*Methods of
Experimental Physics*

VOLUME 28

STATISTICAL METHODS FOR PHYSICAL SCIENCE

# METHODS OF EXPERIMENTAL PHYSICS

Robert Celotta and Thomas Lucatorto, *Editors-In-Chief*

*Founding Editors*

L. MARTON
C. MARTON

Volume 28

# Statistical Methods for Physical Science

*Edited by*

John L. Stanford
*Department of Physics and Astronomy*
*Iowa State University*
*Ames, Iowa*

and

Stephen B. Vardeman
*Department of Statistics*
*Department of Industrial and Manufacturing Systems Engineering*
*Iowa State University*
*Ames, Iowa*

ACADEMIC PRESS

San Diego   New York   Boston   London   Sydney   Tokyo   Toronto

This book is printed on acid-free paper. ∞

Copyright © 1994 by ACADEMIC PRESS, INC.

All Rights Reserved.

No part of this publication may be reproduced or transmitted in any form or by any means, electronic or mechanical, including photocopy, recording, or any information storage and retrieval system, without permission in writing from the publisher.

Academic Press, Inc.
A Division of Harcourt Brace & Company
525 B Street, Suite 1900, San Diego, California 92101-4495

*United Kingdom Edition published by*
Academic Press Limited
24-28 Oval Road, London NW1 7DX

Library of Congress Cataloging-in-Publication Data

Statistical methods for physical science / edited by John L. Stanford
    and Stephen B. Vardeman.
        p.      cm.  -- (Methods of experimental physics (v. 28)
      Includes bibliographical references and index.
      ISBN: 0-12-475973-4
        1. Physical sciences--Experiments--Statistical methods.
    I. Stanford, John., date.   II. Vardeman, Stephen B.
    III. Series.
    Q182.3.S7    1994
    500.2'0724--dc20                                              93-182
                                                                    CIP

PRINTED IN THE UNITED STATES OF AMERICA
 94   95   96   97   98   99   BC   9   8   7   6   5   4   3   2   1

# CONTENTS

CONTRIBUTORS . . . . . . . . . . . . . . . . . . . . . . xiii
PREFACE . . . . . . . . . . . . . . . . . . . . . . . . . xv
LIST OF VOLUMES IN TREATISE . . . . . . . . . . . . . . xvii

### 1. Introduction to Probability Modeling
by WILLIAM R. LEO

1.1. Probability in Experimental Physics . . . . . . . . . 1
1.2. Defining *Probability* . . . . . . . . . . . . . . . . 4
1.3. Elements of Probability Theory . . . . . . . . . . . 8
1.4. Modeling Measurement . . . . . . . . . . . . . . . 27
References . . . . . . . . . . . . . . . . . . . . . 34

### 2. Common Univariate Distributions
by LAURENT HODGES

2.1. Introduction . . . . . . . . . . . . . . . . . . . . 35
2.2. Discrete Probability Mass Functions . . . . . . . . 35
2.3. Continuous Probability Distributions . . . . . . . . 46
References . . . . . . . . . . . . . . . . . . . . . 61

### 3. Random Process Models
by CHRISTOPHER CHATFIELD

3.1. Introduction . . . . . . . . . . . . . . . . . . . . 63
3.2. Probability Models for Time Series . . . . . . . . . 64

3.3. Spectral Properties of Random Processes. . . . . . . . 78
3.4. Point Process Models . . . . . . . . . . . . . . . 83
3.5. Further Reading. . . . . . . . . . . . . . . . . . 91
     References . . . . . . . . . . . . . . . . . . . 92

## 4. Models for Spatial Processes
by NOEL CRESSIE

4.1. Introduction . . . . . . . . . . . . . . . . . . . 93
4.2. Geostatistical Models. . . . . . . . . . . . . . . 95
4.3. Lattice Models . . . . . . . . . . . . . . . . . 103
4.4. Spatial Point Processes . . . . . . . . . . . . . . 112
4.5. Some Final Remarks . . . . . . . . . . . . . . . 122
     References . . . . . . . . . . . . . . . . . . 123

## 5. Monte Carlo Methods
by PETER CLIFFORD

5.1. Introduction . . . . . . . . . . . . . . . . . . 125
5.2. Continuous Distributions . . . . . . . . . . . . . 129
5.3. Discrete Distributions . . . . . . . . . . . . . . 134
5.4. Multivariate Distributions . . . . . . . . . . . . 136
5.5. Monte Carlo Integration . . . . . . . . . . . . . 140
5.6. Time Series . . . . . . . . . . . . . . . . . . 142
5.7. Spatial Processes . . . . . . . . . . . . . . . . 146
5.8. Markov Random Fields . . . . . . . . . . . . . 150
5.9. Point Processes. . . . . . . . . . . . . . . . . 152
5.10. Further Reading . . . . . . . . . . . . . . . . 153
     References . . . . . . . . . . . . . . . . . . 153

## 6. Basic Statistical Inference
by JOHN KITCHIN

6.1. Introduction . . . . . . . . . . . . . . . . . . 155

6.2. Point Estimation. . . . . . . . . . . . . . . . . 160

6.3. Interval Estimation. . . . . . . . . . . . . . . . 165

6.4. Statistical Tests . . . . . . . . . . . . . . . . . 179

6.5. Beyond Basic Statistical Inference . . . . . . . . . . 185

References . . . . . . . . . . . . . . . . . . . 185


## 7. Methods for Assessing Distributional Assumptions in One- and Two-Sample Problems
by VIJAYAN N. NAIR and ANNE E. FREENY

7.1. Introduction . . . . . . . . . . . . . . . . . . 187

7.2. Chi-Squared Tests . . . . . . . . . . . . . . . . 188

7.3. Quantile–Quantile (Q-Q) Plots . . . . . . . . . . . 192

7.4. Formal Test Procedures. . . . . . . . . . . . . . 201

7.5. Extensions to Censored Data. . . . . . . . . . . . 204

7.6. Two-Sample Comparisons . . . . . . . . . . . . . 207

References . . . . . . . . . . . . . . . . . . . 209


## 8. Maximum Likelihood Methods for Fitting Parametric Statistical Models
by WILLIAM Q. MEEKER and LUIS A. ESCOBAR

8.1. Introduction . . . . . . . . . . . . . . . . . . 211

8.2. Data from Continuous Models . . . . . . . . . . . 214

8.3. General Method and Application to the Exponential Distribution (a One-Parameter Model) . . . . . . . . 221

8.4. Fitting the Weibull with Left-Censored Observations (a Two-Parameter Model) . . . . . . . . . . . . . . . . 230

- 8.5. Fitting the Limited Failure Population Model (a Three Parameter Model) . . . . . . . . . . . . . . . . . . 235
- 8.6. Some Other Applications . . . . . . . . . . . . . . 239
- 8.7. Other Topics and Sources of Additional Information . . . 240
- References . . . . . . . . . . . . . . . . . . . . . 243

## 9. Least Squares
by GEORGE A.F. SEBER and CHRISTOPHER J. WILD

- 9.1. Statistical Modeling . . . . . . . . . . . . . . . . 245
- 9.2. The Error Process Viewed Statistically. . . . . . . . . 252
- 9.3. Least Squares Fitting . . . . . . . . . . . . . . . . 254
- 9.4. Statistical Properties of Least Squares Estimates . . . . . 256
- 9.5. Statistical Inference . . . . . . . . . . . . . . . . 261
- 9.6. Diagnostics . . . . . . . . . . . . . . . . . . . . 266
- 9.7. Errors in the Regressors . . . . . . . . . . . . . . . 277
- References . . . . . . . . . . . . . . . . . . . . . 280

## 10. Filtering and Data Preprocessing for Time Series Analysis
by WILLIAM J. RANDEL

- 10.1. Filtering Time Series . . . . . . . . . . . . . . . . 283
- 10.2. Data Preprocessing for Spectral Analysis . . . . . . . . 297
- 10.3. Imperfectly Sampled Time Series . . . . . . . . . . . 305
- References . . . . . . . . . . . . . . . . . . . . . 310

## 11. Spectral Analysis of Univariate and Bivariate Time Series
by DONALD B. PERCIVAL

- 11.1. Introduction . . . . . . . . . . . . . . . . . . . 313
- 11.2. Univariate Time Series . . . . . . . . . . . . . . . 316

11.3. Bivariate Time Series . . . . . . . . . . . . . . . . 340

References. . . . . . . . . . . . . . . . . . . . . . . 347

## 12. Weak Periodic Signals in Point Process Data
by DAVID A. LEWIS

12.1. Introduction . . . . . . . . . . . . . . . . . . . . . 349

12.2. White Noise and Light Curves . . . . . . . . . . . 351

12.3. Tests for Uniformity of Phase . . . . . . . . . . . . 352

12.4. dc Excess vs. Periodic Strength . . . . . . . . . . . 361

12.5. Frequency Searches . . . . . . . . . . . . . . . . . 364

12.6. Multiple Data Sets . . . . . . . . . . . . . . . . . . 367

References. . . . . . . . . . . . . . . . . . . . . . . 372

## 13. Statistical Analysis of Spatial Data
by DALE ZIMMERMAN

13.1. Introduction . . . . . . . . . . . . . . . . . . . . . 375

13.2. Sulfate Deposition Data . . . . . . . . . . . . . . . 376

13.3. The Geostatistical Model . . . . . . . . . . . . . . 378

13.4. Estimation of First-Order Structure . . . . . . . . . 381

13.5. Estimation of Second-Order Structure . . . . . . . . 388

13.6. Spatial Prediction (Kriging) . . . . . . . . . . . . . 396

13.7. Extensions and Related Issues . . . . . . . . . . . . 398

References. . . . . . . . . . . . . . . . . . . . . . . 400

## 14. Bayesian Methods
by Harry F. Martz and Ray A. Waller

14.1. Bayesian Statistical Inference . . . . . . . . . . . . 403
14.2. The Prior Distribution . . . . . . . . . . . . . . . . 407
14.3. Bayesian Estimation. . . . . . . . . . . . . . . . . . 412
14.4. Examples . . . . . . . . . . . . . . . . . . . . . . . 413
14.5. The Gibbs Sampler . . . . . . . . . . . . . . . . . . 426
References. . . . . . . . . . . . . . . . . . . . . . 430

## 15. Simulation of Physical Systems
by John M. Hauptman

15.1. Introduction . . . . . . . . . . . . . . . . . . . . . 433
15.2. Basic Techniques in Simulation . . . . . . . . . . . 436
15.3. Finding Nonalgebraic Solutions . . . . . . . . . . . 440
15.4. Simulation of Experiments . . . . . . . . . . . . . 444
15.5. Validity Testing and Analysis . . . . . . . . . . . . 450
15.6. Improbable Events and Small Effects. . . . . . . . 453
15.7. Simulations within Simulations . . . . . . . . . . . 454
References. . . . . . . . . . . . . . . . . . . . . . 455

## 16. Field (Map) Statistics
by John L. Stanford and Jerald R. Ziemke

16.1. Introduction . . . . . . . . . . . . . . . . . . . . . 457
16.2. Field Statistic Assessment by Monte Carlo Simulation. . 462
16.3. Example One: Atmospheric Temperature Fields . . . . 466
16.4. Example Two: Global Ozone Data Fields . . . . . . . 471
16.5. Example Three: Cross Correlation between Ozone and Solar Flux Time Series . . . . . . . . . . . . . . 475

16.6. Higher Dimensions . . . . . . . . . . . . . . . . . 478

16.7. Summary . . . . . . . . . . . . . . . . . . . . . 478

References. . . . . . . . . . . . . . . . . . . . . 478

## 17. Modern Statistical Computing and Graphics
by FREDERICK L. HULTING and ANDRZEJ P. JAWORSKI

17.1. Introduction . . . . . . . . . . . . . . . . . . . 481

17.2. Statistical Computing Environments . . . . . . . . . 482

17.3. Computational Methods in Statistics . . . . . . . . 488

17.4. Computer-Intensive Statistical Methods . . . . . . . 491

17.5. Application: Differential Equation Models . . . . . . 501

17.6. Graphical Methods . . . . . . . . . . . . . . . . 509

17.7. Conclusion . . . . . . . . . . . . . . . . . . . . 516

References. . . . . . . . . . . . . . . . . . . . . 517

TABLES . . . . . . . . . . . . . . . . . . . . . . . . . 521

INDEX . . . . . . . . . . . . . . . . . . . . . . . . . . 530

# CONTRIBUTORS

Numbers in parentheses indicate the pages on which the authors' contributions begin.

CHRISTOPHER CHATFIELD (63), *School of Mathematical Sciences, University of Bath, Bath, Avon BA2 7AY, United Kingdom*

PETER CLIFFORD (125), *Department of Statistics, Oxford University, 1 South Parks Road, Oxford OX1 3TG, United Kingdom*

NOEL CRESSIE (93), *Statistics Department, Iowa State University, Ames, Iowa 50011*

LUIS A. ESCOBAR (211), *Department of Experimental Statistics, Room 159A, Agricultural Administration Building, Louisiana State University, Baton Rouge, Louisiana 70803-5606*

ANNE E. FREENY (187), *AT&T Bell Laboratories, MH 2C-258, 600 Mountain Avenue, Murray Hill, New Jersey 07974*

JOHN M. HAUPTMAN (433), *Department of Physics and Astronomy, Iowa State University, Ames, Iowa 50011*

LAURENT HODGES (35), *Department of Physics and Astronomy, Iowa State University, Ames, Iowa 50011*

FREDERICK L. HULTING (481), *Applied Mathematics and Computer Technology Division, Alcoa Technical Center, D-AMCT, 100 Technical Drive, Alcoa Center, Pennsylvania 15069-0001*

ANDRZEJ P. JAWORSKI (481), *Applied Mathematics and Computer Technology Division, Alcoa Technical Center, D-AMCT, 100 Technical Drive, Alcoa Center, Pennsylvania 15069-0001*

JOHN KITCHIN (155), *Semiconductor Operations, Digital Equipment Corporation, 77 Read Road, HL02-3/J13, Hudson, Massachusetts 01749-9987*

WILLIAM R. LEO (1), *ASTRAL, rue Pedro Meylan 7, 1211 Geneva 17, Switzerland*

DAVID A. LEWIS (349), *Department of Physics and Astronomy, Iowa State University, Ames, Iowa 50011*

HARRY F. MARTZ (403), *Group A-1 MS F600, Los Alamos National Laboratory, Los Alamos, New Mexico 87545*

WILLIAM Q. MEEKER (211), *Statistics Department, Iowa State University, Ames, Iowa 50011*

VIJAYAN N. NAIR (187), *University of Michigan, Ann Arbor, Michigan 48109*

DONALD B. PERCIVAL (313), *Applied Physics Laboratory, University of Washington, Seattle, Washington 98105*

WILLIAM J. RANDEL (283), *National Center for Atmospheric Research, P.O. Box 3000, Boulder, Colorado 80307*

GEORGE A.F. SEBER (245), *Department of Statistics, Auckland University, Private Bag, Auckland, New Zealand*

JOHN L. STANFORD (457), *Department of Physics and Astronomy, Iowa State University, Ames, Iowa 50011*

RAY A. WALLER (403), *Energy and Technical Directorate, Los Alamos National Laboratory, Los Alamos, New Mexico 87545*

CHRISTOPHER J. WILD (245), *Department of Statistics, Auckland University, Private Bag, Auckland, New Zealand*

JERALD R. ZIEMKE (457), *Department of Physics and Astronomy, Iowa State University, Ames, Iowa 50011*

DALE ZIMMERMAN (375), *Department of Statistics and Actuarial Science, University of Iowa, Iowa City, Iowa 52242*

# PREFACE

This volume is an introduction to probability and statistics for experimental physical scientists—for those physical scientists whose success depends on the wise use of empirical data. Statistics is the study of efficient methods for the collection and analysis of data in a framework that explicitly recognizes and allows for the reality of variation or randomness. Probability is the branch of mathematics that provides tools for the description of randomness; it is thus essential background for the enterprise of statistics, and is basic to the understanding of nondeterministic phenomena.

As the phenomena studied by physical scientists become more complex, statistics and probability become more important to progress in the physical sciences. Larger sets of increasingly complex and noisy data are needed to help answer basic scientific questions. In this book we provide a "source of first resort" for physical scientists who need probabilistic and statistical tools in their research. Physical scientists can and do develop statistical tools of their own, but given an appropriate entry into the statistical literature such as this volume, their work will be easier, more efficient, and more productive. They will be able to quickly find out what is already available (and in the process avoid pitfalls encountered by others) and will be able to make use of the best existing statistical technology.

This volume is not a text on mathematical theory, although a level of mathematical sophistication commensurate with graduate training in the physical sciences is assumed. Instead, it is a readable, self-contained introduction to a variety of methods (old and new) that seem to us most widely applicable and important in the physical sciences. Authors have illustrated their discussions with real physical sciences and also data sets. In all of the chapters, the authors have supplied helpful bibliographies for further reading for those who need to learn more than can be presented in a single volume like this.

In recruiting authors for this book, we (a physicist and a statistician) looked for a creative mix of statistically literate physical scientists and for statisticians with a real interest in the physical sciences. Therefore, the volume reflects a broad understanding of the real modeling and data analysis needs of the modern physical sciences, and also of the best existing probabilistic and statistical methodology. As editors, our hope is that it will serve as a catalyst for interaction between physical scientists and statisticians, reaching far beyond the particular effort that produced it. In the book, physical scientists will find both tools to support their research and evidence that statisticians are interested in crafting such tools. Statisticians will find fascinating problems of genuine scientific importance, easily adequate to occupy their "tool-making" and collaboration efforts full time.

The structure of the book is as follows. It consists of four rather distinct sections. Chapters 1 through 5 provide an introduction to probability modeling for the physical sciences, including discussions of time series and spatial models, and an introduction to Monte Carlo (probabilistic simulation) methods. The second section consists of Chapters 6 through 9, which provide the basics of probability-based statistical inference, including discussions of goodness of fit and maximum likelihood methods and inferences from least squares analyses. Chapters 10 through 13 discuss the important topics of statistical inference from time series and spatial data. The last section, consisting of Chapters 14 through 17, discusses some specialized (but nevertheless important) topics, most of which are only recently available for application through advances in scientific computing. In addition to Bayesian methods, this section contains discussions of applications of Monte Carlo methods to both particle and atmospheric physics and a chapter discussing the capabilities of modern statistical computing systems.

We expect that most physical science readers will use this volume a chapter at a time as they need particular types of statistical methods. Nevertheless, as readers examine Chapters 6 through 17 for data analysis methods, we also expect that they will spend time in Chapters 1 through 5 familiarizing themselves with the modeling background that supports and motivates the statistical methods discussed in the last 12 chapters.

As we come to the end of our work on this volume, we thank the chapter authors for their excellent cooperation in bringing this book together. They have not only written several drafts of their own chapters (reacting with good humor to editorial meddling), but also provided extremely helpful reviews of others' drafts and kept to the ambitious schedule that we set when recruiting them. It has been a pleasure to work with them on this project, and we believe that the reader will find their contributions both engaging and illuminating.

John Stanford<br>
Steve Vardeman

# METHODS OF EXPERIMENTAL PHYSICS

*Editors-in-Chief*
*Robert Celotta and Thomas Lucatorto*


Volume 1. Classical Methods
*Edited by* Immanuel Estermann

Volume 2. Electronic Methods, Second Edition (in two parts)
*Edited by* E. Bleuler and R. O. Haxby

Volume 3. Molecular Physics, Second Edition (in two parts)
*Edited by* Dudley Williams

Volume 4. Atomic and Electron Physics—Part A: Atomic Sources and Detectors; Part B: Free Atoms
*Edited by* Vernon W. Hughes and Howard L. Schultz

Volume 5. Nuclear Physics (in two parts)
*Edited by* Luke C. L. Yuan and Chien-Shiung Wu

Volume 6. Solid State Physics—Part A: Preparation, Structure, Mechanical and Thermal Properties; Part B: Electrical, Magnetic and Optical Properties
*Edited by* K. Lark-Horovitz and Vivian A. Johnson

Volume 7. Atomic and Electron Physics—Atomic Interactions (in two parts)
*Edited by* Benjamin Bederson and Wade L. Fite

Volume 8. Problems and Solutions for Students
*Edited by* L. Marton and W. F. Hornyak

Volume 9. Plasma Physics (in two parts)
*Edited by* Hans R. Griem and Ralph H. Lovberg

Volume 10. Physical Principles of Far-Infrared Radiation
*By* L. C. Robinson

Volume 11. Solid State Physics
*Edited by* R. V. Coleman

Volume 12. Astrophysics—Part A: Optical and Infrared Astronomy
*Edited by* N. Carleton
Part B: Radio Telescopes; Part C: Radio Observations
*Edited by* M. L. Meeks

Volume 13. Spectroscopy (in two parts)
*Edited by* Dudley Williams

Volume 14. Vacuum Physics and Technology
*Edited by* G. L. Weissler and R. W. Carlson

Volume 15. Quantum Electronics (in two parts)
*Edited by* C. L. Tang

Volume 16. Polymers—Part A: Molecular Structure and Dynamics; Part B: Crystal Structure and Morphology; Part C: Physical Properties
*Edited by* R. A. Fava

Volume 17. Accelerators in Atomic Physics
*Edited by* P. Richard

Volume 18. Fluid Dynamics (in two parts)
*Edited by* R. J. Emrich

Volume 19. Ultrasonics
*Edited by* Peter D. Edmonds

Volume 20. Biophysics
*Edited by* Gerald Ehrenstein and Harold Lecar

Volume 21. Solid State: Nuclear Methods
*Edited by* J. N. Mundy, S. J. Rothman, M. J. Fluss, and L. C. Smedskjaer

Volume 22. Solid State Physics: Surfaces
*Edited by* Robert L. Park and Max G. Lagally

Volume 23. Neutron Scattering (in three parts)
*Edited by* K. Sköld and D. L. Price

Volume 24. Geophysics—Part A: Laboratory Measurements; Part B: Field Measurements
*Edited by* C. G. Sammis and T. L. Henyey

Volume 25. Geometrical and Instrumental Optics
*Edited by* Daniel Malacara

Volume 26. Physical Optics and Light Measurements
*Edited by* Daniel Malacara

Volume 27. Scanning Tunneling Microscopy
*Edited by* Joseph Stroscio and William Kaiser

Volume 28. Statistical Methods for Physical Science
*Edited by* John L. Stanford and Stephen B. Vardeman

# 1. INTRODUCTION TO PROBABILITY MODELING

William R. Leo

Astral
Geneva, Switzerland

*To understand probability, I think you need to be older.*
—M. Schöunburg

## 1.1 Probability in Experimental Science

Historically, the invention of probability theory is generally attributed to Pascal and Fermat, who first developed it to treat a narrow domain of problems connected with games of chance. Progress in the ensuing centuries, however, has widened its applications such that today almost all disciplines, be they the physical sciences, engineering, the social sciences, economics, etc., are touched in some way or another. In physics, for example, probability appears in an extremely fundamental role in quantum theory and in statistical mechanics, where it is used to describe the ultimate behavior of matter. Notwithstanding its importance to modern physical theory, however, probability also plays an almost equally fundamental role on the experimental side—for it can be said that probability is one of the truly basic tools of the experimental sciences, without which many results would not have been possible.

Modern science, of course, is based on active experimentation; that is, observation and measurement. This is the fundamental means by which new knowledge is acquired. But it becomes evident to anyone who has ever performed experiments that all measurements are fraught with uncertainty or "errors," which limit the conclusions that can be drawn. (The term *errors* here is the one used most often by scientists to refer to experimental uncertainties, and its meaning should not be confused with that of "making a mistake.") The questions that inevitably arise then are these: How is one to handle these uncertainties? Can one still draw any real conclusions from the data? How much confidence can one have in the results? Moreover, given the data, how does one compare them to results from other experiments? Are they different or consistent? Here the problem is not so much the final result itself but the errors incurred. This implies that a common quantitative measure and "language" are necessary to describe the errors. Finally, given that there are errors in all measurements, is it possible to plan and design one's experiment so that it *will* give meaningful results? Here again, a quantitative procedure is necessary.

The answer to these questions comes from recognizing that measurement and observational errors are in fact *random* phenomena and that they can be modelled and interpreted using the theory of probability. The consequence is that "uncertainty," an intuitive notion, can now be quantified and treated with the mathematical apparatus already developed in probability and in statistics. Ultimately, this allows the confidence one can have in the result, and therefore the meaningfulness of the experiment, to be gauged in a quantitative way. At the same time, the mathematical theory provides a standard framework in which different measurements or observations of the same phenomenon can be compared in a consistent manner.

The notion of probability is familiar to most people, however, the mathematical theory of probability and its application are much less so. Indeed, nowhere in mathematics are more errors (i.e., mistakes!) in reasoning made more often than in probability theory (see, for example, Chapters 1–2 of [1] and Section VI, [2]). Applying probability in a consistent manner requires, in fact, a rather radical change in conceptual thinking.

History gives us some interesting illustrations of this, one of which is the work of Tobias Mayer, who in the mid-18th century successfully treated the problem of the libration of the moon. The term *libration* refers to slow, oscillatory movements made by the moon's body such that parts of its face appear and disappear from view. There are, in fact, several movements along different axes that, over an extended period of time, allow about 60% of the moon's surface to be seen from the earth. The causes of these effects were known at the time of Tobias Mayer, but the problem was to account for these movements either by a mathematical equation or a set of lunar tables. Indeed, a solution was of particular commercial and military value for detailed knowledge of the moon's different motions could serve as a navigational aide to ships at sea.

Thus, in 1748, Johann Tobias Mayer, an already well-known cartographer and practical astronomer, undertook the study of the libration problem. After making observations of the moon for more than a year, Mayer came up with a method for determining various characteristics of these movements. A critical part of this solution was to find the relationship between the natural coordinate system of the moon as defined by its axis of rotation and equator and the astronomer's system defined by the plane of the earth's orbit about the sun (ecliptic). To do this, Mayer focused his observations on the position of the crater Manilius and derived a relation between the two reference systems involving six parameters, three of which were unknown constants and three of which varied with the moon's motion, but could be measured. To solve for these unknowns, therefore, Mayer needed only to make three day's worth of observations. He did better than this, however, and made observations over a period of 27 days.

Keeping in mind that the discovery of the least squares principle (see Chapter 9) was still a half-century off, Mayer was thus faced with the dilemma of having 27 *inconsistent* equations for his three unknowns; that is, if he attempted to solve for the unknowns by just taking three of the equations, different results would be obtained depending on which three equations he chose. The inconsistencies, of course, were due to the observational errors, an effect well known to the scientists and mathematicians of the time, but for which no one had a solution. It was here that Mayer came up with a remarkable idea. From his own practical experience, Mayer knew intuitively that adding observations made under similar conditions could actually reduce the errors incurred. Using this notion, he divided his equations into three groups of nine similar equations, added the equations in each group and solved for his unknowns. Simple[1] as that!

The point, however, is that Mayer essentially thought of his errors as random phenomena, although he had no clear concept of probability, so that adding data could actually lead to a cancellation of the errors and a *reduction* in the overall uncertainty. This was totally contrary to the view of the times, which focused on the *maximum possible error* (as opposed to the *probable* error) that could occur when data were manipulated. From this viewpoint, adding data would only increase the final errors, since it would just cumulate the maximum possible error of each datum. Such a large final error is possible under the probabilistic view but highly *improbable*.

Interestingly enough history also provides a "control experiment" to prove the point. Only one year prior to Mayer's work, Leonhard Euler, undoubtedly one of the greatest mathematicians of all time, reported his work on the problem of "inequalities" in the orbits of Jupiter and Saturn. These effects were thought to be due to the two planets' mutual attraction, an instance of what today is called the *three-body problem.* Formulating a mathematical equation for the longitudinal position of Saturn that took into account the attraction of both the sun and Jupiter, Euler sought to provide some test of his calculation by using some actual data (not of his own taking). After making a number of approximations to linearize his formula,

---

[1]Actually more skill was involved here than has been implied. To get the precise results that he did, the groups had to be chosen judiciously in order to maximize the differences between the three final equations. As well, Mayer's selection of the crater Manilius was very well chosen, indicating perhaps an intuitive feeling for experimental design.

he ended up with 75 equations for six unknowns. He thus faced a problem analytically similar to Tobias Mayer's, although on a somewhat larger scale.

Here, however, Euler failed to make the conceptual jump that Mayer made. Indeed, whereas Mayer made his own measurements and thus had an intuitive feeling for the uncertainties involved, Euler, a *pure* mathematician dealing with data that he did not take, had no basis for even imagining the random nature of measurement errors. Euler was thus left to grope for a solution that he never found. His theoretical work, was nevertheless a significant contribution to celestial mechanics, and in recognition, he was awarded the 1748 prize offered by the Academy of Sciences in Paris. Further details concerning Tobias Mayer's and Leonhard Euler's work may be found in [3].

These historical accounts illustrate the paradigm shift that the application of probability theory required in its early stages, but even today dealing with probability theory still requires a conceptual change that is underestimated by most people. Indeed, to think probabilistically essentially requires giving up *certainty* in everything that is done, as we will see later on. This is also complicated by a good deal of confusion over the meaning of certain terms—beginning with the expressions *probability* and *probability theory* themselves. We will attempt to clarify these points in the next sections.

## 1.2 Defining *Probability*

In applying probability theory it is necessary to distinguish between two distinct problems: the modeling of random processes in mathematical terms and the interpretation and definition of initial probabilities. The first concerns the use of what we have been calling the *theory of probability*. This term refers to the mathematical theory and formalism that one uses to describe and analyze random processes. In this context, the notion of probability is left undefined insofar as its relation to real world processes is concerned. Instead, it is treated as a given mathematical object with certain properties and only the mathematical relationships between these probability objects are considered. These general relationships then allow the manipulation of probabilities to eventually solve specific problems.

The second part concerns the way in which probability is interpreted in terms of the real world and how to relate it to events in the model. For most people, probability is a vague notion that is unconsciously viewed as one single concept, usually a mix of mathematical and physical definitions. However, when analyzed, one will find that *probability* has many different interpretations, and which interpretation is used depends on the problem at hand.

Let us look at this more closely. For example, if the problem concerns a game of chance, say to calculate the odds of obtaining *heads* on the toss of a fair coin, most people would say 1/2. The reasoning used would be that there are two possible outcomes for the toss, each of which is equally likely, so that the probability for *heads* is 1/2. Thus, *probability* here is related to and actually defined as the ratio of the number of outcomes meeting the event criteria to the total number of possible outcomes. This is known as the *classical* view of probability.

If the question now is the probability of a photon scattering off an electron, one would resort to a quantum mechanical calculation, the result of which would be a complex amplitude and the square of which would be the probability.

Obviously, *probability* here has a different interpretation. The *probability* used in quantum mechanics is an expression of fundamental limits in nature, and it is manifested as a complex wave function. In the classical case, however, *probability* is an expression of our lack of complete knowledge about the system, knowledge that in theory exists, and with which one could make a solid prediction about the outcome rather than just the *chances* of such and such an outcome.

The difference between quantum probability and classical probability, of course, is well known, although not always well elucidated in the literature. However, even in the classical world, not all probability questions can be addressed by the same "classical" probability. For example, suppose the question is: What is the probability that a light bulb coming off the manufacturing line of company X will fail on first use? Most people would think of taking a large sample of light bulbs and seeing what fraction actually fail. This would then provide an estimate of the true probability under the implicit assumption that the larger is the sample the better the estimate. Probability here is thus equated to the frequency of occurrences in an infinitely large sample rather than to a ratio of outcomes.

Consider also questions involving things such as sports events (What is the probability that your local football team will win this weekend?) or political questions (What is the probability that the Soviet Union would not have collapsed in 1991?), etc. These questions cannot be answered by calculating a ratio of outcomes nor with a repeated number of trials. Instead, most people would appeal to personal judgment. Probability here, thus takes on a subjective character, so that we have yet another type of probability.

Traditionally, modern science has been concerned with the *classical* and *frequency* interpretations since they are defined by precise mathematical statements. However, as we have noted previously, they do not define *all* of probability, and depending on the situation it may be necessary to consider other types as well (see, for example, Chapter 14 on Bayesian methods).

### 1.2.1 Classical Probability

In the following two sections we will briefly elaborate on these two interpretations.

The development of classical probability was originally tied to the modeling of games of chance. These problems prompted the so-called classical view of probability:

**Definition:** *If an event can occur in N possible mutually exclusive and equally likely ways, and there are $n_A$ outcomes with the attribute A, then the probability that an outcome with attribute A will occur is $n_A/N$.*

For example, if the event is the rolling of one die, the possible outcomes are 1, 2, . . . , 6. With a fair die, each of these outcomes is equally likely, so that the probability of tossing a *2* is 1/6. Similarly the probability of rolling a *2* or *3* is 2/6 = 1/3, since there are two outcomes that satisfy the attribute of *2 or 3*. From this, more complex processes can be described, for example, the rolling of two dice. The number of possible outcomes in this case is 36. To find the probability of rolling, say, a *5,* we count the number of outcomes for which this is satisfied. Thus, a *5* is obtained if we have either (1, 4), (2, 3), (3, 2) or (4, 1), where the first number in the parentheses is the outcome of the first die and the second number the result from the second die. This gives us 4/36 = 1/9 as the probability. More examples of this type are certainly familiar to the reader.

The point to note, however, is that one must be able to reduce the given process to a simpler process or processes with *equally likely outcomes*. The justification for declaring equal probabilities is symmetry, a concept dear to every physicist. For a single fair die, the six faces are "identical" in terms of the relevant parameters for the problem, so there is no reason to assume any difference in the probability of obtaining one face over the other. For a deck of cards, all 52 cards are "identical," so again there is no reason to assume that one will be more probable than the other.

The classical interpretation, however, has limitations. For example, it would be impossible to apply it to the case of a loaded die. Here obviously there is no way of reducing the problem to a symmetrical situation. In this same category are problems such as the probability that a certain chip on a circuit board will fail in 100 hours or the probability that one will have an automobile accident in the next five years. It would be extremely difficult to break these down into simpler elements, and even if one could, it would not necessarily result in a situation of symmetry.

## 1.2.2 Frequency Probability

One solution to the weaknesses of the classical view is the frequency approach. Consider our loaded die in the previous example, which we could not treat using the classical interpretation. Suppose we toss the die many times and, after each throw, note the frequency of occurrence of each face obtained; i.e., the number of times each face has appeared divided by the total number of throws. At first ratios will fluctuate about, but as the number of throws increases, the frequency of each face settles down and approaches a definite number. In the limit of an infinite number of rolls, these frequencies will then approach the true probabilities for each face. In practice, of course, it would be impossible to obtain the exact probability values in this manner since one would have to carry out an infinite number of trials; however, an approximation would be possible and the larger is the sample the better the approximation. Thus, the probability is inferred inductively. This should be contrasted to the classical view where probabilities are inferred *deductively*.

The frequency interpretation has a tremendous importance as it provides the link to many real world processes that otherwise could not be treated under the classical definition. Indeed, if the discipline of statistics is so closely related to probability theory today, it is through frequency probability!

One technical point that we have glossed over is the justification for assuming that, in the limit of an infinite number of trials, the frequency will approach the true probabilities for the process. Although this may appear intuitive, mathematical rigor demands a formal theorem of some sort. This is provided by the *law of large numbers,* which will be demonstrated in Sect. 1.3.8.

While the frequency view solves many of the difficulties of the classical definition, it also has its own limits. Quite obviously, the frequency approach requires that the process be repeatable, at least in a conceptual way. Single events such as the meltdown of a nuclear reactor core or earthquakes, for example, are events that are not very susceptible to analysis by repeated trials, even conceptually. Subjective probability based on expertise knowledge could be applied here; however, science traditionally resists elements of a subjective character so that its use in modern science is controversial to say the least.

Fortunately, for the experimental sciences, the frequency interpretation is usually adequate, for, if anything, scientific experiments should be repeatable! This is the view that we will assume throughout this chapter, then, unless otherwise specified.

## 1.3 Elements of Probability Theory

The theory of probability provides a general structure and formalism for describing and analyzing random processes. This is usually approached from an axiomatic point of view in which probability is simply postulated to be a mathematical object satisfying well-defined rules. From these simple properties, the general laws and theorems that make up the mathematical apparatus of the theory are then derived.

Since its origins are in classical probability, the axiomatic definition of probability is sometimes taken as *the* definition of classical probability. However, the mathematical definition can in fact accommodate many different real world interpretations of probability including quantum probability [2]. Those probabilities satisfying the mathematical axioms of the theory, then, have the entire mathematical apparatus at their disposal.

In this section we will attempt to outline the basic theory and try to give some feeling as to how this is used to model real processes.

### 1.3.1 Sample Space

For a given random process or more precisely an "experiment" involving a random process, the theory of probability begins by defining a *sample space* in which all possible outcomes of the experiment are represented as points in the space. The sample space then forms an exhaustive set of points $S$.

**Example 1.1:** Consider the "experiment" of rolling a single die. The possible outcomes are 1, 2, ..., 6, so that the sample space consists of six points. We can express this using set theory notation as $S = \{1, 2, \ldots, 6\}$ where the curly brackets indicate the "set" of elements contained within.

**Example 1.2:** Let the experiment now be the rolling of two dice at the same time. The possible outcomes in this case are now $S = \{(1, 1), (1, 2), \ldots, (6, 6)\}$ where the first number in the parentheses represents the face of the first die and the second number that of the second die. The sample space thus contains $6 \times 6 = 36$ points.

**Example 1.3:** Consider the experiment of tossing a coin until a head appears. The possible outcomes are $S = \{H, TH, TTH, TTTH, \ldots\}$, where $T$ represents tails and $H$ heads. The number of points in the sample space is infinite.

Depending on the experiment, a sample space may be discrete as in the preceding examples or continuous as in the following example. From a purely mathematical point of view, the change from a discrete to a continuous space requires some justification making use of measure theory, however, the conceptual change is straightforward.

**Example 1.4:** Consider the experiment of measuring the lifetime of a silicon chip. The outcome can be any positive number, so the sample space is continuous.

**Example 1.5:** Consider the experiment of choosing two points at random in the interval 0 to $L$. The outcomes will be a pair of numbers $(x,y)$, where $0 \leq x, y \leq L$. This suggests visualizing the sample space as all points within a square of side $L$.

Using the sample space framework, we can now define the notion of an event. An event is any collection of outcomes of the experiment that satisfies certain specific criteria. For example, we can define the event of throwing a 7 with two dice or the event of drawing three aces from a deck of playing cards. In the sample space, the event $A$ corresponds to the subset of points for which the outcomes satisfy the criteria defining $A$. Thus, if the event $A$ is the throwing of a 7 with two dice, the subset of points satisfying this criterion would be $A = \{(1, 6), (2, 5), (3, 4), (4, 3), (5, 2), (6, 1)\}$. If any of these outcomes occurs, then the event is said to have occurred. Similarly, if the event $B$, is the drawing of three aces, the relevant points are $B = \{(A\heartsuit, A\spadesuit, A\diamondsuit), (A\heartsuit, A\spadesuit, A\clubsuit), \ldots\}$.

Since the sample space is a set, this also allows us to use the notions of set theory for further development. We assume the reader is familiar with these and will only summarize the basic operations through the Venn diagrams in Fig. 1. For the moment we will need two notions to help us define the axioms for probability. First, the complement of the event $A$ is defined as that set of points in the sample space which are not in $A$. This is denoted as $A'$ and by definition $A \cup A' = S$. Second, two events $A$ and $B$ are said to be *mutually exclusive* or *disjoint* if the intersection $A \cap B = \emptyset$, that is, no points in the subset $A$ are in common with the points in subset $B$. Quite obviously, $A$ and $A'$ are mutually exclusive.

We can now define the mathematical notion of probability. Associated with each subset or event $A$ is a *probability function*, $P(A)$, which satisfies the following rules:

1. $P(A)$ is a real number such that $P(A) \geq 0$ for every $A$ in $S$.
2. $P(S) = 1$.

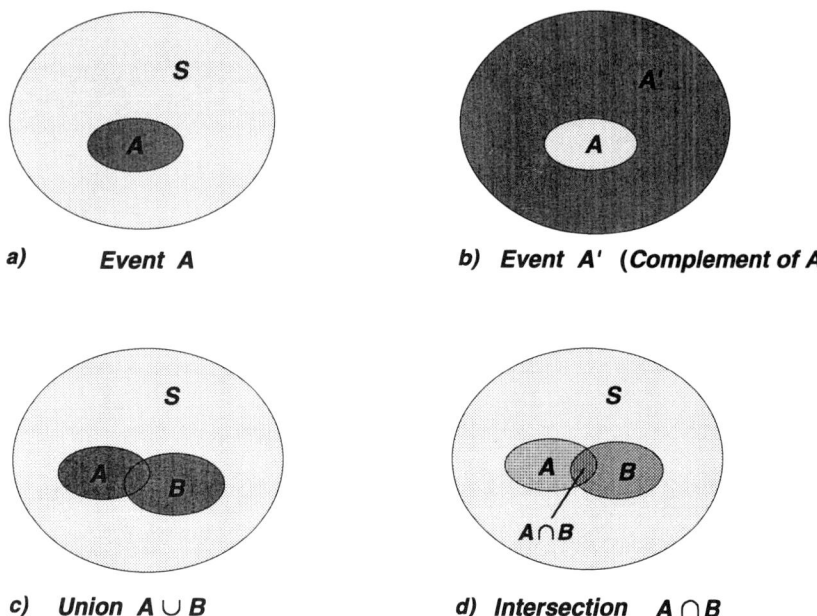

FIG. 1. Venn diagrams illustrating the sample space and its principal subspaces.

3. If $B_1$, $B_2$, $B_3$, . . . is a sequence of mutually exclusive events in $S$; i.e., $B_i \cap B_j = \emptyset$ for all $i \neq j$, then the probability

$$P(B_1 \cup B_2 \ldots) = P(B_1) + P(B_2) + \ldots \quad (1.1)$$

These are the three basic axioms[2] with which the formalism "defines" probability and from which other laws are ultimately derived.

Axiom 2, for example, states that the probability of any of the possible outcomes occurring is 1. This essentially defines *certainty* as being probability equal to 1. Similarly, by considering the probability of the null set, we can show that *impossibility* is equivalent to probability 0. Using all three axioms together leads to a number of other properties such as $0 \leq P(A) \leq 1$, which sets the scale of values for probability. Similarly, it is easy to show that if $P(A)$ is the probability that the event $A$ will occur, then, the probability that it will not occur is $P(A') = 1 - P(A)$.

---

[2]This set of basic axioms are but one of several sets that can be chosen. All eventually lead to the same rules, however.

Axiom 3 expresses the additivity of probabilities for *mutually exclusive* events. In words, what this says is that the probability of either $B_1$ or $B_2$ or $B_3$, etc., is the sum of the individual probabilities for $B_1$, $B_2$, $B_3$ and so on. In general, however, if $A$ and $B$ are any two events in $S$, mutually exclusive or not, then

$$P(A \cup B) = P(A) + P(B) - P(A \cap B) \qquad (1.2)$$

The reason for the last term can be seen by referring to Fig. 1d. If only the simple sum is considered, then any points in the intersection will be counted twice. Thus, we must subtract the intersection region once. In this manner, Eq. (1.2) may be easily extended to more than two events. The case of three events, for instance, leads to

$$P(A \cup B \cup C) = P(A) + P(B) + P(C) - P(A \cap B) \\ - P(A \cap C) - P(B \cap C) + P(A \cap B \cap C) \qquad (1.3)$$

which the reader can verify.

### 1.3.2 Conditional Probability and Independence

A frequently encountered problem is to know the probability of an event $A$ given that another event $B$ has already occurred. This is the conditional probability, written as $P(A \mid B)$ and defined as

$$P(A \mid B) = \frac{P(A \cap B)}{P(B)} \qquad (1.4)$$

where $P(A \cap B)$ is the probability of the event $A$ *and* $B$ occurring. For example, suppose an ace is drawn from a deck of playing cards. What is the probability that this card is also a heart? Clearly, $P(A \cap B) = P(\heartsuit \cap ace)$ and $P(B) = P(ace) = 4/52$, so that, $P(A \mid B) = 1/52 \div 4/52 = 1/4$, which is what is expected intuitively. Note that Eq. (1.4) holds only if the denominator is nonzero. If $P(B) = 0$ then the conditional probability is undefined.

The effect of the condition is thus to limit the sample space to the smaller subspace of events $B$. The conditional probability is then the probability of $A$ relative to this subspace. This should be distinguished from the probability $P(A \cap B)$, which is relative to the entire sample space.

Equation (1.4) also defines the multiplicative rule for probabilities; i.e., the probability of an event $A$ *and* $B$ is equal to the probability of $B$ *times* the conditional probability of $A$ given $B$ (or vice-versa),

$$P(A \cap B) = P(A \mid B)P(B) = P(B \mid A)P(A) \qquad (1.5)$$

This allows us to introduce the concept of *independence*. If the event $A$ does not depend on $B$ in any way, then, clearly,

$$P(A \mid B) = P(A)$$
$$P(B \mid A) = P(B) \qquad (1.6)$$

The multiplicative law for two independent events thus becomes

$$P(A \cap B) = P(A)P(B) \qquad (1.7)$$

And in general for $n$ independent events, we would have

$$P(A_1 \cap A_2 \cap \ldots A_n) = P(A_1)P(A_2) \ldots P(A_n) \qquad (1.8)$$

**Example 1.6** An urn contains three red balls and two white balls. If three balls are drawn from the urn, what is the probability of having three red balls? There are several ways of attacking this problem. The straightforward way is to enumerate the sample space. Since only one of these points satisfies the criteria for the event, the probability is just $1/N$, where $N$ is the number of points in the sample space. Using combinatorial arithmetic, we find that the number of combinations is $N = 10$, so that the probability of selecting three red balls is 1/10.

A second way is to use the multiplicative law. Here we are seeking the probability $P(A \cap B \cap C)$, where $A$ is the event that the first ball is red, $B$ the event that the second is red and $C$ that the third ball is red. Moreover, the individual events are not independent since the selection of a ball removes a ball from the urn, thus affecting subsequent draws. Therefore,

$$P(A \cap B \cap C) = P(C \mid B \cap A)P(A \cap B) = P(C \mid B \cap A)P(B \mid A)P(A)$$

Now let us assign values to the probabilities. The probability that the first ball is red is clearly $P(A) = 3/5$. Given that the first ball is red, the conditional probability that the second ball is red is then $P(B \mid A) = 2/4$. Similarly, the conditional probability that the third ball is red is $P(C \mid B \cap A) = 1/3$. Substituting then gives us the same value as earlier

$$P(A \cap B \cap C) = 1/3 \times 2/4 \times 3/5 = 6/60 = 1/10$$

### 1.3.3 Compound Events

The additive and multiplicative laws allow one to calculate the probabilities of more complicated events from simpler events. Indeed, as the reader may have noticed, the probabilistic description of many "experiments" can be decomposed into sums and products of the probabilities of simpler events. This, of course, allows a clearer view of the essential processes. Events that are made up of simpler events are known as *compound events*, while the basic events from which others may be constructed are called the *elementary events*.

**Example 1.7:** Consider the throwing of the three dice. Constructing the sample space in a straightforward manner gives us the set $S = \{(1, 1, 1), (1, 1, 2), (1, 1, 3), \ldots, (6, 6, 6)\}$ consisting of 216 points. Assuming all three dice are true, the probability of each occurrence is then 1/216. This can be decomposed into the process of throwing three dice separately or one single die three times. The elementary event is thus the throwing of one die for which the probabilities are 1/6 for each face. The probability $P(A \cap B \cap C)$, where $A$ is the first die, $B$ the second and $C$ the third is then given by the multiplicative law

$$P(A \cap B \cap C) = P(A)P(B)P(C) = \frac{1}{6} \times \frac{1}{6} \times \frac{1}{6} = \frac{1}{216}$$

### 1.3.4 Random Variables and Probability Distributions

The representation of the outcomes of a random event as points in a sample space suggests the definition of a variable $X$ to symbolize these points algebraically. The variable $X$ is then known as a *random variable* whose admissible values range over the points in the sample space. The term *variate* is also sometimes used in place of random variable. Depending on the process, a random variable may be discrete or continuous. Similarly, any function of $X$ is also a random variable over the same sample space.

If the outcomes in the space are numerical values, $X$ may take on these same values, otherwise symbolic numerical assignments may be made to each event in a one-to-one correspondence. For example, if the process is the rolling of a single die, the values for $X$ may simply be designated as $X = 1, 2, \ldots, 6$. In the case of flipping a coin, however, the possible outcomes are *heads* or *tails*. Here one can assign the value 1 to *heads* and 0 to *tails*, for example, or −1 for *heads* and +1 for *tails* or vice-versa or some other scheme just as long as it is consistent.

The probability function for a variate $X$ at $X = x$, which we write as $p(x)$, is now known as the *probability distribution* of $X$. For example, if $X$ is the outcome of rolling a single fair die, then, $p(x) = 1/6$ for $x = 1, 2, \ldots, 6$. Similarly, if $Y$ is a random variable function of $X$ with values

$$y = \begin{cases} 1 & x \geq 5 \\ 0 & x < 5 \end{cases}$$

then, its distribution is

$$p(y) = \begin{cases} 1/3 & y = 1 \\ 2/3 & y = 0 \end{cases}$$

In the case of a continuous variable, however, the interpretation of the probability distribution must be modified somewhat. Here, only the probability of finding $X$ in an interval $x$ to $x + dx$ has meaning. This is handled by defining a continuous probability density, $f(x)$, such that the probability of finding $X$ between $x$ and $x + dx$ is given by $f(x)dx$. Note that the density must be multiplied by the differential element $dx$ to obtain the probability. In this manner the probability depends on the size of the interval as one would expect. Probabilities for events, therefore, correspond to areas under the probability density curve.

Very often also, it is required to know the probability of finding $X$ between certain limits, for example, $P(a < X \leq b)$ or $P(X \leq x)$. The latter example is known as the *cumulative distribution* and is defined as

$$F(x) = P(X \leq x) = \sum_{x_i \leq x} p(x_i)$$

and

$$F(x) = P(X \leq x) = \int_{-\infty}^{x} f(x')dx' \qquad (1.9)$$

in the discrete and continuous cases, respectively. The probability $P(a < X \leq b)$ can now be expressed as $F(b) - F(a)$.

**Joint, Marginal and Conditional Distributions.** In cases where events can be described by more than one attribute or criterion, more than one random variable may be used. For example, if the experiment is to choose a playing card at random, we could use two random variables $X$ and $Y$, to represent the denomination and suit of the chosen card respectively. We would thus have a two-dimensional random variable "vector" $(X, Y)$ to represent the outcome. The corresponding probability distribution then becomes a multivariate distribution with $p(x, y) = 1/52$. Similarly, returning to Example 1.5, the outcome of choosing two numbers in the interval $0 \leq x, y \leq L$ could also be represented by the pair $(X, Y)$ and the probability density given by $f(x, y) = 1/L^2$. In both cases, the distribution is known as the *joint distribution* or *joint density* of $X$ and $Y$.

Given the joint distribution, many other questions may be asked of it. For instance, in the first example above: What is the probability distribution of drawing an ace regardless of the suit of the card? Or more generally, what is the distribution of the variate $X$ regardless of the other variate $Y$? This can be obtained by summing over all possible values of the suit variable $Y$,

$$p_X(x) = \sum_{\text{all } y} p(x, y) \qquad (1.10)$$

which in this specific case equals 4/52. Equation (1.10) is more commonly known as the *marginal distribution* of $X$. And in general, given any density or distribution involving more than one variate, one may find the marginal distribution for any subset of the random variables by integrating over all values of the remaining variates.

Using the marginal distribution, the conditional distribution may now also be defined as

$$p_{X|Y}(x \mid y) = \frac{p(x, y)}{p_Y(y)} \qquad (1.11)$$

where $p_{X|Y}(x \mid y)$ is the conditional distribution of $X$ when $Y$ is equal to $y$.

If the conditional distribution of $X$ does not involve the variate $Y$, then the variates are said to be independent. In general, $n$ random variables are said to be independent if their joint probability distribution is equal to the product of the marginal probability distribution of each variate

$$p(x_1, x_2, \ldots, x_n) = p_{X_1}(x_1) p_{X_2}(x_2) \cdots p_{X_n}(x_n) \qquad (1.12)$$

for all values of $X_1, X_2, \ldots, X_n$.

**Example 1.8:** When a subatomic particle such as an electron or a photon enters into matter, it suffers different types of reactions (such as, Coulomb scattering, nuclear reactions, etc.) with the atomic and nuclear constituents of the material (see Chapter 15, for example, or [4], Chapter 2). In nuclear and particle physics, the interaction of subatomic particles is described by the differential cross section, $d\sigma$ $(E, \theta, \phi, \ldots)$, which essentially expresses the quantum probability for a reaction to occur. This is a function of the kinematic variables used to describe the reaction; i.e., the energy, scattering angles and so on. If the differential cross section is integrated over all scattering angles, then one has the total cross section, $\sigma_0$, which describes the probability for a reaction to occur (at energy $E$) regardless of the scattering angle. Each type of reaction has a corresponding cross section. These interactions occur randomly as the particle traverses the material, so that one can ask: What is the probability $P_S(x)$ for it to survive a certain distance $x$ before suffering an interaction? This, of course, depends on the type of particle, its energy, the density of the material, etc. We can, however, derive the form of the distribution for this process without detailed knowledge of these factors.

Suppose $wdx$ is the probability that the particle will suffer an interaction in the interval $x$ to $x + dx$. Then the probability of *not* having an interaction at a point between $x$ and $x + dx$ is $1 - wdx$. If $P_S(x)$ is defined as previously, then the probability that the particle will survive a distance $x + dx$ is

$$P_S(x + dx) = P_S(x)(1 - wdx)$$

Expanding, we obtain

$$P_S(x) + dP_S(x) = P_S(x) - P_S(x)wdx$$

$$\frac{dP_S}{P_S} = -wdx$$

Integrating this and requiring that $P_S(0) = 1$, we have that

$$P_S(x) = e^{-wx} \qquad (1.13)$$

Thus the survival probability for the particle is given by an exponential distribution. It can be shown [4] furthermore that the parameter $w$ is given by $N\sigma_0$, where $N$ is the density of the relevant interacting constituents in the material and $\sigma_0$ the total cross section for an interaction of any kind; i.e., the sum of the total cross sections for each possible type of reaction.

**Example 1.9:** Consider a process in which there are only two outcomes, for example, *heads* or *tails, yes* or *no, success* or *failure*. (To simplify the discussion, we will use *success* and *failure* here as the generic terms for the two outcomes.) A common "experiment" is to perform a repeated number of independent trials and record the total number of *successes* and *failures* in the series regardless of their order of occurrence. Such repeated trials are sometimes called *Bernoulli trials*, after James Bernoulli, who first derived the probability formula for this process.

The sample space for one trial consists of two points: *success* and *failure*. Let $p$ be the probability for a success, then the probability of the opposite outcome will be $1 - p$. If $n$ trials are made, what is the probability of $r$ ($< n$) successes? Suppose the first $r$ trials are successes and the remaining failures. Then, using Eq. (1.8), the joint probability for this would be

$$P(r \text{ successes followed by } (n - r) \text{ failures}) = p^r(1 - p)^{n-r}$$

This is also the probability for any other combination leading to $r$ total *successes*. From combinatorial mathematics, we know that the total number of possibilities is

ELEMENTS OF PROBABILITY THEORY

$$\frac{n!}{r!(n-r)!}$$

so that the total probability is

$$p(r) = \frac{n!}{r!(n-r)!} p^r(1-p)^{n-r} \quad (1.14)$$

Equation (1.14) is known as the *binomial distribution*, and its use is widespread in physics and other sciences. The characteristics of this distribution will be discussed in more detail in Chapter 2.

**Example 1.10:** Consider the scattering of a particle beam from a target. As we saw in Example 1.8, the probability for this is given by the quantum mechanical cross section for the interaction. The probability for a particle to be scattered into a small angular interval between $\theta$ and $\theta + \Delta\theta$ as measured with respect to the incident direction is then

$$p(\theta) = \int_{\theta}^{\theta+\Delta\theta} \sigma'(\theta')d\theta' \approx \sigma'(\theta)\Delta\theta$$

where $\sigma'(\theta)$ is the cross section, and we have used the fact that the interval is small. Suppose now that in the beam there are $N_0$ incident particles per unit time. What is the probability of $r$ scatterings into this angular range within a time period $\Delta t$? Here we clearly have a Bernoulli trials situation, so that the probability will be given by the binomial distribution in Eq. (1.14) with $p = \sigma'(\theta)\Delta\theta$ and $n = N_0\Delta t$.

However, Eq. (1.14) is not very convenient to use, since it involves very large and very small numbers. Indeed, taking high-energy proton–proton scattering as an example, a typical situation might have $N_0$ on the order of $10^6$ or more and $p$ on the order of $10^{-4}$! As will be seen in Chapter 2, in the limit of large $n$ and small $p$ such that at all times the product $Np$ approaches a finite constant, the binomial distribution can be approximated as

$$p(r) = \frac{(np)^r e^{-np}}{r!} \quad (1.15)$$

This is more commonly known as the *Poisson* distribution, which is a distribution in its own right and not just an approximation. It has the advantage of being dependent only on the product $np$, which can be identified as the average number of successes, rather than the individual values of $n$ and $p$. In these types of problems, of course, this quantity is much more accessible to experiment and easier to handle.

Like the binomial distribution, the Poisson distribution has widespread applicability in the sciences and elsewhere. The number of disintegrations in a given period of time in a radioactive sample, for instance, is another well-known example of a Poisson distributed process. Similarly, the probability of finding $r$ bit errors on a given magnetic disk is given by a Poisson distribution as is the probability of $r$ snow flakes falling on a given square centimeter of ground. Other examples are given in Chapter 2.

**Example 1.11:** Many phenomena and, in particular, measurement errors obey a distribution known as the *normal* or *Gaussian* distribution:

$$f(x) = \frac{1}{\sqrt{2\pi\sigma^2}} e^{-(x-\mu)^2/2\sigma^2} \tag{1.16}$$

where the parameters are the mean and standard deviation of the distribution. These terms will be defined in Section 1.3.6.

The Gaussian distribution is a continuous distribution and is perhaps the most widely applied distribution of all. It is the subject of the central limit theorem which will be covered in Section 1.3.9 and thus holds a special place in probability theory. More details on this distribution can be found in Chapter 2.

## 1.3.5 Expected Values

An important notion is the expected value or expectation of a random variable. If $X$ is a random variable distributed as $p(x)$, the expected value of $X$, which we will denote as $E(X)$, is defined as

$$E(X) = \sum_i x_i p(x_i) \tag{1.17}$$

where the summation is over the entire sample space of $X$; i.e., over all admissible values of $X$. If the space is continuous then the summation is replaced by the integral

$$E(X) = \int x f(x) dx \tag{1.18}$$

where $f(x)$ is the probability density. Similarly if $g(X)$ is a function of $X$, the expected value of $g(X)$ is

$$E[g(X)] = \int g(x) f(x) dx \tag{1.19}$$

where we have given the expression for a continuous sample space. To avoid repetition, we will, henceforth, present only one or the other of the cases, under the assumption that the other case is obtained by the appropriate substitution of an integral or summation. The expectation value is thus the probability weighted average of a random variable.

From its definition, we can see immediately that the expectation operation is linear; i.e.,

$$E(aX_1 + bX_2) = aE(X_1) + bE(X_2) \qquad (1.20)$$

where $a$ and $b$ are constants. This is true whether $X_1$ and $X_2$ are independent or not. In the case where $X_1$ and $X_2$ are indeed independent, then

$$E(X_1 X_2) = \sum_{i,j} x_{1_i} x_{2_j} p(x_{1_i}, x_{2_j}) = \sum_{i,j} x_{1_i} x_{2_j} p_{X_1}(x_{1_i}) p_{X_2}(x_{2_j})$$
$$= E(X_1)E(X_2) \qquad (1.21)$$

where we have used Eq. (1.12)

### 1.3.6 Moments and the Mean and Variance

Using the expectation operator, we can define the moments of the random variable $X$. The $r$th moment of $X$ about a point $x_0$ is given by the expectation value of $(X - x_0)^r$, where $r$ is an integer,

$$\mu' = E[(X - x_0)^r] \qquad (1.22)$$

The analogy here with the moments of a mass distribution in mechanics becomes evident. Indeed, it can be shown that the set of all moments, $\mu'$, completely describes the distribution $p(x)$. This is useful for theoretical purposes, but in more practical work, only the very lowest moments have any particular significance and, indeed, in most applications it is usually sufficient to know only the first two moments.

The first moment about zero, usually designated simply as $\mu$, is then given by

$$\mu = E(X) = \int x f(x) dx \qquad (1.23)$$

This moment is generally known as the *mean* of the distribution and essentially locates the "center of mass" of the distribution. The second moment about the mean,

$$V(X) = E[(X - \mu)^2] = \int (x - \mu)^2 f(x) dx \qquad (1.24)$$

is known as the *variance* of the distribution. Expanding the binomial in Eq. (1.24) and evaluating the expectation of each of the individual terms, we also obtain the useful formula

$$V(X) = E(X^2 - 2\mu X + \mu^2) = E(X^2) - 2\mu E(X) + \mu^2$$
$$= E(X^2) - \mu^2 \tag{1.25}$$

The positive square root of the variance,

$$\sigma_X = \sqrt{V(X)} \tag{1.26}$$

is known as the *standard deviation*. This parameter is essentially a measure of the width or dispersion of the distribution about its mean value. Thus, large values of the standard deviation relative to the mean indicate that $X$ varies widely about $\mu$ while a smaller $\sigma_X$ would indicate an $X$ that tends to be clustered about the mean. When cited in conjunction with the result of a measurement, as we will see later, the standard deviation is interpreted as the *error* or *precision* of the measurement.

**Example 1.12:** Let us calculate the mean value of throwing two dice. Let $X$ be the sum of the two dice, so that $X = D_1 + D_2$, where $D_1$ and $D_2$ are random variables representing the outcomes of the two dice, respectively. Assume the dice are fair so that $p(d_1) = p(d_2) = 1/6$ for all values of $d_1$ and $d_2$. Then

$$E(X) = E(D_1) + E(D_2) = 2 \sum_{d=1}^{6} d \times \frac{1}{6} = 2 \times 3.5 = 7$$

Similarly, the variance of $X$ is

$$V(X) = E(X^2) - \mu_X^2$$
$$= E(D_1^2 + 2D_1 D_2 + D_2^2) - \mu_X^2$$
$$= 2E(D^2) + 2E^2(D) - \mu_X^2$$

where we have used the fact that $D_1$ and $D_2$ are independent. Evaluating the individual terms then yields

$$V(X) = 2\frac{91}{6} + 2\frac{49}{4} - 49 = \frac{35}{6} = 5.8333$$

**Example 1.13:** Let us discuss sampling and the expectation and variance of the sample mean. A common procedure in experiments is to make a series of repeated measurements and to take the average value of the individual measurements to obtain a more precise result. This, in fact, is an example of the statistical technique known as *sampling*. The notion of sampling is familiar to most people as the random selection of cases from a population too large to measure in its entirety in order to obtain an estimate of the parameters describing the population. For later use in this volume, we will define sampling here in more formal terms.

**Definition:** *Suppose the random variables $X_1, X_2, \ldots, X_n$ have the joint distribution or density*

$$g(x_1, x_2, \ldots, x_n) = f(x_1)f(x_2) \ldots f(x_n)$$

*where the density of each $X_i$ is $f(x_i)$. Then the $X_1, X_2, \ldots, X_n$ are said to be a random sample of size n from the population with density $f(x)$.*

In the case of $n$ repeated measurements, then, each measurement may be thought of as the outcome of $n$ separate random variables having the same distribution so that we have a random sample. Let us see moreover how this leads to a better estimate of the true mean.

Let the random variables $X_i$ ($i = 1, 2, \ldots, n$) represent the outcomes of the $n$ measurements. Each $X_i$ has a mean $\mu$ and variance $\sigma^2$. The average value of the outcomes is then represented by the variate

$$\overline{X} = \frac{1}{n} \sum_i^n X_i \tag{1.27}$$

Now let us calculate the expected value of $\overline{X}_i$. Using Eqs. (1.20) and (1.23), we obtain

$$E(\overline{X}_n) = \frac{1}{n} \sum_i^n E(X_i) = \frac{1}{n} \sum_i^n \mu = \mu \tag{1.28}$$

Thus, the expected value of the sample mean is the true mean of the probability distribution, which is what one would hope to see. Note that Eq. (1.28) contains no reference to the form of the probability distribution, so that it is a general result subject only to the condition that the mean exist.

Let us go further now and also calculate the variance of $\overline{X}$. Using Eq. (1.24),

$$V(\overline{X}) = E[(\overline{X} - \mu)^2] = E\left[\left\{\left(\frac{1}{n}\sum_i X_i\right) - \mu\right\}^2\right]$$

$$= \frac{1}{n^2} E\left[\left(\sum X_i - n\mu\right)^2\right] \tag{1.29}$$

$$= \frac{1}{n^2} E\left[\left\{\sum (X_i - \mu)\right\}^2\right]$$

where we have brought the constant term $\mu$ into the sum. The square of the sum can now be expanded as

$$\left\{\sum (X_i - \mu)\right\}^2 = \sum\sum_{i \neq j} (X_i - \mu)(X_j - \mu) + \sum_i (X_i - \mu)^2 \qquad (1.30)$$

where all terms with $i = j$ have been separated out explicitly. Taking the expectation of the individual terms, we find that the cross term vanishes, so that

$$\begin{aligned} V(\overline{X}) &= \frac{1}{n^2} E\left[\sum (X_i - \mu)^2\right] \\ &= \frac{1}{n^2} \sum E\left[(X_i - \mu)^2\right] \\ &= \frac{n}{n^2} V(X) \\ &= \frac{V(X)}{n} \end{aligned} \qquad (1.31)$$

or

$$\sigma_{\overline{X}} = \frac{\sigma_X}{\sqrt{n}} \qquad (1.32)$$

Thus the error on the sample mean is reduced relative to the single measurement error ($n = 1$) by a factor which goes inversely as the square root of the number of samples. This then is the mathematical demonstration of Tobias Mayer's intuitive feeling that adding similar data actually helps reduce the inherent errors! Again, Eq. (1.31) is a general result since no reference is made to the form of the distribution. These results will be used again when we consider the law of large numbers and measurement errors later in this chapter.

### 1.3.7 Covariance

In the case of a *multivariate distribution,* the mean and variance of each separate random variable $X$, $Y,$, . . . are calculated in the same way except the summation or integration is now over all random variables. In addition, a third quantity can also be defined: the *covariance,*

$$\begin{aligned} \text{Cov}(X, Y) = \text{Cov}(Y, X) &= E[(X - \mu_X)(Y - \mu_Y)] \\ &= E(XY) - \mu_X \mu_Y \end{aligned} \qquad (1.33)$$

where $\mu_X$ and $\mu_Y$ are the means of $X$ and $Y$, respectively. A covariance is defined for each pair of random variables in the distribution. Thus for three random variables, $X$, $Y$ and $Z$, there will be three covariances: Cov($X$, $Y$), Cov($X$, $Z$) and Cov($Y$, $Z$). And in general, for $p$ variates there will be $p(p-1)/2$ covariances.

The covariance provides a measure of the correlation between the different random variables in the distribution. A nonzero value will thus indicate a relation of some sort between the two variables, while the sign of the value will indicate the sense of the relation; i.e., positive or negative. In contrast, if the variables are independent, the covariance is zero. This can easily be shown by using the result of Eq. (1.21) in Eq. (1.33). The converse of this last statement, however, is not true; if the covariance is zero, the variables will not necessarily be independent. Consider, for example, a random variable $X$ that is distributed symmetrically about zero. Let $Y = X^2$. The covariance between these two variables is then given by Eq. (1.33). Because of symmetry, $\mu_X = 0$. For the same reason, $E(XY) = E(X^3) = 0$, so that Cov($X$, $Y$) = 0, even though $X$ and $Y$ are related!

In many applications, the covariance is often expressed as the *correlation coefficient*, $\rho$, defined as

$$\rho_{XY} = \frac{\text{Cov}(X, Y)}{\sigma_X \sigma_Y} \qquad (1.34)$$

In this form, the correlation coefficient can take on values between $-1$ and $+1$, the sign of which indicates the sense of the correlation. And quite obviously, the correlation of a variable with itself is $\rho_{XX} = 1$.

If a number of random variables, $X_1, X_2, \ldots$, are under consideration, it is convenient to collect the variables and covariances into a single matrix the *covariance* or *error matrix*:

$$\mathbf{C} = \begin{pmatrix} \sigma_1^2 & \rho_{12}\sigma_1\sigma_2 & \rho_{13}\sigma_1\sigma_3 & \cdot \\ \rho_{21}\sigma_1\sigma_2 & \sigma_2^2 & \rho_{23}\sigma_2\sigma_3 & \cdot \\ \cdot & \cdot & \sigma_3^2 & \cdot \\ \cdot & \cdot & \cdot & \cdot \end{pmatrix} = (\text{Cov}(X_i, X_j)) \qquad (1.35)$$

The diagonal elements are the variances of the individual variables while the off-diagonal elements are the covariances between the various pairs. Note that the matrix is symmetric, and if the variables are mutually independent, the covariance matrix becomes a simple diagonal matrix. The utility of the covariance matrix will become evident in later chapters.

**Example 1.14:** Let us consider the propagation of errors. A common practical problem in the experimental sciences is to determine the error on a quantity calculated from measurements or other parameters containing errors. In probabilistic terms, the problem is to calculate the variance $\sigma_W^2 = V(W)$ of a random variable function $W = g(X_1, X_2, \ldots, X_n)$ in terms of the variances of the variables $X_1, X_2, \ldots, X_n$. Depending on the complexity of the function, this may not always be an easy task. An approximate general formula based on a first order expansion can be derived, however.

For simplicity, let us consider only two random variables $X$ and $Y$, and let $\mu_X, \sigma_X^2$ and $\mu_Y, \sigma_Y^2$ represent the mean and variance of $X$ and $Y$, respectively. To calculate $\sigma_W^2$, we will use Eq. (1.25). First, expand $g(X, Y)$ to first order about the means $(\mu_X, \mu_Y)$, so that

$$W \approx g(\mu_X, \mu_Y) + \left(\frac{\partial g}{\partial X}\right)_\mu (X - \mu) + \left(\frac{\partial g}{\partial Y}\right)_\mu (Y - \mu_Y) \quad (1.36)$$

All derivatives here are understood to be evaluated at the point $\mu = (\mu_X, \mu_Y)$. Taking the expectation of Eq. (1.36) then yields an approximation for the mean of $W$:

$$\mu_W \approx g(\mu_X, \mu_Y) \quad (1.37)$$

Now squaring Eq. (1.36) and using the previous approximation gives us

$$W^2 \approx \mu_W^2 + 2\mu_W \left[\left(\frac{\partial g}{\partial X}\right)_\mu (X - \mu_X) + \left(\frac{\partial g}{\partial Y}\right)_\mu (Y - \mu_Y)\right]$$
$$+ \left[\left(\frac{\partial g}{\partial X}\right)_\mu (X - \mu_X) + \left(\frac{\partial g}{\partial Y}\right)_\mu (Y - \mu_Y)\right]^2 \quad (1.38)$$

so that

$$\sigma_W^2 \approx E[W^2] - \mu_W^2 \approx E\left\{\left[\left(\frac{\partial g}{\partial X}\right)_\mu (X - \mu_X) + \left(\frac{\partial g}{\partial Y}\right)_\mu (Y - \mu_Y)\right]^2\right\}$$

Expanding the expression in brackets and taking the expected value of each term then results in the propagation of errors formula:

$$\sigma_W^2 \approx \left(\frac{\partial g}{\partial X}\right)_\mu^2 \sigma_X^2 + \left(\frac{\partial g}{\partial Y}\right)_\mu^2 \sigma_Y^2 + 2\left(\frac{\partial g}{\partial X}\right)_\mu \left(\frac{\partial g}{\partial Y}\right)_\mu \mathrm{Cov}(X, Y) \quad (1.39)$$

In the general case of $n$ variables, it is easy to show that

$$\sigma_W^2 \approx \sum_i \sum_j \left(\frac{\partial g}{\partial X_i}\right)_\mu \left(\frac{\partial g}{\partial X_j}\right)_\mu \mathrm{Cov}(X_i, X_j) \quad (1.40)$$

The interesting point to note is that the errors are added quadratically and that, depending on the sign of the covariance, the final error can actually be made to decrease. In principle, therefore, if the variables in an experiment can be arranged so as to be negatively correlated, very small errors on the final result can be obtained. Because of the complexity of most experiments, however, this is much easier said than done!

### 1.3.8 The Law of Large Numbers

In this and the following section, we will discuss the two most important theorems in modern probability theory. The first is the *law of large numbers*. This theorem was first derived by James Bernoulli in a limited form (for a historical account, see [3], Chapter 2, for example) and essentially justifies the intuitive notion that, if an experiment is repeated many times, the fraction of the trials in which the event $A$ occurs approaches the true probability for the event $A$. It thus provides one of the first mathematical justifications for the frequency view of probability.

To demonstrate the theorem, we will make use of a result known as *Chebyshev's inequality*. Let $X$ be a random variable with a probability distribution $p(x)$, then the expected value of $X^2$ is

$$E(X^2) = \sum_i x_i^2 \, p(x_i) \tag{1.41}$$

Suppose we now limit the sum to terms for which $|x_i| \geq \epsilon$, where $\epsilon$ is some positive number. Then Eq. (1.41) becomes an inequality

$$E(X^2) \geq \sum_{|x_i| \geq \epsilon} x_i^2 \, p(x_i) \tag{1.42}$$

Moreover if we substitute the minimum value $\epsilon$ for all the remaining $x_i$,

$$E(X^2) \geq \epsilon^2 \sum_{|x_i| \geq \epsilon} p(x_i) \tag{1.43}$$

The sum in Eq. (1.43) can now be recognized as the probability $P(|X| \geq \epsilon)$. Substituting this back into Eq. (1.43) and rearranging we obtain

$$P(|X| \geq \epsilon) \leq \frac{1}{\epsilon^2} E(X^2) \tag{1.44}$$

which is Chebyshev's inequality.

The law of large numbers, now, is but an instance of Chebyshev's inequality. Suppose $X_1, X_2, \ldots, X_n$ are a random sample of size $n$ from a distribution $f(x)$ with mean $\mu$ and finite variance $\sigma^2$, and that $\overline{X}_n$ is the sample mean as defined in Eq. (1.27). Now set $X$ in Eq. (1.43) to be $\overline{X}_n - \mu$, so that

$$P(|\overline{X}_n - \mu| \geq \epsilon) \leq \frac{1}{\epsilon^2} E[(\overline{X}_n - \mu)^2] \qquad (1.45)$$

The numerator on the righthand side, now, is just the variance of $\overline{X}_n$ as we saw in Eq. (1.31), so that

$$P(|\overline{X}_n| - \mu| \geq \epsilon) \leq \frac{1}{n\epsilon^2} \sigma^2 \qquad (1.46)$$

For a given constant $\epsilon > 0$, then, we have that

$$\lim_{n \to \infty} P(|\overline{X}_n - \mu| \geq \epsilon) = 0 \qquad (1.47)$$

otherwise known as the *weak law of large numbers.*

What this law says is that, for any given constant $\epsilon > 0$, the probability that the sample mean will deviate from the true mean by an amount greater than $\epsilon$ can be made arbitrarily small by choosing an appropriately large sample size $n$. Note that this statement concerns the *probability* of the deviation to exceed the value $\epsilon$ and not the deviation itself. Thus, even in a *large* but finite sample, there is still a chance that a deviation larger than $\epsilon$ can occur. The law of large numbers allows us to conclude only that such occurrences are rare.

An additional point is that the weak law assumes the existence of the mean and variance of the distribution so that it is not applicable in cases where these expectations do not exist. Fortunately, it can still be shown that even in the case of a nonexistent variance, the general limit still holds (see, for example, Sections 10.1 and 10.7, [5] for a discussion).

### 1.3.9 The Central Limit Theorem

The second theorem is the *central limit theorem,* which is one of the most remarkable theorems in all of mathematics. We will not prove this theorem here, since it requires more advanced theoretical notions that are not generally necessary for applied probability. However, the interested reader may find a proof in most probability theory books, such as Section 8.4, [6] or [7].

In words, the theorem essentially states that, if $S$ is the sum of $n$ random variables, $S = X_1 + X_2 + \ldots + X_n$, then under rather mild conditions, as $n$ increases the distribution of $S$ will approach the Gaussian distribution with mean $\mu_S = \Sigma \mu_x$ and variance $\sigma_S^2 = \Sigma \sigma_{X_i}^2$, *regardless* of the original distributions of the $X_i$. To make a more mathematical statement, let $Y$ be a random variable defined as

$$Y = \frac{S - \mu_S}{\sigma_S} \tag{1.48}$$

where $S$, $\mu_S$ and $\sigma_S$ are defined as earlier. Then, as $n \to \infty$,

$$F(S \leq y) \to \Phi(y) = \int_{-\infty}^{y} \frac{1}{\sqrt{2\pi\sigma^2}} e^{-x^2/2\sigma^2} dx \tag{1.49}$$

where $\Phi(y)$ is the cumulative distribution of the Gaussian. The definition of the variable $Y$ is just a "normalizing" transformation of the variable $S$ such that new variable will be described by a Gaussian distribution with mean 0 and variance 1.

The central limit theorem essentially explains why Gaussian distributions are so ubiquitous, for many random phenomena are, in fact, the sum of many different random factors. Instrumental error is one particular example that will be discussed in the next section.

One possible practical application of the central limit theorem is the computer generation of random numbers. This is necessary for Monte Carlo simulations (see Chapter 15), where numbers distributed according to Gaussian and other distributions are needed to simulate various physical processes. These numbers must all be derived from a single source of uniformly distributed pseudo-random numbers. To obtain Gaussian distributed numbers, therefore, one could sum a sufficient number $n$ of the uniformly distributed numbers and adjust this with appropriate normalization factors. With $n = 12$, for example, a good approximation is already obtained [8]. (From a computer programming point of view, this method is somewhat inefficient because of the large number of operations required per Gaussian number generated. More efficient algorithms are described in Chapter 5.)

## 1.4 Modeling Measurement

We now turn to the application of probability theory to the measurement process and the uncertainties that occur. To use probability efficiently, we must model the actual random process or processes involved such that probabilities can be realistically assigned or inferred. In any modeling, of course, some idealization will be necessary in order to simplify the problem, with the consequent risk of oversimplification. On the other hand, modeling too much detail can lead to an overly complicated model, which is incalculable or which will lead to data analysis problems afterwards. Modeling an experiment, therefore, requires not only a knowledge of the processes involved but a feeling for the relative magnitudes of the probabilities. Only then can appropriate simplifications or approximations be made to allow a meaningful result to be obtained. At all times, however, checks should be made as to how realistic the model is.

At the foundation of all experiments are the measurements that are obtained using instruments of some type. In general, the uncertainties on these measurements will consist of two parts: a *systematic* error and a *random* error.

### 1.4.1 Systematic Errors

Systematic errors concern the possible biases that *may* be present in an observation. A common example is the zeroing of a measuring instrument such as a balance or a voltmeter. Clearly, if this is not done properly, all measurements made with the instrument will be offset or *biased* by some constant amount. However, even if the greatest of care is taken, one can never be certain that the instrument is *exactly* at the zero point. Indeed, various physical factors such as the thickness of the scale lines, the lighting conditions under which the calibration is performed, and the sharpness of the calibrator's eyesight will ultimately limit the process, so that one can say only that the instrument has been "zeroed" to within some range of values, say $0 \pm \delta$. This uncertainty in the "zero value" then introduces the possibility of a bias in all subsequent measurements made with this instrument; i.e., there will be a certain nonzero probability that the measurements are biased by a value as large as $\pm \delta$.

More generally, systematic errors arise whenever there is a comparison between two or more measurements. And indeed, some reflection will show that all measurements and observations involve comparisons of some sort. In the preceding case, for example, a measurement is referenced to the "zero point" (or some other calibration point) of the instrument. Similarly, in detecting the presence of a new particle, the signal must be compared to the background events that could simulate such a particle, etc. Part of the art of experimentation, in fact, is to ensure that systematic errors are sufficiently small for the measurement at hand, and indeed, in some experiments how well this uncertainty is controlled can be *the* key success factor.

One example of this is the measurement of parity violation in high-energy electron–nucleus scattering. This effect is due to the exchange of a $Z^0$ boson between electron and nucleus and manifests itself as a tiny difference between the scattering cross sections for electrons that are longitudinally polarized parallel ($d\sigma_R$) and antiparallel ($d\sigma_L$) to their line of movement. This difference is expressed as the *asymmetry* parameter, $A = (d\sigma_R - d\sigma_L)/(d\sigma_R + d\sigma_L)$, which has an expected value of $A \approx 9 \times 10^{-5}$ [9].

To perform the experiment, a longitudinally polarized electron beam is scattered off a suitable target, and the scattering rates are measured for beam polarization parallel and antiparallel. To be able to make a valid comparison of these two rates at the desired level, however, it is essential to maintain *identical* conditions for the two measurements. Indeed, a tiny change in any number of parameters, for example, the energy of the beam, could easily create an artificial difference between the two scattering rates, thereby masking any real effect. The major part of the effort in this experiment, therefore, is to identify the possible sources of systematic error, design the experiment so as to minimize or eliminate as many of these as possible and monitor those that remain!

Systematic errors are distinguished from random errors by two characteristics. First, in a series of measurements taken with the same instrument and calibration, *all* measurements will have the same systematic error. In contrast, the *random errors* in these same data will fluctuate from measurement to measurement in a completely *independent* fashion. Moreover, the random errors may be decreased by making repeated measurements as shown by Eq. (1.32). The systematic errors, on the other hand, will remain constant no matter how many measurements are made and can be decreased only by changing the method of measurement. Systematic errors, therefore, *cannot* be treated using probability theory, and indeed there is no general procedure for this. One must usually resort to a case by case analysis, and as a general rule, systematic errors should be kept separate from the random errors.

A point of confusion, which sometimes occurs, especially when data are analyzed and treated in several different stages, is that a random error at one stage can become a systematic error at a later stage. In the first example, for instance, the uncertainty incurred when zeroing the voltmeter is a random error with respect to the zeroing process. The "experiment" here is the positioning of the pointer exactly on the zero marking and one can easily imagine doing this process many times to obtain a distribution of "zero points" with a certain standard deviation. Once a zero calibration is made, however, subsequent measurements made with the instrument will all be referred to that particular zero point and its error. For *these* measurements, the zero-point error is a systematic error. Another similar example is the least-squares (see Chapter 9) fitted calibration curve. Assuming that the calibration is a straight line, the resulting slope and intercept values for this fit will *contain* random errors due to the calibration measurements. For all subsequent measurements referred to this calibration curve, however, these errors are not random but systematic.

## 1.4.2 Random Errors

Random errors, as we have noted throughout this chapter, can be treated using the theory of probability. The source of these uncertainties can be instrumental imprecisions or the inherent random nature of the process being measured or both. Mathematically, both types are handled in the same way as we will see.

**Instrumental Errors.** Instrumental errors arise from the inability of the observer and the measuring instrument to control all the variables that can affect the measurement. For example take the simple act of measuring the distance between two points with a ruler. If a series of repeated measurements is made reading the values as accurately as possible, the observer will remark that a spread of values is obtained. This might be due to changes in the lighting that affect the observer's reading of the scale, or changes in the temperature that cause the ruler to slightly expand or contract, or the failure of the observer to place the beginning of the ruler at exactly the same point each time or changes in the reading due to parallax effects, or play in any joints that the ruler might have, etc. The point is that there are many little factors which the experimenter does not always control that can vary haphazardly from measurement to measurement. Of course, the more the experimenter brings these factors under control, the less variation there will be. The instrument or the measurement is then said to be more precise. It is clear, though, that one can never control all factors so that there will always be some imprecision in the final result.

Let us try to model this simple situation. Suppose $x$ is the true value that we are seeking to measure and suppose we make $n$ repeated measurements, $Y_i$, with the same instrument. What is observed, then, are values $y_i$:

$$y_i = x + e_i \qquad (i = 1, 2, \ldots, n) \qquad (1.50)$$

where $e_i$ is a random error distributed according to some probability density $f(e)$. The value $x$, therefore, is essentially nonobservable, at least directly, in that it is always accompanied by the random value $e$. And indeed all that can be measured directly is the distribution $f(e)$, offset by the value $x$. The measurements, $Y_i$, then, are just a random sample from this distribution.

For most instruments, we can argue, in fact, that $f(e)$ is a Gaussian with mean 0 and a variance $\sigma^2$. Indeed, if $e$ represents the sum of many random factors, then, the central limit theorem ensures that $f(e)$ will tend towards a normal distribution, regardless of the distributions of the individual factors. The expected value of $Y_i$ is then $E(Y_i) = x$, which is the true value being sought.

As we have already mentioned, the standard deviation, $\sigma$, is usually interpreted as the *precision* of the instrument. This view is also consistent mathematically. Consider our series of measurements $Y_i$. To estimate the true value $x$, we would take the average $\overline{Y}$, and from Eq. (1.32), we saw that the error on this average was $\sigma/\sqrt{n}$. Thus, as $n$ increases, the error on the measurement decreases, as one would expect. If now a single measurement is made, i.e., $n = 1$, the error would be $\sigma$, which corresponds to what is normally called the *precision* of the instrument.

In the preceding discussion, we ignored the possibility of a systematic error. To account for this, we must modify Eq. (1.50) to

$$y_i = x + s + e_i \qquad (1.51)$$

where $s$ is the systematic error. As we have noted, this is the same for each measurement so that it bears no subscript. It thus enters into the equation in a manner different to $e$. The expected value of $Y$ is then $E(Y) = x + s$, which demonstrates why this error must be treated differently.

**Statistical Errors.** A second source of random error arises when the process being observed or measured is itself a random process. This is the case in nuclear and particle physics experiments, which almost always involve the counting of events such as the number of radioactive decays in a given period of time or the number of particles scattered at a certain angle. As we saw in Example 1.10, however, the number of reactions occurring in a given unit of time is actually described by a Poisson distribution. What is of interest, then, is the mean value of the distribution, which is usually the parameter containing the physics. Clearly, the act of measurement, here, is nothing more than the act of sampling from these distributions in order to estimate the mean. Similarly, the measurement uncertainty incurred can be identified as the sampling error. Because they result from the randomness of the process itself, these are also called *statistical* errors. From a probability and statistical point of view, however, we have the same situation as with instrumental errors.

As an example, let us try to model a simple but realistic experiment, say the counting of the number of disintegrations in a period $\Delta t$ from a radioactive source with a detector. To keep the example simple let us assume that the source emits only one type of radiation, say, gamma rays. The observable variable in this case will be the number of counts $N_r$ registered by the detector.

Now, in the experiment, there will be two sources of events: one from the radioactive source itself and the other from background sources such as cosmic rays, radioactive elements in the surroundings, or radioactive contaminants in the source itself. As a first simplification let us assume that there is only one $\gamma$-ray line, so that the energy does not vary. Let $N_s$ be the number of events emitted by the radioactive source in time $\Delta t$ and $N_b$, the number from background sources in the same period. From Example

1.10, we can argue that the number of these events, in both cases, is governed by the Poisson distribution, Eq.(1.15), so that

$$p_{\text{src}} = \frac{(\lambda_s \Delta t)^{n_s} e^{-\lambda_s \Delta t}}{n_s!}$$

$$p_{\text{bkg}} = \frac{(\lambda_b \Delta t)^{n_b} e^{-\lambda_b \Delta t}}{n_b!} \qquad (1.52)$$

where $\lambda_s$ and $\lambda_b$ are the mean decay rates per unit time for the source and the background, respectively. This, however, is not what is observed since the detector has an efficiency, $\epsilon < 1$. This efficiency includes the geometric solid angle subtended by the detector relative to the source as well as its *intrinsic* detection efficiency, the latter being the probability that the detector will actually register an event when a gamma ray enters into its active volume. As a second simplification, let us assume $\epsilon$ to be a constant. Clearly, the probability that the detector will register $n_r$ counts when $n_s + n_b$ are emitted is a "Bernoulli trials" situation, so that the binomial distribution in Eq. (1.14) applies. Thus,

$$p_{\text{det}}(n_r \mid n_s, n_b) = \frac{(n_s + n_b)!}{n_r!(n_s + n_b - n_r)!} \epsilon^{n_r} (1 - \epsilon)^{n_s + n_b - n_r} \qquad (1.53)$$

This provides us now with the relevant quantities needed to describe the experiment. Using Eq. (1.11), the joint probability is then

$$p(n_r, n_s, n_b) = p_{\text{det}}(n_r \mid n_s, n_b) p_{\text{src}}(n_s) p_{\text{bkg}}(n_b) \qquad (1.54)$$

which allows us to calculate the marginal probability for observing $n_r$ registered counts, i.e.,

$$p_{\text{obs}}(n_r) = \sum_{n_s} \sum_{n_b} p(n_r, n_s, n_b) \qquad (1.55)$$

This then is our mathematical probability model describing the experiment.

Let us have a look at some of the expected values. The expected number of counts is then

$$\begin{aligned}E(N_r) &= \sum_{n_r} n_r p_{\text{obs}}(n_r) = \sum_{n_r} \sum_{n_s} \sum_{n_b} n_r p_{\text{det}}(n_r \mid n_s, n_b) p_{\text{src}}(n_s) p_{\text{bkg}}(n_b) \\ &= \sum_{n_s} \sum_{n_b} \epsilon(n_s + n_b) p_{\text{src}}(n_s) p_{\text{bkg}}(n_b) = \epsilon(\lambda_s + \lambda_b) \Delta t\end{aligned} \qquad (1.56)$$

which is just the sum of the mean decay rates times the efficiency of the detector. Similarly, the variance is

$$V(N_r) = E(N_r^2) - E^2(N_r) = \sum_{n_r}\sum_{n_s}\sum_{n_b} n_r^2 p_{\text{obs}}(n_r)$$
$$- [\epsilon(\lambda_s + \lambda_b)\Delta t]^2$$
$$= \sum_{n_s}\sum_{n_b} \epsilon(n_s + n_b)[(n_s + n_b - 1)\epsilon \quad (1.57)$$
$$+ 1]p_{\text{src}}(n_s)p_{\text{bkg}}(n_b) - [\epsilon(\lambda_s + \lambda_b)\Delta t]^2$$
$$= \epsilon(\lambda_s + \lambda_b)\Delta t$$

Note that the variance is equal to the mean, which, as shown in Chapter 2, is characteristic of Poisson processes.

By making a series of measurements we can thus estimate $E(N_r)$ and its error. To obtain the mean now, one must know the efficiency of the detector $\epsilon$ and the background rate $\lambda_b$. From our model, however, it is clear that these two bits of information must be provided independently, for there is no way that they can be obtained from a single measurement. If these are not already known, then two additional independent experiments must be made under the same conditions, or in some cases such as those outlined later, a calculation such as a Monte Carlo simulation might be necessary to obtain this information. In any case, the final value of $\lambda_s$ will contain the random errors from these values as well as the error from the first experiment as given by the propagation of errors formula in Eq. (1.40). Moreover, a systematic error will be incurred, since there will be an uncertainty over whether identical conditions were actually maintained between the different experiments.

Still another source of uncertainty is in the model itself. For instance, the efficiency of the detector is generally a function of radiation type, its energy, the total count rate, as well as the specific geometry of the experimental setup (see Chapter 5, [4], for example). In our simple model, this was assumed to be constant, which in many experiments is adequate. There may also be correlations between the source and the background rate such that certain background events occur only when the source is in place. In this case, a background measurement would not be indicative of the actual background present during the real measurement, so that any background subtraction would be incorrect. In case of doubt, further experimental checks might be in order, therefore, to see if in fact the model actually corresponds to the experiment. Alternatively, changes in the experimental design might be made to better conform to the model!

## References

1. Hamming, Richard W. (1991). *The Art of Probability for Scientists and Engineers.* Addison-Wesley Publishing Company, Redwood City, CA.
2. Ballantine, L. E. (1986). "Probability Theory in Quantum Mechanics." *Am. J. Phys.* **54,** 883.
3. Stigler, Steven M. (1986). *The History of Statistics,* p. 16. The Belknap Press of Harvard University Press, Cambridge, MA.
4. Leo, W. R. (1987). *Techniques for Nuclear and Particle Physics Experiments.* Springer-Verlag, New York.
5. Feller, William. (1968). *An Introduction to Probability Theory and Its Applications,* **1,** 3rd ed. John Wiley & Sons, New York.
6. Feller, William. (1966). *An Introduction to Probability Theory and Its Applications,* **2.** John Wiley & Sons, New York.
7. Kendall, M. G., and Stuart, M. A. (1963). *The Advanced Theory of Statistics.* Charles Griffen and Co., London.
8. Abramovitz, M. A., and Stegun, I. (1964). *Handbook of Mathematical Functions.* National Bureau of Standards, Washington DC.
9. Leader, Elliot, and Pedrazzi, Enrico. (1983). *An Introduction to Gauge Theories and the New Physics,* p. 92–97. Cambridge University Press, Cambridge.

# 2. COMMON UNIVARIATE DISTRIBUTIONS

## Laurent Hodges
Iowa State University, Ames

## 2.1 Introduction

Although in theory there is no limit to the number of possible univariate probability distributions, a few distributions occur repeatedly in scientific work. In this chapter we will describe seven such distributions. Three of these are discrete distributions (the binomial, the Pascal or negative binomial, and the Poisson distributions) and four of them are continuous distributions (the Gaussian or normal, the lognormal, the exponential, and the Weibull distributions).

## 2.2 Discrete Probability Mass Functions

### 2.2.1 The Binomial Distribution

The binomial probability mass function is a very common discrete probability mass function that has been studied since the 17th century. It applies to many experiments in which there are two possible outcomes, such as heads–tails in the tossing of a coin or decay–no decay in radioactive decay of a nucleus. More generally, it applies whenever the possible outcomes can be divided into two groups.

Formally, the binomial probability mass function applies to a *binomial experiment,* which is an experiment satisfying these conditions:

- The experiment consists of $n$ identical and independent trials, where $n$ is chosen in advance.
- Each trial can result in one of two possible outcomes, success ($S$) or failure ($F$), with the probability $p$ of success being a constant from trial to trial.

An example is the tossing of a fair coin $n$ times, with success defined as "heads up": the experiment consists of $n$ identical tosses, the tosses are independent of one another, there are two possible outcomes (heads =

success and tails = failure), and the probability of success $p = 1/2$ is the same for every trial. In practice, an experiment with three or more possible outcomes can be considered to be a binomial experiment if one focuses on one possible outcome, referring to it as success and all other outcomes as failure. For example, in elementary particle physics, in studying the decay of a kaon ($K$ particle), which can occur in many different ways, one might regard just its decay into two pions as a success.

The analysis of a binomial experiment is straightforward. If we denote success by $S$ and failure by $F$, then the possible outcomes for $n$ trials are easily enumerated for small values of $n$:

$n = 1$: *F, S*

$n = 2$: *FF, FS, SF, SS*

$n = 3$: *FFF, FFS, FSF, FSS, SFF, SFS, SSF, SSS*

$n = 4$: *FFFF, FFFS, FFSF, FFSS, FSFF,*

*FSFS, FSSF, FSSS, SFFF, SFFS, SFSF,*

*SFSS, SSFF, SSFS, SSSF, SSSS*

It is readily seen that there will always be exactly $2^n$ possible outcomes. Any single outcome consisting of $x$ successes and $n - x$ failures has a probability of $p^x(1 - p)^{n-x}$. For example, if an unbiased coin ($p = 0.5$) is tossed $n$ times, the probability of any particular outcome is $(0.5)^x(1 - 0.5)^{n-x} = (0.5)^n = 1/2^n$; in this case all $2^n$ outcomes are equally likely.

It is often the case that the exact sequence of successes and failures is not important to the problem under consideration, but that the number of successes and failures is important. Then we group the possible outcomes into those with 0 successes, 1 success, 2 successes, . . . , $n$ successes. The number of possible outcomes with $x$ successes is just the number of combinations of $n$ objects taken $x$ at a time:

$$\frac{n!}{x!(n-x)!}, \quad \text{usually denoted} \quad \binom{n}{x}$$

This may be checked for the previous case $n = 4$:

$\binom{4}{0} = 1$ (corresponding to the outcome *FFFF*)

$\binom{4}{1} = 4$ (corresponding to the outcomes *FFFS, FFSF, FSFF, SFFF*)

$\binom{4}{2} = 6$    (corresponding to the outcomes *FFSS, FSFS, FSSF, SFFS, SFSF, SSFF*)

$\binom{4}{3} = 4$    (corresponding to the outcomes *SSSF, SSFS, SFSS, FSSS*)

$\binom{4}{4} = 1$    (corresponding to the outcome *SSSS*)

This analysis of the binomial experiment provides us with a succinct formula for the binomial probability mass function $b(x; n, p)$ for $x$ successes in $n$ trials, with $p =$ the probability of success in each trial; it is

$$b(x; n, p) = \binom{n}{x} p^x (1 - p)^{n-x}$$

$$= \frac{n!}{x!(n - x)!} p^x(1 - p)^{n-x} \quad (2.1)$$

$$\text{for } x = 0, 1, 2, \ldots, n$$

$$= 0 \quad \text{for other values of } x$$

For the case $n = 4$ this gives

$$b(0; 4, p) = (1 - p)^4$$
$$b(1; 4, p) = 4p(1 - p)^3$$
$$b(2; 4, p) = 6p^2(1 - p)^2$$
$$b(3; 4, p) = 4p^3(1 - p)$$
$$b(4; 4, p) = p^4$$

Figure 1 shows this probability mass function for the case $p = 0.5$.

The mean of the binomial probability mass function is $E(X) = np$, and its variance is $V(X) = np(1 - p) = npq$, where $q = 1 - p$.

There is no simple expression for the cumulative distribution function of the binomial distribution. The probability that the number of successes is $\leq x$ is denoted $B(x; n, p)$, and it equals

$$B(x; n, p) = \sum_{y=0}^{x} b(y; n, p) \quad \text{for } x = 0, 1, 2, \ldots, n \quad (2.2)$$

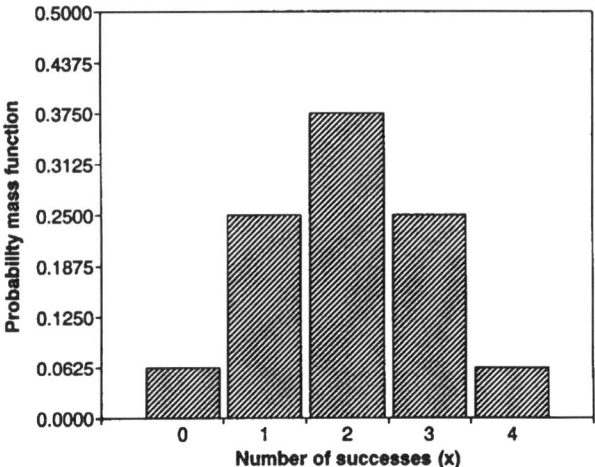

FIG. 1. The binomial distribution $b(x; 4, 0.5)$.

Figure 2(a) shows the binomial distributions for $p = 0.05$, 0.5, and 0.9 in the case of $n = 50$ trials; for all three values of $p$ the distribution is nonzero at all points, but it is often too small to appear in this figure. Figure 2(b) shows the cumulative binomial distributions for the same three cases. These are easily calculated on a computer.

**Example 2.1:** Consider the decimal digits of $\pi$. An unanswered question in mathematics is whether the decimal digits of $\pi = 3.14159 \ldots$ are "randomly" distributed according to the results of a binomial experiment. Suppose we choose a digit (0, 1, ..., 9) and count the number of times it appears in 100 consecutive significant decimal figures of $\pi$. What would we expect if the count were binomially distributed? The probability that the digit would occur $x$ times is $b(x; 100, 0.1)$. We would expect this to be largest for values of $x$ near $np = 10$, and this is borne out by the theoretical binomial distribution shown by the solid line in Figure 3. What do we actually find? If we count the number of times 7 appears in the first 100 significant figures of $\pi$, then the next 100, and so forth, for 250 "trials," we find the distribution represented by the histogram in Figure 3. The fit is excellent. Note that although 10 out of 100 is the most common, occasionally the count is considerably larger or smaller than 10. Tests carried out on billions of digits of $\pi$ show excellent agreement with the predictions of the binomial distribution; we might expect eventually to find 100 7s in a

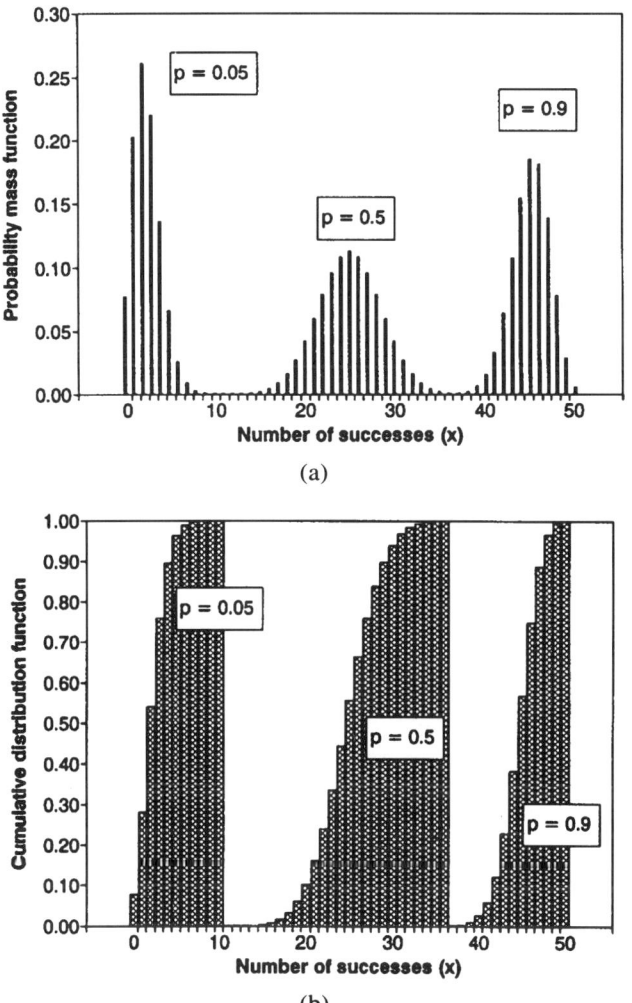

FIG. 2. (a) The binomial probability mass function $b(x; 50, p)$ for $p = 0.05$, 0.5, and 0.9. (b) The binomial cumulative distribution function $B(x; 50, p)$ for $p = 0.05$, 0.5, and 0.9.

row, though this might entail determining more digits than there are atoms in the universe!

**Comments.** For large values of $n$ the binomial distribution can sometimes be approximated by either the discrete Poisson distribution or the

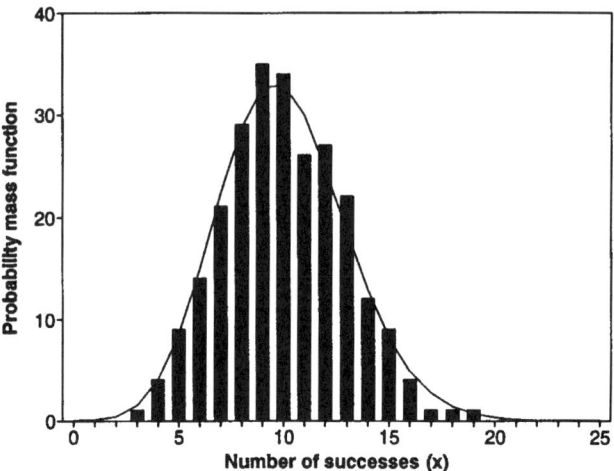

FIG. 3. Solid curve: the binomial distribution $b(x; 100, 0.1)$. Histogram: the distribution of numbers of 7s in 100-digit groups in the first 25,000 decimal digits of $\pi = 3.1415926535\ldots$

continuous normal distribution. These possibilities are discussed in later sections dealing with those distributions.

### 2.2.2 The Pascal or Negative Binomial Distribution

The Pascal or negative binomial distribution is the discrete probability mass function characterizing a binomiallike experiment (a sequence of identical, independent trials, each of which has a probability $p$ of success) that continues until a total of $r \geq 0$ successes have been observed. Thus it differs from the binomial distribution in that it is not the number of trials that is fixed, but the number of successes.

For example, in the case $r = 1$ the possible outcomes are *S, FS, FFS, FFFS,* and so forth; if the probability of success is $p$ then these outcomes have probabilities $p$, $(1 - p)p$, $(1 - p)^2 p$, $(1 - p)^3 p$, and so forth. In the case $r = 2$, the possible outcomes are *SS, FSS, SFS, FFSS, FSFS, SFFS, FFFSS, FFSFS, FSFFS, SFFFS,* and so forth, with probabilities $p^2$, $(1 - p)p^2$, $(1 - p)p^2$, $(1 - p)^2 p^2$, etc.

This distribution is usually expressed in terms of the number $x$ of failures preceding the $r$th success; the possible values of $x$ are then the nonnegative integers 0, 1, 2, . . . . This corresponds to the $r$th success occurring on the $(r + x)$th trial. The Pascal or negative binomial probability mass function is easily determined by combinatorial arguments to be

$$P(x;\ r,\ p) = \binom{x+r-1}{r-1} p^r (1-p)^x \quad \text{for } x = 0, 1, 2, \ldots$$
$$= 0 \quad \text{for other values of } x \tag{2.3}$$

As an example, in the case $x = 2,\ r = 2$ we have

$$P(2;\ 2,\ p) = \binom{3}{1} p^2 (1-p)^2 = 3p^2 (1-p)^2$$

This corresponds to the fact that there are three outcomes consisting of two failures and two successes, ending with a success; namely, *FFSS, FSFS,* and *SFFS.*

The mean of the Pascal distribution is $E(X) = r(1-p)/p$ and its variance is $V(X) = r(1-p)/p^2 = E(X)/p$.

As with the binomial probability mass function, there is no simple formula for the cumulative distribution function. The probability that the number of failures is $\leq x$ is

$$\sum_{y=0}^{x} P(y;\ r,\ p) \quad \text{for } x = 0, 1, 2, \ldots$$

The cumulative distribution function equals $p^r$ for $x = 0$ and approaches 1 as $x \to \infty$; it typically rises sharply over a relatively narrow range of $x$.

Figure 4(a) shows examples of the Pascal distribution for the case $r = 10$ when $p = 0.3$ and $0.7$. Figure 4(b) shows the corresponding cumulative distribution functions as well as that for $p = 0.5$.

**Example 2.2:** A study is being carried out in a large metropolitan area to compare the winter energy consumption in single-family homes heated by natural gas furnaces with those heated electrically. The electric and gas utilities have agreed to provide the energy records if the homeowners give their consent. It is decided to use a random telephone survey to locate homeowners. The telephone company knows that for the exchanges used in that city there is a 52% chance that a number called at random will be a residence. It is also known that about 70% of the dwelling units are single-family homes and that about 60% of these use natural gas furnaces, while 35% use electric heating. Survey experts predict that about 75% of the eligible respondents can be expected to give their approval for the investigators to obtain their energy records. For statistical purposes the investigators want at least 300 homes in each of the two groups. How many random telephone numbers should be selected for calling to have a 95% chance of having at least 300 in each group?

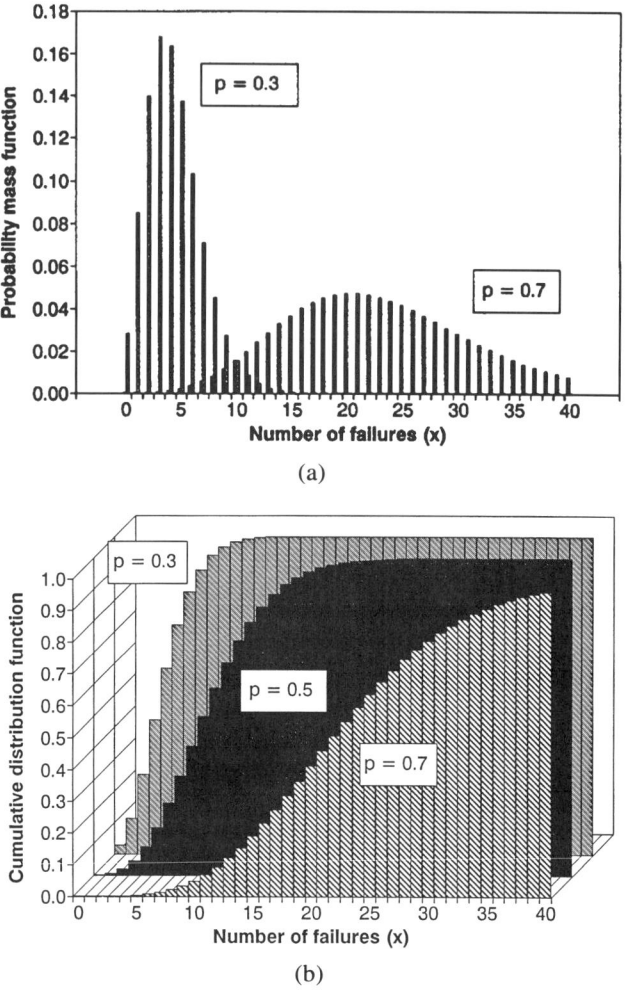

FIG. 4. (a) The Pascal (negative binomial) distribution $P(x; 10, p)$ for $p = 0.3$ and 0.7. (b) The Pascal (negative binomial) cumulative distribution function for $p = 0.3, 0.5,$ and 0.7.

Suppose we assume that the probabilities (of a home being heated by a given fuel, of the homeowners agreeing to participate, etc.) are independent of one another. From the numbers quoted, the chance that a number called will reach a willing homeowner in a single-family home heated with natural gas is $0.75 \times 0.52 \times 0.70 \times 0.60 = 0.1638$, while the chance that it will reach a willing homeowner in an electrically heated single-family home is

$0.75 \times 0.52 \times 0.70 \times 0.35 = 0.09555$. Since the electrically heated homes are scarcer, the number of random phone numbers needed will be determined by them. On the average, $300/0.09555 = 3140$ random phone numbers will have to be called to ensure having 300 electrically heated homes. To have a 95% chance of finding this number of electrically heated homes is equivalent to using the negative binomial distribution for $r = 300$ and $p = 0.09555$, and finding the value of $x$ for which the negative binomial cumulative distribution exceeds 0.95. The number of random telephone numbers that must be drawn for the survey is then $x + r = x + 300$, where $x$ is found from

$$0.95 = \sum_{y=0}^{x} P(y; 300, 0.09555)$$

$$= \sum_{y=0}^{x} \binom{y + 299}{299} (0.09555)^{300}(0.90445)^y$$

(2.4)

A computer calculation shows that $x = 3129$, so a total of $x + 300 = 3429$ randomly selected telephone numbers would be needed. From the cumulative probability mass functions shown in Figure 4(b), we would guess that replacing 0.95 by a different number would not make a drastic difference in the number of phone numbers needed. In fact, computer calculations show that to have a 10% chance of finding the 300 electrically heated homes would require $x = 2621$, while to have a 99% chance would require $x = 3255$.

### 2.2.3 The Poisson Distribution

The Poisson distribution is another common discrete distribution that is defined as follows. The Poisson probability distribution for $x$ occurrences of a phenomenon when the average number of occurrences equals $\lambda$ is

$$p(x; \lambda) = \frac{e^{-\lambda} \lambda^x}{x!} \quad \text{for } x = 0, 1, 2, \ldots \text{ and some } \lambda > 0$$

$$= 0 \quad \text{for other values of } x$$

(2.5)

The Poisson distribution occurs in several important situations. It is the limit of the binomial distribution $b(x; n, p)$ when one lets $n \to \infty$ and $p \to 0$ in such a way that $np$ remains fixed at the value $\lambda > 0$. In practice, the Poisson distribution is an adequate approximation to the binomial distribution if $n \geq 100$ and $p \leq 0.05$.

The Poisson distribution also occurs in a more fundamental context, as a distribution associated with a "Poisson process," which is a process satisfying the following conditions:

1. The number of occurrences in an interval of time $[t, t + \Delta t]$ is independent of the number of occurrences in any other interval that does not overlap with this interval.
2. For small $\Delta t$, the probability of success in the time interval is approximately proportional to $\Delta t$.
3. If the time interval $\Delta t$ is small enough, the probability of more than one occurrence in the interval is small compared to $\Delta t$.

For a Poisson process, the mean number of occurrences in an interval $\Delta t$ can thus be expressed as $\gamma \Delta t$, and $1/\gamma$ is the mean time between occurrences. (See Chapter 3 for more information on the Poisson process.)

An example of a Poisson process is the decay of radioactive nuclei in a sample containing a large number of these nuclei. Over 100 radioactive nuclei occur naturally (such as carbon-14, potassium-40, and uranium-238), plus many others are produced in laboratories. It has been observed, to a very high degree of accuracy, that every radioactive nucleus has a fixed probability of decaying in any particular time interval, and that this probability is independent of how long the nucleus has already been in existence. As an example, the probability that a uranium-238 nucleus will decay in 1 second is $4.87 \times 10^{-18}$; for one mole = $6.022 \times 10^{23}$ nuclei of uranium-238, the mean number decaying in 1 second is thus $6.022 \times 10^{23} \times 4.87 \times 10^{-18} = 2.93 \times 10^6$ nuclei.

The Poisson distribution has a particularly simple mean, $E(X) = \lambda$, and variance, $V(X) = \lambda$.

The Poisson cumulative distribution function does not have a simple form, though it can be easily calculated using a computer; it is

$$P(X \leq x) = \sum_{y=0}^{x} p(y; \lambda) \quad \text{for } x = 0, 1, 2, \ldots, n \quad (2.6)$$

Figure 5(a) shows the Poisson distributions for two values of $\lambda$, while Figure 5(b) shows the cumulative Poisson distributions for three values of $\lambda$.

**Example 2.3:** Next, we consider radioactive decay. One curie (Ci) of a radioactive material is defined as the quantity of the material that leads to an average of $3.7 \times 10^{10}$ decays per second. A more convenient unit in many circumstances is 1 picocurie (pCi) = $10^{-12}$ Ci. What are the Poisson distributions for the number of decays of a 1-pCi sample in 1-sec and 1-min intervals?

The rate of decay of 1 pCi is 0.037/sec or 2.22/min. Setting $\lambda = 0.037$ and $\lambda = 2.22$ leads to the Poisson distributions shown in Table I. Note that

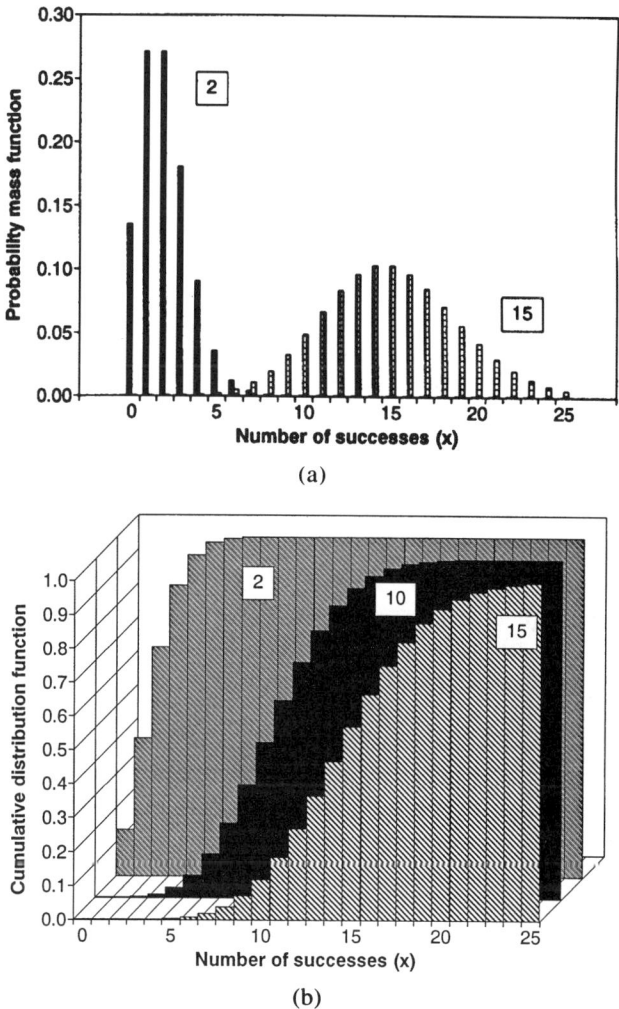

FIG. 5. (a) The Poisson probability mass functions for $\lambda = 2$ and 15. (b) The cumulative Poisson distribution functions for $\lambda = 2$, 10, and 15.

there would be no decays in most 1-sec intervals, while in 1-min intervals the most common result would be two decays.

**Example 2.4:** In a certain state with a stable population in which the annual number of suicides has averaged 300 per year over the past two decades, there are 330 suicides one year. The newspapers report that this

TABLE I. The Poisson Probability Mass Function and Cumulative Distribution Function for $\lambda = 0.037$ and $\lambda = 2.22$, Corresponding to the Radioactive Decays per Second and per Minute for a 1-pCi Sample of a Radioactive Isotype

| $x$ | $\lambda = 0.037$ | | $\lambda = 2.22$ | |
|---|---|---|---|---|
| 0  | 0.963676 | 0.963676 | 0.108609 | 0.108609 |
| 1  | 0.035656 | 0.999332 | 0.241112 | 0.349721 |
| 2  | 0.000660 | 0.999992 | 0.267635 | 0.617356 |
| 3  | 0.000008 | 1.000000 | 0.198050 | 0.815405 |
| 4  | 0.000000 | 1.000000 | 0.109918 | 0.925323 |
| 5  | 0.000000 | 1.000000 | 0.048803 | 0.974126 |
| 6  | 0.000000 | 1.000000 | 0.018057 | 0.992184 |
| 7  | 0.000000 | 1.000000 | 0.005727 | 0.997910 |
| 8  | 0.000000 | 1.000000 | 0.001589 | 0.999500 |
| 9  | 0.000000 | 1.000000 | 0.000392 | 0.999892 |
| 10 | 0.000000 | 1.000000 | 0.000087 | 0.999979 |

is a 10% increase over the previous year, and a higher number than has occurred in the past 20 years, and speculate about the social or economic conditions that might be causing such a large increase. But is it so unusual? How often would the suicide rate be 330 or more? To answer this question, let us make the reasonable assumption that suicide is governed by a Poisson process. A computer calculation shows that for $\lambda = 300$, $P(X \leq 329) = 0.954 = 1 - 0.046$. That means that there is a 4.6% probability that, given an average rate of 300 suicides per year, the number of suicides in a given year is 330 or more. Apparently, the 330 total is not particularly unusual, and it would be advisable to watch the suicide rate for a few more years rather than speculate on the causes for the increase.

## 2.3. Continuous Probability Distributions

### 2.3.1 The Normal (Gaussian) and Standard Normal Distributions

Many probability distributions that occur in science have the shape of the familiar bell-shaped curve shown in Figure 6. This is the standard normal (or Gaussian) probability distribution:

$$f(z; 0, 1) = \frac{1}{\sqrt{2\pi}} e^{-z^2/2} \quad \text{defined for } -\infty < z < \infty \quad (2.7)$$

This continuous distribution is most easily recognized as the limiting shape of the discrete binomial distribution as the number of trials $n \to \infty$, as may be seen by comparing Figure 6 with Figure 2(a).

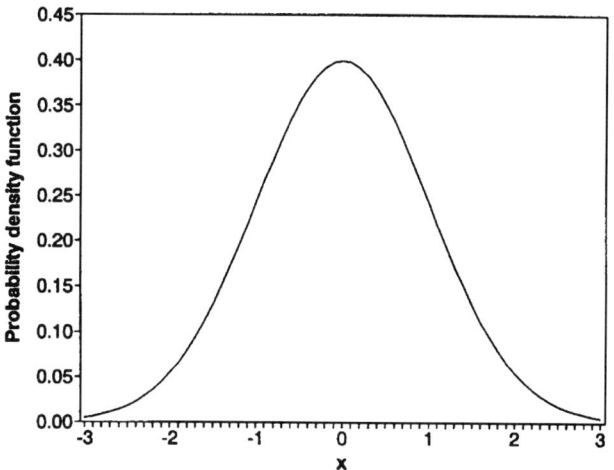

FIG. 6. The standard normal density function $f(x; 0, 1)$.

The standard normal distribution shown in Figure 6 has mean $\mu = 0$ and variance $\sigma^2 = 1$. However, it can be generalized to the probability density function for a normal random variable $X$ with parameters $\mu$ and $\sigma$, where $-\infty < \mu < \infty$ and $\sigma > 0$, which has the form

$$f(x; \mu, \sigma) = \frac{1}{\sqrt{2\pi}\,\sigma} e^{-(x-\mu)^2/2\sigma^2} \quad \text{where } -\infty < x < \infty \quad (2.8)$$

The constant in front of the exponential function, known as the *normalization constant*, is chosen to ensure that the integral of $f(x; \mu, \sigma)$ from $-\infty$ to $+\infty$ is exactly 1. The mean of the normal distribution is $E(X) = \mu$ and its variance is $V(X) = \sigma^2$, which is why the two parameters are given the symbols $\mu$ and $\sigma$.

Tables for the normal distribution are given in terms of the standard normal random variable $Z$ with mean 0 and variance 1, whose probability density function is the standard normal density function given in Eq. (2.7). Any normal random variable $X$ can be converted to a standard normal random variable by the substitution

$$Z = \frac{X - \mu}{\sigma}$$

It is then possible to make calculations of functions of, and integrals involving, $X$ by making the substitutions

$$X \rightarrow \sigma Z + \mu \quad \text{and} \quad dX \rightarrow \sigma dZ$$

The cumulative distribution function for the standard normal distribution is an important function is its own right:

$$P(Z \leq z) = \int_{-\infty}^{z} f(y; 0, 1) \, dy = \frac{1}{\sqrt{2\pi}} \int_{-\infty}^{z} e^{-y^2/2} \, dy \equiv \Phi(z) \quad (2.9)$$

The cumulative distribution function for the normal distribution with mean $\mu$ and variance $\sigma^2$ is then found (using the previous substitutions) to be

$$P(X \leq x) = \Phi\left(\frac{x - \mu}{\sigma}\right)$$

Since the normal density function is symmetric in $x - \mu$, $\Phi(0) = 0.5$.

Table II shows some important special cases of this cumulative distribution function. Approximately 68% of the values in any normal population lie within one standard deviation ($\sigma$) of the mean $\mu$, approximately 95% lie with two standard deviations of $\mu$, and approximately 99.7% lie within three standard deviations of $\mu$.

The term *percentile*, as in 50th percentile or 95th percentile, is often used with normal distributions; it denotes the value of the random variable

TABLE II. Special Values of the Normal and Standard Normal Cumulative Distribution Functions

| $x - \mu$ | $z$ | $\Phi(z)$ |
|---|---|---|
| $-3.08\sigma$ | $-3.08$ | 0.0010 |
| $-3\sigma$ | $-3$ | 0.0013 |
| $-2.58\sigma$ | $-2.58$ | 0.0050 |
| $-2.33\sigma$ | $-2.33$ | 0.0100 |
| $-2\sigma$ | $-2$ | 0.0227 |
| $-1.96\sigma$ | $-1.96$ | 0.0250 |
| $-1.645\sigma$ | $-1.645$ | 0.0500 |
| $-1.28\sigma$ | $-1.28$ | 0.1000 |
| $-\sigma$ | $-1$ | 0.1587 |
| 0 | 0 | 0.5000 |
| $+\sigma$ | $+1$ | 0.8413 |
| $+1.28\sigma$ | $+1.28$ | 0.9000 |
| $+1.645\sigma$ | $+1.645$ | 0.9500 |
| $+1.96\sigma$ | $+1.96$ | 0.9750 |
| $+2\sigma$ | $+2$ | 0.9773 |
| $+2.33\sigma$ | $+2.33$ | 0.9900 |
| $+3\sigma$ | $+3$ | 0.9987 |
| $+3.08\sigma$ | $+3.08$ | 0.9990 |

such that the cumulative distribution function at that value is 0.50, or 0.95. Thus the $100(1 - \alpha)$th percentile, for $0 < \alpha < 1$, is the value $z_\alpha$ (or its corresponding $x_\alpha$) for which $\Phi(z_\alpha) = 1 - \alpha$. From Table II it can be seen that the 90th percentile corresponds to $z = 1.28$ (or $x = \mu + 1.28\sigma$), the 99th percentile to $z = 2.33$, and so forth.

Figure 7(a) shows two normal density functions for the same mean ($\mu = 0$) but different variances ($\sigma = 0.8$ and $0.4$), while Figure 7(b) shows the corresponding normal cumulative distribution functions.

**Normal Approximation to the Binomial Distribution.** It is often convenient to approximate a discrete binomial distribution by a continuous normal distribution. Let $X$ be a binomial random variable based on $n$ trials with success probability $p$. Then if the binomial probability histogram is not too skewed, $X$ has approximately a normal distribution with $\mu = np$ and $\sigma = \sqrt{npq}$. In particular, where $x$ is a possible value of $X$, the cumulative binomial probability distribution is

$$P(X \leq x) = B(x; n, p) \approx \Phi\left(\frac{x + 0.5 - np}{\sqrt{npq}}\right) \qquad (2.10)$$

which is the area under the normal curve to the left of $x + 0.5$. In practice, this approximation is adequate, provided that both $np \geq 5$ and $nq \geq 5$. Figure 8 shows the binomial distribution for $n = 20$, $p = 0.4$, $q = 1 - p = 0.6$ together with its normal approximation, using $\mu = np = 8$ and $\sigma = \sqrt{npq} = 2.19$.

Normal distributions are often encountered in scientific experiments. Some examples are the following:

1. Repeated physical measurements, such as measurements of the length of an object, are usually normally distributed. If there are no systematic errors in the measurements, the mean of the measurements ought to be the real value of the physical quantity, and the variance is an indicator of the precision of the measurement method (the more precise the method, the smaller the variance).
2. The results of random samples from many large populations, such as the heights of adults in the general population, the weights of first-grade children, student grades on an exam, and so forth, are normally distributed. (The key here is that often such populations have relative frequency distributions that are approximately Gaussian.)
3. Any binomial distribution that is not too skewed is approximately normal, as was previously discussed.

**Example 2.5:** A typical example of an experimental distribution that closely resembles a normal distribution is shown in Figure 9. The energy

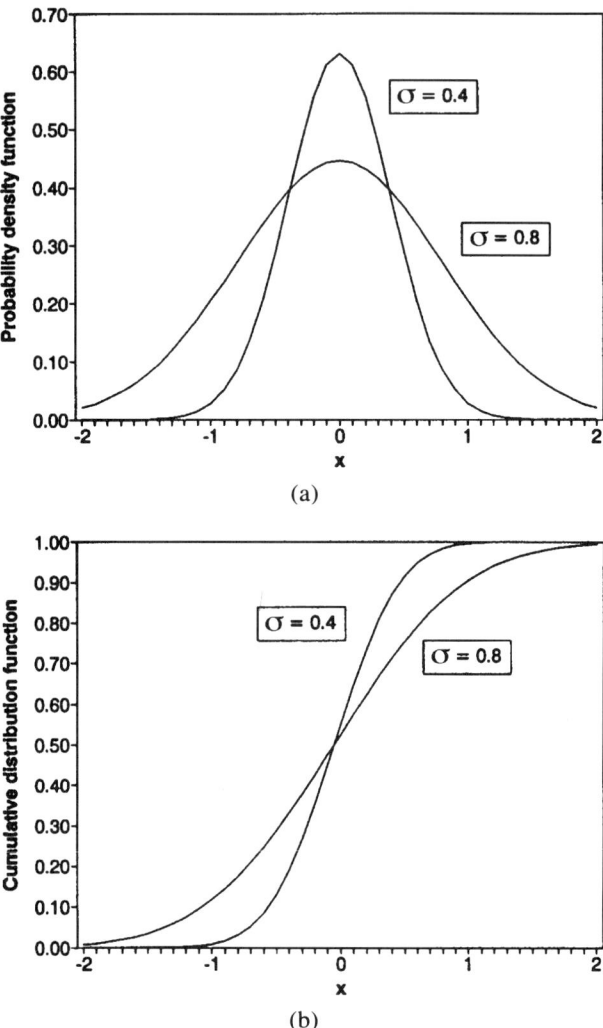

FIG. 7. (a) The normal probability density functions for $\mu = 0$ and $\sigma = 0.4$ and 0.8. (b) The normal cumulative distribution functions for $\mu = 0$ and $\sigma = 0.4$ and 0.8.

records for a three-month period were obtained for 368 single-family homes in a large metropolitan area, the total energy consumption was determined by converting all fuels to the same energy units, and this energy consumption was then divided by the number of heating degree-days for the period

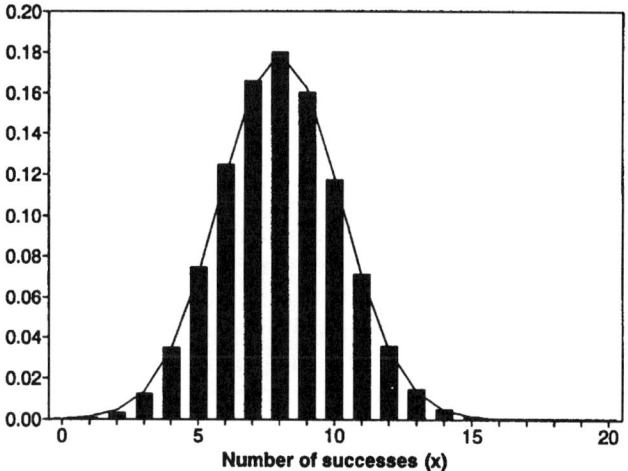

FIG. 8. Solid curve: normal probability density function for $\mu = 8$, $\sigma^2 = np(1 - q) = 4.8$. Histogram: binomial distribution for $n = 20$, $p = 0.4$.

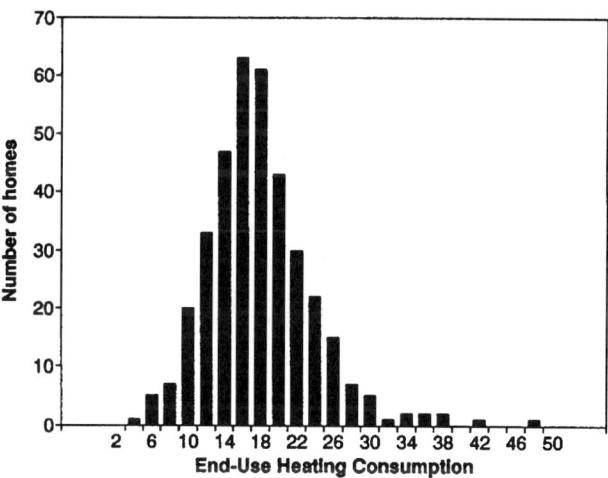

FIG. 9. Histogram of end-use heating consumption values (units: 1000 Btu/degree-day) for 368 homes in central Iowa.

covered by the energy records. The resulting quantity, called the *end-use heating consumption* (EHC), has units of Btus (British thermal units) per degree-day. It is related to the energy-efficiency of the homes, the homes with smaller EHC being the more energy-efficient.

### 2.3.2 The Lognormal Distribution

The normal distribution is very commonly encountered in science. It is obvious that if a random variable $X$ is normally distributed, functions of that random variable (such as $X^2$, $X^3$, log $X$, and $\sqrt{1-X}$) will usually not be normally distributed. Sometimes it happens that the random variable arising most naturally in a scientific situation is not itself normally distributed, but some function of it is. For example, it is often found that, while the concentrations of an air or water pollutant (such as sulfur dioxide in the air or phenol in river water samples) are not normally distributed, their logarithms are.

If the logarithm of a random variable $X$ is normally distributed, then $X$ is said to be lognormally distributed. We let $\ln(x)$ represent the natural (base $e = 2.71828\ldots$) logarithm of $x$. The probability density function of this lognormally distributed random variable $X$ is

$$f(x; \mu, \sigma) = \frac{1}{\sqrt{2\pi}\sigma x} e^{-[\ln(x)-\mu]^2/2\sigma^2} \quad \text{for } x \geq 0$$
$$= 0 \quad \text{for } x \leq 0 \quad (2.11)$$

where $\mu$ and $\sigma^2$ are the mean $E(\ln(X))$ and variance $V(\ln(X))$ of the normally distributed random variable $\ln(X)$.

It should be noted that $\mu$ and $\sigma^2$ are *not* the mean and variance of the lognormally distributed variable $X$. The variable $X$ will have mean $E(X) = e^{\mu + \sigma^2/2}$ and variance $V(X) = e^{2\mu + \sigma^2} \cdot (e^{\sigma^2} - 1)$.

The mean of the variable $X$ is called its *arithmetic mean*. The quantity $e^\mu$ is an important quantity, known as the *geometric mean* of $X$; the geometric mean of $X$ is clearly always less than the arithmetic mean of $X$.

The cumulative distribution function of a lognormal random variable can be readily expressed in terms of the cumulative distribution function $\Phi(z)$ of a standard normal random variable. For $x \geq 0$, the cumulative distribution function is

$$P(X \leq x) = F(x; \mu, \sigma) = \Phi\left(\frac{\ln x - \mu}{\sigma}\right) \quad \text{for } x > 0$$
$$= 0 \quad \text{for } x \leq 0 \quad (2.12)$$

Figure 10 shows the lognormal and cumulative probability functions for the cases $(\mu, \sigma) = (1, 1)$ and $(2, 1)$. Compared to a normal variable, a lognormal variable has a greater probability in the right tail: very large values of a lognormal variable are much more common than they are for a normal variable.

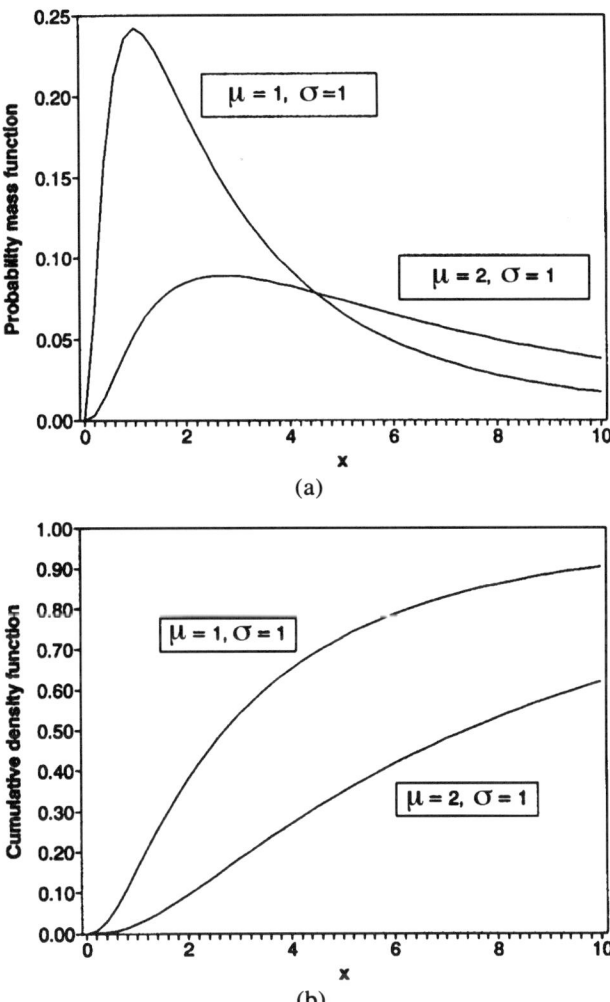

FIG. 10. (a) Lognormal probability density functions for $(\mu, \sigma) = (1, 1)$ and $(2, 1)$. (b) Lognormal cumulative distribution functions for the same two sets of parameters.

**Example 2.6:** Figure 11(a) shows the distribution of nearly 7000 measurements of the radon concentrations in the air in the basements of homes in Iowa. It is clear from the skewness of the distribution that the radon concentrations are not normally distributed. However, Figure 11(b) shows

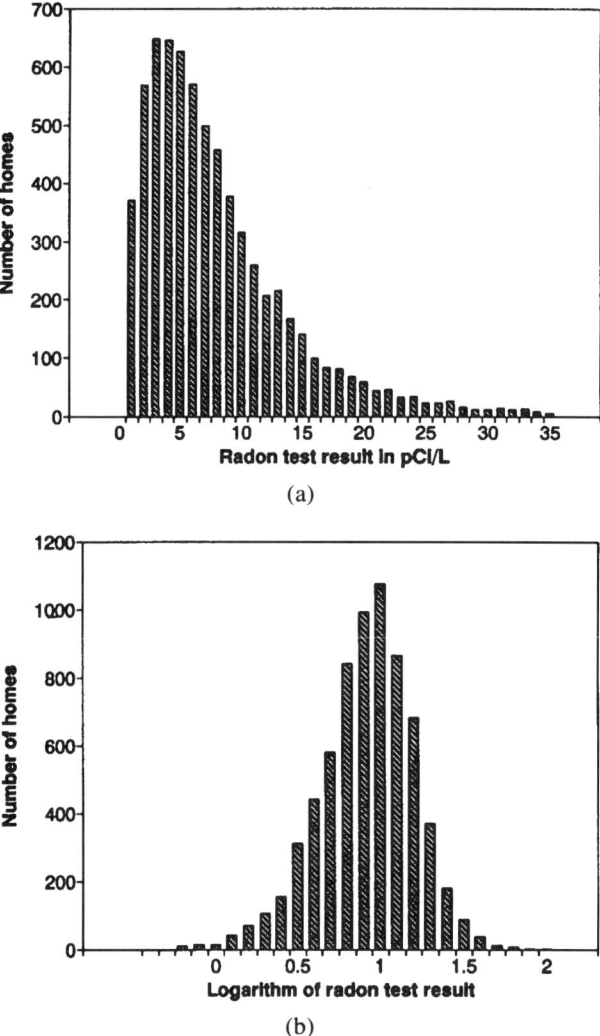

FIG. 11. (a) Histogram of radon concentrations found in 6926 basements in Iowa. (b) Histogram of the logarithms of the same radon concentrations.

the distribution of the logarithms of the radon concentrations; this distribution does resemble a normal distribution.

A consequence of the fact that the radon concentrations are approximately lognormally distributed is that the geometric mean of the radon concentrations is a more meaningful distribution summary than the arithmetic mean. For example, a single extra-large radon concentration found in one home may have a drastic effect on the arithmetic mean, but it will have little effect on the geometric mean.

### 2.3.3 The Exponential Distribution

Another commonly encountered continuous probability distribution is the exponential distribution. It is defined by the exponential density function

$$f(x; \lambda) = \lambda e^{-\lambda x} \quad \text{for } x \geq 0$$
$$= 0 \quad \text{for } x < 0 \quad (2.13)$$

The mean of this distribution is $E(X) = 1/\lambda$ and the variance is $V(X) = 1/\lambda^2$.

The cumulative distribution function for the exponential distribution is easy to calculate by direct integration. The probability that the random variable has a value $\leq x$ is

$$P(X \leq x) = F(x; \lambda) = 1 - e^{-\lambda x} \quad \text{for } x \geq 0$$
$$= 0 \quad \text{for } x \leq 0 \quad (2.14)$$

Figure 12(a) shows two exponential density functions corresponding to $\lambda = 1$ and 2, while Figure 12(b) shows the corresponding cumulative distribution functions.

Some of the most common applications of this distribution involve time as the random variable, in which case the random variable is generally denoted $T$ instead of $X$. One such application is to radioactive decay, which we first described in connection with the Poisson distribution. When radioactive decay was introduced there, we were considering a large number of nuclei and interested in the number of decays observed in finite time intervals, such as 1 sec or 1 min. Now let us look instead at the probability that an individual nucleus in existence at time $t = 0$ will decay during some time interval $t_1 < t < t_2$, such as during the first second.

Radioactive decay has been found to be a purely stochastic process, not a deterministic process. There is no way to predict exactly when a given nucleus will decay. However, there is a certain probability $p$ that the nucleus will decay during the first second, so the probability that it survives the first second is $1 - p$. If it does survive the first second, then it has exactly

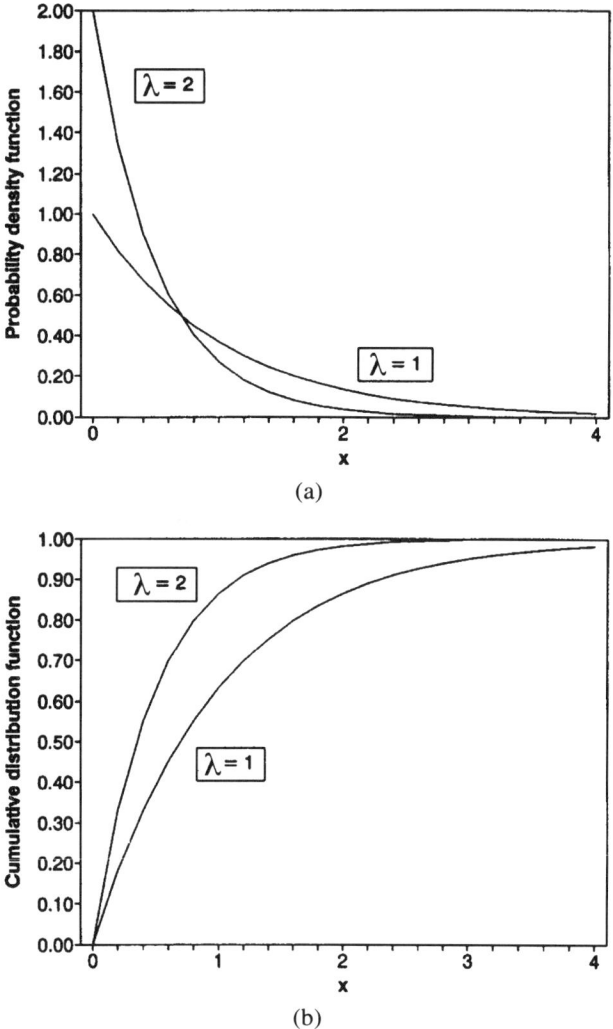

FIG. 12. (a) Exponential probability density functions for $\lambda = 1$ and $\lambda = 2$. (b) Exponential cumulative distribution functions for $\lambda = 1$ and $\lambda = 2$.

the same probability $p$ of decaying during the next second, so the probability of surviving two seconds is $(1 - p)^2$. In general, the probability of surviving $N$ seconds is $(1 - p)^N$. The survival function that has this form is an exponential: $S(t) = e^{-\lambda t}$.

If we focus not on the survival function $S(t)$ but on the function $1 - S(t)$, the latter has the form $1 - e^{-\lambda t}$. This is the probability that the nucleus will have decayed by time $t$. This function has exactly the form of the exponential cumulative distribution function of Eq. (2.14). The exponential density function $\lambda e^{-\lambda t}$ thus represents the probability density function for the time of decay of the individual nucleus. Note that this function is proportional to the survival function $S(t)$ introduced earlier.

The parameter $\lambda$ that governs exponential decay is related to the half-life of the radioactive material, which is defined as the time by which an individual nucleus will have a 50% chance of having decayed. Denoting the half-life by $t_{1/2}$, we must have

$$\frac{1}{2} = 1 - e^{-\lambda t_{1/2}} \tag{2.15}$$

from which it follows that $t_{1/2} = \dfrac{(\ln 2)}{\lambda} = 0.693/\lambda$.

To the extent that radioactive decay can be regarded as a Poisson process, the exponential distribution occurs in another way: in a large sample, with many nuclei, the distribution of elapsed time between the occurrence of two successive decays is exponential with the same parameter $\lambda$. Poisson processes are discussed further in Chapter 3.

An important property of the exponential distribution is its "memoryless" feature: the shape of the probability density function is the same beginning at any value of $x$, which means the distribution has no "memory" of where (or when) it started. Mathematically, this can be expressed by stating that the conditional probability

$$P(X \geq t + t_0 | X > t_0) = P(X \geq t)$$

This condition leads directly to the exponential distribution, which is thus the *only* distribution with the property of being memoryless. In the case of radioactive decay, the memoryless property corresponds to the fact that the decay rate (the probability of decay per unit time interval) is independent of the age of the nuclei. If the failure rate of a component is modeled by an exponential distribution, the memoryless property corresponds to the failure being independent of age, i.e., the component shows no wear and tear due to its age. Failure rates that do not have the memoryless property are discussed in the section on the Weibull distribution.

**Example 2.7:** Practically all the naturally occurring atoms of uranium on earth are the uranium-235 and uranium-238 isotopes. The uranium-235 isotope has a half-life of $7.1 \times 10^8$ yr and an abundance of 0.711%, while the uranium-238 isotope has a half-life of $4.51 \times 10^9$ yr and an abundance of 99.283%. Both isotopes have been decaying since the earth formed approximately $4.6 \times 10^9$ years ago. How much of each isotope originally present has decayed, and what were the relative abundances of the two isotopes when the earth was young?

Converting the known half-lives into rate parameters $\lambda$, we find that $\lambda = 9.76 \times 10^{-10}$/yr for uranium-235 and $\lambda = 1.54 \times 10^{-10}$/yr for uranium-238. The values of $e^{-\lambda t}$ for $t = 4.6 \times 10^9$ yr are, respectively, 0.0112 and 0.493. Thus only 1.12% of the original uranium-235 is still present on earth, but 49.3% of the original uranium-238. Consequently, while the current ratio of uranium-235 to uranium-238 is 0.711 to 99.283, or 1 to 140, the ratio when the earth was young must have been 0.711/0.0112 to 99.283/0.493, or 1 to 3.2.

### 2.3.4 The Weibull Distribution

Sometimes an experiment is characterized by a continuous random variable whose probability distribution is best fit by a function involving two or more parameters, which allows greater freedom in fitting the experimental results. A useful distribution of this type is the Weibull distribution, a generalization of the exponential distribution that was introduced by Weibull [1].

The Weibull distribution is most easily introduced in terms of its cumulative distribution function, which has the form

$$F(x; \beta, \alpha) = 1 - e^{-(x/\alpha)^\beta} \quad \text{for } x \geq 0$$
$$= 0 \quad \text{for } x \leq 0 \quad (2.16)$$

where the two parameters $\alpha$ and $\beta$ are positive. This function is easily seen to be a generalization of the cumulative distribution function for the exponential distribution, which corresponds to the choice of $\beta = 1$.

The probability density function for the Weibull distribution can then be obtained as the derivative of $F(x; \beta, \alpha)$ with respect to $x$:

$$f(x; \beta, \alpha) = \beta \alpha^{-\beta} x^{\beta-1} e^{-(x/\alpha)^\beta} \quad \text{for } x \geq 0$$
$$= 0 \quad \text{for } x \leq 0 \quad (2.17)$$

When $\beta = 1$, the Weibull distribution reduces to the exponential distribution with the identification of the exponential parameter $\lambda$ with $1/\alpha$.

The parameter $\beta$ is a shape parameter affecting the shape of the distribution, while $\alpha$ is a scaling parameter affecting the scale. It is also possible to change the location of the Weibull distribution by replacing the variable $x$ by $x - x_0$.

The mean and variance of the Weibull distribution have complicated forms:

$$E(X) = \alpha\Gamma\left(1 + \frac{1}{\beta}\right) \quad \text{and} \quad V(X) = \alpha^2\left\{\Gamma\left(1 + \frac{2}{\beta}\right) - \Gamma^2\left(1 + \frac{1}{\beta}\right)\right\}$$

where $\Gamma(\beta) = \int_0^\infty x^{\beta-1} e^{-x} dx$ is the gamma function.

The Weibull distribution is widely used in modeling failure times, because a great variety of shapes of probability curves can be generated by different choices of the two parameters, $\beta$ and $\alpha$. Three examples of Weibull distributions are shown in Figure 13. Weibull distributions range from exponential distributions to curves resembling the normal distribution. The exponential distribution limit corresponds to a "memoryless" failure rate (the failure rate of an individual item is independent of its current age), while the other Weibull distributions correspond to distributions of failure times that are peaked at certain ages and skewed in different fashions.

In fitting a Weibull distribution to experimental data it is worth noting that from the cumulative distribution function $F(x; \beta, \alpha)$ of Eq. (2.16) we can easily derive the relation

$$\ln\ln\left(\frac{1}{1 - F(x; \beta, \alpha)}\right) = \beta \ln(x/\alpha) = -\beta \ln \alpha + \beta \ln x \quad (2.18)$$

This gives a straight line when $\ln \ln[1 - F(x; \beta, \alpha)]^{-1}$ is plotted against $\ln x$, a useful method of determining whether the Weibull distribution is appropriate and, if so, determining the parameters $\beta$ and $\alpha$ from the intercept and slope of the straight line. This plot will also identify an exponential function by yielding $\beta = 1$.

A special property of the Weibull distribution is that the natural logarithm of a Weibull variable has the "smallest extreme value" distribution, discussed further in Chapter 7.

**Example 2.8:** An example of a distribution well fit by a Weibull distribution is plotted in Figure 14. A large number of 3.3-microfarad solid tantalum capacitors were operated continuously at high temperature to determine their failure rate [2]. It was found that 5% had failed after 4 hr, 10% after 13 hr, 15% after 26 hr, 20% after 36 hr, 25% after 51 hr, 30% after 75 hr, 35% after 100 hr, 40% after 111 hr, 45% after 162 hr, and 50% after 174 hr. Setting $x = 4, 13, \ldots, 174$ and $F = 0.05, 0.10, \ldots, 0.50$ and plotting

$$\ln \ln[1 - F(x; \beta, \alpha)]^{-1} \quad \text{versus} \quad \ln x$$

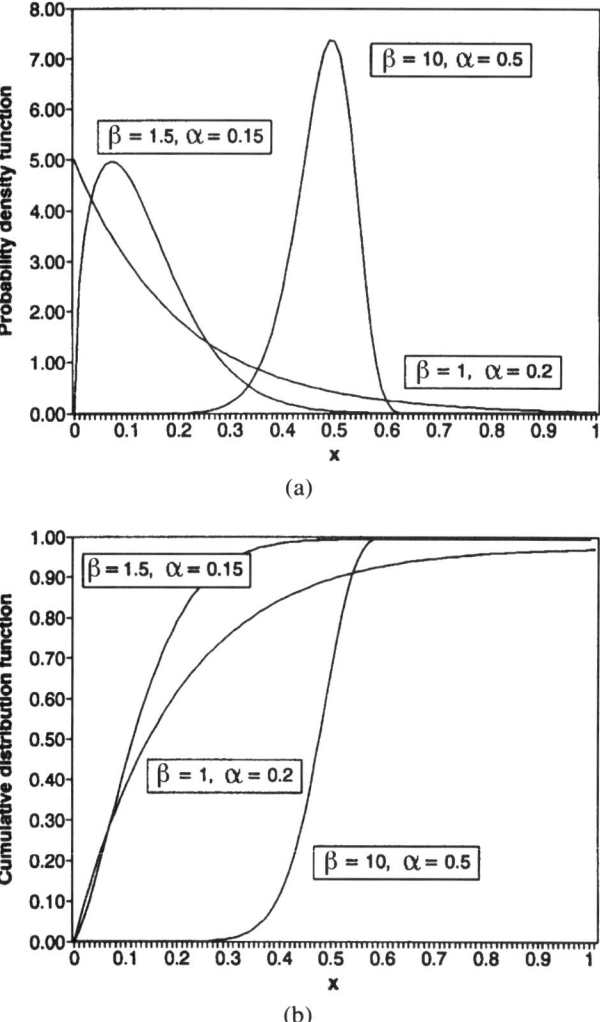

FIG. 13. (a) Weibull probability density functions for three sets of parameters: $(\beta, \alpha) = (1, 0.2), (1.5, 0.15)$, and $(10, 0.5)$. (b) Weibull cumulative distribution functions for the same three sets of parameters.

gives the values plotted in Figure 14. The fact that the plotted values are well fit by the straight line (which corresponds to (2,18) with $\beta = 0.70$ and $\alpha = 310$) shows that the Weibull distribution is appropriate to model the failure rate of these capacitors.

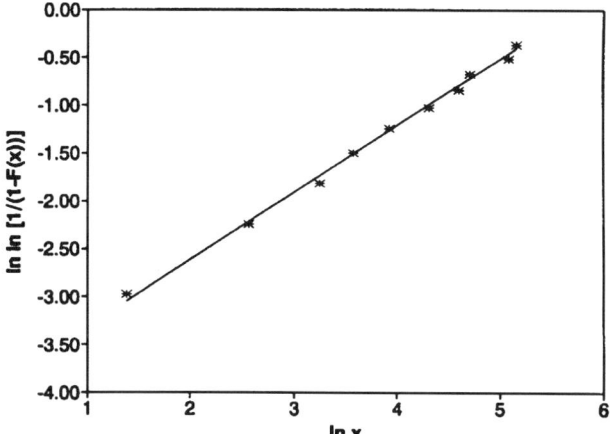

FIG. 14. Plot of capacitor failure data [2]. The random variable is hours to failure. The straight line corresponds to choosing the Weibull distribution parameters as $\beta = 0.70$ and $\alpha = 310$.

## References

1. Weibull, W. (1951). "A Statistical Distribution Function of Wide Applicability." *J. Appl. Mechanics* **18,** 293–297.
2. Berrettoni, J. N. (1964). "Practical Applications of the Weibull Distribution." *Industrial Quality Control* (August), 71–78.

# 3. RANDOM PROCESS MODELS

## Christopher Chatfield
University of Bath, United Kingdom

## 3.1 Introduction

Most physical processes evolve through time according to probabilistic laws. As such they are *stochastic*, rather than *deterministic*, in that the future is only partly determined by past values. Such phenomena are collectively called *random processes*, though some writers prefer the synonym *stochastic processes*. This chapter describes various probability models for two broad classes of random process; namely, *time series* and *point processes*.

A *time series* is a collection of observations made sequentially through time. Examples include air temperature measured at regular intervals at a particular site, and measurements of brain activity of a particular person (an EEG). An example of a time series is shown in Fig. 1. A graph such as this, which plots the observations against time, is called a *time plot*. Plotting such a graph is the first and sometimes the most important step in a time-series analysis, as it will show up features such as *trend* (i.e., a long-term drift in the underlying mean), *seasonal* or *cyclic variation*, sudden *discontinuities* and unexpected observations called *outliers*. For example in Fig 1 we see little or no long-term trend but there is a clear seasonal cycle (summers are hotter!). There are no obvious discontinuities or outliers in this case. It is always worth spending a few minutes looking carefully at the time plot.

Plotting a time series is not as easy as it sounds and many computer-drawn plots are of poor quality. Care is needed to choose suitable scales, the size of the intercept, and the plotting symbol, as well as to label axes carefully and give a clear self-explanatory title.

A *point process* consists of a series of events occurring randomly through time. Examples include the times of earthquakes and arrival times of radio-active particles at a geiger counter. An example of a realization of a point process is shown in Fig. 2. Note that the time between successive events can sometimes be quite short, giving apparent clustering, even when the events are in fact "random."

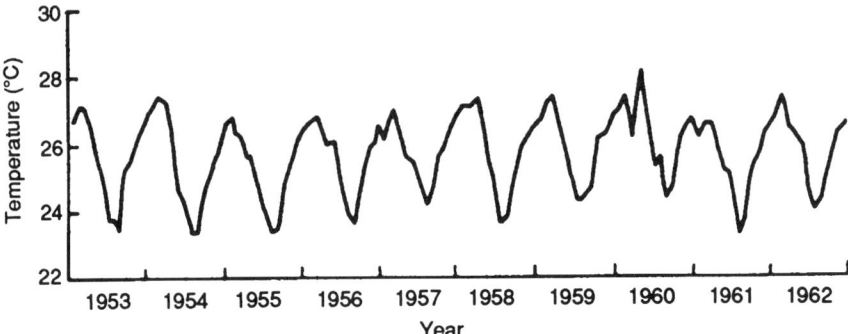

FIG. 1. Average air temperature at Recife, Brazil, in successive months from 1953 to 1962.

The two types of process are related but distinct. For a time series, the modelling procedure aims to describe the way in which the observed variable evolves through time, while for a point process interest centers on the placement of the points (or the distances between them) along the time axis.

## 3.2 Probability Models for Time Series

A time series may be regarded as an ordered sequence of dependent random variables. This sequence of variables will evolve through time according to probabilistic laws. This section introduces various classes of probability model that can be used to describe time series. For a more thorough treatment see, for example, Chatfield [3], Papoulis [11], and Priestley [12].

### 3.2.1 General Terminology

First we introduce some general terminology and some notation. A time series is said to be *continuous* when observations are made continuously

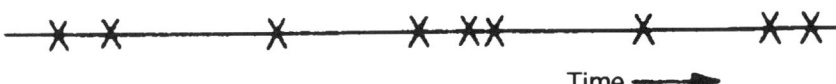

FIG. 2. A realization of a point process, where $X$ denotes an event.

through time, as for example when temperature is recorded on a continuous trace. A time series is said to be *discrete* when observations are taken at discrete times, such as every hour or every day. The intervals are usually equally spaced. Note that the term *discrete* refers to the spacing between observations rather than the values of the observations themselves, which can be continuous or discrete valued (among other possibilities). Also note that a continuous series in the form of a continuous trace is often *digitized* at equal intervals of time so that the analysis is carried out on the discrete-time sampled series.

If a time series is regarded as a sequence of random variables, we denote the random variable at time $t$ by $X_t$, if time is discrete (e.g., $t = 0, 1, 2, \ldots$) and by $X(t)$ if time is continuous (e.g., $0 < t < \infty$).

Most statistical problems are concerned with estimating the properties of a population from a sample. Time-series analysis raises special problems because, although it may be possible to make an observed time series longer, it is usually impossible to make more than one observation at any given time. Thus there is usually only a single realization of the process and a single observation on the random variable at time $t$. [An exception to this is the *repeated measures* problem (Diggle [8], Chapter 5), where a set of time series is recorded for a group of "similar" experimental units.] The problem of having only a single realization is overcome by regarding the observed time series as just one example of the population of time series that might have been generated by the same underlying probability model. This notional population is called the *ensemble*.

One way of trying to describe this ensemble, or the underlying probability mechanism, is to specify the joint probability distribution of the random variables that make up the time series. In general this involves specifying the joint distribution of $X(t_1), X(t_2), \ldots, X(t_n)$ for any set of times $t_1, t_2, \ldots, t_n$ and any value of $n$. However, this is generally impractical. Instead the first and second moments of the process are specified. These will be defined in discrete time, with similar definitions applying in continuous time.

**Mean.** The mean function $\mu(t)$ is defined by

$$\mu(t) = E[X_t] = \text{Expected value of } X_t \tag{3.1}$$

**Variance.** The variance function $\sigma^2(t)$ is defined by

$$\begin{aligned}\sigma^2(t) &= \text{Var}[X_t] \\ &= \text{Variance of } X_t \\ &= E\{[X_t - \mu(t)]^2\}\end{aligned} \tag{3.2}$$

**Autocovariance.** The more general second moment (of which the variance function is a special case when $t_1 = t_2$) is the autocovariance function of $X_{t_1}$ and $X_{t_2}$ that is given by

$$C(t_1, t_2) = \text{Covariance }(X_{t_1}, X_{t_2}) \\ = E\{[X_{t_1} - \mu(t_1)] [X_{t_2} - \mu(t_2)]\} \quad (3.3)$$

If the series contains a trend, then $\mu(t)$ will change through time. The variance function is often assumed to be constant, though this is an unreasonable assumption when the series exhibits turbulence or appears to get more variable as the mean level increases. The interpretation of the autocovariance function will become clearer in the next subsection.

### 3.2.2 Stationary Processes and the Autocorrelation Function

An important class of time series are those that are *stationary*. Put simply, a time series is stationary if there is no systematic increase or decrease in the mean (i.e., no trend) and no systematic change in variance (no turbulence).

There are two main ways of defining stationarity; namely, *strict* and *second-order stationarity*. The latter is easier to understand as well as being more useful in practice and so will be given here. A time series is called *second-order stationary* (or weakly stationary) if its first and second moments do not change through time. This implies that the mean and variance of the process are constant and furthermore that the autocovariance function $C(t_1, t_2)$ depends only on the value of $(t_2 - t_1) = \tau$ called the *lag*, and it may be written as

$$C(\tau) = \text{Covariance }[X_t, X_{t+\tau}] \quad (3.4)$$

Chapters 4 and 14 introduce a function called the *variogram*, which for stationary processes is closely related to the autocovariance function by

$$\gamma(\tau) = C(0) - C(\tau)$$

The value of $C(\tau)$ depends on the units in which $X_t$ is measured. Thus it is usually helpful to standardize $C(\tau)$ to produce a function lying between $\pm 1$ called the *autocorrelation* function, which is given by

$$\rho(\tau) = C(\tau)/C(0) \quad (3.5)$$

This measures the (auto)correlation between $X_t$ and $X_{t+\tau}$. As such it can be shown that it satisfies the "usual" property of correlation; namely, that $|\rho(\tau)| \leq 1$. The sign of $\rho(\tau)$ can also be interpreted in the "usual" way. For example, a positive autocorrelation at lag 1 implies that if $X_t$ is above its mean, then $X_{t+1}$ will also tend to be above the mean. Short-term positive autocorrelation is often observed in practice in that a series of values above the mean may be followed by a series of values below the mean, and so on.

Another useful property of $\rho(\tau)$ is that it can be shown to be an *even* function of lag, so that $\rho(\tau) = \rho(-\tau)$.

One important class of random processes are *normal* processes, where the joint distribution of any set of observations follows a multivariate version of the normal distribution called the *multivariate normal* distribution. Such processes are completely characterized by their first and second moments. It is also worth noting that a given stochastic process has a unique autocorrelation function, but the reverse is not in general true. It is usually possible to find many normal and nonnormal processes with the same $\rho(\tau)$, and this means that the sample estimate of $\rho(\tau)$ (called the *sample autocorrelation function* or the (sample) *correlogram*) cannot be uniquely interpreted. Even for stationary normal processes, a special condition, called the *invertibility condition* (see Section 3.2.5), is required to ensure that there is a unique model for a given $\rho(\tau)$.

### 3.2.3 A Purely Random Process or White Noise

Having established some general terminology, the next seven subsections introduce a range of different probability models that are useful in describing time series. The most basic type of model is the *purely random process* that is occasionally used for modelling in its own right, but is more often used as a "building block" in a more complicated model such as a moving average process (see Section 3.2.5). For this reason, it will have its own special notation, namely, $\{Z_t\}$, and this notation will be used throughout to denote a purely random process.

A random process in discrete time is called a *purely random process* if it consists of a sequence of random variables, say, $\{Z_1, Z_2, \ldots, Z_t, \ldots\}$, which are mutually independent and which have a common probability distribution. The measured noise amplitude on a radio at noon on successive days may perhaps follow such a model. As the values of $Z$ are identically distributed, it follows that the mean and variance of the process must be constant. As the values of $Z$ are independent, it follows that the autocovariance function must be zero (except at lag zero) so that

$$C(k) = \text{Covariance } (Z_t, Z_{t+k})$$
$$= \begin{cases} \text{Variance } (Z_t) & k = 0 \\ 0 & k = \pm 1, \pm 2, \ldots \end{cases} \quad (3.6)$$

Note that we use $k$ rather than $\tau$ to denote the lag, when, as here, it is integer valued. As the first and second moments do not depend on $t$, the process is second-order stationary (and also strictly stationary in this case). It follows that the autocorrelation function is given by

$$\rho(k) = \begin{cases} 1 & k = 0 \\ 0 & k = \pm 1, \pm 2, \ldots \end{cases}$$

A purely random process is sometimes called *white noise*, particularly by engineers.

Note that attempts to define a purely random process in continuous time are fraught with mathematical difficulty. Priestley [12], Section 3.7.1, says "no such process exists, except in a highly degenerate sense."

### 3.2.4 Autoregressive Processes

Very few time series are completely random. Rather the observation in a particular time period can be expected to depend on one or more observations that immediately precede it. Thus a natural class of models to consider are those called *autoregressive models*, which may be defined as follows. A time series $\{X_t\}$ is said to be an autoregressive process of order $p$ if an observation depends (linearly) on the preceding $p$ observations so that

$$X_t = \alpha_1 X_{t-1} + \alpha_2 X_{t-2} + \ldots + \alpha_p X_{t-p} + Z_t \tag{3.7}$$

where $\{Z_t\}$ denotes a purely random process with zero mean and variance $\sigma_z^2$, and $\alpha_1, \alpha_2, \ldots, \alpha_p$ are constants. This is rather like a multiple regression model, except that $X_t$ is regressed, not on separate explanatory variables, but on past values of $X_t$; hence, the prefix *auto*. An autoregressive process of order $p$ will be abbreviated as an AR($p$) process.

The first-order model, AR(1), is of particular interest and may be written as

$$X_t = \alpha X_{t-1} + Z_t \tag{3.8}$$

In order to examine the properties of this model, it is helpful to introduce the *backward shift* operator, $B$, such that $BX_t = X_{t-1}$. Then $B^2 X_t = B(BX_t) = BX_{t-1} = X_{t-2}$. More generally the operator $B^j$ shifts the variable back $j$ steps so that $B^j = X_{t-j}$. Then Eq. (3.8) may be written as

$$X_t = \alpha B X_t + Z_t$$

or

$$(1 - \alpha B) X_t = Z_t$$

or

$$X_t = Z_t/(1 - \alpha B)$$
$$= (1 + \alpha B + \alpha^2 B^2 + \ldots)Z_t \quad (3.9)$$
$$= Z_t + \alpha Z_{t-1} + \alpha^2 Z_{t-2} + \ldots$$

This result could also be obtained directly from Eq. (3.8) by successive substitution. Note that the series on the righthand-side of Eq. (3.9) converges only if $|\alpha| < 1$. From Eq. (3.9) it can be shown that

$$E(X_t) = 0$$
$$\text{Variance}(X_t) = \sigma_z^2(1 + \alpha^2 + \alpha^4 + \ldots)$$
$$= \sigma_z^2/(1 - \alpha^2)$$

again provided that $|\alpha| < 1$. In fact it turns out that the process is stationary provided that $|\alpha| < 1$. This result is intuitively sensible in that a value of $\alpha$ greater than 1 will tend to give larger and larger values and make the series "blow up." If $-1 < \alpha < 0$, the series will tend to oscillate about its mean, while if $0 < \alpha < 1$, the series will tend to exhibit runs of observations on the same side of the mean. If $\alpha$ is exactly equal to 1, then the model becomes a *random walk*—see Section 3.2.7.

When the AR(1) process is stationary (i.e., $|\alpha| < 1$), it can be shown that the autocorrelation function depends only on the lag and has the simple form

$$\rho(k) = \alpha^k \quad k = 0, 1, 2, \ldots \quad (3.10)$$

An even function can be defined for all $k$ by

$$\rho(k) = \alpha^{|k|} \quad k = 0, +1, +2, \ldots$$

Thus if $\alpha = 0.5$, for example, the correlation between successive observations (where the lag, $k$, is 1) is also 0.5. For observations two time intervals apart (where $k$ is 2), the correlation goes down to $0.5^2 = 0.25$. There will be very little correlation between observations that are several lags apart.

The more general $p$th-order model in Eq. (3.7) may also be rewritten using the backward shift operator as

$$X_t = (\alpha_1 B + \alpha_2 B^2 + \ldots + \alpha_p B^p)X_t + Z_t$$

or

$$\phi(B)X_t = Z_t \quad (3.11)$$

where $\phi(B) = 1 - \alpha_1 B - \alpha_2 B^2 \ldots - \alpha_p B^p$ is a polynomial in $B$ of order $p$. It can be shown that the process is stationary provided that the roots of the equation

$$\phi(B) = 0 \qquad (3.12)$$

lie outside the unit circle. Then the autocorrelation function turns out to be the sum of exponentially decreasing terms such as that in Eq. (3.10) or a mixture of damped sine and cosine waves. Either way the values of $\rho(k)$ tend to 0 as $k$ increases. Further properties of AR processes can be found in many books including the book that popularized such models, Box and Jenkins [2].

As well as being useful for summarizing and modelling the properties of a given time series, AR models are also useful for computing linear predictors for forecasting future values of a time series. They can also be used for estimating the spectrum of a stationary process—see Section 3.3.5.

### 3.2.5 Moving Average Processes

A time series $\{X_t\}$ is said to be a moving average process of order $q$ (abbreviated to a MA($q$) process) if

$$X_t = Z_t + \beta_1 Z_{t-1} + \ldots + \beta_q Z_{t-q} \qquad (3.13)$$

where $\beta_1, \beta_2, \ldots, \beta_q$ are constants. At first sight this looks rather like an AR process, but the values of $\{Z_t\}$ cannot be observed directly and can be inferred only from the values of $\{X_t\}$. Thus it is generally a more tricky class of models to handle. If the $\{Z_t\}$ are regarded as a series of random "shocks" to the system, then the value of $X_t$ depends on the previous $q$ "shocks".

It can be shown that an MA process is always stationary, and the mean, variance, and autocorrelation function of Eq. (3.13) are given by

$$E(X_t) = 0$$

$$\text{Variance }(X_t) = \sigma_z^2 \sum_{i=1}^{q} \beta_i^2$$

$$\rho(k) = \begin{cases} 1 & k = 0 \\ \sum_{i=0}^{q-k} \beta_i \beta_{i+k} / \sum_{i=0}^{q} \beta_i^2 & k = 1, 2, \ldots, q \\ 0 & k > q \end{cases}$$

where $\beta_0$ is taken to be 1 for mathematical convenience. Note that $\rho(k)$ "cuts off" at lag $q$, which is a special feature of MA processes (in contrast to AR processes for which $\rho(k)$ generally decreases exponentially).

Equation (3.13) may be rewritten using the backward shift operator, $B$, as

$$X_t = \theta(B)Z_t \tag{3.14}$$

where $\theta(B) = 1 + \beta_1 B + \ldots + \beta_q B^q$. Although no restrictions are required on the values of $\{\beta_i\}$ to make a MA process stationary, it is usual to impose a condition, called the *invertibility condition* (e.g., see Chatfield [3], p. 34) to ensure that there is a unique MA process for a given autocorrelation function. This is done by requiring the values of $\{\beta_i\}$ to be such that the roots of the equation

$$\theta(B) = 0 \tag{3.15}$$

lie outside the unit circle. This also ensures that $X_t$ can be expressed as a converging linear combination of the past values $X_{t-1}, X_{t-2}, \ldots$.

Further details about MA processes may be found in many books, including Box and Jenkins [2], but this class of models is not recommended for the beginner.

### 3.2.6 ARMA Processes

An even more general class of models may be obtained by combining AR and MA processes. A time series $\{X_t\}$ is said to be a mixed autoregressive–moving average process of order $(p, q)$ (abbreviated to an ARMA $(p, q)$ process) if it contains $p$ AR terms and $q$ MA terms. Then

$$\begin{aligned}X_t = \alpha_1 X_{t-1} + \ldots + \alpha_p X_{t-p} + Z_t + \beta_1 Z_{t-1} \\ + \ldots + \beta_q X_{t-q}\end{aligned} \tag{3.16}$$

Using the backward shift operator, $B$, this may be rewritten in the form (see Eqs. (3.11) and (3.14))

$$\phi(B)X_t = \theta(B)Z_t \tag{3.17}$$

where $\phi(B)$, $\theta(B)$ are polynomials in $B$ of orders $p$, $q$, respectively. As for an AR process, the values of $\{\alpha_i\}$ that make the process stationary must be such that the roots of Eq. (3.12) lie outside the unit circle. As for an MA process the values of $\{\beta_i\}$ that make the process invertible must be such that the roots of Eq. (3.15) lie outside the unit circle.

It is straightforward in principle, though algebraically rather tedious, to calculate the autocorrelation function of a given ARMA process. Like that of an AR process, it will typically decay exponentially rather than "cut off" as for an MA process.

The importance of the mixed ARMA class of models arises because many time series can be described by such a model with fewer parameters than would be needed if a pure AR model were to be fitted.

### 3.2.7 The Random Walk

In practice, many observed time series are nonstationary. In order to fit a stationary model, it is first necessary to remove any nonstationary sources of variation. The simplest type of nonstationarity is a long-term drift in the mean, which is often fairly small compared with other sources of variation and which is often approximately linear with time. Such drift can easily be subtracted from a series to give a better approximation to stationary data.

A completely different type of nonstationarity arises when a series is locally smooth but where the local mean seems to go up and down in a fairly random way. The simplest type of model for describing such behavior is the so-called random walk model, for which

$$X_t = X_{t-1} + Z_t \tag{3.18}$$

where $\{Z_t\}$ again denotes a purely random process.

Many observed time series can be approximated by a random walk, including share prices on successive days for which we typically find

share price on day $t$ = share price on day $(t - 1)$ + "random shock"

It can readily be shown that a random walk is not stationary (as the variance increases through time), but that the first differences of the process, namely,

$$\nabla X_t = X_t - X_{t-1} = Z_t \tag{3.19}$$

*do* form a stationary process.

### 3.2.8 Integrated ARMA Processes

The idea of differencing a time series in order to make it stationary can be extended to other classes of model including ARMA processes. If $X_t$ in Eq. (3.16) is replaced by $\nabla X_t = (X_t - X_{t-1})$, then we have what is called an *integrated ARMA* (or ARIMA) model. Such a model is called *integrated* because a stationary ARMA model is fitted to the differenced data, and this has to be summed or "integrated" in order to provide a model for the original nonstationary data. If a series has to be differenced, not once, but $d$ times, and then $p$ AR terms and $q$ MA terms are fitted to the stationary differenced series, then the model is called an ARIMA($p$, $d$, $q$) model. Of particular interest is the ARIMA (0, 1, 1) model, which has been used to model the frequency fluctuations of atomic clocks. It turns out that a well-known forecasting method called *exponential smoothing* is optimal for this model (e.g., Chatfield [3], Chapter 5). We will not pursue this topic here.

### 3.2.9 State-Space Models

This class of models was originally developed by control engineers but is now finding applications in many areas including statistics. This section provides a brief introduction. Further details may be found in Chatfield [3], Chapter 10, and (for a viewpoint from control engineering) Anderson and Moore [1] and Maybeck [10].

When a scientist tries to measure any sort of signal, it will typically be contaminated by noise, so that the actual observation, $X_t$, is given (in words) by

$$\text{observation} = \text{signal} + \text{noise} \tag{3.20}$$

In state-space models the signal is assumed to be a linear combination of a set of variables, called *state variables,* which constitute what is called the *state vector* at time $t$. This vector describes the state of the system at time $t$ and is sometimes called the *state of nature.* Unfortunately the state vector cannot be observed directly, and we have to use the actual observations to make inferences about the signal, and hence about the state vector.

Assume that the state vector, $\theta_t$, consists of $m$ state variables and so is an $(m \times 1)$ vector. Denote the "noise" or "observation error" by $n_t$. Then we write Eq. (3.20) as

$$X_t = \mathbf{h}_t^T \theta_t + n_t \tag{3.21}$$

where the $(m \times 1)$ vector $\mathbf{h}_t$ is assumed to be a vector of known quantities.

Although $\theta_t$ cannot be observed directly, it is often reasonable to assume that we know how it changes through time, and this updating equation can be represented in the general form

$$\theta_t = G_t \theta_{t-1} + \mathbf{w}_t \tag{3.22}$$

where the $(m \times m)$ matrix, $G_t$, is assumed known and $\mathbf{w}_t$ denotes a vector of $m$ random shocks or deviations. Note that if $G_t$ is independent of $t$, then the process becomes a vector-valued AR(1) model.

Equations (3.21) and (3.22) together constitute the general form of the (univariate) state-space model. Equation (3.21) is customarily called the *observation* equation (because it includes the observed variable), while Eq. (3.22) is called the *transition* (or *system*) equation. The "error" terms in the observation and transition equations are generally assumed to be uncorrelated with each other and also to be uncorrelated through time. The model can readily be generalized to the situation where $X_t$ is a vector by choosing $h_t$ to be a matrix of appropriate size and by making $n_t$ a vector of appropriate length.

This general formulation of a state-space model looks quite complicated at first sight so let us look at a simple example. The so-called steady model, has a state vector that consists of a single state variable, say $\mu_t$, which can be described as the local level and is assumed to follow a random walk. Thus the observation equation is given by

$$X_t = \mu_t + n_t \tag{3.23}$$

and the transition equation by

$$\mu_t = \mu_{t-1} + w_t \tag{3.24}$$

This model involves two "error" terms, $n_t$ and $w_t$, which are usually assumed to be purely random processes with respective variances $\sigma_n^2$ and $\sigma_w^2$. One of the "error" terms, $n_t$, is the observation error, and the other is the "error" in the transition (signal) equation. The ratio of their two variances, namely, $\sigma_w^2/\sigma_n^2$, is often called the *signal-to-noise ratio*. Comparing with Eqs. (3.21) and (3.22) we see that $\theta_t$ is just a scalar and equals $\mu_t$, while $h_t$ and $G_t$ are also scalars taking the constant value of unity. The steady model may be a reasonable approximation for data showing no long-term trend and no seasonality, but some short-term correlation.

The steady model only depends on one state variable. Another important, slightly more general, model is the *linear growth model*, which depends on *two* state variables, the local level $\mu_t$ and the local trend $\beta_t$. This model is specified by three equations:

$$X_t = \mu_t + n_t \tag{3.25}$$

which is the observation equation, and

$$\mu_t = \mu_{t-1} + \beta_{t-1} + w_{1,t} \tag{3.26}$$

$$\beta_t = \beta_{t-1} + w_{2,t} \tag{3.27}$$

which are the two transition equations. We see that the local level, $\mu_t$, changes at each time interval by the value of the local trend, $\beta_t$, while the local trend in turn also changes through time as a random walk. Comparing with Eqs. (3.21) and (3.22), we see that $\theta_t^T = (\mu_t, \beta_t)$, while $\mathbf{h}_t^T = (1, 0)$ and

$$G_t = \begin{pmatrix} 1 & 1 \\ 0 & 1 \end{pmatrix}$$

are both constant through time. Of course, if $w_{1,t}$ and $w_{2,t}$ have zero variance, then the trend is constant and we have what is called a *global* linear trend model. However modern writers often prefer the extra flexibility allowed in letting the trend evolve through time. Although the trend may be approximately linear over short time spans, it can evolve in a highly nonlinear way over longer time spans. Thus the linear growth model is an example of what is called a *local* linear trend model.

Many other models can also be put into state-space form, including models for trend and seasonality, ARMA models, regression models, and the *dynamic linear models* of Bayesian forecasting (West and Harrison [17]).

Often the prime objective is to estimate the signal in the presence of noise. In other words we want to estimate the state vector, $\theta_t$, given observations on the measured variable, $X$, up to time $t$. A general method of doing this is provided by a recursive numerical procedure called the *Kalman filter*. This provides optimal estimates of the current state of a dynamic system as well as expressions for the variances of these estimates. Indeed one advantage of expressing a model in state-space form is that the Kalman filter can then be implemented. However, we do not have space to describe it here—see, for example, Chatfield [3], Section 10.2.

### 3.2.10 Harmonic Models

Short-term correlation effects are usually best described by the autocorrelation function and by fitting some form of ARIMA or state-space model. Such models are sometimes called *time-domain models*. However, physical scientists are often more interested in periodic variation such as can be described by sinusoidal-type functions. Such models, sometimes called *frequency-domain models,* are introduced in Section 3.3.

### 3.2.11 Bivariate Processes

Thus far we have been concerned with modelling a single time series. This section considers the case where we have observations on *two* time series, say, $X_t$ and $Y_t$, and are interested in the relationship, if any, between them.

We assume for simplicity that both time series are stationary (or have already been made stationary by differencing or some other trend-removal technique). In the univariate case, we started by defining the autocorrelation function. Here we start by defining the cross-covariance and cross-correlation functions. The *cross-covariance* function of a (discrete-time) bivariate stationary process $\{X_t, Y_t\}$, may be defined by

$$C_{xy}(k) = \text{Covariance } (X_t, Y_{t+k}) \qquad (3.28)$$

This function differs from the autocovariance function in that it is generally *not* an even function of lag. The size of the cross-covariance depends on the units in which $X_t$ and $Y_t$ are measured. Thus it is often useful to standardize the cross-covariance function to produce a function called the *cross-correlation* function, $\rho_{xy}(k)$, which is given by

$$\rho_{xy}(k) = C_{xy}(k)/(\sigma_x \sigma_y) \qquad (3.29)$$

where $\sigma_x$, $\sigma_y$ are the standard deviations of $X$ and $Y$, respectively. This function measures the correlation between $X_t$ and $Y_{t+k}$ and has the properties that

(a) $\rho_{xy}(k) = \rho_{yx}(-k)$  note the reversal of subscripts

(b) $|\rho_{xy}(k)| \leq 1$.

Whereas autocorrelations are always 1 at lag zero, note that this is generally *not* true for $\rho_{xy}(0)$. The latter can take any value in the range $(-1, 1)$ and is 0 for two uncorrelated processes.

Models for bivariate processes fall into two classes. In one class, the two processes "arise on an equal footing," as for example when seismic signals are recorded at two different sites. Here the observation on each variable could involve lagged values of both variables as well as a "random shock" term. One example could be

$$X_t = \alpha_1 X_{t-1} + \beta_1 Y_{t-2} + Z_{1,t}$$
$$Y_t = \alpha_2 Y_{t-1} + Z_{2,t}$$

where $\alpha_1$, $\beta_1$, $\alpha_2$ are constants and $Z_{1,t}$, $Z_{2,t}$ are independent purely random processes.

The second class of models are those where the two series are causally related, so that one series, say, $X_t$, is regarded as the input to a (usually linear) system and the other series is regarded as the output. Then a typical model would be

$$Y_t = \sum_{k=0}^{\infty} h_k X_{t-k} + \text{noise} \qquad (3.30)$$

where the set of weights $\{h_k\}$ constitute what is called the *impulse response function* of the system. The Fourier transform of the latter function, called the *frequency response function* or *transfer function* of the system, is an alternative and complementary way of describing a linear system in the frequency domain. The *identification of linear systems* is a major topic in control engineering, and we do not have space to discuss it here (see Chatfield [3], Section 9.4. The problem is relatively straightforward when data are collected in what is called an *open-loop* system (where the output has no effect on the input), but much more difficult when some form of feedback control is being applied so that the data arise from what is called a *closed-loop* system.

### 3.2.12 Model Building

This chapter is primarily about probability models, and we do not have space to discuss detailed procedures for fitting such models to data. However, a few general comments on model building can usefully be made. In addition some references will be given.

The three main stages of *model building* (see Box and Jenkins [2], Section 1.3; Chatfield [3], Sections 4.8 and 5.2.4) are

1. *Formulating* (or specifying) an appropriate type of model,
2. *Estimating* the parameters of that model,
3. *Checking* that the model really does give a good fit to the data.

Textbooks often concentrate on estimation. This is unfortunate because model formulation is often much the harder problem. If one knows for example that the data may be described by an AR(1) process, then many packages are available to fit such a model. The real problem lies in knowing that an AR(1) model is the right one to fit in the first place.

Model formulation is helped by getting background knowledge, looking at a time plot of the data (is there a trend? cyclic variation? and so on) and looking at various statistics such as the sample autocorrelation function or sample *correlogram*. The latter is usually defined by

$$r_k = \text{sample autocorrelation coefficient at lag } k \quad (3.31)$$
$$= \sum_{t=1}^{N-k} (x_t - \bar{x})(x_{t+k} - \bar{x}) / \sum_{t=1}^{N} (x_t - \bar{x})^2$$

where the observed series is donated by $x_1, x_2, \ldots, x_N$. The quantity $r_k$ provides an estimate of $\rho(k)$ for a stationary series (the original data may need to be detrended). The experienced analyst can use the correlogram, along with other sample statistics (such as the *partial autocorrelation function*), to identify a suitable model (e.g., Chatfield [3], Chapter 4). For example, if the sample correlogram decays exponentially as in Eq. (3.10), then an AR(1) model is indicated. However model formulation is not easy, and the choice of model is often not clear-cut.

After fitting a model, various diagnostic checks including a residual analysis (e.g., Chatfield [3], Section 4.7) can check whether the correct model has been fitted. If the checks cast doubt on the model, then the analyst should go back and try alternative models until one is found that is consistent with all aspects of the data. Thus there may be several cycles of model fitting as the model is modified in response to diagnostic checks or to additional data. This emphasizes that model building is an iterative, interactive procedure.

For bivariate data, $\{(x_i, y_i); i = 1, 2, \ldots, N\}$, model identification depends partly on the bivariate analogue of Eq. (3.31); namely,

$r_{xy}(k) =$

$$\begin{cases} \sum_{t=1}^{N-k} (x_t - \bar{x})(y_{t+k} - \bar{y})/s_x s_y & k = 0, 1, \ldots, N - 1 \\ \sum_{t=1-k}^{N} (x_t - \bar{x})((y_{t+k} - \bar{y})/s_x s_y & k = -1, -2, \ldots, -(N - 1) \end{cases} \quad (3.32)$$

where $s_x$, $s_y$ are the respective standard deviations of the $x$ and $y$ series. This is called the *sample cross-correlation function*. Unfortunately this sample function is even harder to interpret than the (univariate) sample correlogram and may in particular show apparent cross-correlations that are actually induced by trends within each series.

## 3.3 Spectral Properties of Random Processes

The autocorrelation function is one way of describing a stationary random process. An analysis based on this function and on models such as AR models is sometimes called an analysis in the *time domain*. An alternative complementary function for describing a stationary random process is provided by the so-called spectral density function. In brief this function is the Fourier transform of the autocorrelation function so that the two functions are mathematically equivalent. What they do is to present the same information in completely different ways, and an analysis based on the spectral density function is called *an analysis in the frequency domain* or *spectral analysis*. Spectral analysis, with its emphasis on identifying periodic behavior, is often more important to the physical scientist than the estimation of short-term correlation effects using time-domain models such as ARMA models. The basic aim is to decompose the variance of the process into components ascribable to different frequencies. This section provides a brief introduction to spectral ideas (see also Chatfield [3], Chapters 6 and 8, and Priestley [12]).

First we clarify the idea of *frequency*. A time series that consists of a single sinusoidal curve may be written

$$X_t = R \cos(\omega t + \theta) \quad (3.33)$$

where $\omega$ is called the *frequency* of the sinusoidal variation, $R$ is called the *amplitude*, and $\theta$ is called the *phase*. Here the frequency, $\omega$, is measured in radians per unit time (where $\pi$ radians $= 180°$) and is sometimes called the *angular* frequency. An alternative form of frequency, namely, $f = \omega/2\pi$, represents the number of cycles per unit time and is often easier to interpret from a physical point of view. The period of a sinusoidal cycle, called the *wavelength*, is equal to $2\pi/\omega$ or $1/f$.

### 3.3.1 The Spectrum

More generally the variation in a time series may be caused by variation at a range of different frequencies, and a function called the *power spectral density function* or *power spectrum* may be defined for stationary random processes. This function describes how the variance of the process is attributed to different frequencies. If strictly deterministic sinusoidal components are excluded and the autocovariances come down to zero quickly enough, then it turns out that the power spectrum is the Fourier transform of the autocovariance function, $C(k)$; namely,

$$f(\omega) = \text{power spectrum at frequency } \omega \quad (3.34)$$

$$= \sum_{k=-\infty}^{\infty} C(k)\, e^{-i\omega k}/\pi$$

$$= [C(0) + 2 \sum_{k=1}^{\infty} C(k)\, \cos(\omega k)]/\pi \quad (3.35)$$

for $0 < \omega < \pi$. Note the upper limit, $\pi$, for the allowed range of frequencies. This upper limit, sometimes called the *Nyquist frequency*, arises because observations taken at unit intervals of time give no direct information about variation at frequencies higher than 1/2 cycle per unit time, which corresponds to an (angular) frequency of $\omega = \pi$ (radians per unit time). Rather any variation in the series at a frequency higher than the Nyquist frequency will be *aliased* (or confused with or cannot be distinguished from) a corresponding frequency below the Nyquist frequency. Details will not be given here.

Also note the lower limit of the allowed range of frequencies is zero frequency. Some writers prefer an alternative definition of the power spectrum that gives an even function of frequency defined over the range $(-\pi, \pi)$. The introduction of negative frequencies has some mathematical advantages but some readers may find the idea of a negative frequency rather unappealing from a practical point of view. Negative frequencies are just a mathematical convenience. Note that Chapter 11 uses the notation $S(f)$ to denote the (power) spectral density function, with frequency measured in cycles per unit time and with negative frequencies allowed. Thus the allowed range for $f$ is $(-1/2, +1/2)$ rather than $(-\pi, \pi)$.

The power spectrum defined previously may be interpreted in the following way. The area underneath the curve between two particular frequencies represents the contribution to the variance of the time series due to sinusoidal components with frequencies in the given range. In particular it can be shown that the overall variance of the series is equal to the total area under the curve, so that

$$\text{Var}(X_t) = C(0) = \int_0^\pi f(\omega)d\omega \qquad (3.36)$$

This is just a particular case of the inverse Fourier transform of Eq. (3.34); namely,

$$C(k) = \int_0^\pi \cos(\omega k) f(\omega) d\omega$$

when $k = 0$. We do not have space to give full mathematical justification for these formulae or show in detail how to deal with series containing deterministic sinusoidal components—see Section 3.3.3 and Chatfield [3], Chapter 6.

Statisticians sometimes prefer to work with a dimensionless version of the spectrum, called the *normalized spectrum,* which is equal to the power spectrum divided by the variance of the process. The area underneath this curve between two particular frequencies will then represent the *proportion* of variance accounted for by frequencies in the given range. More formally, we have

$$\begin{aligned}f^*(\omega) &= \text{normalized spectrum at frequency } \omega \\ &= f(\omega)/C(0) \qquad (3.37) \\ &= \left[1 + 2\sum_{k=1}^\infty \rho(k)\cos(\omega k)\right]/\pi\end{aligned}$$

using Eq. (3.35). Thus $f^*(\omega)$ is the Fourier transform of the autocorrelation function.

### 3.3.2 The Shape of the Spectrum

The shape of the spectrum of a particular random process is a useful guide to its properties. If the spectrum is "large" at low frequencies but "small" at high frequencies, then the variance of the process is concentrated at low frequencies indicating a smooth process that varies rather slowly about its mean. Such a spectrum is called a *low-frequency spectrum.* In contrast a spectrum that is relatively large at high frequencies is called a *high-frequency spectrum,* and this indicates a process that tends to oscillate rather quickly about its mean. These remarks can be illustrated by the AR(1) process

$$X_t = \alpha X_{t-1} + Z_t \tag{3.38}$$

for which the autocorrelation function is given by Eq. (3.10). Using Eq. (3.37) we find, after some algebra, that the normalized spectrum is given by

$$f^*(\omega) = 1/\pi(1 - \alpha^2)(1 - 2\alpha \cos\omega + \alpha^2) \tag{3.39}$$

The shape of this spectrum depends on the value of $\alpha$. If $\alpha$ is negative, then it is clear from Eq. (3.38) that values of $X_t$ will tend to alternate in sign. This means they tend to oscillate at high frequency. As expected, we then find that Eq. (3.39) gives a high-frequency spectrum (i.e., $f^*(\omega)$ is relatively large at high frequencies). In contrast, if $\alpha$ is positive, then power will be concentrated at low frequencies, as might intuitively be expected from Eq. (3.38).

The spectra of other random processes can also be worked out using Eqs. (3.35) or (3.37) or perhaps by some other means. For example the normalized spectrum of the MA(1) process

$$X_t = Z_t + \beta Z_{t-1} \tag{3.40}$$

can be shown to be

$$f^*(\omega) = (1 + 2\beta \cos\omega + \beta^2)/\pi(1 + \beta^2) \tag{3.41}$$

This gives a low-frequency spectrum when $\beta > 0$, and a high-frequency spectrum when $\beta < 0$, as may be intuitively expected from Eq. (3.40).

The spectrum of a purely random process is also of interest. Using Eqs. (3.6) and (3.34), it can easily be shown that the spectrum will be *constant* in the range $(0, \pi)$. This indicates, as might be expected, that variation is not caused by components at any particular frequency, but rather that all frequencies contribute equally in some sense.

### 3.3.3 Deterministic Periodic Variation

The spectral density function can be used to describe a stationary process having no deterministic components. However, there are many series where it is known that a deterministic component *does* exist. For example, temperature has a 24-hour cycle through the day. An important class of models may be obtained from Eq. (3.33) by adding a noise term so that

$$X_t = R \cos(\omega t + \theta) + Z_t \tag{3.42}$$

This equation defines a stationary process when $\theta$ is regarded as a random variable uniformly distributed over the interval $[-\pi, +\pi]$. However, $\theta$ will be fixed for a single realization and so this is something of a mathematical "fiddle." If $Z_t$ is a purely random process or white noise, then the spectrum of $X_t$ consists of two parts. The first part due to $Z_t$ is a constant function over the frequency range $(0, \pi)$, while the second part due to the deterministic sinusoidal component consists of an infinite "spike" at frequency $\omega$. It turns out that the spectral density function is not defined at frequency $\omega$ and that one has to work with a function called the *integrated spectrum*, or *spectral distribution function*, $F(\omega)$. This is actually a more fundamental function than the power spectrum as it always exists for a stationary process and may be defined as the contribution to the variance of the process accounted for by frequencies in the range $(0, \omega)$. When the power spectrum $f(\omega)$ exists, the integrated spectrum is the integral of $f(\omega)$ up to a given frequency (or conversely $f(\omega)$ is the derivative of $F(\omega)$). For the model in Eq. (3.42), it can be shown that $F(\omega)$ has a discrete jump at frequency $\omega$ that is proportional to the square of the amplitude of the deterministic component, namely, $R^2$. Because of this discontinuity, the integrated spectrum cannot be differentiated at the particular frequency $\omega$ and that is why the power spectrum does not exist there.

The model described by Eq. (3.42) can be further generalized by allowing $Z_t$ to be *colored* noise, rather than white noise (i.e., to be some stationary process that is not necessarily a sequence of uncorrelated random variables) or by adding further deterministic sinusoidal terms at different specified frequencies, say $\omega_1, \omega_2, \ldots$.

The detection of deterministic components is very important for determining long-term behavior. It can be difficult when, as is usually the case, such variation is mixed up with nondeterministic variation that will be present over a continuous range of frequencies. Thus the detection of deterministic components is an important aspect of spectral analysis and various tests exist for detecting deterministic spectral peaks—see Chapter 12.

### 3.3.4 The Cross Spectrum

For a stationary *bivariate* process, it is possible to define what is called the *cross spectrum* of the process as the Fourier transform of the cross-covariance function (or of the cross-correlation function for the normalized cross spectrum). More formally, we have

$$f_{xy}(\omega) = \left[ \sum_{k=-\infty}^{\infty} C_{xy}(k) \, e^{-i\omega k} \right] / \pi \qquad (3.43)$$

over the range $0 < \omega < \pi$. (The sum converges provided that $C_{xy}(k)$ damp down to zero quickly enough). Unlike the (auto) spectrum, which is a real-valued function, the cross spectrum is generally a complex-valued function and is therefore much harder to understand and assess. One possibility is to look at the real and complex parts of the cross spectrum separately (they are called the *cospectrum* and *quadrature spectrum,* respectively). An alternative possibility is to look at the *amplitude* and *phase* of the cross spectrum. Further details are given in Chapter 11.

### 3.3.5 Spectral Analysis

The *estimation* of spectra and cross spectra, called *spectral analysis,* raises many interesting statistical problems that will be discussed later in Chapters 10 and 11—see also Chatfield [3], Chapters 7 and 8, and Priestley [12]. Data must first be made stationary by some sort of filtering or preprocessing particularly to remove trend. The (univariate) spectrum and the cross spectrum are then essentially estimated by taking a weighted Fourier transform of the corresponding sample correlation function, namely, $r_k$ or $r_{xy}(k)$, respectively, or by smoothing, over neighboring frequencies, the squared modulus of the raw Fourier transform of the data (univariate or bivariate as appropriate) in an appropriate way. The raw squared modulus is usually called the *periodogram.* The computational calculations can be made more efficient by utilizing an algorithm called the *fast Fourier transform* or FFT.

An alternative parametric approach to spectral analysis is also possible using autoregressive (AR) models. This is called *autoregressive spectrum estimation.* An AR model of sufficiently high order is fitted to the data, and the estimated value of $p$, the order of the process, and the estimated values of the parameters $\alpha_1, \ldots, \alpha_p$ of Eq. (3.7) are inserted into the expression for the spectrum of an AR process—see Chatfield [3], page 198. This approach is parametric (whereas the usual Fourier approach is nonparametric) and has the advantage that it produces an AR linear predictor as a byproduct at the same time.

## 3.4 Point Process Models

The idea of a *point process* was briefly introduced in Section 3.1. It consists of a series of events occurring randomly through time, such as machine failures, accidents, and earthquakes. More generally we would like to have a range of models to describe different mechanisms for the placement of points or events along the time axis. The events are often called *arrivals* or *failures,* depending on the context. It turns out—see Chapter 4—that the models can also be applied to the placement of points in some region of *space,* usually one or two dimensional, but for convenience we refer to placement through time throughout.

Observed data from a point process could, for example, consist of the sequence of dates of earthquakes in a particular region, the arrival times of customers joining a queue, and (a spatial example) the positions of a particular type of plant in a field. This section is concerned primarily with probabilistic aspects of point processes, rather than with statistical aspects, and the reader is referred, for example, to Cox and Lewis [6] for guidance on the latter topic.

There are many possible features of interest in a point process, with the emphasis depending on the particular application. Sometimes the main interest lies in estimating the overall mean rate of occurrence of events, while in other applications we may be more interested in seeing if there is a trend in the event rate. Alternatively we may be interested in seeing if events are in some sense "random" or perhaps come in clusters or tend to recur at periodic intervals (see Chapter 12) or whatever.

There is a wide range of functions for describing different properties of point processes, and this can be rather bewildering for the new reader, especially when the notation also varies considerably between different applications. For example in reliability problems, the "events" are typically breakdowns or failures of some kind and interest may center on evaluating the probability that a system will function properly for a specified time. In queueing theory the "events" are typically the arrival times of customers joining a queue, and interest may center on the probability distribution of the queue length. For simplicity we will refer throughout to "events" or "failures," but the reader should bear in mind that the results apply more generally.

### 3.4.1 Some Definitions

This section introduces a variety of functions for describing point processes as well as the notation that will be used.

There are broadly speaking two different ways of looking at a point process. The first approach concentrates on the *number of events occurring in a fixed time interval,* while the second approach concentrates on the *time intervals between events.* The two approaches are equivalent and complementary for processes in time (though the time-interval approach does not generalize so well to spatial data) and there is a fundamental duality between the two random variables. For example if the number of events in a particular time interval of length $T$ is 0, then the time interval between events that includes this interval must be at least of length $T$. In reliability, typical questions for the two approaches could be "How many failures are likely to occur in the warranty period?" or, for the second approach, "How long will it be until the first failure occurs?" Of course, we may sometimes be interested in answering both sorts of question, which is why a mathematical description in terms of "numbers of events" is complementary to a description in terms of "length of time intervals."

We assume throughout that events cannot occur simultaneously. We also assume that time is continuous, although many results can be adapted to the discrete-time case, as for example when a system is inspected once an hour rather than continuously.

We begin by looking at the distribution of the number of events in a fixed time interval. Let $N(t)$ denote the number of events in the interval $(0, t]$, where the square bracket indicates that the upper end-point, $t$, *is* included in the interval, whereas the round bracket indicates that the lower end-point 0 is *not* included. Clearly $N(t)$ is a discrete random variable as it takes nonnegative integer values. It is an increasing step function through time, with steps of unit size provided there are no simultaneous arrivals.

More generally, let $N(t_1, t_2)$ denote the number of events that occur in $(t_1, t_2]$. Clearly $N(t_1, t_2) = N(t_2) - N(t_1)$. A point process may formally be specified by the joint distributions of collections of random variables of the form $N(t_1, t_2)$, where $t_1$ and $t_2$ lie in a specified range. A point process is said to be *stationary* or *homogeneous* if translating the whole process through time by a constant amount has no effect on these joint distributions. In particular this would imply that the distribution of $N(t, t + k)$ does not depend on $t$.

The expected value of $N(t)$, sometimes called the *renewal function*, may be of particular interest and is defined by

$$M(t) = \text{Expected value of } [N(t)] \tag{3.44}$$

Like $N(t)$, $M(t)$ is a nondecreasing function but will generally be continuous rather than a step function. If it is a nicely behaved function whose derivative exists, then the *arrival (or failure) rate of the process* is defined by this derivative (with respect to time); namely,

$$\mu(t) = M'(t) \tag{3.45}$$

If simultaneous events cannot occur, it can be shown that $\mu(t)$ is equal to another function, sometimes called the *intensity function of the process*, which is defined by

$$\lambda(t) = \lim_{\Delta t \to 0} \frac{\text{Prob [1 event in } (t, t + \Delta t)]}{\Delta t} \tag{3.46}$$

The intensity function is unfortunately sometimes confused with the hazard function (see Eq. (3.50)), which relates to the distribution of the time between events rather than to the occurrence of events in a fixed time interval. The two functions are generally not the same, although they do happen to be identical for the Poisson process (see Section 3.4.2). Note that $\mu(t)$ and $\lambda(t)$ are both constant for stationary (or homogeneous) processes; that is $\mu(t) = \mu$ and $\lambda(t) = \lambda$, say.

We now turn our attention to the alternative way of looking at a point process; namely, quantities related to the distribution of time between events. Suppose we start to observe the process at time 0, and let $Y_1, Y_2, \ldots,$ denote successive event times, so that $Y_r =$ (time at which the $r$th event occurs). Let $X_1$ denote the time until the first event and $X_r$ denote the time between the $(r - 1)$th and $r$th events. Thus $X_1 = Y_1, X_2 = (Y_2 - Y_1), \ldots,$ $X_r = (Y_r - Y_{r-1}), \ldots$ . Whereas the number of events, $N(t)$, is a discrete random variable, it is clear that $\{X_i\}$ and $\{Y_j\}$ are in general continuous random variables.

For a stationary process it may be possible to assume that the time to the first event, $X_1$, has the same distribution as the subsequent $\{X_i\}$. This is true if the time origin is chosen to be at an event, in which case $X_1$ will also represent a time between events. It is also the case for the Poisson process (see Section 3.4.2), where it can be shown that the history of the process is irrelevant so that it does not matter if there is an event at the time origin or not. However, it is also worth stressing that the properties of some processes are quite different if one starts from an arbitrary time point rather than from an arbitrary event.

There are various equivalent ways of describing the *distribution of time between events,* say, $X$, for a stationary process. Here, $X$ is a nonnegative random variable that is generally continuous and hence may be described by its cumulative distribution function (c.d.f.), $F(x)$, defined by

$$F(x) = \text{Prob } (X \leq x) \tag{3.47}$$

or equivalently by its probability density function (p.d.f.), $f(x)$, defined by

$$f(x) = \frac{dF(x)}{dx} \tag{3.48}$$

For some purposes it is more convenient to describe the distribution of $X$ in terms of the *survivor* or *reliability* function, $R(x)$, defined by

$$\begin{aligned} R(x) &= \text{Prob}(X > x) \\ &= 1 - F(x) \end{aligned} \tag{3.49}$$

or by the *hazard function* defined by

$$h(x) = f(x)/R(x) \tag{3.50}$$

In reliability work, this function is also known as the *age-specific failure rate* or the conditional failure rate, but it should not be called just the *failure rate* as this term can be mixed up with the process failure rate defined in Eq. (3.45).

The hazard function may not be easy to understand at first. It is best explained in terms of failure times, where it can be shown that $h(x)\Delta x$ is the probability that an event (or failure) will occur in the interval from $x$ to $x + \Delta x$ *given that* the failure time is at least as large as $x$. Conditional probabilities of this kind are often more meaningful than unconditional probabilities. For example the proportion of automobiles that are scrapped between the ages of 15 and 16 years, say, may be found from the p.d.f., $f(x)$, of the time-to-failure (scrapping) distribution and will be small because few automobiles survive to the age of 15 years. However, if you have a 15 year-old automobile, there is a much higher probability that it will fail in the coming year, and as the current owner you will be interested in only probabilities *conditional* on having reached this age. In fact a scrapped automobile is an example of a *nonrepairable* system. While it is possible to represent such a system as a very simple point process, in which there is a maximum of just one event, this is not generally helpful and such systems are generally studied by looking at lifetime distributions (e.g., Thompson [14]). Point process theory is usually applied in reliability problems only to *repairable systems* or to systems where a failed component is immediately replaced by an identical new component. Of course, in most applications outside reliability (e.g., times between earthquakes), the crucial distinction between repairable and nonrepairable systems is not relevant and there is no difficulty about the meaning of an *inter-event time*.

As well as the distribution of inter-event times, we may also wish to examine the *distribution of the time until the rth event occurs;* namely, $Y_r = \sum_{i=1}^{r} X_i$. Other related quantities are the *backward recurrence time*, defined as the length of time measured backwards from time $t$ to the most recent event before $t$, and the *forward recurrence time* defined in an analogous way.

As an example of how we can relate statements concerning the numbers of events in a fixed time interval to times between events, we note that the number of events in $(0, t]$, $N(t)$, is equal to $r$ if and only if $Y_r < t < Y_{r+1}$. Thus probability statements can readily be found linking the two (dual) approaches. For example, Prob $[N(t) \geq r]$ = Prob $[Y_r \leq t]$. (If this is not obvious, think about it!).

### 3.4.2 The Poisson Process

A relatively simple type of point process, and arguably the most important, is one in which events occur completely "at random" in a sense now to be defined. In fact a Poisson process may be defined in at least three mutually equivalent ways, as follows. Let $\lambda$ denote the average number of events per unit time. Note that it is assumed to be constant. The first specification defines the process in terms of the count variables, $N(t)$, and so is sometimes called the *counting specification.*

**Definition 1:** *A series of events form a Poisson process of rate $\lambda$ over the range $(0, T)$ if (i) for any $(t_1, t_2)$, such that $0 < t_1 < t_2 < T$, $N(t_1, t_2)$ is a Poisson random variable with mean $\lambda(t_1 - t_2)$; (ii) the numbers of events in disjoint intervals are mutually independent.*

It follows immediately from this definition that $N(0, t)$ is a Poisson random variable with mean $\lambda t$; hence, the name of the process.

The second specification is called the *interval specification,* as it depends on the intervals between events.

**Definition 2:** *A series of events form a Poisson process if the time to the first event, $X_1$, and the the times between subsequent events, $X_2, X_3, \ldots$, are independent exponential random variables with mean $1/\lambda$.*

The third specification, sometimes called the *intensity specification,* is perhaps the most fundamental and involves explicit probability statements.

**Definition 3:** *A series of events form a Poisson process if there exists a constant $\lambda$ such that, for all $t > 0$, (i) Prob $[N(t, t + \Delta t) = 1] = \lambda \Delta t + o(\Delta t)$; (ii) Prob $[N(t, t + \Delta t) = 2] = o(\Delta t)$; so that (iii) Prob $[N(t, t + \Delta t) = 0] = 1 - \lambda \Delta t + o(\Delta t)$. Moreover, all these probability statements do not depend on what has happened before time t so that Prob $[N(t, t + \Delta t) = 1 \mid$ history of event times up to time $t] = $ Prob $[N(t, t + \Delta t) = 1]$ and the process is said to be memoryless.*

In this third definition, $o(\Delta t)$ is the usual mathematical notation for terms that go to zero faster than $\Delta t$ itself, and we see from (ii) that the chance of simultaneous events is negligible.

The three definitions may be hard to take in at first and need a bit of thought. We will not demonstrate that they are equivalent. The reader is referred to any textbook on stochastic process (e.g., Cox and Miller [7]). The interrelationships between the three specifications is a recurring theme in the study of a Poisson process. While the first definition is perhaps the easiest to understand, the second one is more useful for constructing a realization of the process, and the third specification is often the easiest one to derive results from. We will also state several additional results without proof.

1. The Poisson process is clearly stationary and is sometimes called the *homogeneous* Poisson process to distinguish it from more complicated variants of the Poisson process, some of which will be briefly introduced.
2. It makes no difference to any properties of the Poisson process if there is assumed to be an event at time 0 or not as the process has no memory. (The memoryless property of the process means that it is an example of a general class of processes called *Markov processes*). Thus $X_1$ may be the time to the first event from an arbitrary origin or from an arbitrary event. It follows that the forward recurrence time has the same distribution as the time between events; namely, exponential with mean $1/\lambda$.
3. The average number of events per unit time, $\lambda$, is equal to various other quantities describing the properties of a Poisson process, including the arrival rate of the process (see Eq. 3.45), the intensity function of the process (see Eq. 3.46), and the hazard function of the distribution of interevent times. Indeed the parameter $\lambda$ is known by many descriptions including event rate, rate, and intensity.
4. It can be shown that the time until the $r$th event, $Y_r$, has a gamma distribution with parameters $\alpha = r$ and $\lambda$ in the usual notation (see Eq. 3.51). The mean value of $Y_r$ is $r/\lambda$.

The basic (homogeneous) Poisson process just introduced can be generalized and extended in various ways. These include

1. the Poisson process in space, which may be one, two, or three dimensional;
2. the time-dependent or nonhomogeneous Poisson process where the rate parameter $\lambda(t)$ varies through time;
3. the combination or superposition of several independent Poisson processes;
4. a Poisson process in which multiple events at the same time are allowed.

We briefly consider the first two of these possibilities. Further details about 3 and 4 may be found for example in Thompson [15], Chapter 7, and Cox and Isham [5], page 49, respectively.

A *Poisson process in space* may be developed in a straightforward way by varying the assumptions of the basic Poisson process in an appropriate way. Consider two-dimensional space as an example. Let $N(\Delta S)$ denote the number of events in an arbitrary small area of size $\Delta S$. Then Definition 3 for example may be varied by requiring that Prob $[N(\Delta S) = 1] = \lambda \Delta S + o(\Delta S)$. Then the number of events in any given area can be shown to have a Poisson distribution whose mean is equal to that area multiplied by $\lambda$. Further details and extensions may be found for example in Cox and Isham [5], Chapter 6, as well as in Chapter 4.

A *nonhomogeneous* or *(inhomogeneous) Poisson process* may be developed by allowing the rate parameter or intensity function, $\lambda(t)$, to vary through time. The process is then nonstationary or time dependent, but the adjective *nonhomogeneous* is generally used instead. For the homogeneous process, the expected value of $N(t)$ is given by $M(t) = \lambda t$. In the nonhomogeneous case, this is no longer true. Instead we find that $N(t)$ still has a Poisson distribution but with mean $M(t) = \int_0^t \lambda(u)du$. We do not have space to develop the properties of this interesting class of processes here, and the reader is referred, for example, to Cox and Miller [7], Section 4.2, and Thompson [14], Section 4, as well as to Chapter 4, which introduces inhomogeneous Poisson processes, the Cox process, and the Neyman-Scott process among others in a spatial context.

### 3.4.3 Renewal Processes

A completely different way of generalizing the Poisson process is to consider a series of events where the times between events, $\{X_i\}$, are independent but have some distribution other than the exponential. Such a process is called a *renewal process*. This description arises from their application to the failure and replacement of components such as light bulbs.

Although the definition of a renewal process looks like a variant of Definition 2 of the Poisson process, the numbers of events will now generally *not* be Poisson and the other two definitions of the Poisson process will not apply. If there is an event at time 0, so that $X_1$ is also a time between events, then the process is called an *ordinary* renewal process. Because the $\{X_i\}$ are independent, then every time an event occurs, it is as if the process starts all over again from the beginning. If there is not necessarily an event at time 0, so that $X_1$ may have a different distribution to subsequent $\{X_i\}$, then we have what is called a *modified* renewal process. Further details may be found in many references including Cox and Isham [5], Cox and Miller [7], Section 9.2, and Thompson [15], Chapter 5, though it is still hard to beat the classic work of Cox [4].

The simplest special case of a renewal process is the Poisson process itself, for which the $\{X_i\}$ have an exponential distribution. More generally, we can allow the $\{X_i\}$ to come from some other family such as the *gamma* distribution, for which the probability density function is given, for $x > 0$, by

$$f(x) = \lambda(\lambda x)^{\alpha-1} e^{-\lambda x}/\Gamma(\alpha) \tag{3.51}$$

where $\Gamma$ denotes the gamma function and $\alpha$ and $\lambda$ are nonnegative parameters. It can be shown that the distribution has mean $\alpha/\lambda$. When the parameter

$\alpha$ is an integer, this distribution is sometimes called the *special Erlangian* distribution.

An alternative possibility is to allow the $\{X_i\}$ to follow a distribution called the *Weibull* (or *extreme-value*) distribution—see Chapter 2 and Cox [4]—for which the survivor function is given, for $x > 0$, by

$$R(x) = \exp\{-(\lambda x)^\alpha\} \tag{3.52}$$

We may often be interested in finding the distribution of the time to the $r$th event, $Y_r = X_1 + \ldots + X_r$. If the values of $X$ are exponential with mean $1/\lambda$, then it can be shown that $Y_r$ is gamma with parameters $\alpha = r$ and $\lambda$. If instead the values of $X$ are gamma with parameters $\alpha$ and $\lambda$, then it can be shown that $Y_r$ will also be gamma with parameters $r\alpha$ and $\lambda$. More generally, if each $X_i$ has mean $\mu$ and variance $\sigma^2$, then it can be shown that $Y_r$ will be asymptotically normally distributed with mean $r\mu$ and variance $r\sigma^2$. The exact distribution can be investigated if desired by using moment generating functions or Laplace transforms—see Cox [4], Chapter 1.

Let us now turn attention to the number of events in $(0, t]$; namely, $N(t)$. Its distribution is related to that of $Y_r$ as noted at the end of Section 3.4.1, and a variety of results are available about $N(t)$, about the expected value of $N(t)$, namely, $M(t)$, the renewal function (Eq. 3.44), and about the derivative of $M(t)$ with respect to time, which is often called the *renewal density* in the context of renewal processes. Lack of space prevents us giving more details, and the reader is referred to the references given previously.

## 3.5 Further Reading

An introduction to time-series analysis is given by many authors including Chatfield [3], Kendall and Ord [9], and Wei [16]. A more advanced treatment is given by Priestley [12].

An introduction to point processes is given by Thompson [15] and also in various general texts on stochastic processes such as Cox and Miller [7]. More advanced treatments are given by Cox and Isham [5] and Snyder [13] The early book by Cox and Lewis [6] is still a valuable source regarding methods of statistical analysis for point processes. Additional references on both probabilistic and statistical aspects are given by Thompson [15], page 113.

More details on a variety of spatial processes are given in Chapter 4.

## References

1. Anderson, B. D. O., and Moore, J. B. (1979). *Optimal Filtering*. Prentice-Hall, Englewoods Cliffs, NJ.

2. Box, G. E. P., and Jenkins, G. M. (1976). *Time Series Analysis, Forecasting and Control,* rev. ed. Holden-Day, San Francisco.
3. Chatfield, C., (1989). *The Analysis of Time Series,* 4th ed. Chapman and Hall, London.
4. Cox, D. R. (1962). *Renewal Theory.* Methuen, London.
5. Cox, D. R., and Isham, V. (1980). *Point Processes.* Chapman and Hall, London.
6. Cox, D. R., and Lewis, P. A. W. (1965). *The Statistical Analysis of Series of Events.* Methuen, London.
7. Cox, D. R., and Miller, H. D. (1965). *The Theory of Stochastic Processes.* John Wiley and Sons, New York.
8. Diggle, P. J. (1990). *Time Series: A Biostatistical Introduction.* Oxford University Press, Oxford.
9. Kendall, M. G., and Ord, J. K. (1990). *Time Series,* 3d ed. Arnold, Sevenoaks, U.K.
10. Maybeck, P. S. (1979). *Stochastic Models, Estimation and Control.* Academic Press, New York.
11. Papoulis, A. (1984). *Probability, Random Variables, and Stochastic Processes,* 2nd ed. McGraw-Hill, New York.
12. Priestley, M. B. (1981). *Spectral Analysis and Time Series,* **1** and **2.** Academic Press, London.
13. Snyder, D. L. (1975). *Random Point Processes.* John Wiley and Sons, New York.
14. Thompson, W. A., Jr., (1981). "On the Foundations of Reliability." *Technometrics,* **23,** 1–13.
15. Thompson, W. A., Jr., (1988). *Point Process Models with Applications to Safety and Reliability.* Chapman and Hall, London.
16. Wei, W. W. S. (1990). *Time Series Analysis: Univariate and Multivariate Methods.* Addison-Wesley, Reading, MA.
17. West, M., and Harrison, J. (1989). *Bayesian Forecasting and Dynamic Models.* Springer-Verlag, Berlin.

# 4. MODELS FOR SPATIAL PROCESSES

Noel Cressie

Department of Statistics, Iowa State University, Ames

## 4.1 Introduction

Much of the statistical inference currently taught to scientists and engineers has its roots in the methods advocated by Fisher [1] in the 1920s and 1930s, which were developed for agricultural and genetic experiments. Statistics and its applications have grown enormously in recent years, although at times it has been difficult for the new developments to fit into Fisher's framework, namely, blocking, randomization, and replication within a well designed experiment. Contrast data from an agricultural field trial with the time series of Wolf's sunspot numbers, which are often used exogenously in models of global climate. In the most extreme case, there is just one experimental unit and hence no real replication; here the term *observational study* is used to describe the investigation. (On occasions, it may be possible to obtain replication when the observational study is one of several like studies that are available, but those occasions are rare.)

With one of the underpinnings of the classical framework missing, it may seem that statistical methods would have little to say about an observational study. However, statistics is a vital, evolving discipline that is highly adaptive to new scientific and engineering problems. At the very least, statistical design (and analysis) forces one to think about important factors, use of a control, sources of variability, and statistical models; it reveals new structure and so prompts conjectures about that structure; and it is a medium for productive multidisciplinary communication. It may even be possible to substitute observational studies in the field with well-designed experiments in the laboratory. Indeed, in the *computer* laboratory, large simulation experiments have been a very effective statistical tool.

Spatial statistics is a relatively new development within statistics; ironically, Fisher was well aware of potential complications from the spatial component in agricultural experiments. In [2] he wrote: "After choosing the area we usually have no guidance beyond the widely verifiable fact that patches in close proximity are commonly more alike, as judged by the yield of crops, than those which are far apart." Fisher went to great lengths to neutralize the effect of spatial correlation by randomizing treatment assignments. (Just as important, randomization also controls for bias in the way treatments are assigned.) In 1938, Fairfield Smith [3] wrote an article showing empirically that as the plot size increased, the variance of the average yield decreased but that beyond a particular plot size, the decrease was negligible. However, this is impossible if the data are uncorrelated, so that Fairfield Smith's study indirectly recognized the presence of spatial correlation. Statistical models for such phenomena did not begin to appear until Whittle's article in the 1950s [4].

The most general model considered in this chapter is spatio-temporal, which is written as

$$\{Z(\mathbf{s}; t) : \mathbf{s} \in D(t), t \in T\} \tag{4.1}$$

where $\mathbf{s}$ and $t$ index the location and time of the random observation $Z$ (which could be real-valued, a random vector, or even a random set). The model (4.1) achieves a great deal of generality by allowing index sets $\{D(t) : t \in T\}$ and $T$ to be possibly random sets themselves; see Section 4.4. A particular case of (4.1) is the purely *spatial model:*

$$\{Z(\mathbf{s}) : \mathbf{s} \in D\} \tag{4.2}$$

where $D$ is an index set (possibly random) in the Euclidean space $\Re^d$; typically $d = 2$ or 3. For example, Figure 1 shows a predicted surface $\{\hat{Z}(\mathbf{s}) : \mathbf{s} \in D\}$ (in pounds of wheat per rectangular plot, where a plot is 1/500 acre) of (4.2) over a 1-acre rectangular region, based on data in [5].

In the last 30 years, a number of texts and monographs on models for spatial processes have appeared. Preparation of this chapter has benefited from material in Matern [6], Bartlett [7], Journel and Huijbregts [8], Cliff and Ord [9], Ripley [10], and Upton and Fingleton [11]. Nevertheless, the approach taken is a personal one and draws considerably from my recent book (Cressie [12]), where complete references can be found to all the topics covered in this chapter.

In the three sections that follow, $D$ will take different forms. Section 4.2 deals with geostatistical models where $D$ is a continuous region; Section 4.3 deals with lattice models where $D$ is a finite or countable set of sites; and Section 4.4 deals with point process models where $D$ is a random set made up of random events. These three cases cover the majority of spatial statistical models, although the sections do not go into very much detail on statistical inference for the models presented. A full discussion of these issues can be found in Cressie [12], along with applications to real data.

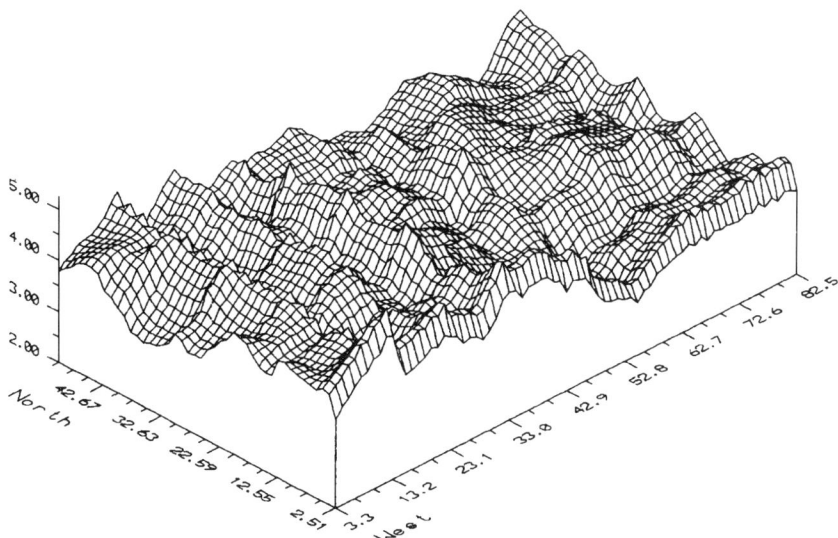

FIG. 1. A predicted surface of wheat yields (in pounds per 1/500 acre) over a 1-acre rectangular region. (Copyright © 1991. Reprinted by permission of John Wiley & Sons, Inc.)

Finally, in Section 4.5, a discussion of further developments in spatial modeling is given.

## 4.2 Geostatistical Models

The prefix *geo* in geostatistics originally implied statistics pertaining to the earth (see Matheron [13]). However, more recently, geostatistics has been used in a variety of disciplines ranging from agriculture to zoology. (Within meteorology, a virtually identical theory called *objective analysis* was developed by Gandin [14].) Its flexibility, in being able to incorporate the known action of physical, chemical, and biological processes along with uncertainty represented by spatial heterogeneity, makes it an attractive tool.

Geostatistics is mostly concerned with spatial prediction, but there are other important areas, such as model selection, effect of aggregation, and spatial sampling design, that offer fruitful open problems. More details can be found in Part I of Cressie [12]. The emphasis in this section will be on a spatial-prediction method known as *kriging*. Matheron [13] coined the term in honor of D. G. Krige, a South African mining engineer (see [15] for an account of the origins of kriging). Chapter 13 should be consulted for applications of kriging.

### 4.2.1 The Variogram

First, a measure of the (second-order) spatial dependence exhibited by spatial data is presented. A model-based function known as the *variogram* will be defined; its estimate provides such a measure. Physicists are used to dealing with the power spectrum and the mathematically equivalent autocovariance function. It is demonstrated here that the class of processes with a variogram contains the class of processes with an autocovariance function and that optimal linear prediction can be carried out on a wider class of processes than the one traditionally used in statistics.

Let $\{Z(\mathbf{s}) : \mathbf{s} \in D \subset \Re^d\}$ be a real-valued stochastic process defined on a domain $D$ of the $d$-dimensional space $\Re^d$, and suppose that differences of variables lagged $\mathbf{h}$-apart vary in away that depends only on $\mathbf{h}$. Specifically, suppose

$$\text{var}[Z(\mathbf{s} + \mathbf{h}) - Z(\mathbf{s})] = 2\gamma(\mathbf{h}), \quad \text{for all } \mathbf{s}, \mathbf{s} + \mathbf{h} \in D \quad (4.3)$$

The quantity $2\gamma(\cdot)$, which is a function of only the *difference* between the spatial locations $\mathbf{s}$ and $\mathbf{s} + \mathbf{h}$, is called the *variogram* (or sometimes the structure function). The variogram must satisfy the conditional-negative-semidefiniteness condition, $\sum_{i=1}^{k} \sum_{j=1}^{k} \alpha_i \alpha_j 2\gamma(\mathbf{s}_i - \mathbf{s}_j) \leq 0$, for any finite number of spatial locations $\{\mathbf{s}_i : i = 1, \ldots, k\}$, and real numbers $\{\alpha_i : 1, \ldots, k\}$ satisfying $\sum_{i=1}^{k} \alpha_i = 0$. This condition is extremely important because it guarantees that all model-based variances are nonnegative (e.g., [12], sec. 2.5.2). Thus, one must be very careful to choose such valid variogram models. When $2\gamma(\mathbf{h})$ can be written as $2\gamma^\circ(\|\mathbf{h}\|)$, for $\mathbf{h} = (h_1, \ldots, h_d)' \in \Re^d$, and $\|\mathbf{h}\| \equiv (h_1^2 + \ldots + h_d^2)^{1/2}$ the variogram is said to be *isotropic;* otherwise it is said to be *anisotropic,* in which case the process $Z$ is also referred to as *anisotropic.* Figure 2 shows a commonly used isotropic (semi)variogram model. Note the discontinuity at the origin (notated as $c_0$ in the models that follow), which is important in capturing micro-scale spatial variation.

Variogram models that depend on only a few parameters $\theta$ can be used as summaries of the spatial dependence and as an important component of optimal linear prediction (kriging). Three basic isotropic models, given here in terms of the semivariogram (half the variogram), are as follows. The *linear* model (valid in $\Re^d$, $d \geq 1$)

$$\gamma(\mathbf{h};\theta) = \begin{cases} 0 & \mathbf{h} = \mathbf{0} \\ c_0 + b_l \|\mathbf{h}\| & \mathbf{h} \neq \mathbf{0} \end{cases}$$

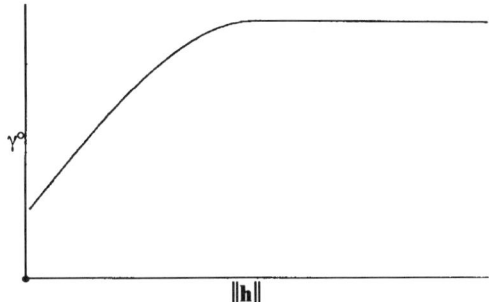

FIG. 2. Semivariogram model $\gamma$ plotted as a function of the lag distance. Shown is the spherical model. (Copyright © 1991. Reprinted by permission of John Wiley & Sons, Inc.)

where $\boldsymbol{\theta} = (c_0, b_l)'$, $c_0 \geq 0$, $b_l \geq 0$. The *spherical* model (valid in $\Re^1$, $\Re^2$, and $\Re^3$)

$$\gamma(\mathbf{h}; \boldsymbol{\theta}) = \begin{cases} 0 & \mathbf{h} = \mathbf{0} \\ c_0 + c_s \left[ \frac{3}{2}(\|\mathbf{h}\|/a_s) - \frac{1}{2}(\|\mathbf{h}\|/a_s)^3 \right] & 0 < \|\mathbf{h}\| \leq a_s \\ c_0 + c_s & \|\mathbf{h}\| \geq a_s \end{cases}$$

where $\boldsymbol{\theta} = (c_0, c_s, a_s)'$, $c_0 \geq 0$, $c_s \geq 0$, $a_s \geq 0$. This is the model shown in Figure 2. The *exponential* model (valid in $\Re^d$, $d \geq 1$)

$$\gamma(\mathbf{h}; \boldsymbol{\theta}) = \begin{cases} 0 & \mathbf{h} = \mathbf{0} \\ c_0 + c_e[1 - \exp(-\|\mathbf{h}\|/a_e)] & \mathbf{h} \neq \mathbf{0} \end{cases}$$

where $\boldsymbol{\theta} = (c_0, c_e, a_e)'$, $c_0 \geq 0$, $c_e \geq 0$, $a_e \geq 0$.

A semivariogram model that exhibits negative correlations caused by periodicity of the process is the *wave* (or *hole-effect*) model (valid in $\Re^1$, $\Re^2$, and $\Re^3$):

$$\gamma(\mathbf{h}; \boldsymbol{\theta}) = \begin{cases} 0 & \mathbf{h} = \mathbf{0} \\ c_0 + c_w \dfrac{1 - a_w \sin(\|\mathbf{h}\|/a_w)}{\|\mathbf{h}\|} & \mathbf{h} \neq \mathbf{0} \end{cases}$$

where $\boldsymbol{\theta} = (c_0, c_w, a_w)'$, $c_0 \geq 0$, $c_w \geq 0$, $a_w \geq 0$.

A further condition that a variogram model must satisfy is

$$2\gamma(\mathbf{h})/\|\mathbf{h}\|^2 \to 0, \qquad \text{as } \|\mathbf{h}\| \to \infty$$

In fact, the *power* semivariogram model,

$$\gamma(\mathbf{h}; \boldsymbol{\theta}) = \begin{cases} 0 & \mathbf{h} = \mathbf{0} \\ c_0 + b_p \|\mathbf{h}\|^\lambda & \mathbf{h} \ne \mathbf{0} \end{cases}$$

where $\boldsymbol{\theta} = (c_0, b_p, \lambda)'$, $c_0 \ge 0$, $b_p \ge 0$, $0 \le \lambda < 2$, is a valid semivariogram model in $\mathfrak{R}^d$, $d \ge 1$.

When the process $Z$ is anistropic, the variogram is no longer purely a function of distance between two spatial locations. Anisotropies are caused by the underlying physical process evolving differentially in space. Sometimes the anisotropy can be corrected by an invertible linear transformation of the lag vector $\mathbf{h}$. That is,

$$2\gamma(\mathbf{h}) = 2\gamma^\circ(\|A\mathbf{h}\|), \qquad \mathbf{h} \in \mathfrak{R}^d$$

where $A$ is a $d \times d$ matrix and $2\gamma^\circ$ is a function of only one variable. The preceding paragraphs demonstrate that variogram models are very flexible. In the paragraphs that follow, it will be seen that they are even more general than autocovariance models.

Replacing (4.3) with the stronger assumption,

$$\text{cov}[Z(\mathbf{s} + \mathbf{h}), Z(\mathbf{s})] = C(\mathbf{h}), \qquad \text{for all } \mathbf{s}, \mathbf{s} + \mathbf{h} \in D \qquad (4.4)$$

and specifying the mean function to be constant; i.e.,

$$E[Z(\mathbf{s})] = \mu, \qquad \text{for all } \mathbf{s} \in D \qquad (4.5)$$

defines the class of *second-order* (or wide-sense) *stationary* processes in $D$, with auto covariance function $C(\cdot)$. Time series analysts often assume (4.4) and work with the autocorrelation function $\rho(\cdot) \equiv C(\cdot)/C(\mathbf{0}).$

Conditions (4.3) and (4.5) define the class of *intrinsically stationary* processes, which is now shown to contain the class of second-order stationary processes. Assuming only the existence of the autocovariance function given by (4.4),

$$\gamma(\mathbf{h}) = C(\mathbf{0}) - C(\mathbf{h}) \qquad (4.6)$$

That is, the semivariogram given by (4.3) exists and is related very simply to the covariance function, which establishes the containment we set out to prove. An example of a process for which $2\gamma(\cdot)$ exists but $C(\cdot)$ does not is a one-dimensional standard Wiener process $\{W(t) : t \ge 0\}$. Here, $2\gamma(h) = |h|$ (where $-\infty < h < \infty$) but $\text{cov}(W(t), W(u)) = \min(t, u)$, which is not a function of $|t - u|$; an analogous result is true in $\mathfrak{R}^d$. Thus, the class of intrinsically stationary processes *strictly* contains the class of second-order stationary processes.

Now consider estimation of the variogram from data $\{Z(\mathbf{s}_i) : i = 1, \ldots, n\}$. Suppose these are observations on an intrinsically stationary process (i.e., a process that satisfies (4.3) and (4.5)), taken at the $n$ spatial locations $\{\mathbf{s}_i : i = 1, \ldots, n\}$. Because of assumption (4.5), $\text{var}[Z(\mathbf{s} + \mathbf{h}) - Z(\mathbf{s})] = E[Z(\mathbf{s} + \mathbf{h}) - Z(\mathbf{s})]^2$. Hence, a natural (method-of-moments) estimator of the variogram $2\gamma(\mathbf{h})$ is

$$2\hat{\gamma}(\mathbf{h}) \equiv \sum_{N(\mathbf{h})} [Z(\mathbf{s}_i) - Z(\mathbf{s}_j)]^2 / |N(\mathbf{h})|, \quad \mathbf{h} \in \Re^d \qquad (4.7)$$

where the average in (4.7) is taken over $N(\mathbf{h}) \equiv \{(\mathbf{s}_i, \mathbf{s}_j) : \mathbf{s}_i - \mathbf{s}_j = \mathbf{h}\}$, and $|N(\mathbf{h})|$ is the number of distinct elements in $N(\mathbf{h})$. For irregularly spaced data, $N(h)$ is usually modified to $\{(\mathbf{s}_i, \mathbf{s}_j) : \mathbf{s}_i - \mathbf{s}_j \in T(h)\}$, where $T(\mathbf{h})$ is a tolerance region in $\Re^d$ surrounding $\mathbf{h}$. Other estimators, more robust than (4.7), are discussed in [12], sec. 2.4. Parametric models $2\gamma(\cdot; \boldsymbol{\theta})$ can be fit to the estimator (4.7) by various means; as a compromise between efficiency and simplicity, one could minimize the weighted sum of squares,

$$\sum_{k=1}^{K} \left\{ \frac{2\hat{\gamma}[\mathbf{h}(k)]}{2\gamma[\mathbf{h}(k); \boldsymbol{\theta}]} - 1 \right\}^2 |N[\mathbf{h}(k)]|$$

with respect to variogram model parameters $\boldsymbol{\theta}$. The sequence $\mathbf{h}(1), \ldots, \mathbf{h}(K)$ denotes the "lags" at which an estimator (4.7) was obtained, which are less than half the maximum possible $\|\mathbf{h}\|$ and satisfy $|N[\mathbf{h}(k)]| \geq 30$; $k = 1, \ldots, K$ (e.g., [8], p. 194, eq. III.42). Minimizing the weighted sum of squares usually performs well and never does poorly against other more complicated competitors; for more details, see [12], sec. 2.6.

### 4.2.2 Kriging

For the purposes of this section, assume that the variogram is known; in practice, variogram parameters are estimated from the spatial data $\mathbf{Z} \equiv [Z(\mathbf{s}_1), \ldots, Z(\mathbf{s}_n)]'$ observed at known locations $\{\mathbf{s}_1, \ldots, \mathbf{s}_n\}$. Suppose it is desired to predict $Z(\mathbf{s}_0)$ at some unsampled spatial location $\mathbf{s}_0$ using a linear function of the data:

$$\hat{Z}(\mathbf{s}_0) = \sum_{i=1}^{n} \lambda_i Z(\mathbf{s}_i) \qquad (4.8)$$

It is sensible to look for coefficients $\{\lambda_i : i = 1, \ldots, n\}$ for which (4.8) is uniformly unbiased (e.g., see (4.9)) and that minimize the mean-squared prediction error $E[Z(\mathbf{s}_0) - \hat{Z}(\mathbf{s}_0)]^2$. More generally, one could try to minimize $E(L[Z(\mathbf{s}_0), p(\mathbf{Z})])$ with respect to predictor $p(\mathbf{Z})$, where $L$ is a given loss function. For example, the loss function,

$$L[Z(\mathbf{s}_0), p(\mathbf{Z})] = b\{\exp[a(Z(\mathbf{s}_0) - p(\mathbf{Z}))] - a(Z(\mathbf{s}_0) - p(\mathbf{Z})) - 1\},$$
$$a \in \Re, \quad b > 0$$

allows overprediction to incur a different loss than underprediction. Minimizing mean-squared prediction error results from using

$$L[Z(\mathbf{s}_0), p(\mathbf{Z})] = b[Z(\mathbf{s}_0) - p(\mathbf{Z})]^2, \quad b > 0$$

which is the squared-error loss function. In all that is to follow, squared-error loss is used.

Assuming a constant mean (4.5), the *uniform unbiasedness* condition imposed on the predictor (4.8) becomes $E[\hat{Z}(\mathbf{s}_0)] = \mu = E[Z(\mathbf{s}_0)]$, for all $\mu \in \Re$. This is equivalent to

$$\sum_{i=1}^{n} \lambda_i = 1 \tag{4.9}$$

If the process is second-order stationary (i.e., (4.4) and (4.5) hold) and (4.9) is assumed,

$$E[Z(\mathbf{s}_0) - \sum_{i=1}^{n} \lambda_i Z(\mathbf{s}_i)]^2 = C(\mathbf{0}) - 2 \sum_{i=1}^{n} \lambda_i C(\mathbf{s}_i - \mathbf{s}_0) \\ + \sum_{i=1}^{n} \sum_{j=1}^{n} \lambda_i \lambda_j C(\mathbf{s}_i - \mathbf{s}_j) \tag{4.10}$$

If the weaker assumption is made that the process is intrinsically stationary (i.e., (4.3) and (4.5) hold) and (4.9) is assumed,

$$E[Z(\mathbf{s}_0) - \sum_{i=1}^{n} \lambda_i Z(\mathbf{s}_i)]^2 = 2 \sum_{i=1}^{n} \lambda_i \gamma(\mathbf{s}_i - \mathbf{s}_0) \\ + \sum_{i=1}^{n} \sum_{j=1}^{n} \lambda_i \lambda_j \gamma(\mathbf{s}_i - \mathbf{s}_j) \tag{4.11}$$

Using differential calculus and the method of Lagrange multipliers, optimal coefficients $\boldsymbol{\lambda} = (\lambda_1, \ldots, \lambda_n)'$ can be found that minimize (4.11) subject to (4.9). Define $\boldsymbol{\gamma} \equiv [\gamma(\mathbf{s}_1 - \mathbf{s}_0), \ldots, \gamma(\mathbf{s}_n - \mathbf{s}_0)]'$, $\mathbf{1} \equiv (1, \ldots, 1)'$, and the $n \times n$ matrix $\Gamma$ to be the symmetric matrix with $(i, j)$th element $\gamma(\mathbf{s}_i - \mathbf{s}_j)$, which is assumed to be invertible. Then the optimal coefficients are given by

$$\lambda = \Gamma^{-1}\left[\gamma + \frac{(1 - \mathbf{1}'\Gamma^{-1}\gamma)\mathbf{1}}{\mathbf{1}'\Gamma^{-1}\mathbf{1}}\right] \quad (4.12)$$

and the minimized value of (4.11) (sometimes called the *kriging variance*) is

$$\sigma_k^2(\mathbf{s}_0) = \gamma'\Gamma^{-1}\gamma - \frac{(1 - \mathbf{1}'\Gamma^{-1}\gamma)^2}{\mathbf{1}'\Gamma^{-1}\mathbf{1}} \quad (4.13)$$

The kriging predictor given by (4.8) and (4.12) is appropriate if the process Z contains no measurement error. If measurement error is present, then a "noiseless version" of Z should be predicted; see [12], sec. 3.2.1.

Thus far, kriging has been derived under the assumption (4.5) of a constant mean. More realistically, assume

$$Z(\mathbf{s}) = \mu(\mathbf{s}) + \delta(\mathbf{s}), \quad \mathbf{s} \in D \quad (4.14)$$

where $E[Z(\mathbf{s})] = \mu(\mathbf{s})$, for $\mathbf{s} \in D$, and $\delta(\cdot)$ is a zero-mean, intrinsically stationary stochastic process with $\text{var}[\delta(\mathbf{s} + \mathbf{h}) - \delta(\mathbf{s})] = \text{var}[Z(\mathbf{s} + \mathbf{h}) - Z(\mathbf{s})] = 2\gamma(\mathbf{h})$, $\mathbf{h} \in \Re^d$. In (4.14), the "large-scale variation" $\mu(\cdot)$ and the "small-scale variation" $\delta(\cdot)$ are modeled as deterministic and stochastic processes, respectively, but with no unique way of identifying either of them. What is one person's mean structure could be another person's correlation structure. Often this problem is resolved by relying on scientific or habitual reasons for determining the mean structure.

Suppose $\mu(\mathbf{s}) = \sum_{j=0}^{p} x_j(\mathbf{s})\beta_j$, a linear combination of known variables that could include trend-surface terms or other explanatory variables thought to influence the behavior of the large-scale variation. Thus, assume the model

$$Z(\mathbf{s}) = \mathbf{x}(\mathbf{s})'\boldsymbol{\beta} + \delta(\mathbf{s}), \quad \mathbf{s} \in D \quad (4.15)$$

where $\boldsymbol{\beta} \equiv (\beta_0, \ldots, \beta_p)'$ are unknown parameters and $\delta(\cdot)$ is intrinsically stationary (i.e., satisfies (4.3) and (4.5)) with zero mean. Although the model has changed, the problem of predicting $Z(\mathbf{s}_0)$ using an unbiased linear predictor (4.8) is still of interest. The uniform unbiasedness condition is now equivalent to the condition

$$\lambda'X = \mathbf{x}_0' \quad (4.16)$$

where $\mathbf{x}_0 \equiv [x_0(\mathbf{s}_0), \ldots, x_p(\mathbf{s}_0)]'$ and $X$ is an $n \times (p + 1)$ matrix whose $(i, j)$th element is $x_{j-1}(\mathbf{s}_i)$. Then, provided (4.9) is implied by (4.16), minimizing the mean-squared prediction error subject to (4.16) yields the *universal kriging* predictor

$$\hat{Z}_U(s_0) = \lambda'_U Z \tag{4.17}$$

where

$$\lambda_U = \Gamma^{-1}[\gamma + X(X'\Gamma^{-1}X)^{-1}(x_0 - X'\Gamma^{-1}\gamma)] \tag{4.18}$$

and the (universal) kriging variance is

$$\sigma_U^2(s_0) = \gamma'\Gamma^{-1}\gamma - (X'\Gamma^{-1}\gamma - x_0)'(X'\Gamma^{-1}X)^{-1}(X'\Gamma^{-1}\gamma - x_0) \tag{4.19}$$

Another way to write the equations (4.16) and (4.17) is

$$\hat{Z}_U(s_0) = v'_1\gamma + v'_2 x_0 \tag{4.20}$$

where $v_1$ (an $n \times 1$ vector) and $v_2$ (a $(p + 1) \times 1$ vector) solve

$$\begin{aligned}\Gamma v_1 + X v_2 &= Z \\ X' v_1 &= 0\end{aligned} \tag{4.21}$$

Equations (4.20) and (4.21) are known as the *dual-kriging equations*, since the predictor is now expressed as a linear combination of the elements of $(\gamma', x'_0)$. Interestingly, these are also the equations used for *thin-plate spline smoothing*, where the basis function (or variogram) used to define $\Gamma$ and $\gamma$ is prespecified (see [12], sec. 3.4.5). Kriging has the advantage that in practice the data are used first to estimate the basis function, so adapting to the quality and quantity of spatial dependence in the data. Furthermore, kriging produces a mean-squared prediction error, given by (4.19), that quantifies the degree of uncertainty in the predictor. Cressie [12], sec. 5.9, presents these two methods of spatial prediction along with 12 others, obtained from both stochastic and nonstochastic considerations.

### 4.2.3 Change of Support

The support of $Z(B) \equiv (1/|B|) \int_B Z(u) du$ is simply $B$, the region over which $Z$ is averaged. The concept is an extremely important one in modeling a physical process; one could think of $|B|$ as the spatial resolution at which the phenomenon is being studied, although data often come as $Z = [Z(s_1), \ldots, Z(s_n)]'$, at point support. Kriging adapts very easily to accommodate the change from predicting $Z(s_0)$ with point support $s_0$ to predicting $Z(B)$ with block support $B$. For example, in (4.12) and (4.13), $\gamma$ is modified to $\gamma(B) \equiv [(1/|B|) \int_B \gamma(s_1 - u) du, \ldots, (1/|B|) \int_B \gamma(s_n - u) du]'$ and, in (4.13), $\sigma_k^2(B)$ has the extra term $-(1/|B|^2) \int_B \int_B \gamma(u - v) du dv$.

Suppose now that we wish to predict $g[Z(B)]$, where $g$ is a nonlinear function. Then the quantity of greatest interest is the conditional distribution $\Pr[Z(B) \leq z|Z]$, because any optimal predictor is a function of it. Both disjunctive kriging and indicator kriging attempt to estimate this quantity based on bivariate distributional properties of the (possibly transformed) process; see [12], sec. 5.2. However, these and other methods have their drawbacks and so optimal prediction of $g[Z(B)]$ remains an open problem.

## 4.3 Lattice Models

This section is concerned with the situation where the index set $D$ of the spatial process (4.2) is a countable collection of spatial sites at which data are observed. The set $D$ of such sites is called a *lattice*. For example, the Ising model in statistical mechanics (see, e.g., [16]) is defined for binary data on a two-dimensional lattice of sites:

$$D \equiv \{(u, v) : u = \ldots, -1, 0, 1, \ldots ;$$
$$v = \ldots, -1, 0, 1, \ldots \} \quad (4.22)$$

It is used as a model for ferromagnetism; each site is thought of as possessing a spin, which may be either an "up" spin or a "down" spin.

The set of sites is supplemented with neighborhood information that define (conditional) dependencies between sites. When the sites are located on a regular grid, the resulting *regular* lattice has neighborhoods that are often defined as adjoining sites or sets of adjoining sites. For example, in the Ising model, we could define the neighborhood of site $(u, v)$ to be $\{(u, v + 1), (u, v - 1), (u + 1, v), (u - 1, v)\}$. By contrast, *irregular* lattices have neighborhoods that are often (but do not have to be) defined by Euclidean distances to nearby sites. For example, $\mathbf{s}_i$ and $\mathbf{s}_j$ are neighbors if $\| \mathbf{s}_i - \mathbf{s}_j \| \leq r$. The formal definition of neighborhood, in terms of the conditional dependence structure, is given by (4.25) later.

### 4.3.1 Markov Random Fields

For temporal data, Markov models have been a popular way of introducing statistical dependence. Such models assume that the "present," conditioned on the "past," in fact depends only on the "immediate past." A well-known example is the random walk process.

For spatial data, an analogous notion can be formulated. As in Section 4.2, the data are

$$\mathbf{Z} \equiv [Z(\mathbf{s}_1), \ldots, Z(\mathbf{s}_n)]'$$

located at known sites

$$\{\mathbf{s}_1, \ldots, \mathbf{s}_n\} \quad (4.23)$$

(In fact, it is not necessary to know exactly the locations (4.23), but simply which sites are neighbors of each other.) Assume that the data are jointly distributed according to

$$\Pr[Z(\mathbf{s}_1), \ldots, Z(\mathbf{s}_n)] \tag{4.24}$$

from which the conditional probabilities, $\Pr[Z(\mathbf{s}_i)|\{Z(\mathbf{s}_j) : j \neq i\}]$; $i = 1, \ldots, n$, can be calculated.

The *neighborhood* $N_i$ of the $i$th site is defined to be the collection of all other sites $\mathbf{s}_j$ ($j \neq i$) such that,

$$\Pr[Z(\mathbf{s}_i)|\{Z(\mathbf{s}_j) : j \neq i\}] = \Pr[Z(\mathbf{s}_i)|\{Z(\mathbf{s}_j) : j \in N_i\}];$$
$$i = 1, \ldots, n \tag{4.25}$$

In words, (4.25) says that the $i$th site, conditioned on all other sites, in fact depends only on its neighboring values $\{Z(\mathbf{s}_j) : j \in N_i\}$. In this sense, (4.25) is the spatial analog of the temporal Markov property.

When the right-hand sides of (4.25) define (4.24), the stochastic process **Z** is called a *Markov random field*. A very attractive feature of working with a Markov random field is that modeling can be carried out at the *local* level by specifying the neighbors,

$$\{N_i : i = 1, \ldots, n\} \tag{4.26}$$

and the conditional probabilities,

$$\{\Pr[Z(\mathbf{s}_i)|\{Z(\mathbf{s}_j) : j \in N_i\}] : i = 1, \ldots, n\} \tag{4.27}$$

site by site. However, such specifications have to be made consistently so that at a *global* level the joint measure (4.24) is well defined. In addition, parameter estimation by maximum likelihood (see Chapter 8) requires an expression for (4.24) in terms of (4.27), since parameters are specified at the local level but estimated from the joint distribution (i.e., the likelihood).

Much of the material presented in this section follows closely the notation and development of Besag [17], sec. 3, and can also be found in [12], Part II. Define $\zeta \equiv \{\mathbf{z} : \Pr(\mathbf{z}) > 0\}$ and $\zeta_i \equiv \{z(\mathbf{s}_i) : \Pr[z(\mathbf{s}_i)] > 0\}$; $i = 1, \ldots, n$. The *positivity condition* is said to be satisfied if $\zeta = \zeta_1 \times \ldots \times \zeta_n$, which says that if each of $z(\mathbf{s}_i)$ individually has positive probability, then jointly they have positive probability. In all that follows, this condition will be assumed to hold.

Recall from (4.25) the definition of neighbor and neighborhood. Then a *clique* is defined to be a set of sites that consists either of a single site or of sites that are all neighbors of each other.

Without loss of generality, assume that $z(\mathbf{s}_i) = 0$ can occur at each site; i.e., assume $\mathbf{0} \in \zeta$. Define

$$Q(\mathbf{z}) \equiv \log\{\Pr(\mathbf{z})/\Pr(\mathbf{0})\}; \quad \mathbf{z} \in \zeta \tag{4.28}$$

and call $Q(\cdot)$ the *negpotential function*. The important part of (4.28) to note is that it involves a *ratio* of probabilities; nothing is special about choosing the event **0** in the denominator except that, because $Q(\mathbf{0}) = 0$, it makes some of the subsequent expressions shorter. Hence,

$$\Pr(\mathbf{z}) = \exp(Q(\mathbf{z}))/\sum_{\mathbf{y}\in\zeta}\exp[Q(\mathbf{y})]; \quad \mathbf{z} \in \zeta \qquad (4.29)$$

(Note that, for continuous data, the summation in the denominator of (4.29) is replaced with integration.)

Let $0(\mathbf{s}_i)$ denote the event "$Z(\mathbf{s}_i) = 0$," and write $\mathbf{z}_i \equiv [z(\mathbf{s}_1), \ldots, z(\mathbf{s}_{i-1}), 0, z(\mathbf{s}_{i+1}), \ldots, z(\mathbf{s}_n))'$. Then, from [17], $Q(\cdot)$ can be expanded uniquely on $\zeta$ as

$$Q(\mathbf{z}) = \sum_{1\le i\le n} z(\mathbf{s}_i)G_i[z(\mathbf{s}_i)] + \sum\sum_{1\le i\le j\le n} z(\mathbf{s}_i)z(\mathbf{s}_j)G_{ij}[z(\mathbf{s}_i), z(\mathbf{s}_j)]$$

$$+ \ldots + z(\mathbf{s}_1)z(\mathbf{s}_2) \ldots \qquad (4.30)$$

$$z(\mathbf{s}_n)G_{12\ldots n}[z(\mathbf{s}_1), z(\mathbf{s}_2), \ldots, z(\mathbf{s}_n)]; \quad \mathbf{z} \in \zeta$$

From the joint probability function $\Pr(\mathbf{z}); \mathbf{z} \in \zeta$, conditional probabilities can be calculated. Furthermore, and most important, the summands in (4.30) can be expressed in terms of conditional probabilities. For example,

$$z(\mathbf{s}_i)G_i[z(\mathbf{s}_i)] = \log[\Pr(z(\mathbf{s}_i)|\{0(\mathbf{s}_j) : j \ne i\})/\Pr(0(\mathbf{s}_i)|\{0(\mathbf{s}_j) : j \ne i\})] \qquad (4.31)$$

and

$$z(\mathbf{s}_i)z(\mathbf{s}_j)G_{ij}[z(\mathbf{s}_i), z(\mathbf{s}_j)] =$$

$$\log\left[\frac{\Pr(z(\mathbf{s}_i)|z(\mathbf{s}_j), \{0(\mathbf{s}_k) : k \ne i, j\})}{\Pr(0(\mathbf{s}_i)|z(\mathbf{s}_j), \{0(\mathbf{s}_k) : k \ne i, j\})} \cdot \frac{\Pr(0(\mathbf{s}_j)|\{0(\mathbf{s}_k) : k \ne i\})}{\Pr(z(\mathbf{s}_j)|\{0(\mathbf{s}_k) : k \ne i\})}\right] \qquad (4.32)$$

From similar considerations, expressions for three-way and higher-way interaction $G$-functions can also be derived.

The derivation of the expression (4.32) involves an arbitrary choice of the $i$th site or the $j$th site to be featured in the (conditional) probability statement. Because either could have been chosen, it is clear that the conditional-probability consistency conditions must ensure that this choice is immaterial. In fact, conditional probabilities,

$$\Pr[z(\mathbf{s}_i) \mid \{z(\mathbf{s}_j) : j \ne i\}]; \quad i = 1, \ldots, n \qquad (4.33)$$

for which $G$-functions are well defined through (4.31), (4.32), etc., yield a Markov random field given by (4.29) and (4.30). For example, in [18] it is shown that if the specified probabilities (4.33) are such that each of the $G$-functions, *defined* by (4.31), (4.32), etc., is invariant under permutations of its respective subset of variables, then the negpotential function $Q(\cdot)$, *defined* by (4.30), defines a Markov random field through (4.29), provided $\sum_{\mathbf{y} \in \zeta} \exp[Q(\mathbf{y})] < \infty$.

**Hammersley–Clifford Theorem.** Suppose that the random process on $\zeta$ is a Markov random field with neighborhood structure $\{N_i : i = 1, \ldots, n\}$. Then the negpotential function $Q$ given by (4.30) must satisfy the following property:

$$\text{if sites } i, j, \ldots, s \text{ do not form a clique, then } G_{ij\ldots s}(\cdot) \equiv 0 \quad (4.34)$$

A proof is given in [17]. The original result, by Hammersley and Clifford in 1971, was proven in a less general setting and the paper was never published.

By assuming that all three-way and higher way interaction $G$-functions in (4.30) are identically zero, Besag [17] shows how the one-parameter exponential family of conditional probabilities,

$$\Pr[z(\mathbf{s}_i) \mid z(N_i)] = \exp[A_i(z(N_i))B_i(z(\mathbf{s}_i)) + C_i(z(\mathbf{s}_i)) + D_i(z(N_i))]; \quad i = 1, \ldots, n \quad (4.35)$$

can be used to define a Markov random field. (The notation $z(N_i)$ is used to denote $\{z(\mathbf{s}_j) : j \in N_i\}$.) He shows that, up to an additive constant,

$$Q(z) = \sum_{i=1}^{n} \{\alpha_i B_i[z(\mathbf{s}_i)] + C_i[z(\mathbf{s}_i)]\} + \sum\sum_{1 \le i < j \le n} \theta_{ij} B_i[z(\mathbf{s}_i)] B_j[z(\mathbf{s}_j)] \quad (4.36)$$

where $\theta_{ji} = \theta_{ij}$, $\theta_{ii} = 0$, and $\theta_{ik} = 0$ for $k \notin N_i$. Special cases of (4.35) will now be used to generate both discrete and continuous Markov random fields.

### 4.3.2 Markov Random-Field Models for Discrete Data

Suppose that the data $Z(\mathbf{s}_1), \ldots, Z(\mathbf{s}_n)$ come as *counts*. A natural model to choose for the conditional distributions (4.35) is the Poisson distribution:

$$\Pr[z(\mathbf{s}_i) \mid z(N_i)] = \exp\{-\theta_i[z(N_i)]\} \{\theta_i[z(N_i)]\}^{z(\mathbf{s}_i)} / z(\mathbf{s}_i)!; \quad i = 1, \ldots, n \quad (4.37)$$

Then,

$$\theta_i[z(N_i)] + E[Z(\mathbf{s}_i) \mid z(N_i)] = \exp\left\{\alpha_i + \sum_{j \in N_i} \theta_{ij} z(\mathbf{s}_j)\right\} \quad (4.38)$$

which is of loglinear form. Equation (4.36) gives the negpotential function $Q$, where $B_i[z(\mathbf{s}_i)] \equiv z(\mathbf{s}_i)$ and $C_i[z(\mathbf{s}_i)] \equiv -\log[z(\mathbf{s}_i)!]$. The joint probability mass function is then given by (4.29), provided $\theta_{ij} \leq 0$ for all $j \in N_i$ and $i = 1, \ldots, n$ (which guarantees summability of the denominator of (4.29)). This spatial model is often called the *auto-Poisson* model.

When data are *binary* (taking values 0 or 1) and all three-way and higher way interaction $G$-functions in (4.30) are identically zero, then

$$Q(\mathbf{z}) = \sum_{i=1}^{n} \alpha_i z(\mathbf{s}_i) + \sum\sum_{1 \leq i < j \leq n} \theta_{ij} z(\mathbf{s}_i) z(\mathbf{s}_j)$$

where $\theta_{ij} = \theta_{ji}$, $\theta_{ii} = 0$, and $\theta_{ik} = 0$ for $k \notin N_i$. Consequently,

$$\Pr[z(\mathbf{s}_i) \mid z(N_i)] = \frac{\exp\{\alpha_i z(\mathbf{s}_i) + \sum_{j \in N_i} \theta_{ij} z(\mathbf{s}_i) z(\mathbf{s}_j)\}}{1 + \exp\{\alpha_i + \sum_{j \in N_i} \theta_{ij} z(\mathbf{s}_j)\}};$$
$$z(\mathbf{s}_i) = 0, 1$$
(4.39)

This spatial model is often called an *autologistic model*.

In both (4.38) and (4.39), statistical independence in the models is achieved by putting $\theta_{ij} \equiv 0$. Further, any large-scale inhomogeneities can be modeled via $\boldsymbol{\alpha} \equiv (\alpha_1, \ldots, \alpha_n)'$. For example, put $\boldsymbol{\alpha} = X\boldsymbol{\beta}$ where $X$ is an $n \times q$ matrix of known explanatory variables and $\boldsymbol{\beta}$ is a $q \times 1$ vector of unknown regression coefficients. When $\boldsymbol{\alpha} = (\alpha, \ldots, \alpha)'$, the model is homogeneous in large-scale effects, leaving the spatial dependence parameters $\{\theta_{ij}\}$ to model any spatial inhomogeneities.

The *Ising model* is defined by (4.39) on the infinite lattice (4.22); in (4.39), $\alpha_{(u,v)} = \alpha$ and $N_{(u,v)} = \{(u - 1, v), (u + 1, v), (u, v - 1), (u, v + 1)\}$. Thus, the exponent in the numerator of (4.39) becomes

$$\exp[z(u,v)\{\alpha + \gamma_1(z(u - 1, v) + z(u + 1, v))$$
$$+ \gamma_2(z(u, v - 1) + z(u, v + 1))\}]$$

The model was originally presented by Ising [19] under the added assumption that $\gamma_1 = \gamma_2$. Because the lattice $D$ is infinite (but countable), a delicate mathematical question remains as to whether (4.39) yields a well-defined probability measure on $D$. There may be regions of the parameter space where two different probability measures give rise to the same conditional probabilities (4.39) (physicists call this phenomenon *phase transition*). Further, for $\gamma_1 = \gamma_2 = \gamma$, critical values of $\gamma$ exist, below which the limiting nearest-neighbor correlation is zero, but above which the process exhibits long-range dependence [20].

Binary data can be viewed as a black-and-white picture, where $z(\mathbf{s}_i) = 1$ corresponds to the $i$th site being colored black and $z(\mathbf{s}_i) = 0$ corresponds to the $i$th site being colored white. When more than two colors (or categories) are allowed, the autologistic model can be generalized. Let $z(\mathbf{s}_i) \in \{0,$

$1, \ldots, c\}$, the set of $c + 1$ colors available. Assuming homogeneity and pairwise-only dependence between sites,

$$Q(\mathbf{z}) = \sum_{i=1}^{n} z(\mathbf{s}_i) G_1[z(\mathbf{s}_i)] + \sum\sum_{1 \leq i < j \leq n} z(\mathbf{s}_i)z(\mathbf{s}_j) G_{1,2}[z(\mathbf{s}_i), z(\mathbf{s}_j)]$$

$$= \sum_{r=1}^{c} m_r u_r + \sum_{r=1}^{c}\sum_{s=1}^{c} n_{rs} v_{rs}$$

where $u_r \equiv rG(r)$, $v_{rs} \equiv rsG(r, s)$; $r, s = 1, \ldots, c$. The data appear through $m_r \equiv$ number of sites having color $r$ and $n_{rs} \equiv$ number of pairs of neighboring sites where one has color $r$ and the other has color $s$.

The conditionally specified spatial models presented in the preceding section have proved particularly useful in the analysis of images, pattern recognition, and classification from satellite data. The maximum a posteriori (MAP) estimate of an image, given the degraded observations, can be computed when such spatial models are used as a "prior distribution" of the image. See Geman and Geman [21] and Cressie [12], sec. 7.4, for more details on statistical image analysis and remote sensing and Chapter 14 for more details on the use of prior distributions and the related Bayesian analysis.

### 4.3.3 Markov Random-Field Models for Continuous Data

Suppose that the data $Z(\mathbf{s}_1), \ldots, Z(\mathbf{s}_n)$ are continuous variables. A natural model to choose for the conditional distributions (4.35) is the Gaussian (abbreviated as Gau) distribution:

$$\Pr[z(\mathbf{s}_i)|z(N_i)] = (2\pi\tau_i^2)^{-1/2} \exp[-\{z(\mathbf{s}_i) - \theta_i[z(N_i)]\}^2/2\tau_i^2];$$

$$i = 1, \ldots, n \qquad (4.40)$$

Then,

$$\theta_i[z(N_i)] = E[Z(\mathbf{s}_i)|z(N_i)] = \mu(\mathbf{s}_i) + \sum_{j \in N_i} c_{ij}[z(\mathbf{s}_j) - \mu(\mathbf{s}_j)] \qquad (4.41)$$

where $c_{ij}\tau_j^2 = c_{ji}\tau_i^2$, $c_{ii} = 0$, $c_{ik} = 0$ for $k \notin N_i$, and $\mu(\mathbf{s}_i) \equiv E[Z(\mathbf{s}_i)]$; $i, j = 1, \ldots, n$. Furthermore, provided $(I - C)^{-1}$ is positive-definite, the joint probability density function is Gaussian given by

$$\mathbf{Z} \sim \text{Gau}[\boldsymbol{\mu}, (I - C)^{-1}M] \qquad (4.42)$$

where $C$ is an $n \times n$ matrix whose $(i, j)$th element is $c_{ij}$, $M \equiv \text{diag}(\tau_1^2, \ldots, \tau_n^2)$, and $\boldsymbol{\mu} \equiv [\mu(\mathbf{s}_1), \ldots, \mu(\mathbf{s}_n)]'$. Call (4.40) an *auto-Gaussian model* or, equivalently, a *conditionally specified Gaussian* (CG) *model*.

When scientific interest centers on the large-scale effects, the idea is to use a few extra small-scale (spatial dependence) parameters so that the large-scale parameters are estimated more efficiently. Suppose that the data **Z** are modeled according to a CG model and that the large-scale variation is modeled as

$$\boldsymbol{\mu} = X\boldsymbol{\beta} \tag{4.43}$$

where $X$ is an $n \times q$ matrix of explanatory variables and $\boldsymbol{\beta}$ is a $q \times 1$ vector of regression coefficients. The classical Gaussian regression model is $\mathbf{Z} \sim \text{Gau}(X\boldsymbol{\beta}, \tau^2 I)$. Now, (4.42) and (4.43) allow one to write

$$\mathbf{Z} = X\boldsymbol{\beta} + \boldsymbol{\delta} \tag{4.44}$$

where the error $\boldsymbol{\delta}$ satisfies

$$\boldsymbol{\delta} \sim \text{Gau}[\mathbf{0}, (I - C)^{-1} M]$$

When $C = 0$ and $M = \tau^2 I$, the CG regression model (4.44) reduces to the usual Gaussian regression model.

The poor performance of various *classical* statistical inference procedures when $C \neq 0$ (e.g., [9], ch. 7; [12], sec. 1.3) should convince scientists analyzing spatial data that their classical regression models need to be generalized to include spatial dependence terms. If $C = 0$ is inferred from the data, the classical regression model is fitted. If it appears that $C \neq 0$, there may be a missing explanatory variable (varying spatially) causing it. By modeling its presence through spatial dependence parameters, the spatial model is more resistant to misspecification errors. Another possible cause of a nonzero $C$ is the cumulative effect of many small-scale, spatially varying components.

When $M$ and $C$ are unknown, as is usually the case, they are often expressed in terms of a few (small-scale) parameters, at least one of which captures the spatial dependence. Thus, in general, $C \neq 0$.

The CG regression model (4.44) is compared with the SG (simultaneously specified Gaussian) regression model in Section 4.3.4. There is no difference in computational complexity but there is a difference in interpretation of the spatial-dependence parameters.

For the CG model, spatial dependence is characterized through the conditional expectation and the inverse covariance matrix is modeled. This is in contrast to Section 4.2, where the covariance matrix is modeled directly (often through the variogram). Indeed, a zero in the off-diagonal elements of $C$ corresponds to conditional independence. On the regular lattice (4.22), spatial dependence in a homogeneous process (i.e., $\boldsymbol{\mu} = (\mu, \ldots, \mu)'$) is expressed through

$$E[Z(u, v) \mid \{z(k, l) : (k, l) \neq (u, v)\}]$$
$$= \mu + \sum_i \sum_j \gamma(i, j)[z(u - i, v - j) - \mu]$$

where $\gamma(i, j) = \gamma(-i, -j)$ and $\gamma(0, 0) = 0$. The first-order dependence model is the special case: $\gamma(1, 0) = \gamma(-1, 0) = \gamma_1$, $\gamma(0, 1) = \gamma(0, -1) = \gamma_2$, and all other $\gamma(i, j)$s are zero. For this model, it can be seen that the correlation between two observations decays rather slowly with increasing distance between their lattice sites.

For the irregular lattice, one could model the $\{c_{ij}\}$ in (4.41) as some function of distance $d_{ij} \equiv \|\mathbf{s}_i - \mathbf{s}_j\|$ between site $i$ and site $j$. Note that, although $C \equiv (c_{ij})$ depends only on $\{d_{ij} : 1 \leq i < j \leq n\}$, the covariance, cov$[Z(\mathbf{s}_i), Z(\mathbf{s}_j)]$, does *not* in general depend on $\mathbf{s}_i - \mathbf{s}_j$, and hence the finite-lattice process is generally not stationary.

Care must be taken with parameterizing the spatial-dependence matrix $C$. It is not difficult to show that corr$^2$ $[Z(\mathbf{s}_i), Z(\mathbf{s}_j) \mid \{z(\mathbf{s}_k) : k \neq i, j\}) = c_{ij}c_{ji}$, which implies that $0 \leq c_{ij}c_{ji} \leq 1$. When data in $\Re^2$ occur at a potentially infinite number of sites (such as in (4.22)), construction of a Gaussian Markov random field proceeds by constructing a finite process on a torus (a donut-shaped surface that is the two-dimensional analog of the circle) lattice and then allowing its dimensions to tend to infinity. This can also be generalized to $d$-dimensional lattices, as discussed in [12], p. 438.

When the model is no longer auto Gaussian, the normalizing constant, $\int_\zeta \exp[Q(\mathbf{y})]d\mathbf{y}$, in (4.29) is typically intractable. A large class of non-Gaussian models is generated by members of the exponential family (4.35), whose distributions are absolutely continuous. For example, assuming pairwise-only dependence between sites, an *autogamma* process is defined by

$$\Pr[z(\mathbf{s}_i) \mid z(N_i)] = [\Gamma(\lambda_i)]^{-1} \beta_i^{\lambda_i} z(\mathbf{s}_i)^{\lambda_i - 1} e^{-\beta_i z(\mathbf{s}_i)} I[z(\mathbf{s}_i) > 0]$$

where

$$\lambda_i[z(N_i)] = \alpha_i + \sum_{j \in N_i} \theta_{ij} \log [z(\mathbf{s}_j)]$$

$\theta_{ij} = \theta_{ji}$, $\theta_{ii} = 0$, and $\theta_{ik} = 0$ for $k \notin N_i$; $i, j = 1, \ldots, n$. From (4.36), the negpotential function is, up to an additive constant,

$$Q(\mathbf{z}) = \sum_{i=1}^n \alpha_i \log[z(\mathbf{s}_i)] + \sum\sum_{1 \leq i < j \leq n} \theta_{ij} \log[z(\mathbf{s}_i)] \log[z(\mathbf{s}_j)]$$
$$- \sum_{i=1}^n [\beta_i z(\mathbf{s}_i) + \log z(\mathbf{s}_i)]$$

The real hurdle to obtaining the joint probability density function remains the normalizing constant, $\int_\zeta \exp[Q(\mathbf{y})] \, d\mathbf{y}$.

### 4.3.4 Spatial Autoregressive Models

This section develops lattice models based on the simultaneous interaction of spatial data at various sites. For ease of presentation, Gaussian models will be considered. Suppose that the data $\mathbf{Z} \equiv [Z(\mathbf{s}_1), \ldots, Z(\mathbf{s}_n)]'$ satisfy

$$(I - B)\mathbf{Z} = \boldsymbol{\epsilon} \tag{4.45}$$

where $\boldsymbol{\epsilon} \equiv [\epsilon(\mathbf{s}_1), \ldots, \epsilon(\mathbf{s}_n)]'$ is a vector of independent and identically distributed Gau(0, $\sigma^2$) errors and $B$ is an $n \times n$ matrix $(b_{ij})$ whose diagonal elements $\{b_{ii}\}$ are zero. Another way to write (4.45) is

$$Z(\mathbf{s}_i) = \sum_{j=1}^{n} b_{ij} Z(\mathbf{s}_j) + \epsilon(\mathbf{s}_i) \tag{4.46}$$

which is why the model is sometimes referred to as a *spatial autoregressive model*.

Clearly,

$$\mathbf{Z} \sim \text{Gau}[\mathbf{0}, (I - B)^{-1}(I - B')^{-1} \sigma^2] \tag{4.47}$$

Because of (4.45) and (4.47), the model will be called a *simultaneously specified Gaussian* (SG) *model*. More general models, on a regular lattice with a countable number of sites, are considered in [4].

There are at least two possible ways to include large-scale regression parameters in the model, although they have sometimes been confused in the literature. I now give the model I prefer and make brief mention of the other.

If it is desired to interpret large-scale effects $\boldsymbol{\beta}$ through $E(\mathbf{Z}) = X\boldsymbol{\beta}$ (where the columns of $X$ might be treatments, spatial trends, factors, etc.), then (4.45) should be modified to

$$(I - B)(\mathbf{Z} - X\boldsymbol{\beta}) = \boldsymbol{\epsilon} \tag{4.48}$$

Thus,

$$\mathbf{Z} \sim \text{Gau}[X\boldsymbol{\beta}, (I - B)^{-1}(I - B')^{-1} \sigma^2] \tag{4.49}$$

Define $\boldsymbol{\delta} \equiv \mathbf{Z} - X\boldsymbol{\beta}$, which is the error process. Then,

$$\mathbf{Z} = X\boldsymbol{\beta} + \boldsymbol{\delta} \tag{4.50}$$

where $\boldsymbol{\delta}$ satisfies

$$\boldsymbol{\delta} = B\boldsymbol{\delta} + \boldsymbol{\epsilon} \tag{4.51}$$

Times series typically exhibit *serial* correlation, which can be modeled through the error process $\boldsymbol{\delta}$.

The alternative (undesirable) model is

$$(I - B)Z = X\beta + \epsilon \quad (4.52)$$

which has the same variance matrix as in (4.49) but

$$E(Z) = (I - B)^{-1} X\beta \quad (4.53)$$

Thus, the regression is with respect to explanatory variables $(I - B)^{-1}X$, so confounding large-scale and small-scale effects.

A detailed comparison of CG models (Section 4.3.3) and SG models (Section 4.3.4) is given in [12], sec. 6.3. The most important difference between the two models is in the correct interpretation of spatial dependence parameters ($\{c_{ij}\}$ for the CG model versus $\{b_{ij}\}$ for the SG model).

## 4.4. Spatial Point Processes

In this section, the index set $D$ consists of the random locations of events in the Euclidean space $\Re^d$. Thus, the spatial models differ markedly from those in Sections 4.2 and 4.3 (where $D$ is a fixed subset of $\Re^d$) and, indeed, they have a longer history. Student [22] was concerned with the distribution of particles throughout a liquid; by counting the number of yeast cells in each of 400 squares of a hemocytometer, he concluded that the number of cells per square followed a Poisson distribution. In problems of cosmology, complete spatial randomness of the locations of galaxies in space was obviously inappropriate; Neyman and Scott [23] suggested a Poisson cluster process as a way to model the clustering of galaxies.

Ecologists, studying the way plants interact both within and between species, have made considerable use of point-pattern analysis but less of point-process modeling. Their statistical analyses have traditionally concentrated on *quadrat* (counting the number of plants in nonoverlapping regions) and distance-based (measuring the distance between plants) methods. The former does not retain precise spatial information, while the latter does not capture patterns over widely diverse spatial scales. These problems were largely overcome in the 1980s with the popularization of a statistic called the *K*-function [10], which is discussed later.

Informally speaking, a spatial point process is any random mechanism for generating the locations of a countable number of events in $\Re^d$. More formally, a spatial point process $N$ can be defined through the collection of probabilities,

$$\{\Pr[N(A_i) = n_i : i = 1, \ldots, k] :$$
$$n_i = 0, 1, 2, \ldots; \quad A_i \in \mathcal{A};$$
$$i = 1, \ldots, k; \quad k = 1, 2, \ldots\}$$

where $N(A_i)$ denotes the number of events in the bounded set $A_i$ and $\mathcal{A}$ is the set of all such sets. Figure 3 shows the spatial point pattern of locations of longleaf pine trees in an old-growth forest in southern Georgia. This may then be assimilated to a spatial point process model; see [12], ch. 8.

The homogeneous Poisson process is the simplest point process, and it is the null model against which spatial point patterns are frequently compared. Its realizations are said to exhibit *complete spatial randomness* (CSR). It should be noted that others have used the term *CSR* to describe realizations from an inhomogeneous Poisson process; that convention will be avoided here.

A spatial point process $N$ is a *homogeneous Poisson process* with intensity $\lambda$ if

1. The number of events in any bounded region $A \in \mathcal{A}$ is Poisson distributed with mean $\lambda|A|$:

$$\Pr\{N(A) = n\} = e^{-\lambda|A|}(\lambda|A|)^n/n!$$

   where $|A|$ denotes the $d$-dimensional volume of $A$.
2. Given that there are $n$ events in $A$, those events form an independent random sample from a uniform distribution on $A$.

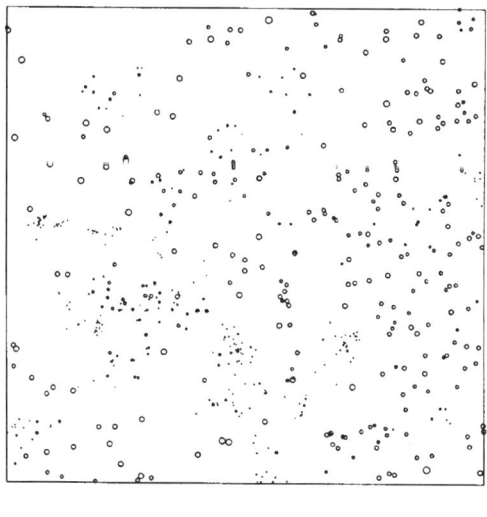

FIG. 3. Map of locations and relative diameters (at breast height) of all longleaf pines in a 4 ha study region of an old-growth forest in southern Georgia. (Copyright © 1991. Reprinted by permission of John Wiley & Sons, Inc.)

Note that the distribution of $N(A)$ depends only on the volume of $A$ and not on its shape or location. If $A_1, \ldots, A_k$ are disjoint sets, then under a Poisson process $N(A_1), \ldots, N(A_k)$ are independent Poisson distributed random variables.

Departure from CSR is often toward either the *clustering* of events or the *regularity* of events. Clustering may result from either the clustering of events around parents, from spatial inhomogeneity caused by interactions with a heterogeneous environment, or from positive interactions among events. Regularity may result from negative interactions among events.

Because the data usually consist of a single realization of a spatial point process, the additional assumptions of stationarity and sometimes isotropy are often made to reduce the parameter space and allow parameter estimation. Let $A \oplus \mathbf{s} \equiv \{\mathbf{a} + \mathbf{s} : \mathbf{a} \in A\}$ denote the translation of $A$ by the vector $\mathbf{s}$. A spatial point process $N$ is *stationary* if, for every collection of sets $A_1, \ldots, A_k$, the joint distribution of $\{N(A_1 \oplus \mathbf{s}), \ldots, N(A_k \oplus \mathbf{s})\}$ does not depend on $\mathbf{s} \in \Re^d$. Thus, stationary point processes are invariant under translations. Similarly, a point process $N$ is *isotropic* if it is invariant under rotations about the origin.

The moment measures of a spatial point process are analogous to the moments of random variables or random vectors. For a point process $N$, the number of events $N(A)$ in $A \in \mathcal{A}$ is a random variable with *mean measure*

$$\mu(A) \equiv E\{N(A)\} \tag{4.54}$$

The $k$th *moment measure* of $N$ is defined to be

$$\mu_k(A_1, \ldots, A_k) \equiv E\{N(A_1) \times \ldots \times N(A_k)\} \tag{4.55}$$

Now consider the behavior of the moment measures as the sets $\{A_i\}$ become small. Let $d\mathbf{s}$ and $d\mathbf{u}$ be small regions located at $\mathbf{s}$ and $\mathbf{u}$, respectively. Then, provided the limit exists, the *first-order intensity* at a point $\mathbf{s} \in \Re^d$ is defined to be

$$\lambda(\mathbf{s}) \equiv \lim_{|d\mathbf{s}| \to 0} \mu(d\mathbf{s})/|d\mathbf{s}| \tag{4.56}$$

where recall that $|d\mathbf{s}|$ is the volume of the region $d\mathbf{s}$. Similarly, the *second-order intensity* is defined to be

$$\lambda_2(\mathbf{s}, \mathbf{u}) = \lim_{\substack{|d\mathbf{s}| \to 0 \\ |d\mathbf{u}| \to 0}} \frac{\mu_2(d\mathbf{s}, d\mathbf{u})}{|d\mathbf{s}| |d\mathbf{u}|} \tag{4.57}$$

provided the limit exists.

For a stationary isotropic point process, $\lambda(\mathbf{s}) = \lambda$, and $\lambda(\mathbf{s}, \mathbf{u}) = \lambda°(\| \mathbf{s} - \mathbf{u} \|)$, for all $\mathbf{s}, \mathbf{u} \in \Re^d$. Here, $\| \mathbf{s} - \mathbf{u} \|$ is the Euclidean distance between the points **s** and **u**. In particular, under CSR, $\lambda(\mathbf{s}) = \lambda$ and $\lambda_2(\mathbf{s}, \mathbf{u}) = \lambda^2$.

### 4.4.1 The K-function

The *reduced Palm distribution* is one of the most important concepts in the theory of spatial point processes. Originally applied to an investigation of long-distance telephone queues, Palm [24] defined the function $\psi(t; t_0)$ to be the conditional probability that there are no calls (events) in the interval $(t_0, t_0 + t)$, given there was a call at time $t_0$. Applying this idea of conditioning on the location of an event in space, leads to the following definition for a stationary spatial point process:

$$K(r) \equiv \lambda^{-1} E \text{ (number of extra events within distance } r$$
$$\text{of a randomly chosen event)}, \quad r \geq 0$$

The relationship between the K-function and the second-order intensity function in $\Re^2$ is given by

$$K(r) = 2\pi\lambda^{-2} \int_0^r u\lambda_2^\circ(u)du. \quad (4.58)$$

Under a homogeneous Poisson process in $\Re^2$, $\lambda_2^\circ(r) = \lambda^2$, so that $K(r) = \pi r^2$. Under regularity, $K(r)$ tends to be smaller than $\pi r^2$, while under clustering $K(r)$ tends to be greater than $\pi r^2$.

Estimating the K-function from an observed pattern on $\Re^2$ is complicated by edge effects. As distance $r$ increases, edge effects become more troublesome. From a complete map of events, let $(\mathbf{s}_1, \ldots, \mathbf{s}_n)$ denote the locations of all events in a sample region $A$. An edge-corrected estimator that uses information on events for which $d_i > r$ is

$$\hat{K}_1(r) = \hat{\lambda}^{-1} \sum_{\substack{i=1 \\ j \neq i}}^{n} \sum_{j=1}^{n} w(\mathbf{s}_i, \mathbf{s}_j)^{-1} I(\|\mathbf{s}_i - \mathbf{s}_j\| \leq r)/n \quad (4.59)$$

The weight $w(\mathbf{s}_i, \mathbf{s}_j)$ is the proportion of the circumference of a circle centered at $\mathbf{s}_i$, passing through $\mathbf{s}_j$, which is inside the study region $A$. Another approximately unbiased estimator is

$$\hat{K}_2(r) = \hat{\lambda}^{-2} \sum_{\substack{i=1 \\ j \neq i}}^{n} \sum_{j=1}^{n} \frac{I(\|\mathbf{s}_i - \mathbf{s}_j\| \leq r)}{|(A \oplus \mathbf{s}_i) \cap (A \oplus \mathbf{s}_j)|} \quad (4.60)$$

where again $A \oplus \mathbf{s} \equiv \{\mathbf{a} + \mathbf{s} : \mathbf{a} \in A\}$. For both of these estimators, $\hat{\lambda} \equiv N(A)/|A|$ is used to estimate $\lambda$.

As a method of data summary, the empirical $K$-function has obvious advantages. It does not depend on arbitrary choices of quadrat location, size, or shape. Spatial information is presented at all scales of pattern, and precise locations of events are used in the estimator. The $K$-function also plays a role in model building. Figure 4 shows a graph of

$$\hat{L}(r) \equiv \{\hat{K}_1(r)/\pi\}^{1/2} - r \tag{4.61}$$

versus $r$ for the spatial point pattern of longleaf pine trees shown in Figure 3. We plot $\hat{L}$ to emphasize departures from CSR and to stabilize variances. By simulating (see Chapter 5) 100 times the same number of trees from a CSR process and plotting the envelopes of $\hat{L}$, it can be seen clearly that a CSR model for the pattern in Figure 3 is not appropriate.

### 4.4.2 Inhomogeneous Poisson Process and Cox Process

The *inhomogeneous Poisson process* is perhaps the simplest alternative to CSR and can be used to model realizations resulting from environmental heterogeneity. In contrast to the homogeneous Poisson (or CSR) process,

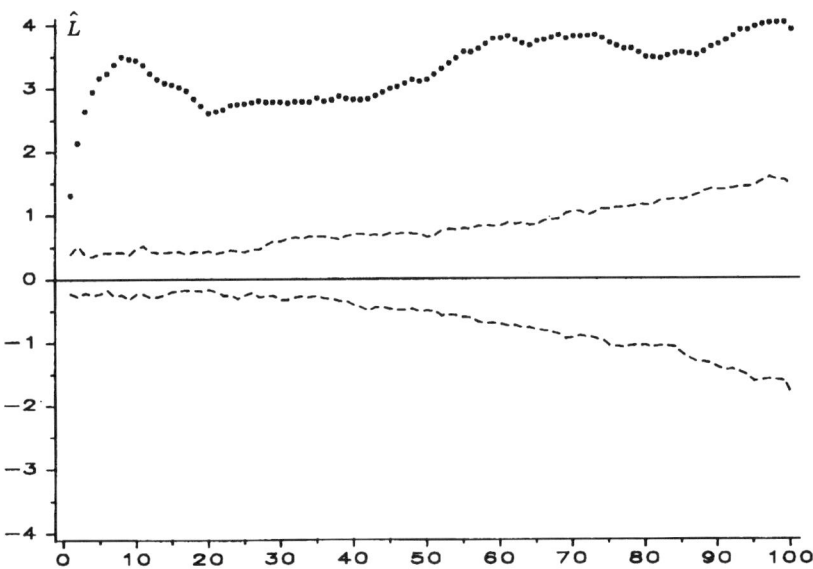

FIG. 4. Plot of $\hat{L}(r)$ versus $r$, and upper and lower envelopes from 100 simulations of CSR (dashed lines). Units on the horizontal and vertical axes are in meters. (Copyright © 1991. Reprinted by permission of John Wiley & Sons, Inc.)

the intensity function of an inhomogeneous Poisson process is a nonconstant function $\lambda(\mathbf{s})$ of spatial location $\mathbf{s} \in \Re^d$.

Let $\lambda(\cdot)$ be a nonnegative function on $\Re^d$. Then an *inhomogeneous Poisson process* on the region $A \subset \Re^d$ can be obtained as follows.

1. Select a random number $n$ from a Poisson distribution with mean $\mu(A) \equiv \int_A \lambda(\mathbf{s})d\mathbf{s}$:

$$\Pr\{N(A) = n\} = e^{-\mu(A)} [\mu(A)]^n/n! \qquad (4.62)$$

2. Sample $n$ event locations from the distribution on $A$ whose probability density is proportional to $\lambda(\mathbf{s})$:

$$f(\mathbf{s}) = \lambda(\mathbf{s})/\mu(A); \qquad \mathbf{s} \in A \qquad (4.63)$$

Note that, by assuming the existence of the intensity function $\lambda(\cdot)$, "pathologies" such as multiple events at the same spatial location have been avoided. An important example of this latter type of process is the *compound Poisson process*, which is obtained by putting "batches" of events of random size $K_i \geq 1$ at Poisson locations $\mathbf{s}_i$; $i = 1, 2, \ldots$, where $K_1, K_2, \ldots$ are independent and identically distributed positive-integer valued random variables.

Under an inhomogeneous Poisson process, regions of high intensity will tend to contain large numbers of events, while regions of low intensity will tend to contain few events. One general class of intensity functions is

$$\lambda(\mathbf{s}; \boldsymbol{\theta}) = g[\boldsymbol{\theta}'\mathbf{x}(\mathbf{s})] \qquad (4.64)$$

where $\boldsymbol{\theta}$ is an unknown parameter, $\mathbf{x}(\mathbf{s})$ is a known vector of concomitant spatial variables, and $g$ is a known nonnegative function. The ability to model $\lambda(\cdot)$ as a function of concomitant spatial variables makes the inhomogeneous Poisson process an attractive model for spatial point patterns. However, the numbers of events in disjoint regions are still independent random variables. So, this process cannot be used to model spatial dependence, as might arise from competitive interactions among events. If the assumption of independence is not tenable, then one could attempt to fit either a Cox process or a Markov point process (Section 4.4.4). Both of these models can also include the effects of concomitant spatial variables.

*Cox processes* were first considered by Lundberg [25] and Cox [26]. A Cox process $N$ is an inhomogeneous Poisson process with random intensity function $\Lambda(\cdot)$. The Cox process can be used to model the effects of rather complex environmental heterogeneity. More formally, a Cox process on $A$ can be defined as follows:

1. Generate a nonnegative random function $\Lambda(\mathbf{s})$; $\mathbf{s} \in A$.
2. Conditional on $\Lambda(\mathbf{s}) = \lambda(\mathbf{s})$; $\mathbf{s} \in A$, generate an inhomogeneous Poisson process with intensity function $\lambda(\mathbf{s})$ (see Section 4.4.1).

The simplest Cox process is the inhomogeneous Poisson process, where $\Lambda(\mathbf{s}) = \lambda(\mathbf{s})$; $\mathbf{s} \in \Re^d$, with probability one. Next, the mixed Poisson process has $\Lambda(\mathbf{s}) = Wf(\mathbf{s})$; $\mathbf{s} \in \Re^d$, where $W$ is a nonnegative random variable and $f(\cdot)$ is a deterministic function on $\Re^d$. If the intensity function of a Poisson process is $\lambda(\mathbf{s}; \boldsymbol{\theta})$, then a Cox process can be obtained by allowing $\boldsymbol{\theta} \in \Theta$ to be a random vector. Another possibility would be to let $\Lambda(\mathbf{s}; \boldsymbol{\theta}) = \exp\{\boldsymbol{\theta}'\mathbf{x}(\mathbf{s}) + \Psi(\mathbf{s})\}$, where $\Psi(\mathbf{s})$ is a random function.

### 4.4.3 Poisson Cluster Process

A special case of the Poisson cluster process was introduced by Neyman and Scott [23]. They suggested that groups of galaxies may have originated from the same points in space, resulting in the clustered point pattern of the galaxies observed today. They defined what is now known as the *Neyman–Scott process:*

1. Generate locations $\{\mathbf{u}_i : i = 1, \ldots, m\}$, in $A$, of parent events from a homogeneous Poisson process with intensity $\rho$.
2. For each parent $i$, independently generate a random number $J_i$ of offspring, from a discrete probability distribution $\{p_j : j = 0, 1, 2, \ldots\}$.
3. For each parent $i$, independently locate the $J_i$ offspring according to a $d$-dimensional density function $f(\mathbf{s} - \mathbf{u}_i)$ centered at $\mathbf{u}_i$.
4. The final process is composed of the superposition of offspring only.

If $f(\cdot)$ is radially symmetric about $\mathbf{0}$, then the process is stationary and isotropic. Note that astronomical evidence collected since Neyman and Scott's article suggests that their model, in fact, does not provide a good fit to the clustering of galaxies.

The more general *Poisson cluster process* is obtained by generalizing condition (1) to allow an inhomogeneous Poisson process, generalizing condition (2) to specify simply that each parent produces a random number of offspring, generalizing condition (3) to allow an arbitrary spatial positioning of offspring, and invoking condition (4). In this section, the properties of the simpler Neyman–Scott process will be summarized.

Now, certain Neyman–Scott processes and certain Cox processes can be identical. Consider a Cox process $N$ with random intensity

$$\Lambda(\mathbf{s}) = \omega \sum_{i=1}^{\infty} f(\mathbf{s} - \mathbf{u}_i); \qquad \mathbf{s} \in \Re^d$$

where ω is a positive constant, and $\mathbf{u}_1, \mathbf{u}_2, \ldots,$ are points of a homogeneous Poisson process with intensity ρ. Then $N$ is identical to a Neyman–Scott process in which a Poisson number (mean ω) of offspring are distributed around their parent events according to the density function $f(\cdot)$ and whose parent process is realized from a Poisson process with intensity ρ. No method of statistical analysis can distinguish between the two processes.

In the following, assume that $f(\cdot)$ is radially symmetric and hence the Neymann–Scott process $N$ is stationary and isotropic. The intensity of $N$ is $\lambda = \rho m_J$, where $m_J$ is the expected number of events per cluster. The density of the distance $r$ between two events in the same cluster is given by the convolution of $f$ with itself: $f_2(r) = f_2(\|\mathbf{v}\|) = \int_{\Re^d} f(\mathbf{w}) \cdot f(\mathbf{w} - \mathbf{v}) d\mathbf{w}$. Then, from (4.58), the $K$-function of $N$ in $\Re^2$ is

$$K(r) = \pi r^2 + E[J(J-1)] \cdot F_2(r)/(\rho m_J^2); \qquad r > 0 \qquad (4.65)$$

where $F_2(\cdot)$ is the distribution function of the distance between any two events in the same cluster. For example, suppose that the number of offspring per parent is Poisson distributed with mean $m_J$ and that $f(\cdot)$ is a bivariate normal density function with mean 0 and variance $\sigma^2 I_2$. Then,

$$K(r; \sigma^2, \rho) = \pi r^2 + \rho^{-1} \{1 - \exp(-r^2/\sigma^2)\}; \qquad r > 0 \qquad (4.66)$$

### 4.4.4 Markov Point Process

The Markov point process provides a flexible framework for modeling interactions among events that may be positive or negative. This point process has a long history in statistical mechanics, since Gibbs [27] introduced it to "explain the laws of thermodynamics on mechanical principles." There, the goal was not to model the locations of molecules, but rather to find a mechanism that would explain the thermodynamic behavior of gases, liquids, etc.

The Markov point process, here on a bounded set $A \subset \Re^d$, is a natural extension of the Markov notion in time. A point process is defined to be *Markov of range* ρ if the conditional intensity at $\mathbf{s}$, given the realization in $A - \{\mathbf{s}\}$, depends only on the events in $b(\mathbf{s}, \rho) - \{\mathbf{s}\}$, where $b(\mathbf{s}, \rho)$ is the ball of radius ρ centered at $\mathbf{s}$.

Consider a point process defined through its *total potential energy* function $U_n(\mathbf{s}_1, \ldots, \mathbf{s}_n)$ that does not depend on the order of the $\{\mathbf{s}_i\}$. The Gibbs point process can be defined as follows.

1. Generate a random number $n$ from the discrete probability distribution $\{p_n : n = 0, 1, 2, \ldots\}$, where

$$p_n = \begin{cases} c^{-1}e^{-|A|}; & n = 0 \\ \dfrac{e^{-|A|}}{cn!}\int_{A^n} \exp\{-U_n(\mathbf{s}_1, \ldots, \mathbf{s}_n)\}d\mathbf{s}_1 \ldots d\mathbf{s}_n; & n = 1, 2, \ldots \end{cases} \quad (4.67)$$

and the normalizing constant $c$ is chosen so that $\Sigma_{n=0}^{\infty} p_n = 1$.

2. Given $n$, generate $\mathbf{s}_1, \ldots, \mathbf{s}_n$ from the probability density function that is proportional to $\exp\{-U_n(\mathbf{s}_1, \ldots, \mathbf{s}_n)\}$:

$$f_n(\mathbf{s}_1, \ldots, \mathbf{s}_n) = \frac{\exp\{-U_n(\mathbf{s}_1, \ldots, \mathbf{s}_n)\}}{\int_{A^n} \exp\{-U_n(\mathbf{u}_1, \ldots, \mathbf{u}_n)\} d\mathbf{u}_1 \ldots d\mathbf{u}_n} \quad (4.68)$$

The function

$$f[(\mathbf{s}_1, \ldots, \mathbf{s}_n), n] = \frac{e^{-|A|}}{cn!} \exp\{-U_n(\mathbf{s}_1, \ldots, \mathbf{s}_n)\} \quad (4.69)$$

is called the *Gibbs grand canonical distribution* and can be thought of as the joint "density" of $n$ and $(\mathbf{s}_1, \ldots, \mathbf{s}_n)$.

The total potential energy of a realization of a Gibbs process can be written in the form

$$U_n(\mathbf{s}_1, \ldots, \mathbf{s}_n) = \sum_{i=1}^{n} g_1(\mathbf{s}_i) + \sum_{i<j} g_{12}(\mathbf{s}_i, \mathbf{s}_j) + \cdots + g_{1\ldots n}(\mathbf{s}_1, \ldots, \mathbf{s}_n) \quad (4.70)$$

Define $\mathbf{s}_i, \mathbf{s}_j \in A$ to be *neighbors* if $\|\mathbf{s}_i - \mathbf{s}_j\| < \rho$, for some $\rho > 0$. As in Section 4.3, a *clique* is defined to be a single event or a set of events, all of which are neighbors of each other. By the Hammersley–Clifford theorem, $g_{1\ldots k}(\mathbf{s}_1, \ldots, \mathbf{s}_k) = 0$, unless the events $\mathbf{s}_1, \ldots, \mathbf{s}_k$ form a clique. Then the point process defined by (4.69) and (4.70) is Markov of range $\rho$.

Frequently, third- and higher order interaction terms in (4.70) are assumed to be negligible and second-order interactions are assumed to depend only on the distances between events. Then the grand canonical distribution defined by (4.69) is

$$\frac{e^{-|A|}}{cn!} \exp\{-\sum_{i=1}^{n} \xi(\mathbf{s}_i) - \sum_{i<j} \psi(\|\mathbf{s}_i - \mathbf{s}_j\|)\} \quad (4.71)$$

when $\xi$ models large-scale effects and the pair-potential function $\psi$ models small-scale interactions between pairs of events. The function $\xi$ may depend on concomitant spatial variables. For distances $r$ such that $\psi(r) > 0$, the model shows inhibition among events, while for $r$ such that $\psi(r) < 0$, the model shows attraction among events. Thus, the pair-potential Markov point process might be used to model clustering as well as aggregation.

Strictly positive pair-potential $\psi$ always yield stable processes, but those with $\psi(r) < 0$ for some $r$ are frequently unstable. A Strauss process [28] is defined by $\xi(s) \equiv -\log \beta$, and

$$\psi_1(r) = \begin{cases} -\log \gamma & r \leq \rho \\ 0 & r > \rho \end{cases}$$

where $\beta > 0$ and $0 \leq \gamma \leq 1$. Other potential functions considered are

$$\psi_2(r) = -\log\{1 + (\alpha r - 1)e^{-\beta r^2}\}; \qquad \alpha \geq 0, \quad \beta > 0$$

$$\psi_3(r) = -\log\{1 + (\alpha - 1)e^{-\beta r^2}\}; \qquad \alpha \geq 0, \quad \beta > 0$$

$$\psi_4(r) = \beta(1/r)^n - \alpha(1/r)^m; \qquad m < n, \quad \alpha \in \Re, \quad \beta > 0$$

In statistical physics, $\psi_4$ is known as the *Lennard–Jones potential*. The potential $\psi_5(r) \equiv \theta|b(\mathbf{0}, R) \cap b(\mathbf{r}, R)|; \theta > 0, R > 0$, where $r = \|\mathbf{r}\|$ has been suggested for modeling plant competition.

The requirement that $U_n(\mathbf{s}_1, \ldots, \mathbf{s}_n)$ in (4.69) does not depend on the order of the $\mathbf{s}_i$ implies that interactions between all pairs of events are symmetric. However, in reality, competitive interactions between events may be asymmetric, with the strong eliminating the weak. Simple inhibition point processes (Section 4.4.5) more closely mimic the asymmetric nature of competitive interactions.

### 4.4.5 Other Models of Point Processes

Apart from the trivial case of regular lattices, the simplest class of point-process models whose realizations exhibit regularity is the class of *simple inhibition processes*. For such models, a minimum permissible distance $\delta$ is imposed; no event may be located within distance $\delta$ of another event. Matern [6] was the first to describe formally the simple inhibition process. Suppose that disks of constant radius $\delta$ are placed sequentially over a finite region $A$. At each stage, the next disk center is chosen at random from a uniform distribution over those points in $A$ for which the new disk would not overlap any previously located disk. The procedure terminates when a prespecified number of disks have been placed or it is impossible to continue. Further details can be found in [12], sec. 8.5.4.

*Thinned point processes* describe the result of randomly removing events from a realization of an initial point process $N_0$; the retained events define the realization of the point process of interest $N_1$. Thinned and related point processes have been used to describe survivorship patterns of trees in natural forests or in plantations, and abandonment patterns of villages in central Europe. The initial spatial point process $N_0$ could be an inhomogeneous Poisson process, a Cox process, a Poisson cluster process, a simple inhibition process, a Markov point process, etc. Further details can be found in [12], sec. 8.5.6.

In addition to their locations, certain measurements or "marks" may be associated with events. For example, marks may be tree diameters (Figure 3), magnitudes of earthquakes, or population sizes of towns. The locations of events and their corresponding marks form a *marked point pattern*. The marks associated with events may provide crucial information for modeling physical processes. In forest ecology, for example, the marked point pattern of trees and their diameters may reflect the frequency of births, patterns of natural disturbances (e.g., lightning, windstorms), and the requirements of seedlings for openings in the forest canopy. Further details can be found in [12], sec. 8.7.

Realized (marked) spatial point patterns are often the result of dynamic processes that occur over time as well as space. For example, the spatial point pattern of trees shown in Figure 3 has evolved over time as new trees are born and old trees die. Most published attempts at point-process modeling have had data available from only a single instant of time. Consequently, many point-process models suffer nonidentifiability problems in the sense that two or more distinct models may yield identical realizations. (An example is given in Section 4.4.3.) No amount of statistical analysis can distinguish between such models. Dynamic space–time models are less susceptible to this problem, since such models attempt to follow the temporal evolution of the spatial point pattern.

## 4.5 Some Final Remarks

This chapter brings together many of the commonly used models in spatial statistics. Very little attention has been given to inference, namely, estimation of model parameters and subsequent predictions based on the fitted models. A full discussion of these issues can be found in Cressie [12], along with several applications to real data. In this volume, Chapter 13, written by Dale Zimmerman, provides an analysis of spatial data using geostatistical models. Special techniques for simulating spatial data are presented by Peter Clifford in Chapter 5.

The future of spatial statistics is in solving problems for space–time data. Dynamic spatial models are needed to represent phenomena distributed through space and evolving in time. Much more of this type of model development remains to be done.

## Acknowledgments

I would like to express my appreciation to Don Percival, Peter Clifford, Chris Chatfield, and Steve Vardeman for their helpful remarks on the first draft of this chapter. Research support came from the Office of Naval Research, the National Science Foundation, and the National Security Agency.

## References

1. Fisher, R. A. (1925). *Statistical Methods for Research Workers.* Oliver and Boyd, Edinburgh.
2. Fisher, R. A. (1935). *The Design of Experiments.* Oliver and Boyd, Edinburgh.
3. Fairfield Smith, H. (1938). "An Empirical Law Describing Heterogeneity in the Yields of Agricultural Crops." *Journal of Agricultural Science (Cambridge* **28**, 1-23.
4. Whittle, P. (1954). "On Stationary Processes in the Plane." *Biometrika* **41**, 434–449.
5. Mercer, W. B., and Hall, A. D. (1911). "The Experimental Error of Field Trials." *Journal of Agricultural Science (Cambridge)* **4**, 107–132.
6. Matern, B. (1960). *Spatial Variation* **49**(5), Meddelanden fran Statens Skogsforskningsitut.
7. Bartlett, M. S. (1975). *The Statistical Analysis of Spatial Pattern.* Chapman and Hall, London.
8. Journel, A. G., and Huijbregts, C. (1978). *Mining Geostatistics.* Academic Press, London.
9. Cliff, A. D., and Ord, J. K. (1981). *Spatial Processes: Models and Applications.* Pion, London.
10. Ripley, B. D. (1981). *Spatial Statistics.* John Wiley and Sons, New York.
11. Upton, G. J. G., and Fingleton, B. (1985). *Spatial Data Analysis by Example,* **1.** *Point Pattern and Quantitative Data.* John Wiley and Sons, Chichester.
12. Cressie, N. (1993). *Statistics for Spatial Data, Revised Edition.* John Wiley and Sons, New York.
13. Matheron, G. (1963). "Principles of Geostatistics." *Economic Geology* **58**, 1246–1266.
14. Gandin, L. S. (1963). *Objective Analysis of Meteorological Fields.* GIMIZ, Leningrad.
15. Cressie, N. (1990). "The Origins of Kriging." *Mathematical Geology* **22**, 239–252.
16. Ruelle, D. (1969). *Statistical Mechanics: Rigorous Results.* Benjamin, New York.
17. Besag, J. E. (1974). "Spatial Interaction and the Statistical Analysis of Lattice Systems." *Journal of the Royal Statistical Society B* **36**, 192–225.
18. Cressie, N., and Lele, S. (1992). "New Models for Markov Random Fields." *Journal of Applied Probability* **29**, 877–884.
19. Ising, E. (1925). "Beitrag zur Theorie des Ferromagnestismus." *Zeitschrift fur Physik* **31**, 253–258.
20. Pickard, D. K. (1987). "Inference for Discrete Markov Fields: The Simplest Nontrivial Case." *Journal of the American Statistical Association* **82**, 90–96.
21. Geman, S., and Geman, D. (1984). "Stochastic Relaxation, Gibbs Distributions and the Bayesian Restoration of Images," *IEEE Transactions on Pattern Analysis and Machine Intelligence* **PAMI-6**, 721–741.

22. Student. (1907). "On the Error of Counting with a Haemacytometer." *Biometrika* **5,** 351–360.
23. Neyman, J., and Scott, E. L. (1958). "Statistical Approach to Problems of Cosmology." *Journal of the Royal Statistical Society B* **20,** 1–43.
24. Palm, C. (1943). "Intensitätsschwankungen im Fernsprechverkehr." *Ericsson Technics* **44,** 1–189.
25. Lundberg, O. (1940). *On Random Processes and Their Application to Sickness and Accident Statistics.* Almqvist and Wiksells, Uppsala.
26. Cox, D. R. (1955). "Some Statistical Methods Related with Series of Events." *Journal of the Royal Statistical Society B* **17,** 129–157.
27. Gibbs, J. W. (1902). *Elementary Principles in Statistical Mechanics, Developed with Especial Reference to the Rational Foundation of Thermodynamics.* Charles Scribner's Sons, New York.
28. Strauss, D. J. (1975). "A Model for Clustering." *Biometrika* **62,** 467–475.

# 5. MONTE CARLO METHODS

Peter Clifford

University of Oxford, United Kingdom

## 5.1 Introduction

Consider a perfectly balanced spinner, spinning on a vertical axis inside a round box, rather like an unmagnetized compass needle. Suppose that the circumference of the box is marked with a scale from 0 to 1, so that when the spinner stops it will point to a value between 0 and 1. Imagine repeatedly spinning the spinner and recording the values when the spinner comes to rest. Under ideal conditions, this experiment will generate a sequence of independent, uniformly distributed random variables.

Most scientific calculators now have a button marked RAND. When it is pressed a number between 0.0000 and 0.9999 appears; the number of decimal places depending on the calculator. Pressing the button repeatedly produces a sequence of numbers all in the same range, with no discernible pattern in the sequence. All high-level computer languages have equivalent functions; RND for example in Basic. What the RAND function on the calculator is trying to do is to imitate the type of data produced by the idealized spinner experiment.

The numbers that are generated on the calculator are said to be pseudo-random variables. They are intended to have the same statistical properties as the numbers generated by the experiment. Most important the numbers should be uniformly distributed between 0 and 1, since we imagine that the spinner is unbiased. Figure 1(a) shows a histogram formed by collecting 1000 values from the RND function in Microsoft QuickBasic. The histogram appears to be fairly flat, and in fact a formal test reveals no significant departure from the uniform assumption. Another property that we expect our idealized spinner to have is that it forgets about the previous value between spins; in other words, successive values should be independent. This can be assessed by plotting successive values against each other as in Figure 1(b). Visually, there does not appear to be any reason to doubt the independence; this is confirmed by a formal statistical test. One could go on and consider the uniformity of triples and other simple tests of randomness

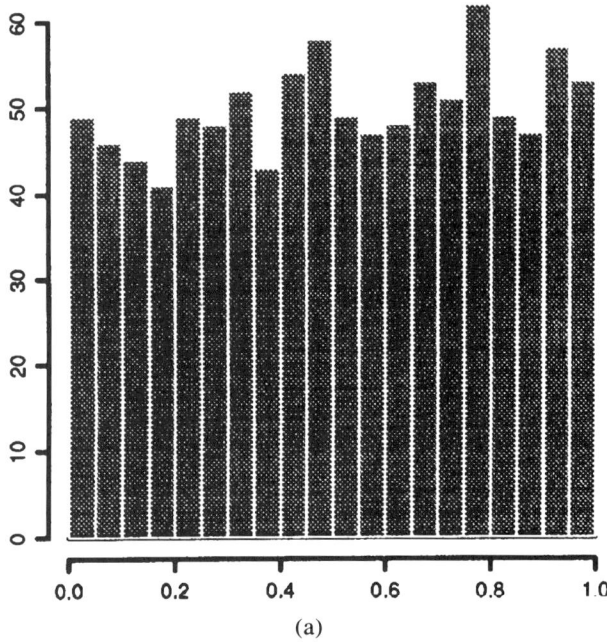

FIG. 1. (a) Histogram of 1000 uniformly distributed pseudo-random variables. (b) Scatter plot of successive pairs.

of the sequence. A good random number generator will pass just the right number of tests. In other words, it will produce a sequence of numbers that is indistinguishable from a sequence of independent uniform random variables. It is unrealistic to hope that a random number generator will pass every test of randomness, since any such test picks out a particular type of coincidence, and we do not want to restrict ourselves to sequences in which every type of coincidence is eliminated. It is only fair to warn the reader that there are deep questions here, relating to the foundations of probability theory. Philosophically we are skating on thin ice, so it seems sensible to move on rapidly to more practical matters.

There is a substantial literature on the art of designing and programming fast random number generators. By this time, in the 1990s, there is every reason to hope that the basic ingredients of a good generator will be included in the software you are using. Such is the case in the S language mentioned in Chapter 17. Ideally there should be detailed information about the built-in generator, which should state explicitly which generator is used and

# INTRODUCTION

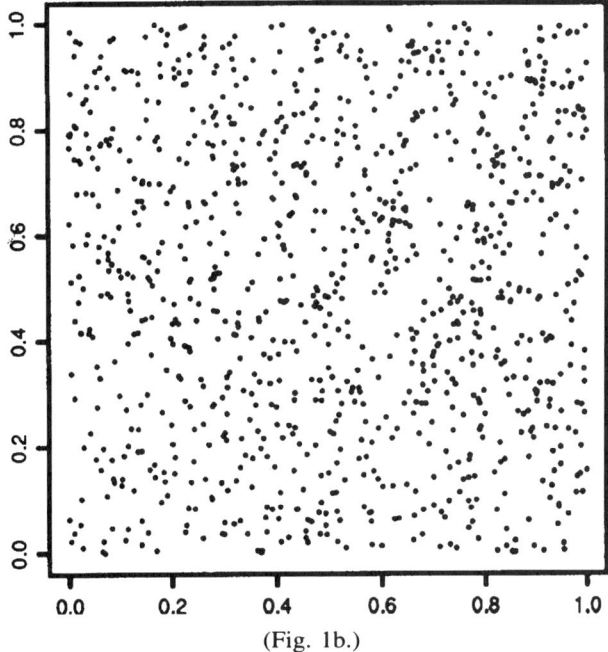

(Fig. 1b.)

document its performance in statistical tests of randomness. Failing this, some confidence in the generator can be built up by running your own tests and by tapping into the experience of other users. Finally there is the possibility of writing your own generator based on published recommendations.

Most random number generators work by iterating a function $\phi$. The function maps the set of integers $0, 1, \ldots, N - 1$ onto itself, so that starting from some initial value $w_0$, called the seed, successive values are given by

$$w_{n+1} = \phi(w_n) \tag{5.1}$$

Numbers in the range 0 to 1 are then formed by calculating $w_n/N$. The idea is to find a function that produces a highly chaotic sequence, so that the numbers $w_n$ skip about in the range $0, 1 \ldots N - 1$ with no obvious pattern. Obviously if the sequence returns to a value it has visited before the sequence will repeat. For this reason the function is chosen so that every one of the values in $0, 1 \ldots N - 1$ is visited in a cycle and $N$ is chosen to be large, say $2^{32}$, to maximise the time between repetition. An example of such a function is the *multiplicative congruential* generator

$$\phi(w) = (69069\ w + 1) \bmod 2^{32} \qquad (5.2)$$

with $N = 2^{32}$. The seed can be specified by the user, which enables the sequence to be reproduced; a facility that is helpful in debugging computer programs. Alternatively, the seed can be calculated as a function of the internal state of the computer, for example the current time.

Note that if such a generator is run for the full cycle length $N$ the histogram will be completely flat, since every value occurs exactly once. However, only $N$ different successive pairs $(w_n, w_{n+1})$ are generated out of $N^2$ possible pairs, so that the set of pairs of successive values forms a relatively small subset. A similar phenomenon occurs with triples of successive values. The structure of these subsets may exhibit highly undesirable properties, for example the points may lie on widely separated planes. For these reasons it is wise to use generators such as the one given previously, with a well-documented performance. For a discussion of these problems, see Ripley [10].

One may ask why it is important to be able to generate *uniform* random variables, since observed distributions are rarely of that form. The reason is that in principle any random phenomenon can be simulated by taking a suitable transformation of a sequence of uniform random variables. The remainder of this chapter will be devoted to such constructions. The aim will be to present the basic tools of Monte Carlo simulation. No attempt will be made to discuss specialized commercial simulation packages. Throughout, we use $U$ with or without subscripts to denote a random variable that is distributed uniformly on the interval $(0, 1)$, and similarly, we use $u$ to denote observed or simulated values from such a distribution.

Computer generated random variables play an increasingly important role in scientific investigation. They are an essential component in the simulation of complicated real-world phenomena. In some applications the aim may be merely to see if the computer model produces results that are in accord with experimental data. Generally as the investigation proceeds and confidence in the model increases the aim will shift to the task of accurately calculating some property of the system. At a final stage the focus will be on optimization of the system with respect to this property.

In principle, properties of a mathematical model can be determined by mathematical analysis but in practice the analysis can easily become bogged down with intractable calculations. It is easy to lose sight of the often simple mechanism that lies behind the model. Computer generated random variables enable the mechanism to be mimicked directly. The properties of the model can thus be investigated empirically on the computer. This "hands on" interaction with the model builds up intuition and may eventually suggest good approximate formulae.

## 5.2 Continuous Distributions

Any random variable can be simulated by transforming a computer generated uniform random variable. The simplest example is the standard *exponential* random variable, with probability density $e^{-x}$, $x > 0$. This can be simulated by the transformation $x = -\log(u)$, where $u$ is a simulated uniform random variable. To see that the value is simulated from the right density we use the change of variable formula, which states that the density of a random variable $X = g(U)$ is given by $\left|\frac{d}{dx}g^{-1}(x)\right|$, for an arbitrary smooth monotone transformation $g$.

In our case $g(U) = -\log(U)$ so that $g^{-1}(x) = e^{-x}$ and $\left|\frac{d}{dx}g^{-1}(x)\right| = e^{-x}$ for $x > 0$, as required. More generally $-\lambda^{-1}\log(u)$ simulates an exponentially distributed random variable with density $\lambda e^{-\lambda x}$, $x > 0$. The exponential distribution occurs frequently throughout the physical sciences and it is fortunate that it is so simple to simulate.

The *normal* distribution can also be simulated in a straightforward manner. First note that, if $Z$ has distribution $N(0, 1)$ then $\mu + \sigma Z$ has distribution $N(\mu, \sigma^2)$. The problem of sampling from $N(\mu, \sigma^2)$ is therefore reduced to that of sampling from $N(0, 1)$ and then taking a simple linear transformation.

The Box–Muller algorithm simulates a pair of independent $N(0, 1)$ variables by making use of the following result. If $(X, Y)$ are two independent $N(0, 1)$ random variables, represented in polar coordinates as $R$, $\Theta$, i.e., $X = R\cos(\Theta)$ and $Y = R\sin(\Theta)$, then $R$ and $\Theta$ are independent. Furthermore $R^2$ has an exponential distribution with mean 2, and $\Theta$ is uniformly distributed on the interval $(0, 2\pi)$. This can be checked by the two-dimensional change of variables formula. The result enables a pair of independent $N(0, 1)$ variables to be simulated, by first simulating $R^2$ and $\Theta$ and then converting to Cartesian coordinates. The algorithm is therefore as follows.

**Algorithm 5.1:** The Box and Muller algorithm simulates a pair $(x, y)$ of independent $N(0, 1)$ variables:

```
u = RND : v = RND
r = sqr(-log(v)) : t = 2 * π * u
x = r * cos(t) : y = r * sin(t)
```

Random variables with the so called $\chi^2$, $t$ and $F$ distributions (used in the development of inference methods based on samples from normal populations) can be constructed by using independent $N(0, 1)$ and exponential variables. By definition, the $\chi^2$ distribution with $n$ degrees of freedom is the distribution of the sum of squares of $n$ independent $N(0, 1)$ variables. A simulated $\chi^2$ variable is therefore given by $z_1^2 + \cdots + z_n^2$, where $z_1, \ldots, z_n$ are sampled from $N(0, 1)$. Alternatively, when $n = 2m$, it can simulated as $-2 \log(u_1) - \cdots - 2 \log(u_m)$, since the sum of two independent squared $N(0, 1)$ variables has an exponential distribution with mean 2. When $n = 2m + 1$ it is necessary to add on only another squared $N(0, 1)$ variable. Similarly, a $t$ variable with $n$ degrees of freedom is given by

$$t = n^{1/2} z_{n+1}/(z_1^2 + \cdots + z_n^2)^{1/2} \qquad (5.3)$$

and an $F$ variable with $m$ and $n$ degrees of freedom is given by

$$F = (n/m)(z_1^2 + \cdots + z_m^2)/(z_{m+1}^2 + \cdots + z_{m+n}^2) \qquad (5.4)$$

The $\chi^2$ distribution is a member of the gamma family of distributions. Fast algorithms have been developed for sampling from the general gamma distribution. These algorithms should be used when undertaking substantial simulation studies. References are given in the standard texts listed at the end of this chapter.

The standard *Cauchy* random variable with density $\pi^{-1}/(1 + x^2)$, $-\infty < x < \infty$ has the same distribution as the ratio of two independent $N(0, 1)$ variables. It can therefore be simulated as $z_1/z_2$. Note that if $z_1 = y$ and $z_2 = x$ as in Algorithm 5.1 then $z_1/z_2 = \tan(2\pi u)$, which turns out to be equivalent to the transformation obtained by inversion described next.

### 5.2.1 The Inversion Method

*Inversion* is a general method for simulating random variables. It makes use of the fact that the transformation $X = F^{-1}(U)$ gives a random variable $X$ with distribution function $F$ provided the inverse function $F^{-1}$ exists. This is a simple consequence of the change of variables formula, this time with $g(U) = F^{-1}(U)$. Since $g^{-1}(x) = F(x)$, the density of $X$ becomes $\frac{d}{dx} F(x) = f(x)$, which is the probability density corresponding to the distribution function $F$. The method is illustrated in Figure 2. Note that applying this procedure in the exponential case, for which $F(x) = 1 - e^{-x}$, $x > 0$, gives $F^{-1}(U) = -\log(1 - U)$ rather than $-\log(U)$ as given previously. However there is no real inconsistency here, since $U$ and $1 - U$ have the

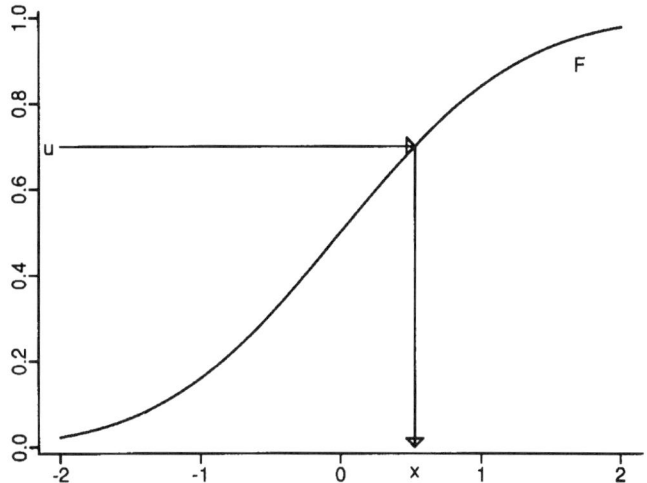

FIG. 2. Generation of simulated value $x$ from the distribution function $F$ by the method of inversion.

same distribution. Another simple example is the *Weibull* distribution, which has distribution function

$$F(x; \alpha, \beta) = 1 - \exp[-(x/\beta)^\alpha], \qquad x > 0, \alpha > 0, \beta > 0 \qquad (5.5)$$

This is simulated by $x = \beta[-\log(1 - u)]^{1/\alpha}$ or equivalently $\beta[-\log(u)]^{1/\alpha}$. The Weibull distribution is used extensively in reliability theory as a model of the distribution of breaking strengths in material testing. Similarly, the *logistic* distribution with distribution function

$$F(x; \alpha, \beta) = \frac{1}{1 + e^{\alpha - \beta x}} \qquad (5.6)$$

is simulated by $x = [\alpha - \log((1 - u)/u)]/\beta$.

To use the inversion method, the inverse function $F^{-1}$ either has to be available explicitly, as in the exponential, Weibull and logistic cases, or has to be computable in a reasonable amount of time. The equality $x = F^{-1}(u)$ is equivalent to $u = F(x)$, so that finding $x$ for given $u$ is equivalent to finding a root of the equation $F(x) - u = 0$. When $F$ is strictly monotone, there is only one root and standard numerical root-finding algorithms can be used, provided of course that $F(x)$ itself is easy to evaluate. If it is required to sample repeatedly from the same distribution, it may be worthwhile devoting some time to the development of an accurate approximation to $F^{-1}$ beforehand. For the standard normal distribution $N(0, 1)$, a numerical approximation to $F^{-1}$ is given in Abramovitz and Stegun [1]. However inversion is usually not the method of choice for the normal distribution since there are other simple methods such as Algorithm 5.1.

The main advantage of the inversion method is that it is generally easy to verify that a computer algorithm using it is written correctly. In this sense the method is efficient; i.e., in saving the time of the programmer. However, there are usually competing methods that will run faster at the expense of mathematical and algorithmic complexity. We proceed to discuss some of these.

### 5.2.2 Rejection and Composition

*Rejection* and *composition* techniques are used extensively in simulation algorithms for both discrete and continuous distributions. Rejection sampling works by repeatedly simulating random variables from the wrong probability density and then rejecting them with a probability that depends on the ratio of the right density to the wrong density, until an acceptable value is obtained.

A very simple example is that of simulating a random variable uniformly distributed on a disk of radius 1. The rejection method samples a pair $(x, y)$ of independent random variables uniformly distributed on $(-1, 1)$ and then rejects the pair unless $x^2 + y^2 < 1$. The rejection probability is either 0 or 1 in this example.

**Algorithm 5.2:** This simulates a pair $(x, y)$ uniformly distributed on the unit disk, by the rejection method:

```
do
   x = 2 * RND - 1 : y = 2 * RND - 1
loop until x^2 + y^2 < 1
```

Marsaglia's polar algorithm converts the pair $(x, y)$ into a simulated pair of independent $N(0, 1)$ variables by computing

```
w = x^2 + y^2 : t = sqr(-2 log(w) / w)
x = x t : y = y t
```

This too is justified by the change of variables formula.

More generally, suppose that it is easy to simulate random variables with density $g$ but we want random variables with density $f$. Provided there is a constant $K$ such that $f(x) \leq Kg(x)$, $-\infty < x < \infty$, the rejection algorithm proceeds as follows.

**Algorithm 5.3:** Here we simulate $x$ from $f$, where $f \leq Kg$, by the rejection method:

```
do
    simulate x from the density g
loop until K * g(x) * RND ≤ f(x)
```

If $f$ is known only up to a multiplicative constant, i.e., $f(x)$ is proportional to a function $a(x)$, then rejection can still be used provided a constant $L$ can be found for which $a(x) \leq Lg(x)$, $-\infty < x < \infty$. The rejection probability for the value $x$ becomes $a(x)/[Lg(x)]$.

The method is justified as follows. The frequency with which the algorithm stops and thereby gives a value between $x$ and $x + \Delta$ is the product of the probability of being in the right range and the conditional probability of stopping; i.e., the product of $g(x)\Delta$ and $a(x)/[Lg(x)]$. The product is proportional to $a(x)\Delta$, which is itself proportional to $f(x)\Delta$. Given that the algorithm has stopped, the value $x$ is therefore a simulated value from the density $f$.

The probability that the algorithm stops on a particular iteration, i.e., that a sample from $g$ is accepted, is

$$\int g(x) \frac{a(x)}{Lg(x)} dx = \frac{1}{L} \int a(x) dx \qquad (5.7)$$

The algorithm stops quickly when $L$ is small. This can be achieved by matching the shape of the function $a$ as closely as possible to that of $f$. Algorithm 5.6 illustrates this principle.

### 5.2.3 Composition

Composition methods exploit the mixture structure inherent in certain probability distributions. Suppose that

$$f(x) = \sum_{k=0}^{\infty} p_k f_k(x) \qquad (5.8)$$

where each of the $f_k$ are probability densities and the $p_k$ are probabilities adding to 1. Such a density $f$ is then said to be a discrete mixture or composition. Continuous mixtures are defined similarly with an integral replacing the sum. Simulation from $f$ proceeds in two stages. First, a value is simulated from the discrete distribution $p_0, p_1, \ldots$. If the simulated value is $n$ then at the next stage $x$ is simulated from $f_n$. Composition plays an important role in the design of sophisticated simulation algorithms. The books referenced in Section 5.10 contain numerous examples of its usage. The method is also particularly suited to the sampling of multimodal densities that can be built up as mixtures of simple unimodal densities.

## 5.3 Discrete Distributions

The simplest discrete random variable takes two values, 0 and 1, with probabilities $1 - p$ and $p$, respectively. It can be simulated by taking $x$ to be 0 if $u < 1 - p$ and 1 otherwise. This is equivalent to splitting the interval $(0, 1)$ into two subintervals of lengths $1 - p$ and $p$; if $u$ falls in the first then $x = 0$, and if it falls in the second, $x = 1$. More generally, if the possible values of the random variable $X$ are $x_0, x_1, \ldots$ in increasing order, and the associated probabilities are $p_0, p_1, \ldots$ with $\Sigma p_k = 1$, then the interval $(0, 1)$ is split up into successive subintervals $I_0, I_1, \ldots$ of lengths $p_0, p_1, \ldots$, respectively. If $u$ falls in the $k$th interval then the simulated value of $X$ is $x_k$. Equivalently $x$ is the smallest value of $x$ for which $F(x) \geq u$, where $F$ is the distribution function of $X$. This is the *inversion* method, in its most general form.

Binary search can be used to put the method into practice. Firstly, the values $S(k) = F(x_k)$, $k = 0, 1, \ldots, m$ are calculated for $m$ sufficiently large that a high percentage of the distribution has been included, e.g., until $F(x_m) > 0.9999$. The algorithm simulates a value $x$ as follows.

**Algorithm 5.4:** This uses binary search for sampling $x$ from a general discrete distribution:

```
u = RND : a = 0 : b = m
do
    x = int((a + b)/2)
    if S(x) < u then
        a = x
    else
        b = x
    endif
loop until a ≥ b - 1
```

Other methods of dealing with the general discrete distribution, are detailed in the specialized texts referenced at the end of this chapter.

### 5.3.1 Binomial and Geometric

As in the continuous case, special methods exist for special distributions. For example, the *binomial* distribution $B(n, p)$ can be sampled by adding together $n$ independent variables each taking the value 0 or 1 with probabilities $1 - p$ and $p$, respectively. For small values of $n$ this will be faster than general methods such as binary search. The *geometric* distribution with $p_i = p(1 - p)^i$, $i = 0, 1, \ldots$ is a discrete version of the exponential. It can be simulated by taking $x = \text{int}[\log(u)/\log(1 - p)]$.

## 5.3.2 Poisson

Properties of the Poisson process can be exploited to simulate from a *Poisson* distribution. Suppose that the intensity of the process is 1, then the times between arrivals are independent and exponentially distributed with mean 1. Furthermore, the number of arrivals in the $\mu$ has a Poisson distribution with mean $\mu$. The strategy is therefore to sum simulated exponential variables until time $\mu$ has elapsed. The number of arrivals at this time will therefore have the required Poisson distribution. Note that checking whether $-\log(u_1) - \cdots - \log(u_n)$ is greater than $\mu$ is equivalent to checking whether $u_1 u_2 \cdots u_n$ is less than $\exp(-\mu)$. The algorithm to produce a simulated value $x$ is therefore as follows.

**Algorithm 5.5:** This simulates a Poisson distribution ($\mu < 20$), using the Poisson process:

```
a = RND : b = EXP (-μ) : x = 0
do while a > b
   a = a * RND
   x = x + 1
loop
```

The algorithm will not perform correctly for large values of $\mu$ because the product $u_1 u_2 \cdots u_n$ is eventually miscalculated as a result of numerical underflow. Although this problem can be remedied by reverting to the test based on the sum of logarithms, Algorithm 5.6 is very much faster and is to be preferred when $\mu > 20$. The algorithm uses a rejection method by comparison with the logistic distribution. Suitably scaled, the logistic density provides an upper bound for the Poisson probabilities. See Figure 3.

**Algorithm 5.6:** Here, the Poisson distribution ($\mu \geq 20$) is simulated by Atkinson's method:

```
b = π / sqr(3 * μ) : a = b * μ
c = .767 - 3.36 / μ : k = log(c / b) - μ
do
   do
      u = RND
      y = (a - log((1 - u) / u)) / b
   loop until y > -.5
   x = int(y + .5)
   z = k + x * log(μ) - logfact(x)
loop until log(RND * u * (1-u)) < z
```

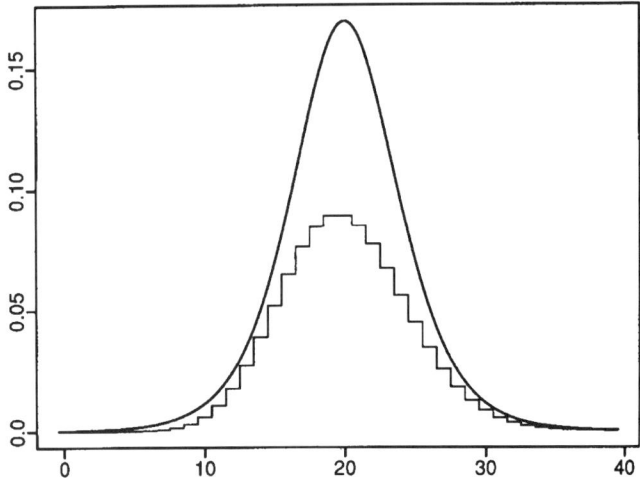

FIG. 3. Comparison of a scaled logistic density and a histogram constructed with bars centered at integer points and with heights equal to the Poisson probabilities ($\mu = 20$).

The function logfact $(x) = \log(x!)$, which appears in the algorithm, can be calculated efficiently using standard numerical routines based on Stirling's approximation.

### 5.4 Multivariate Distributions

A random vector having a *multivariate normal* distribution with mean vector $\mu$ and variance-covariance matrix $V$ can be simulated as follows. First, form the Cholesky decomposition of $V$; i.e., find the lower triangular matrix $L$ such that $V = LL^T$. Next, simulate a vector $z$ with independent $N(0, 1)$ elements. A vector simulated from the required multivariate normal distribution is then given by $\mu + Lz$.

For example, suppose that we want to simulate a three-dimensional normal vector with mean (100, 200, 300) and variance covariance matrix

$$V = \begin{pmatrix} 4 & 8 & 2 \\ 8 & 20 & 6 \\ 2 & 6 & 11 \end{pmatrix} \quad (5.9)$$

The Cholesky decomposition of $V$ is

$$\begin{pmatrix} 4 & 8 & 2 \\ 8 & 20 & 6 \\ 2 & 6 & 11 \end{pmatrix} = \begin{pmatrix} 2 & 0 & 0 \\ 4 & 2 & 0 \\ 1 & 1 & 3 \end{pmatrix} \begin{pmatrix} 2 & 4 & 1 \\ 0 & 2 & 1 \\ 0 & 0 & 3 \end{pmatrix} \quad (5.10)$$

so that the simulated normal vector $(x_1, x_2, x_3)$ is given by

$$x_1 = 100 + 2z_1$$
$$x_2 = 200 + 4z_1 + 2z_2 \quad (5.11)$$
$$x_3 = 300 + z_1 + z_2 + 3z_3$$

where $z_1$, $z_2$ and $z_3$ are values simulated from the $N(0, 1)$ distribution.

The method is practical provided the dimension of the vector **X** is not too large; i.e., smaller than a few hundred. The main computational task is the calculation of $L$. Efficient algorithms for this purpose are included in most numerical packages. Cases in which the dimension of **X** is large occur naturally in time series and image analysis. Special techniques that can deal with these cases are described later in this chapter.

In principle, the sampling of multivariate distributions presents no new problems, since any multivariate probability density $f(x_1, \ldots, x_n)$ can be decomposed into the product of conditional densities; i.e.,

$$f(x_1, \ldots, x_n) = f(x_1)f(x_2|x_1)f(x_3|x_1, x_2) \cdots \\ f(x_n|x_1, \ldots, x_{n-1}) \quad (5.12)$$

To sample the vector $(x_1, \ldots, x_n)$ the algorithm proceeds by firstly sampling $x_1$ from $f(x_1)$ and then $x_2$ from $f(x_2|x_1)$ and so on until finally $x_n$ is sampled from $f(x_n|x_1, \ldots, x_{n-1})$. At each stage the sampling problem is univariate. In practice, the decomposition may be unwieldy and the univariate densities may be difficult to sample from. Special methods exist in special cases; the books listed at the end of the chapter give many examples. If all else fails, it is usually possible to implement the Metropolis algorithm, which is described in the next section.

### 5.4.1 The Metropolis Algorithm

The algorithm of Metropolis, Rosenbluth, Rosenbluth, Teller and Teller [9], commonly known as the *Metropolis algorithm,* is a general method of Monte Carlo simulation. As in rejection sampling, it is necessary to know the probability density $f$ only up to a constant of proportionality. However, unlike the rejection method, there is no requirement that the density should be bounded by a multiple of a more easily sampled density. The Metropolis algorithm produces a sample from $f$ by simulating a specific Markov process, which will be described later. To build some intuition about how the

Metropolis algorithm works we give a simple illustration. Think of a helicopter making a succession of 1-hr flights according to the following rules. Whenever the helicopter lands, the pilot plots a new course involving 1 hour's flying but in a random direction. If the destination is further north than the helicopter's present position, the pilot takes off and flies to the new destination. Otherwise the pilot decides what to do by tossing a coin: *heads* to fly off to the chosen destination and *tails* to stay put, wait an hour and try again with a new random direction and, if necessary, another coin toss. If we plot out several thousand of the landing sites as the helicopter travels endlessly over the globe we will notice that they tend to cluster around the North Pole according to some distribution. If we eventually stop the clock and see where the helicopter lands next, this point will be a random sample from the equilibrium distribution. This is the basis of the Metropolis algorithm. The helicopter course is a Markov process, and we have shown how to sample from the equilibrium distribution associated with it. The Metropolis algorithm is actually much cleverer than this: the trick is to design the Markov process so that it has a specific equilibrium density, $f$.

The Markov process $\{X_n\}$ is constructed as follows. First, a transition density $q(x \to y)$ must be found, defined for all $x, y$ in the sample space of $f$, and having the property that if $q(x \to y)$ is positive then so is $q(y \to x)$. Suppose that at the $n$th time step the Markov process is in state $x$. Now simulate a value $y$ using the transition density $q(x \to y)$. The value $y$ is called the *proposal*. In the case of discrete random variables the proposal density $q$ has the interpretation

$$q(x \to y) = P(Y = y | X_n = x) \tag{5.13}$$

and in the continuous case

$$P(Y \in A | X_n = x) = \int_A q(x \to y) dy \tag{5.14}$$

The next stage of the algorithm is to decide whether $y$ is accepted or not. If it is accepted then $X_{n+1}$, the new state of the Markov process, will be $y$. If not the process remains in state $x$.

The Metropolis algorithm accepts the proposal with probability

$$A(x, y) = \max\left\{1, \frac{f(y)q(y \to x)}{f(x)q(x \to y)}\right\} \tag{5.15}$$

that is, the proposal is accepted if

$$uf(x)q(x \to y) < f(y)q(y \to x) \tag{5.16}$$

To check that this process has the required equilibrium, it is necessary to confirm only that detailed balance holds; i.e.,

$$f(x)q(x \to y)A(x, y) = f(y)q(y \to x)A(y, x) \; \forall x, y \qquad (5.17)$$

The algorithm is widely used in molecular physics applications. For example consider the problem of simulating a three-dimensional molecular system in which spherical molecules are subject to a potential $U$. According to the theory of statistical mechanics, the equilibrium probability density of **x**, the vector of molecular positions, is proportional

$$\exp\{-\beta U(\mathbf{x})\} \qquad (5.18)$$

where $\beta$ is inversely proportional to temperature. In practice simulations are restricted to a few hundred molecules, typically in a cube with periodic boundary conditions. The simplest proposal is to displace each coordinate of each molecule independently by a random amount uniformly distributed in $(-\Delta, \Delta)$, where $\Delta$ is an adjustable parameter. See Figure 4. Because the transition density is constant, the acceptance probability for a new position vector **y** becomes

$$\max\{1, \exp[-\beta U(\mathbf{y})]/\exp[-\beta U(\mathbf{x})]\} \qquad (5.19)$$

In other words, if the new position decreases the energy of the system it is certain to be accepted and if it increases the energy it is accepted with a probability less than 1.

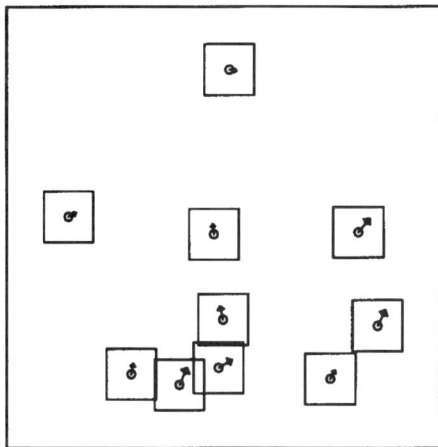

FIG. 4. Molecules are given independent displacements uniformly distributed within the boxes. The proposed vector of positions **y** is accepted with probability $\max\{1, \exp[-\beta U(\mathbf{y})]/\exp[-\beta U(\mathbf{x})]\}$.

The algorithm proceeds by successively generating proposed moves and then moving if the proposals are acceptable. Note that if $\Delta$ is small, the proposals will be accepted with a high probability, but the configuration will need a large number of transitions to change substantially. On the other hand, if $\Delta$ is large then large jumps are possible but the probability of accepting the change will typically be small. A general rule of thumb is to adjust $\Delta$ so that proposals are accepted roughly 50% of the time. After a large number of time steps the Markov process can be expected to settle into equilibrium. It should be stressed that the steps that the process takes do not correspond to a physically realistic evolution; they are just a mechanism for arriving at a specific equilibrium distribution. The method can be contrasted with the numerical solution of the deterministic equations of motion as in molecular dynamics simulation. The book by Allen and Tildesley [2] provides a comprehensive coverage of methods of simulating molecular systems and includes a discussion of how best to determine when equilibrium has been attained. For a recent review of the use of Markov chain Monte Carlo methods in statistics, see Besag and Green [3].

## 5.5 Monte Carlo Integration

The evaluation of a physical constant frequently involves the calculation of a multidimensional integral. Monte Carlo integration is based on a simple idea, namely, that an integral

$$I = \int_0^1 g(x)dx \tag{5.20}$$

can be approximated by

$$\hat{I} = n^{-1} \sum_{i=1}^{n} g(u_i) \tag{5.21}$$

where $u_1, \ldots, u_n$ are simulated uniform $(0, 1)$ variables. The estimate $\hat{I}$ has expected value $I$, and its standard deviation decreases as $n^{-1/2}$. Higher dimensional integrals are dealt with in a similar manner; i.e., if $A$ the domain of integration is scaled to fit within a cube $(0, 1)^d$ in $d$-dimensions then $I = \int_A g(x_1, \ldots, x_d)dx_1 \ldots dx_d$ is approximated by

$$n^{-1} \sum_{i=1}^{n} g_A(u_{1i}, \ldots, u_{di}) \tag{5.22}$$

where $g_A(u_1, \ldots, u_d)$ equals $g(u_1, \ldots, u_d)$ when $(u_1, \ldots, u_d)$ is in $A$ and equals 0 otherwise. As usual, the $u$ terms are simulated independent variables, uniformly distributed on (0, 1). The problem with Monte Carlo integration is that in all dimensions the standard deviation of the estimate decreases as $Kn^{-1/2}$. The best that can be done is to reduce the size of the constant $K$.

### 5.5.1 Importance Sampling

Notationally it is easier to illustrate the method in the case $d = 1$. Importance sampling works by simulating from a probability density $f$ other than the uniform density on (0, 1). The estimate of $I$ is then given by

$$\hat{I} = n^{-1} \sum_{i=1}^{n} g(x_i)/f(x_i) \tag{5.23}$$

where $x_1, \ldots, x_n$ are independently simulated from $f$. Provided $f(x)$ is positive whenever $g(x)$ is nonzero, the estimate is unbiased. The standard error (estimated standard deviation) of $\hat{I}$ can be calculated in the usual way, namely,

$$\frac{1}{\sqrt{n}} \left( \frac{1}{n} \sum_{i=1}^{n} (y_i - \bar{y})^2 \right)^{1/2} \tag{5.24}$$

where $y_i = g(x_i)/f(x_i)$ for $i = 1, \ldots, n$. To make the standard error small, $f$ must be adapted to $g$ so as to reduce the variation in the ratio $g(x)/f(x)$. When $g$ is positive the ideal $f$ will have approximately the same shape; in other words, $f$ will be large when $g$ is large and $f$ will be small when $g$ is small. This is the basic principle of importance sampling: that you should bias the sampling towards those regions where the contributions to the integral are large. When $g$ changes sign, the situation is less clear. It may be possible to find a function $h$ with a known integral $J$ such that $g + h$ is for the most part nonnegative. In this case, importance sampling can be applied by using an $f$ shaped roughly like $g + h$. The resulting Monte Carlo integral is then

$$\hat{I} = n^{-1} \sum_{i=1}^{n} [g(x_i) + h(x_i)]/f(x_i) - J \tag{5.25}$$

where $x_1, \ldots, x_n$ are independently simulated from $f$.

Over the years, a number of further refinements of Monte Carlo integration have been developed. These make use of whatever additional information may be available about the shape of $g$. Hammersley and Handscomb [5] present a succession of techniques of increasing complexity that yield progressively improved estimates of a specific integral. Some or all of the methods may be relevant to a particular practical problem.

## 5.6 Time Series

One of the simplest stochastic processes is the random walk. The position of the walk at time $n$ is given by $X_n = Z_1 + \cdots + Z_n$, where $Z_1, \ldots, Z_n$ are independent, identically distributed random variables constituting the steps of the walk. In general, $Z_n$ can be in any dimension. For physical applications it is typically in three dimensions. The process is simple to simulate since $X_{n+1} = X_n + Z_n$ so that successive positions are obtained by taking the current position of the walk $X_n$ and adding on the next step $Z_n$. When the components of each $Z_n$ are independent $N(0, 1)$ variables, $X_n$ corresponds to the position of *Brownian motion* at time $n$. To interpolate at a time $n + t$ between $n$ and $n + 1$ the *Brownian bridge* can be simulated. Given that $X_n = x_n$ and $X_{n+1} = x_{n+1}$ the distribution of $X_{n+t}$ is normal with mean $x_n(1 - t) + x_{n+1}t$ and the variance of each coordinate of $X_{n+t}$ is $t(1 - t)$. Thus the trajectory of a particle moving according to Brownian motion can be simulated first on a coarse time scale, to determine the general direction of its movement, and subsequently finer details of the trajectory can be filled in at points in time and space that are of special interest. This technique is used extensively in modelling molecular trajectories in studies of reaction kinetics.

The simple random walk is a prototype for the general autoregressive process *AR(p)* that has the following structure

$$X_{n+1} = a_1 X_n + a_2 X_{n-1} + \cdots + a_p X_{n-p} + Z_n \tag{5.26}$$

where the $Z_n$ are again independent and identically distributed. Under certain conditions on the constants $a_1, \ldots, a_p$, the process has a stationary distribution. See Chapter 3. For convenience we will assume that all variables have zero means. The easiest way to simulate the process in this case, is to start with an arbitrary set of initial values $x_1, \ldots, x_p$ and then, using simulated values of $z_n$, calculate

$$x_{n+1} = a_1 x_n + a_2 x_{n-1} + \cdots + a_p x_{n-p+1} + z_n \tag{5.27}$$

for $n = p, p + 1, \ldots$, until equilibrium is obtained. After a period of equilibration, the values of $x_n$ come from the stationary distribution, and the subsequent time series can then be used as a realization of the stationary autoregressive process. This is known as the harvesting phase of the algorithm. The main problem is to determine when equilibration has been completed. In practice, a number of *ad hoc* methods are used, including monitoring the empirical mean and covariances of the series. If computer time is not a constraint, substantial periods of equilibration can be contemplated when in doubt.

When the $Z_n$ are normally distributed, $\{X_n\}$ is a *Gaussian process*. The equilibrium distribution of $X_1, \ldots, X_p$ is multivariate normal with a variance–covariance matrix that can be obtained from the Yule–Walker equations. See Chapter 3. The process can then be started from a simulation of $X_1, \ldots, X_p$ using the methods of Section 5.4 and continued using the recurrence relation as previously. Having started in equilibrium it remains in equilibrium.

The autoregressive moving-average process, ARMA, $(p, q)$ has structure

$$X_{n+1} - a_1 X_n - \cdots - a_p X_{n-p+1} = b_1 Z_n + \cdots + b_q Z_{n-q+1} \qquad (5.28)$$

Again we will restrict attention to the problem of sampling the process in equilibrium, which means that certain restrictions have to placed on the parameters $a_1, \ldots, a_p$ and $b_1, \ldots, b_q$. See Chapter 3.

As before, simulations can be started in an arbitrary state, say $z_1 = \cdots = z_q = 0$, $x_1 = \cdots = x_p = 0$, and then continued through an equilibration period before harvesting values of the process.

Alternatively, in the Gaussian case, the process can be started in equilibrium, although an additional complication is that the initial state must specify both $z_1, \ldots, z_q$ and $x_1, \ldots, x_p$. Rather than go into detail about how to do this for the ARMA $(p, q)$ model we will move on to describe methods for simulating an arbitrary stationary Gaussian process. The methods we describe will also enable us to simulate time series that are defined on a continuous time scale, rather than on the discrete set of times $\ldots, 0, 1, 2, \ldots$.

### 5.6.1 Spectral Sampling

We start by considering processes taking real values at discrete times. A stationary Gaussian process $\{X_n\}$ is specified by its autocovariance function. For simulation purposes, we may as well assume that the mean of the process is zero. In Chapter 3 it was shown that the autocovariance function $C(k)$ has a spectral representation

$$C(k) = E(X_n X_{n+k}) = \int_0^\pi f(\omega) \cos(k\omega) d\omega \qquad (5.29)$$

where $f(\omega)$, is a nonnegative function called the *spectral density*. The variance of the $X_n$ is given by

$$\sigma_Y^2 = \int_0^\pi f(\omega) d\omega \qquad (5.30)$$

Conversely, if the covariance function is known then $f$ is given by

$$f(\omega) = \frac{1}{\pi}\left[1 + 2\sum_{k=1}^{\infty} C(k)\cos(\omega k)\right] \tag{5.31}$$

Since $f$ integrates to $\sigma_Y^2$ the function

$$f^*(\omega) = \frac{f(\omega)}{\sigma_Y^2} \tag{5.32}$$

is a probability density function. It is straightforward to show that if $W$ is sampled from the density $f^*$ and if $U$ is uniformly distributed on $(0, 1)$, then the sequence of random variables

$$\sigma_Y\sqrt{2}\cos(\pi U + nW) \qquad \text{for } n = \ldots, -1, 0, 1, \ldots \tag{5.33}$$

has the same covariance as $\{X_n\}$. The simplest simulation of $\{X_n\}$ is therefore

$$\sigma_Y (2/m)^{1/2} \sum_{i=1}^{m} \cos(\pi u_i + nw_i)$$
$$\text{for } n = \ldots, -1, 0, 1, \ldots \tag{5.34}$$

where $(u_i, w_i)$ are independent simulations of $(U, W)$ and $m$ is large enough to guarantee approximate normality by the central limit theorem. A conservative rule is to chose $m$ to be 10 times the number of time points at which the process is simulated.

If it is difficult to sample from the probability density $f^*$ the following procedure can be used. It uses the fact that when $U$ is uniformly distributed on $(0, 1)$ and when $V$ independently has probability density $k$ then

$$[2f(V)/k(V)]^{1/2}\cos(\pi U + nV)$$
$$\text{for } n = \ldots, -1, 0, 1, \ldots \tag{5.35}$$

has the same covariance function as $\{X_n\}$. Here we have assumed that the support of $f$ is contained in that of $k$. A simulation of the process $\{X_n\}$ is then given by

$$(2/m)^{1/2} \sum_{i=1}^{m} [f(v_i)/k(v_i)]^{1/2}\cos(\pi u_i + nv_i)$$
$$\text{for } n = \ldots, -1, 0, 1, \ldots \tag{5.36}$$

where $(u_i, v_i)$ are independent simulations of $(U, V)$ and $m$ is large enough to guarantee approximate normality. It should be noted that if $k$ is poorly matched to $f$, components of the sum will have a long-tailed distribution and large values of $m$ will be needed to achieve normality.

To illustrate the method we will simulate values from an ARMA(1, 1) process. The general ARMA($p$, $q$) process has spectral density

$$f(\omega) = \frac{\sigma_Z^2}{\pi} \frac{|1 + \sum_{k=1}^{q} b_k e^{ik\omega}|^2}{|1 - \sum_{k=1}^{p} a_k e^{iw k}|^2} \tag{5.37}$$

where $\sigma_Z^2$ is the variance of $Z_n$. For the particular case

$$X_n = aX_{n-1} + Z_n + bZ_{n-1} \tag{5.38}$$

this becomes

$$f(\omega) = \frac{\sigma_Z^2}{\pi} \frac{1 + 2b\cos(\omega) + b^2}{1 - 2a\cos(\omega) + a^2} \tag{5.39}$$

The spectral density with $b = 0.5$, $a = 0.7$ and $\sigma_Z^2 = 1$ is graphed in Figure 5. A simulation of 200 successive values of the ARMA process, $X_1, \ldots, X_{200}$ calculated by sampling from the uniform density $k(\omega) = 1/\pi$, with $m = 2000$, is displayed in the adjoining panel.

A stationary time series in continuous time can be thought of as a one-dimensional spatial process. In the next section, we show how homogeneous Gaussian random fields in an arbitrary number of dimensions can be simulated.

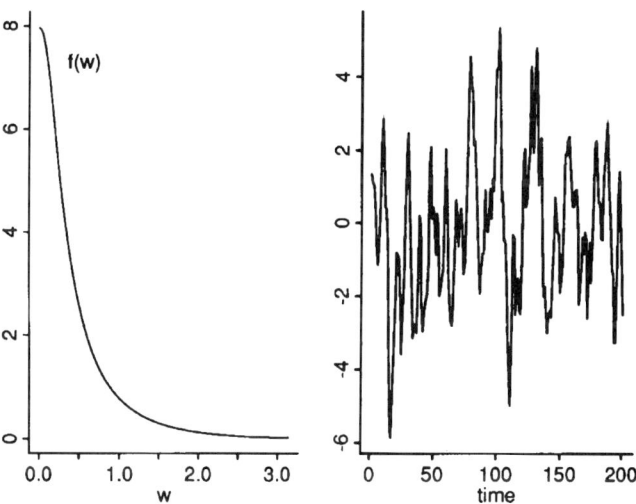

FIG. 5. Spectral density of ARMA(1, 1) process with $b = 0.5$, $a = 0.7$ and $\sigma_Z^2 = 1$. Simulation of process using spectral sampling with $k$ uniform.

## 5.7 Spatial Processes

### 5.7.1 Gaussian Random Fields

The simulation of Gaussian random fields is important in the study of spatially distributed data, both as a means of investigating the properties of proposed models of spatial variation and as a way of constructing goodness-of-fit tests by the Monte Carlo method. When there are only a few spatial locations, the Cholesky decomposition method of Section 5.4 can be used. When a large number of locations is involved, for example in simulating a random surface in a study of frictional effects, we have to employ different techniques. In geostatistical applications, *turning band* simulation, developed by Matheron [8], appears to be the method of choice. The technique has the following advantages: (a) it permits the field to be simulated at irregularly spaced locations; (b) the second-order properties of the simulated process are exactly those specified by the Gaussian model; (c) the multivariate distribution of the simulated process is approximately Gaussian, relying on the central limit theorem for the approximation. However, the technique is restricted to the simulation of fields that by transformation of the spatial coordinates can be made homogeneous and isotropic. Furthermore, the method of turning bands involves repeatedly simulating and superimposing independent copies of an associated stationary one-dimensional process. Each such component of the simulation may therefore by itself involve a substantial computational cost.

The spectral sampling methods for simulating stationary Gaussian time series extend naturally to random fields in higher dimensions. First of all, we generalize the discussion to consider processes defined at all points in a continuum, e.g., at all points along a time axis or all points in the plane or in space in one, two and three dimensions, respectively. With the notation of Chapter 4, we denote a real-valued Gaussian random field by $X(h)$, $h \in R^d$, where $d$ is the dimension of the space. We will assume that the field is homogeneous; i.e., that $X(h_1), \ldots, X(h_m)$ has the same distribution as $X(h_1 + k_1), \ldots, X(h_n + k_n)$ for any displacement $k \in R^d$. We assume that the mean of the process is 0 and we denote the covariance function $E[X(h)X(0)]$ by $C(h)$. The spectral density function $f(\omega)$ of the random field and the covariance function are related by

$$f(\omega) = (2\pi)^{-d} \int_{R^d} C(h) \cos(\omega^T h) dh \qquad (5.40)$$

and

$$C(h) = \int_{R^d} f(\omega) \cos(\omega^T h) d\omega \qquad (5.41)$$

If the covariance function is isotropic, then $f(\omega)$ is a function of $|\omega|$ alone. Furthermore, as Matheron has shown, the covariance function $C$ that is now a function of $r = |h|$ can be expressed as

$$C(r) = \frac{2\Gamma(d/2)}{\sqrt{\pi}\Gamma((d-1)/2)} \int_0^1 C_1(vr)(1 - v^2)^{(d-3)/2} dv \tag{5.42}$$

where $C_1$ is the covariance of a one-dimensional stationary time series.

Note that the spectral density is defined over the whole of $R^d$. In particular $\sigma_X^2$, the variance of $X(h)$ is given by

$$\sigma_X^2 = \int_{R^d} f(\omega) d\omega \tag{5.43}$$

For homogeneous fields that are defined only on integer lattices, i.e., discrete grid points, the spectral density is confined to the cube $(-\pi, \pi)^d$ and the integral in Eq. (5.40) is replaced by summation over the grid points. For the case $d = 1$ this differs slightly from the definitions given in Section 5.6.

### 5.7.2 Turning Bands

Matheron [8] showed that if $V$ is a point chosen randomly on the unit sphere in $R^d$ and $Z(t)$ is a stationary process with covariance $C_1$, then $Z(h^T V)$ has the same second-order properties as $X(h)$. In two dimensions the process $Z(h^T V)$ can be visualized as a corrugated surface whose contours are parallel bands, turned through a random angle. The turning band simulation of $X(h)$ is given by

$$m^{-1/2} \sum_{i=1}^{m} z_i(h^T v_i) \tag{5.44}$$

where $(v_i, z_i(t))$ are independent simulations $(V, Z(t))$. By choosing $m$ sufficiently large the distribution of (5.44) at a finite number of locations is approximately Gaussian by virtue of the central limit theorem.

### 5.7.3 Spectral Sampling

A simple alternative to the method of turning bands is to use spectral sampling as in Eq. (5.36). Suppose that the field is to be simulated at a set of points $h_1, \ldots, h_n$. First of all, we must find a $d$-dimensional density that is easy to simulate from. The density has to have the property that it is positive whenever $f$ is positive. Call this density $k$ and write $V$ for the $d$-dimensional random variable with this distribution. As usual we denote a uniform $(0, 1)$ variable by $U$. The random field is simulated by

$$(2/m)^{1/2} \sum_{i=1}^{m} [f(v_i)/k(v_i)]^{1/2} \cos(\pi u_i + h^T v_i) \quad \text{for } h \in R^d \quad (5.45)$$

where $(u_i, v_i)$ are independent simulations of $(U, V)$ and $m$ is large enough to guarantee approximate normality. The special case when $f \propto k$ leads to some simplification.

Spectral sampling is more widely applicable than the method of turning bands. It is also easier to implement since there is no need to solve the functional Eq. (5.42) to obtain $C_1$, and the problem of simulating the process $Z(t)$ is avoided. Most important, spectral sampling is not restricted to isotropic fields. In particular, it is possible to designate one of the space axes as a time axis and simulate a Gaussian spatial–temporal process using the anisotropic spectral density corresponding to the spatial–temporal covariance function. Furthermore, many processes are most conveniently specified in terms of their spectral density, and in those cases in which only the covariance function is given, Eq. (5.40) can be used to obtain the spectral density by integration.

### 5.7.4 Lattice-Based Gaussian Processes

The treatment of lattice-based spatial processes is similar. The principle difference is that $f$ is constrained to $(-\pi, \pi)^d$. A benefit of this is that there is an obvious simple choice for $k$, namely, the uniform density on $(-\pi, \pi)^d$. The two-dimensional autonormal random field described in Chapter 4 has spectral density

$$f(\omega) = [1 - \alpha \cos(\omega_1) - \beta \cos(\omega_2)]^{-1}$$
$$\text{for } (\omega_1, \omega_2) \in (-\pi, \pi)^2 \quad (5.46)$$

where $|\alpha|$ and $|\beta|$ are less than 0.5. Figure 6 gives a simulation of such a field obtained by using Eq. (5.45) with $k$ the uniform density on $(-\pi, \pi)^2$.

### 5.7.5 Systematic Sampling of Spectral Density

A number of refinements of the basic method (5.45) have been proposed. The usual modification is to replace the simulated variables $v_i$ by values chosen from a predetermined set of points distributed systematically throughout the spectral domain. Effectively this replaces the continuous spectral density by a discrete one. A side effect is that the simulated process becomes periodic, which will distort long range correlations. The main advantage is that when the points are suitably chosen, the fast Fourier transform can be used to do the calculations, resulting in a substantial

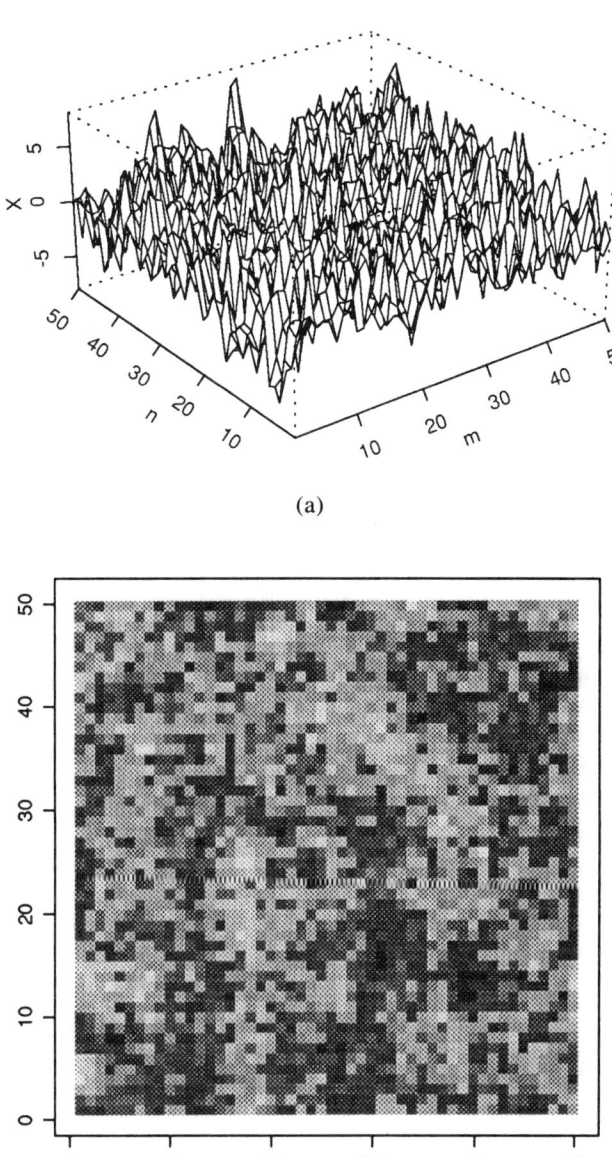

FIG. 6. Simulation of autonormal random surface $\alpha = \beta = 0.49$: spectral sampling with $k$ uniform and $m = 50,000$.

reduction in computing time. For example, placing points at $(\omega_j, \omega_k)$, where $\omega_j = j\pi/N$ for $j, k = -N + 1, \ldots, N$ leads to the simulation

$$x_{mn} = \sum_{j=-N+1}^{N} \sum_{k=-N+1}^{N} \left(\frac{2\pi}{N} f(\omega_j, \omega_k)\right)^{1/2} \cos(u_{jk}\pi + m\omega_j + n\omega_k) \qquad (5.47)$$

for $m, n = \ldots, -1, 0, 1, \ldots$, in the two-dimensional lattice-based case. The best value of $N$ will depend on circumstances. A large value of $N$ ensures that the discrete spectrum yields a covariance function close to that of $f$ at the cost of additional computing time. An added refinement is to include independently simulated exponential random variables under the square root sign. The advantage is that the resulting expression has an exact Gaussian distribution. In practice, this produces no real benefit when $N$ is large since the central limit theorem will ensure approximate normality even without the added variables.

## 5.8 Markov Random Fields

In this section we will focus on methods for the simulation of Markov random fields on graphs and lattices. We will assume as in Chapter 4 that the joint distribution of the values $Z(s_1), \ldots, Z(s_n)$ at the sites $s_1, \ldots, s_n$ is such that at every site $s_k$, the conditional density of $Z(s_k)$ given the rest of the values is of the form $p_k[z(s_k)|z(N_k)]$, where $N_k$ is the set of sites that neighbour $s_k$. When the number of sites is small, it may be feasible to obtain the joint density of $Z(s_1), \ldots, Z(s_n)$ explicitly or at least up to a multiplicative constant. In such cases, standard methods such as rejection sampling can be used. However, most applications of Markov fields involve lattices with large numbers of sites. In image processing, for example, $n$ may be $512 \times 512$; i.e., the number of pixels on the screen. Simulation methods make use of the Markov property of the field. The most straightforward technique is to visit sites cyclically in some prescribed order and at each visit update the site according to a certain probabilistic rule. If this is done correctly the succession of updated configurations forms a stochastic process having $f$ as its equilibrium.

Updating can be achieved in a number of ways. The simplest is to use the Metropolis method on the conditional densities. Suppose that site $s_k$ is in state $z(s_k)$ when it is visited and that the neighbouring sites are in state $z(N_k)$. The *site-wise Metropolis* procedure generates a proposal using a density $g_k(y)$ and then accepts the proposal with probability

$$\max \left\{ 1, \frac{p_k[y|z(N_k)]g_k[z(s_k)]}{p_k[z(s_k)|z(N_k)]g_k(y)} \right\} \qquad (5.48)$$

The density $g_k$ can depend on the site $k$ and the values $z(N_k)$. When the Markov field takes only a finite number of values, $g_k$ can simply be uniform.

Another method of updating that has become popular is the *Gibbs sampler*. In general it is more difficult to implement, and there is some evidence that the convergence to equilibrium is slower than the site-wise Metropolis method. The Gibbs sampler updates a site by simulating a value from the conditional density $p_k[y|z(N_k)]$. When the field takes only two values at each site, e.g., a black-and-white image, it is trivial to apply the Gibbs sampler. However, when the number of possible values at each site is large, use of the Gibbs sampler can be cumbersome and costly in computer time. Under these circumstances the site-wise Metropolis method is to be preferred.

An unresolved problem is that of determining when convergence has been attained. The number of iterative cycles that have to be completed before equilibrium is achieved will depend both on the strength of dependence between site values (long-range order effect) and which properties of the equilibrium distribution are under investigation. Careful monitoring of the evolution of the random field is required.

Recently there has been a great deal of interest in a new method of simulating the Potts model, a particular class of discrete-state Markov random fields in which each site is assigned one of $q$ different values or colours. In its simplest form $p[z(s_1), \ldots, z(s_n)]$ the probability density for the Potts model is taken to be proportional to $\exp[-\beta v(z)]$, where $\beta > 0$ and $v(z)$ is the number of neighbouring pairs of sites $(s_i, s_j)$ for which $z(s_i) \neq z(s_j)$. The distribution is a model for an aggregation process in which neighbouring sites tend to be of the same type. It has been used in image analysis as a prior distribution for the Bayesian reconstruction of a corrupted image. The prior distribution represents a prejudice that neighbouring pixels should tend to be of the same colour. The site-wise Metropolis algorithm converges very slowly for the Potts model, particularly when $\beta$ is large. Quite dramatic improvements can be produced by using a different Markov chain introduced by Swendsen and Wang. See, for example, Besag and Green [3]. At each step in the chain, the algorithm starts by joining each pair of neighbouring sites with a bond. The algorithm then identifies which pairs of neighbouring sites are such that $z(s_i) \neq z(s_j)$. The bonds between such pairs are then broken. The algorithm proceeds by independently breaking each of the unbroken bonds with probability $e^{-\beta}$, then finds which clusters of sites are held together by the remaining unbroken bonds and finally colours each cluster randomly, independently and with equal probability using one of the $q$ possible colours. Computer experiments show that the Swendsen–Wang algorithm reaches equilibrium much more rapidly than the site-wise Metropolis method. Finding Markov chains with such good convergence properties for more general densities than that of the Potts model is a challenging area of statistical research.

## 5.9 Point Processes

The simplest spatial point process is the *Poisson process* and the simplest case is when there is only one space dimension that can equivalently be thought of as time. A one-dimensional Poisson process with intensity $\lambda$ can be simulated most simply by making use of the fact that the times *between* arrivals are independent and exponentially distributed. Alternatively, from the definitions in Section 4.4, it is known that the number of arrivals in an interval $(0, t)$ has a Poisson distribution with mean $\lambda t$, and the times of arrival are then uniformly distributed on $(0, t)$. This leads to the following algorithm.

**Algorithm 5.7:** Here we simulate the arrival times in a Poisson process, during an interval $(0, t)$:

```
simulate x from Poisson (λ t) by
using Algorithm 5.5 or 5.6
for i = 1 to x
t(i) = t * RND
next i
```

The advantage of the first method is that the arrival times are generated as time evolves. In Algorithm 5.7, which is a faster algorithm, the arrival times $t(i)$ are generated in no particular order, so that an additional sorting operation is required if they are needed in an increasing sequence. Algorithm 5.7 extends immediately to the spatial case; i.e., the homogeneous Poisson process with intensity $\lambda$ in a cube $(0, t)^d$ in $d$-dimensions. The only changes that have to be made are that $x$ has to be sampled from the Poisson distribution with mean $\lambda t^d$, the mean number of points in the cube, and the location of a typical point is then uniformly distributed in $(0, t)^d$; i.e., each one of its coordinates is uniformly and independently distributed between 0 and $t$. To simulate the homogeneous Poisson process in a more complicated region, it is necessary only to locate the region within a cube and record just those points that lie in the region of interest.

The inhomogeneous Poisson process can be simulated directly from the construction in Chapter 4. Similarly the constructive definitions of the Cox process and the Poisson cluster process in the same chapter can be used to provide straightforward simulation algorithms. The Markov point process mentioned in Chapter 4 can be simulated either constructively or by making use of a spatial–temporal birth–death process whose equilibrium distribution is of the form required (Ripley [10]).

## 5.10 Further Reading

There are a number of good books on Monte Carlo methods. The monograph by Hammersley and Handscomb [5] is easy to read and describes a wide range of applications. Devroye [4] provides a more recent and compendious catalogue of techniques. Volume II of Knuth's book [7], *The Art of Computer Programming* contains many ingenious algorithms, emphasising good computing practice. Other recent books are those by Ripley [10], which is strong on theory, and Kalos and Whitlock [6], which emphasizes physical applications. It is always worth checking software manuals to see whether random variables can be simulated by built-in algorithms. The S language, mentioned in Chapter 17, has routines for common continuous variables.

## References

1. Abramovitz, M., and Stegun, I. A. (1965). *Handbook of Mathematical Functions with Formulas, Graphs and Mathematical Tables*. Dover, New York.
2. Allen, M. P., and Tildesley, D. J. (1987). *Computer Simulations of Liquids*. Clarendon Press, Oxford.
3. Besag, J., and Green, P. (1993). "Spatial Statistics and Bayesian Computation (with Discussion)." *J. R. Statist. Soc. B* **55.** (1), 25–37.
4. Devroye, L. (1986). *Non-Uniform Random Variate Generation*. Springer Verlag, New York.
5. Hammersley, J. M., and Handscomb, D. C. (1964). *Monte Carlo Methods*. Methuen, London.
6. Kalos, M. H., and Whitlock, P. A. (1986). *Monte Carlo Methods,* **1.** *Basics*. John Wiley and Sons, New York.
7. Knuth, D. E. (1981). *The Art of Computer Programming*. Addison-Wesley, Reading, MA.
8. Matheron, G. (1973). "The Intrinsic Random Functions and Their Applications." *Adv. in Appl. Probab.* **5,** 439–468.
9. Metropolis, N., Rosenbluth, A. W., Rosenbluth, M. N., Teller, A. H., and Teller, E. (1953). "Equations of State Calculations by Fast Computing Machines." *J. Chem. Phys.* **21,** 1087–1092.
10. Ripley, B. D. (1987). *Stochastic Simulation*. John Wiley and Sons, New York.

# 6. BASIC STATISTICAL INFERENCE

John Kitchin

Semiconductor Operations
Digital Equipment Corporation
Hudson, Massachusetts

## 6.1 Introduction

We see in Chapter 1 that observations of physical phenomena are often subject to uncertainty or random "errors" and that probability theory can be useful for modeling and interpreting them. In this chapter we begin to address the problem of making inferences, or more generally, answering questions based on observations subject to randomness. We build a framework for *statistical* inference.

### 6.1.1 Goals of this Chapter

The goals of this chapter are to introduce the basic concepts of classical inference under a parametric model; to illustrate these concepts through simple, though real, examples in a way that demonstrates what is possible with classical statistical inference; and to give pointers to literature that can help the reader carry out statistical inference for more general problem sets. Included are brief discussions of some of the weaknesses of classical statistical inference and reference to other approaches to inference that, to many, are in competition with classical statistical inference.

### 6.1.2 Motivation

Consider the observations in Table I (prearranged for convenience in ascending order). Each is the natural log of the time to failure (in hours)

TABLE I. Natural Logs of Electromigration Failure Times from Accelerated Stress

| | | | |
|---|---|---|---|
| $x_1 = 3.29$ | $x_5 = 3.72$ | $x_9 = 3.99$ | $x_{13} = 4.21$ |
| $x_2 = 3.47$ | $x_6 = 3.80$ | $x_{10} = 3.11$ | $x_{14} = 4.28$ |
| $x_3 = 3.54$ | $x_7 = 3.82$ | $x_{11} = 3.15$ | $x_{15} = 4.53$ |
| $x_4 = 3.55$ | $x_8 = 3.93$ | $x_{12} = 4.17$ | $x_{16} = 4.94$ |

of a microscopic metal conductor line (interconnect) on a individual semiconductor test "chip." Failure is defined as a 10% or greater increase in electrical resistance, signaling end-of-life of the metal line due to electromigration.[1] Though the 16 chips (from the same production lot) were manufactured to be identical and experienced equal temperature and current density accelerations by careful control of test conditions, the results vary over a range of 1.65, equivalent to a factor of $\exp(1.65) \doteq 5$ from shortest to longest time till failure. Each of the 16 metal lines represents (after scaling) the metal interconnect on an individual semiconductor chip under mass production. Our interest is in answering important questions about the entire production lot; for example, does the lot meet a certain reliability requirement?

*Scientific inference* is the act of drawing conclusions or answering questions about an unknown "state of nature" based solely on theory and observation. The methods of science generally exclude expediency, the subjective judgment of the experimenter, or consideration of the consequences of the answers. Classical *statistical inference* was developed in order to pursue scientific inference based on observations that are subject to statistical uncertainty. Table II pairs specific questions we address in the examples of this chapter with more general questions that arise in scientific inference, to illustrate how statistical inference supports scientific inference. To develop classical statistical inference, we must develop a formal way of describing a "state of nature" and posing a question about it coupled with a set of methods for transforming observations subject to uncertainty into an answer to the question. First, however, we briefly examine methods for summarizing the electromigration observations separately from the problem of how to answer questions.

### 6.1.3 Descriptive Statistics and Graphical Methods

Commonly used descriptive statistics that provide a measure of the *location* or "central tendency" of a set of observations are the *sample mean* $\bar{x} = 3.97$ and the *sample median* $\tilde{x} = \frac{1}{2}(x_8 + x_9) = 3.96$, when applied to

---

[1]Electromigration is a directed diffusion phenomena of intense interest in the semiconductor industry because it can cause electrical failure over time in the metal "interconnect" lines of integrated circuits by forming voids or metal extrusions. The causes of randomness in electromigration time to failure are not well-understood, yet questions of design and production reliability have to be answered on the basis of the observations such as $x_1, x_2, \ldots, x_{16}$. Thus, this example demonstrates the need for statistical inference methods. Note, the statistical treatment of these observations in this chapter is simplified. More elaborate statistical models and inference methods are used to account for length scaling and temperature stress.

INTRODUCTION

TABLE II. Electromigration-Specific versus General Questions Answerable with Statistical Inference

| | Electromigration Question | General Question |
|---|---|---|
| I | Is this lot as reliable as the standard requires? | Is this claim about the population true? |
| II | How much does annealing decrease the frequency of grain boundary triple points? | How much does this new treatment improve the process? |
| III | Does electromigration accelerate as the square of the current density? | Do these observations confirm the theory? |
| IV | What lifetime is predicted for this customer's chip? | What observation is predicted for the next experiment? |
| V | Are electromigration failure times Lognormal? | Does this model fit the data? |

the observations of Table I. (See Section 6.2.1 for the general formulas for sample mean and sample median.)

A descriptive statistic called the *range*

$$r = \max(x_1, x_2, \ldots, x_{16}) - \min(x_1, x_2, \ldots, x_{16}) = 1.65 \quad (6.1)$$

is sometimes used to measure the variability among or *scale* of the observations. However, the most commonly used descriptive statistic for scale is the *sample standard deviation*:

$$s = \sqrt{\frac{\sum_{i=1}^{16} (x_i - \bar{x})^2}{16 - 1}} \doteq 0.42 \quad (6.2)$$

(See Section 6.2.1 for the general formulas for sample range and sample standard deviation.)

Figure 1 shows a relative frequency histogram of the 16 observations, revealing a symmetric, unimodal (one-hump) distribution of values around the the two nearly equal location statistics. This situation is generally consistent with a *statistical model* that stipulates that the observations were drawn at random from a normal (Gaussian) population (see Chapter 2). To further support this conjecture, Figure 2 gives a normal *probability plot,* on which a normal cumulative distribution function plots as a straight line.

A much richer collection of descriptive statistics and graphical methods can be found in many texts, including [15] (very readable, though elementary), [7], and [19] (very rich).

We have gained insight into the location and scale of the *population* of electromigration log failure times through application of descriptive statistics

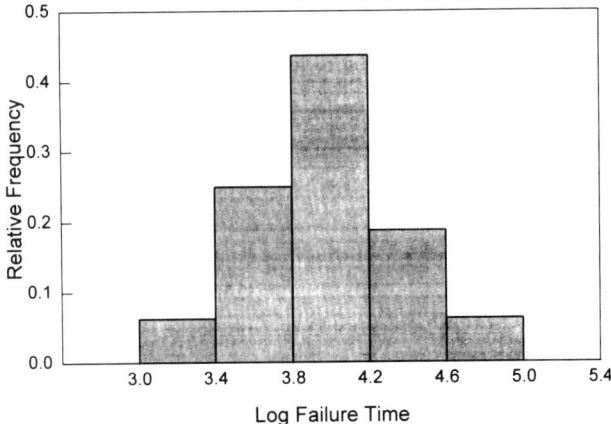

FIG. 1. Histogram of 16 electromigration log failure times.

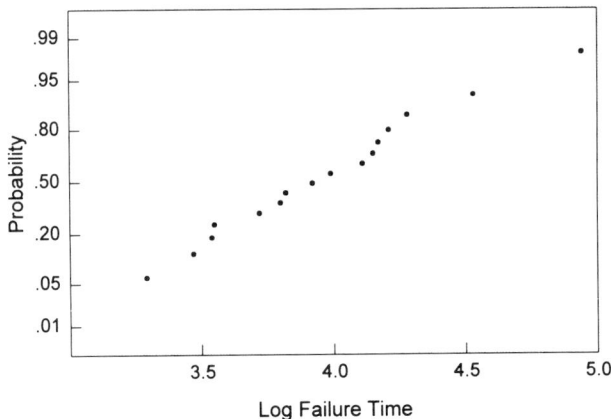

FIG. 2. Normal probability plot of electromigration log failure times.

to the *sample* observations. Also, we have a tentative probability model to describe the relationship between the population and the sample. However, we are not at present equipped to do any formal inference.

### 6.1.4 Statistical Models and Population Parameters

Statistical inference requires a *statistical model*. A formal (parametric) statistical model consists of an observable **X**, a vector of unknown parameters

Θ, and a family of probability distributions indexed by Θ, usually specified by a joint probability mass function $p(\mathbf{x}; Θ)$ or a joint probability density function $f(\mathbf{x}; Θ)$.

Our brief excursion through descriptive statistics applied to the electromigration observations leads us to assume the statistical model

$$\mathbf{X} = X_1, X_2, \ldots, X_n \text{ independent, identically distributed} \quad (6.3)$$

$$n = 16 \quad (6.4)$$

$$Θ = \langle \mu, \sigma \rangle \quad (6.5)$$

$$f(\mathbf{x}; Θ) = \frac{\exp\left(-\frac{(x_1 - \mu)^2}{2\sigma^2}\right)}{\sqrt{2\pi}\sigma} \cdot \frac{\exp\left(-\frac{(x_2 - \mu)^2}{2\sigma^2}\right)}{\sqrt{2\pi}\sigma} \cdot \ldots \cdot \frac{\exp\left(-\frac{(x_{16} - \mu)^2}{2\sigma^2}\right)}{\sqrt{2\pi}\sigma} \quad (6.6)$$

What is being said so formally is that the population of chips in the lot has a one-to-one correspondence with a population of electromigration log failure times that is normally distributed with mean $\mu$ and standard deviation $\sigma$. Testing 16 of the chips to failure gives a realization of the observable $X_1, X_2, \ldots, X_{16}$ that yields information on the unknown population parameters $\mu$ and $\sigma$ through their $N(\mu, \sigma)$ relationship with the observable.

The statistical model describes *all* random samples of size 16 that might be drawn from the population of electromigration log failure times. The 16 observations in Table I are but a single *realization* of the observable (but at present the only realization we have). This distinction between the statistical model for all possible realizations of the observable and a specific realization of the observable is key to the interpretation of classical statistical inference.

Any question about the "state of nature" of the population that is framed in terms of the parameters $\mu$ and $\sigma$ can be answered via statistical inference based on this statistical model. Note, however, that to answer Question V from Table II requires a statistical model more general than that assumed here for the electromigration observable. We defer to Chapter 7 the introduction of methods of statistical inference for questions about how well a parametric statistical model fits the observable.

Given a statistical model we can now develop methods to estimate the population parameters or functions of them (estimation), establish the precision of such estimates (confidence intervals), and formally answer questions about the population (statistical testing).

## 6.2 Point Estimation

In this section we develop methods for obtaining an estimator that yields a "good" point (that is, single number) estimate of a population parameter within the statistical model chosen.

### 6.2.1 Concept of a Statistic

Formally, a statistic is any *function* of the observable. A statistic $T$ maps the observable to a real number $T(\mathbf{X})$ and is therefore a random variable with a probability distribution (see Chapter 1). For example, the descriptive statistics introduced in Section 6.1 to measure location are the sample mean

$$\overline{X} = (1/n) \sum_{i=1}^{n} X_i \qquad (6.7)$$

and the sample median

$$\tilde{X} = \begin{cases} X_{[k+1]} & \text{if } n = 2k+1 \\ (1/2)(X_{[k]} + X_{[k+1]}) & \text{if } n = 2k \end{cases} \qquad (6.8)$$

where $X_{[1]}, X_{[2]}, \ldots, X_{[n]}$ denotes the observable arranged in *increasing order*. Further, statistics to measure scale are the sample range

$$R(\mathbf{X}) = \max(X_1, X_2, \ldots, X_n) - \min(X_1, X_2, \ldots, X_n) \qquad (6.9)$$

and the sample standard deviation

$$S(\mathbf{X}) = \sqrt{\frac{\sum_{i=1}^{n}(X_i - \overline{X})^2}{n-1}} \qquad (6.10)$$

the latter invoked as the preferred estimator of the parameter $\sigma$. Note that the statistic $S$ utilizes the statistic $\overline{X}$.

In general we speak of an *estimator T* as a statistic that yields information on a parameter $\theta$. Several statistics are introduced in the succeeding sections, not only as *estimators*, but also as *limits* for confidence intervals and as *test statistics* for statistical hypothesis tests.

### 6.2.2 Desirable Properties of Estimators

We naturally desire the estimator $T$ to yield *estimates* (realized values) that are in some sense "close" to $\theta$. A common way to measure "close" for a random variable is through the *mean squared error:*

POINT ESTIMATION

$$E(T - \theta)^2 = V(T) + \{E(T) - \theta\}^2 \tag{6.11}$$

where $E$ and $V$ are the expectation and variance operators, respectively (see Chapter 1). The realized values of $T$ will spread about the center of its distribution $E(T)$ with a *precision* measured by its variance $V(T)$. On the other hand, the *accuracy* of $T$ is measured by its *bias* for $\theta$, given by $E(T) - \theta$.

For example, in the estimation of $\mu$ the variance of the sample mean, $V(\bar{X})$ for a sample of size $n$ from a $N(\mu, \sigma)$ population equals $\sigma^2/n$ while its bias, $E(\bar{X}) - \mu$ equals zero; that is, $\bar{X}$ is *unbiased*. In general, the smaller the mean squared error of an estimator, the better it is *on average* at estimating the parameter. It is obvious from Eq. (6.11) that reducing either the variance or the bias of an estimator improves its "closeness" to $\theta$. Hence, *minimum variance* and *unbiasedness* are two desirable properties of estimators.

In comparison to $\bar{X}$ in the $N(\mu, \sigma)$ statistical model, the sample median $\tilde{X}$, though also unbiased, has $V(\tilde{X}) > 7\sigma^2/6n$ for any $n \geq 3$. Hence its mean square error is always at least 7/6 times as large as that of $\bar{X}$, regardless of the value of the standard deviation $\sigma$ (expect in the trivial cases $n = 1$ and $n = 2$ when the sample mean and sample median are identical). This is why the median is not used for inference about $\mu$ in the $N(\mu, \sigma)$ case. In fact, $\bar{X}$ is the *minimum variance unbiased estimator* for $\mu$ in the $N(\mu, \sigma)$ model.

A third desirable property of an estimator $T$ is *consistency*—as the sample size $n$ increases without bound the probability that $T$ yields a value "close" to $\theta$ converges to 1. Any reasonable estimator will be consistent, typically because of the Law of Large Numbers (see Chapter 1). Also, if an estimator's mean squared error converges to 0 as $n$ increases without bound, consistency is assured. For these reasons developing a "good" estimator usually focuses on minimizing variance and bias, balanced against the complexity and computational burden that an estimator or inference based on it may impose.

### 6.2.3 Methods of Obtaining Estimators

#### 6.2.3.1 Method of Moments.
One of the conceptually simplest ways to obtain an estimator is to match sample moments to population moments. (See Chapter 1 on moments.) As a simple case, $\bar{X}$ is the first moment of the sample data and estimates the population first moment $\mu$ (the mean) in the electromigration statistical model. The *sample variance* $S^2$ is the second moment about the mean of the sample data, and it estimates the second moment about the population mean $\sigma^2$ in the electromigration statistical model.

162　　　　　　　　　　　　　BASIC STATISTICAL INFERENCE

TABLE III. Frequency of Grain Boundary Triple Points in Semiconductor Interconnect Lines

| Type of Treatment | Number of Triple Points | Length of Line Examined | Triple Points per 100 $\mu$m |
|---|---|---|---|
| As deposited | 64 | 1140 $\mu$m | 5.61 |
| Annealed | 32 | 1340 $\mu$m | 2.38 |

A more challenging example of the method of moments comes from an electron microscopy study of the microstructure of aluminum interconnect.[2] Grain boundary triple points are counted over two differently treated lengths of 1.3 μm wide semiconductor interconnect (Table III). The study seeks inference about parameters $\rho$ and $\lambda$, the mean number of grain boundary triple points per 100 μm of interconnect in the population of chips as deposited and in the population of chips annealed, respectively. The natural method of moments estimators are the sample means $X/11.4$ for $\rho$ and $Y/13.4$ for $\lambda$, where $X$ and $Y$ are the total number of grain boundary triple points in the 1140 μm and 1340 μm lengths, respectively. Note, this "averaging" is over a continuous variable (length) rather than over a discrete set of values such as the 16 in the electromigration example. This is common in problems dealing with time or space averages, but some extra thought is required to formulate the statistical model, which is done in Section 6.3.1. Under that model both of these method of moments estimators are minimum variance and unbiased.

In general, estimators obtained via the method of moments are easy to compute and have the consistency property. Often they are also unbiased and minimum variance. If no minimum variance unbiased estimator is known and the *maximum likelihood estimator* (see Section 6.2.3.3) is too difficult to obtain, there is no reason not to use the method of moments estimator.

**6.2.3.2 Method of Least Squares.** Like the method of moments, the *method of least squares* is a conceptually simple way of developing an estimator with good properties and predates by a century the systemization of mathematical statistics (with its comparative study of the properties of estimators) that began early in the twentieth century. Though least squares is addressed in full in Chapter 9, we illustrate its rationale and usefulness in statistical inference with an application to *simple linear regression* (SLR).

---

[2] Digital Equipment Corp. internal report by B. Miner and S. Bill.

Many studies of electromigration postulate that the median time to failure equals $cj^{-m}$, where $c$ is an unknown scaling constant, $j$ is the current density in the interconnect, and $m$ is the current density exponent. By a log transformation we have equivalently that the log median failure time equals $\beta_0 + \beta_1 x$, where $x = \log(j)$ and $\beta_1$ is the negative of the current density exponent. In a *Monte Carlo* experiment (see Chapters 5 and 15) we generate 30 independent simulated electromigration log failure times at each of 3 known current densities, plotted in Figure 3. Thus we have $n = 90$ pairs $(x_i, y_i)$, where $y_i$ is a log failure time at log current density $x_i$, with which to estimate the unknown parameters $\beta_0$ and $\beta_1$ in the SLR relating current density to failure time. (Of course, unlike in a physical experiment, we know the values of $\beta_0$ and $\beta_1$ in our Monte Carlo experiment.)

The method of least squares seeks estimators $\hat{\beta}_0$ and $\hat{\beta}_1$ that minimize the *sum of squares for error:*

$$\text{SSE} = \sum_{i=1}^{n} (Y_i - \hat{\beta}_0 - \hat{\beta}_1 x_1)^2 \tag{6.12}$$

(Here we are using a common nomenclature rule that the estimator of a parameter $\theta$ is denoted $\hat{\theta}$.) The solution that minimizes *SSE* (found through elementary calculus) is

$$\hat{\beta}_1 = \frac{\sum_{i=1}^{n} (x_i - \bar{x})(Y_i - \bar{Y})}{\sum_{i=1}^{n} (x_i - \bar{x})^2} \tag{6.13}$$

$$\hat{\beta}_0 = \bar{Y} - \hat{\beta}_1 \bar{x} \tag{6.14}$$

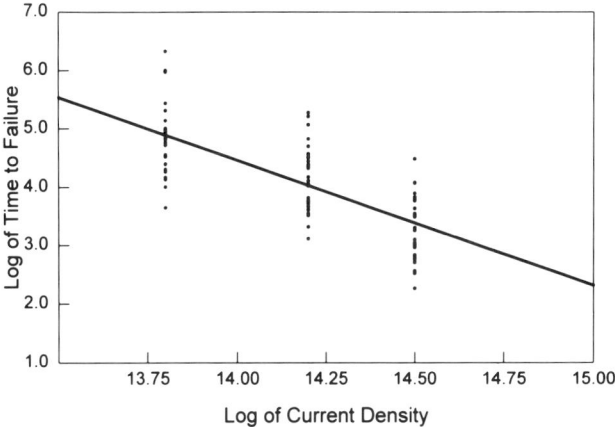

FIG. 3. Linear regression on simulated electromigration log failure times at three current densities.

As expected, $\hat{\beta}_0$ and $\hat{\beta}_1$ are statistics, that is, functions of the observable $(x_1, Y_1), (x_2, Y_2), \ldots, (x_n, Y_n)$, but not of the unknown parameters $\beta_0$ and $\beta_1$.

For the 90 pairs $(x_i, y_i)$ realized in the Monte Carlo experiment, the values of *SSE*, the slope $\hat{\beta}_1$, and intercept $\hat{\beta}_0$ calculated with Eqs. (6.12), (6.13), and (6.14) are 27.97, $-2.14$, and 34.5, respectively. Inserting these realized values of the statistics $\hat{\beta}_1$ and $\hat{\beta}_0$ into $\hat{\beta}_0 + \hat{\beta}_1 x$ yields the fitted regression line drawn in Figure 3.

The method of least squares optimizes estimation by minimizing the sum of the vertical distances (error) of the sample pairs from the fitted regression line. Note that this method does not rely on a statistical model for the pairs. However, least squares estimation in SLR is often carried out under the *normal error model:*

$$Y_i = \beta_0 + \beta_1 x_i + \epsilon_i \qquad (6.15)$$

$$\epsilon_1, \epsilon_2, \ldots, \epsilon_n \text{ are independent } N(0, \sigma) \qquad (6.16)$$

under which $SSE/(n-2)$, $\hat{\beta}_1$, and $\hat{\beta}_0$ are the minimum variance unbiased estimators of $\sigma^2$, $\beta_1$, and $\beta_0$, respectively.

### 6.2.3.3 Method of Maximum Likelihood.
Loosely speaking, *maximum likelihood* estimation for a parameter $\theta$ sets the estimator $T$ equal to the value of $\theta$ that maximizes the *a priori* probability of the realized values of the observable. In our formal statistical model of Section 6.1.4, the *likelihood function L* to be maximized is simply the joint probability density function $f(\mathbf{x}; \Theta)$ (or joint probability mass function $p(\mathbf{x}; \Theta)$) considered as a function $\Theta$ while $\mathbf{x}$ is *fixed* at the realized values of the observable.

For example, the likelihood $L$ under the statistical model in Eqs. (6.3) through (6.6) of our getting the 16 values in Table I is given by

$$L = \frac{\exp\left(-\frac{(3.29 - \mu)^2}{2\sigma^2}\right)}{\sqrt{2\pi}\sigma} \cdot \frac{\exp\left(-\frac{(3.47 - \mu)^2}{2\sigma^2}\right)}{\sqrt{2\pi}\sigma} \cdot \ldots \cdot \frac{\exp\left(-\frac{(4.94 - \mu)^2}{2\sigma^2}\right)}{\sqrt{2\pi}\sigma} \qquad (6.17)$$

which, by simple calculus, is maximum at $\mu = \bar{x} = 3.97$, regardless of the (unknown) value of $\sigma$. If, further, we search for the value of $\sigma$ that maximizes $L$, we get $\sigma = s_{mle} = .41$. In general, $\bar{X}$ and $S_{mle} = (\sqrt{(n-1)/n})S$ are the *maximum likelihood estimators* of $\mu$ and $\sigma$ in a $N(\mu, \sigma)$ statistical model.

Besides being the fruit of the compelling *likelihood principle* of mathematical statistics, a maximum likelihood estimator has (under technical regularity conditions) the consistency property and large-sample versions of the minimum variance and unbiasedness properties.

### 6.2.4 Looking Further on Point Estimation

See Chapters 3 and 4 of [3] for a more complete and mathematically rigorous (yet readable) coverage of methods of developing and comparing estimators for classical inference in parametric models. Included in [3] is a discussion of the *large sample theory* of estimators, of which this chapter's mention of consistency is only the slightest of beginnings.

Chapter 8 gives a much more complete introduction to the method of maximum likelihood and insight into the origin of computational problems using it, along with detailed application to several different statistical models.

A more mathematically focused source for estimators, with references to the original literature, is [11], [12], and [13]. This series, indexed by probability distribution, gives formulas, algorithms, and statistical tables for computing a wide variety of estimators, including method of moments, minimum variance unbiased, maximum likelihood, and best linear unbiased.

## 6.3 Interval Estimation

This section addresses how *precision* in statistical estimation can be obtained. Confidence intervals and other interval and "region" estimation methods based on the confidence interval concept provide a special type of initial precision for inference under uncertainty.

### 6.3.1 Single Parameter Confidence Intervals

It is common in science and engineering to convey the precision for a sample average of repeated measurements as $\bar{x} \pm s/\sqrt{n}$. An important question is: "If the sample is subject to randomness and $\bar{x}$ is meant to estimate the population mean $\mu$, is $\mu$ *always* contained in such an interval?" The answer is, "No." However, classical inference offers the concept of a *confidence interval* that conveys a type of precision in estimation, but one qualified by the level of uncertainty inherent in the estimation.

To illustrate how a simple confidence interval is built, we start with the estimation of $\mu$ based on $\bar{X}$ when the population is $N(\mu, \sigma)$. We know from the tables of the standard normal distribution that the event $-1.96 < (\bar{X} - \mu)/(\sigma/\sqrt{n}) < 1.96$ has probability .95. Simple rearrangement of the inequalities defining this event leads us to

$$P\left(\overline{X} - 1.96\frac{\sigma}{\sqrt{n}} < \mu < \overline{X} + 1.96\frac{\sigma}{\sqrt{n}}\right) = .95 \qquad (6.18)$$

The interval $\overline{X} \pm 1.96\sigma/\sqrt{n}$ defines a 95% *confidence* interval for $\mu$.

The *general* form of a $1 - \alpha$ confidence interval for a single parameter $\theta$ is $(\mathcal{L}, \mathcal{U})$, where $\mathcal{L}$ and $\mathcal{U}$ are statistics specially chosen to allow a statement of $1 - \alpha$ probability for the event $\{\mathcal{L} < \theta < \mathcal{U}\}$ (analogous to that in Eq. (6.18)) under the chosen parametric statistical model. The difference $\mathcal{U} - \mathcal{L}$, the width of the confidence interval, conveys the statistical precision of the estimation of $\theta$ in a way consistent with classical statistical inference.

In Eq. (6.18), $\sigma$ is a *nuisance* parameter, since its value is needed for the confidence interval for $\mu$, but $\sigma$ is not the object of the inference. Often, a nuisance parameter has an estimator that can be substituted for it without changing the basic form of the confidence interval. For the case of estimating a normal mean

$$P\left(\overline{X} - t_{\alpha/2,n-1}\frac{S}{\sqrt{n}} < \mu < \overline{X} + t_{\alpha/2,n}\frac{S}{\sqrt{n}}\right) = 1 - \alpha \qquad (6.19)$$

defines a confidence interval for $\mu$ using the sample standard deviation $S$ to estimate $\sigma$ (and requiring use of the Student $t$-distribution for the quantity $(\overline{X} - \mu)/(S/\sqrt{n})$.) For example, the electromigration date of Table I, the 95% confidence interval for $\mu$ is $3.97 \pm (2.13 \cdot .42)/\sqrt{16}$, equal to (3.75, 4.19).

Our success in manipulating the probability statement about the quantity $(\overline{X} - \mu)/(\sigma/\sqrt{n})$ into a confidence interval follows from the fact that the distribution of $(\overline{X} - \mu)/(\sigma/\sqrt{n})$ is completely specified (either by assigning a value to $\sigma$ or replacing it with its estimator $S$), even though the quantity depends on the unknown parameter $\mu$. In deriving confidence intervals it is common to work with such quantities (called *pivots*), often having the form $(T - \theta)/\sqrt{\hat{V}(T)}$, where $\sqrt{\hat{V}(T)}$ denotes an estimator of the standard deviation of $T$ (also called the *standard error* of $T$).

We turn now from our confidence interval introduction to an example for which no pivot exists—but first the statistical model needs to be developed.

A statistical model for the grain boundary triple point "annealed" results in Table III is had by first thinking of the lines as divided into squares 1.3 $\mu$m on a side. Since over a thousand such squares were examined for each treatment and for each square the chance of it containing a triple point is small (apparently $\leq .01$), we invoke the Poisson probability model of Chapter 1 as the limit of these *binomial trials.* Independence of the trials is required, which here corresponds to assuming that the presence of a grain boundary in any one square does not affect the chance of a grain boundary being present in any other collection of squares.[3] The resulting statistical model is $\mathbf{Y} = Y$ and $\Theta = 13.4\lambda$ with $p(y; \Theta) = \exp(-13.4\lambda)[13.4\lambda]^y/y!$.

The goal of this discussion is to develop a confidence interval for $\lambda$, the frequency of grain boundary triple points under the annealed treatment. First, we seek a confidence interval based on the Poisson distribution of $Y$. Mimicking the probability statement of Eq. (6.18), we seek statistics $\mathcal{L}$ and $\mathcal{U}$, functions of $Y$, for which

$$P(\mathcal{L} < 13.4\lambda < \mathcal{U}) = 1 - \alpha \quad (6.20)$$

Fortunately, a special relationship between the Poisson and $\chi^2$ distributions makes Eq. (6.20) true when

$$\mathcal{L} = \frac{1}{2}\chi^2_{\alpha/2, 2Y} \quad (6.21)$$

$$\mathcal{U} = \frac{1}{2}\chi^2_{1-\alpha/2, 2(Y+1)} \quad (6.22)$$

From Table III the realized value of $Y$ is 32. To apply Eq. (6.21) and (6.22) to yield a 90% confidence interval on $\lambda$ first requires linear interpolation among the $\chi^2$ distribution percentiles in Table 3 of the Appendix to get

$$\chi_{.05,64} = 40.48 + \left(\frac{64-60}{70-60}\right)(48.76 - 40.48) = 43.72 \quad (6.23)$$

$$\chi_{.05,66} = 79.08 + \left(\frac{66-60}{70-60}\right)(95.53 - 79.08) + 88.95 \quad (6.24)$$

Then the 90% confidence interval for $\lambda$ is $(\frac{1}{2}\cdot 43.72/13.4, \frac{1}{2}\cdot 88.95/13.4) = (1.65, 3.31)$, in units of triple points per 100 μm.

Using the same development of a Poisson statistical model, but using numerical computation of the $\chi^2$ quantiles needed, we get (4.20, 6.91) for the 90% confidence interval for ρ, the frequency of grain boundary triple points *as deposited*.

### 6.3.2 Interpreting Confidence Intervals

The advantage of a confidence interval for $\theta$ over a point estimate of $\theta$ is that the interval conveys an *a priori* (*before* the sample is realized) precision in the estimation of $\theta$, though a precision qualified by the interval's

---

[3]The "fit" of the Poisson model for the triple point counts could be assessed via the technique of Chapter 7 applied to a detailed tally of the frequency of triple points in equal-length subsections of the total length examined.

confidence level $1 - \alpha$. For a fixed confidence level, a narrower confidence interval conveys a greater precision for $\theta$ than a wider interval. However, *no* level of *a posteriori* (*after* the sample is realized) precision is imputed to a confidence interval in classical inference. In particular, the 95% confidence interval (3.75, 4.19) from the electromigration data of Table I *either does or does not* contain the unknown mean $\mu$. This is a subtle, and often confusing, point in the interpretation of classical statistical intervals.

To further clarify the interpretation of a confidence interval, Table IV reports the results of a Monte Carlo simulation of repeatedly drawing independent samples of size 16 from a $N(3.9, 0.5)$ population and applying the confidence interval determined by Eq. (6.19) for $\alpha = .05$. Since we know the value of $\mu$ on which the simulation is based, we know which of the 1000 intervals actually contain $\mu$. As expected, the 95% confidence interval *formula* applied repeatedly yields about 950 "correct" intervals. Conversely, there is no way to judge that an interval is correct without knowing the parameter $\mu$ (which if known makes the confidence interval unneeded). Thus "confidence" must come *a priori* from the properties of the confidence interval formula, and not *a posteriori* from the realized confidence interval that our sample yields.

That the interpretation of a confidence interval relies on repeated trials, usually difficult or even conceptually impossible to pursue in practice, is a major source of criticism for classical inference. Alternatives are pointed out in Section 6.5., but confidence intervals are firmly embedded in statistical practice (and software). On the other hand, it is common to loosely (but in truth wrongly) interpret a 95% realized confidence interval, like our previous (3.75, 4.19) interval, as itself having a 95% chance of being correct.

TABLE IV.  Confidence Intervals for 1000 Monte Carlo Samples from a $N(3.9, .5)$ Population

| Trial # | $\bar{x}$ | $s$ | 95% Confidence Interval | Contains $\mu$? |
|---|---|---|---|---|
| *1* | 3.85 | .48 | ( _____ $\bar{x}$ _____ ) | Yes |
| *2* | 3.67 | .34 | ( ___ $\bar{x}$ ___ ) | No |
| *3* | 3.88 | .52 | ( _____ $\bar{x}$ _____ ) | Yes |
| . | | | | . |
| . | | | | . |
| *999* | 3.88 | .58 | ( _____ $\bar{x}$ _____ ) | Yes |
| *1000* | 3.68 | .57 | ( _____ $\bar{x}$ _____ ) | Yes |
| Summary: | | | $\mu = 3.90$ | 946 Yes<br>54 No |

### 6.3.3 Some Useful Options in Applying Confidence Intervals

Since a confidence interval ($\mathcal{L}$, $\mathcal{U}$) for a parameter $\theta$ results from a probability statement about inequalities, any transformation that preserves the inequalities $\mathcal{L} < \theta < \mathcal{U}$ is also a confidence interval on the transformed $\theta$ at the same confidence level.

For example, since the transformation $\exp(x)$ is a monotone function, we can claim that $(\exp(3.75), \exp(4.19)) \doteq (43 \text{ hrs}, 66 \text{ hrs})$ is itself a 95% confidence interval for $\exp(\mu)$, the lognormal median time to failure from electromigration for the chip population under the same acceleration.

Though all confidence intervals presented in this chapter have finite lower and upper bounds, a *one-sided* interval is always available by taking either $\mathcal{L}$ or $\mathcal{U}$ to be infinite. This is desirable when it is not useful to produce inference about how small ($\mathcal{L} = -\infty$) or large ($\mathcal{U} = \infty$) the parameter is.

For example, in reliability testing for electromigration, interest may rest only in the least possible value of the median time to failure, since there is no consequence when the median parameter is larger than the reliability standard. The advantage is that, for a given confidence level, $\mathcal{U}$ is smaller (or $\mathcal{L}$ is larger) in a one-sided interval than for a two-sided interval.

Since a confidence interval has an *a priori* precision, the sample size needed for a given precision (at a specified confidence level) can, in principle, be determined.

For example, the confidence interval for $\exp(\mu)$ derived at the beginning of this section has a relative width (precision) of $66/43 \doteq 1.53$ or about 53%. Suppose a 95% confidence interval with 20% relative precision is desired. This means $\exp(\mathcal{U}) = 1.2 \exp(\mathcal{L})$, or equivalently $\mathcal{U} - \mathcal{L} \doteq .182$. From Equation 6.18 it is clear that $\mathcal{U} - \mathcal{L}$ must also equal $2(1.96)\sigma/\sqrt{n}$. Solving these two constraints for $n$ gives $n = (2\sigma 1.96/.182)^2$, which for $\sigma = .5$ yields a sample size of $n = 116$. Several points need to be made concerning this discussion.

The general sample size formula for a normal mean $(1 - \alpha)$-level confidence interval of *(a priori)* width $W$ is $n = (2\sigma z_{\alpha/2}/W)^2$, where $n$ is rounded up to an integer.

A *planning value* for the nuisance parameter $\sigma$ must be specified to arrive at a unique value for $n$. In practice, the value of $\sigma$ used is often that from a similar previous experiment or a conservative guess based on prior judgment about the population being sampled.

Though a planning value of $\sigma$ may be used, the confidence interval should be based on the estimator $S$ (as in Eq. (6.19)). Thus, the width of the confidence interval is a random variable and on average will be wider than the $W$ from which the sample size is computed. More refined techniques

TABLE V. Statistics from Die Probe Travel Study

| Pin Type | n | $\bar{x}$ | s |
|---|---|---|---|
| Short | 8177 | 0.897 | 0.854 |
| Long | 8180 | 0.870 | 0.610 |

exist that offer some control over the variation in width, at the price of a larger sample size. See Section 8.3.4 ff. of [8].

Though this example deals only with the normal mean case, formulas (or tables and graphs) are available for choosing sample sizes for confidence intervals under many other statistical models. In some cases however (for example a confidence interval for a Poisson mean), a planning value for the parameter of interest (not just nuisance parameters) is required.

### 6.3.4 Comparing Two Independent Samples

Often there is interest in comparing two populations by performing inference on the difference or ratio of their individual parameters. This difference or ratio is itself a single parameter, and hence the methodology developed so far in Section 6.3.1 can be applied. For example, Table V shows sample statistics on measured contact resistance (in ohms) from a study[4] to explain variations in two types of semiconductor die probes, short pin and long pin, with population means $\mu_1$ and $\mu_2$, respectively. Our interest here is in the single *derived* parameter $\mu_1 - \mu_2$ that describes the difference in contact resistance between the two pin types.

Development of a confidence interval proceeds along the lines that the difference of sample moments $\bar{X} - \bar{Y}$ is the minimum variance unbiased estimator of $\mu_1 - \mu_2$. For large sample sizes a combination of the Law of Large Numbers and the Central Limit Theorem (see Chapter 1) ensures that the pivot

$$Z = \frac{\bar{X} - \bar{Y} - (\mu_1 - \mu_2)}{\sqrt{\frac{s_1^2}{m} + \frac{s_2^2}{n}}} \qquad (6.25)$$

has approximately a $N(0, 1)$ distribution. Equation (6.25) can then be unfolded (analogous to the single sample case of Eq. (6.18)) to yield

---

[4]Digital Equipment Corp. internal report by R. Hart, G. Wozniak, J. Ramírez, and B. Cantell.

$$P\left(\overline{X} - \overline{Y} - z_{\alpha/2} \sqrt{\frac{S_1^2}{m} + \frac{S_2^2}{n}} < \mu_1 - \mu_2 < \overline{X} - \overline{Y}\right.$$
$$\left. + z_{\alpha/2} \sqrt{\frac{S_1^2}{m} + \frac{S_2^2}{n}}\right) = 1 - \alpha \tag{6.26}$$

For the die probe study a 90% confidence interval for $\mu_1 - \mu_2$ based on Eq. (6.26) is then

$$(.897 - .870) \pm 1.645 \sqrt{\frac{(.854)^2}{8177} + \frac{(.610)^2}{8180}} = (.008, .046) \tag{6.27}$$

The triple point frequency experiment of Section 6.2.3.1 offers another two-sample inference problem—for the derived parameter $\lambda/\rho$, equal to the relative reduction in triple point frequency due to annealing. (The ratio $\lambda/\rho$ seems a more natural comparison than the difference $\rho - \lambda$ since triple point frequency has no reference value in this experiment.)

A confidence interval for this ratio is difficult to find in standard textbooks. In such a situation a more mathematical reference like [11] can be consulted, where on page 98 is found an approximate confidence interval for the ratio of Poisson means $13.4\lambda/11.4\rho$ based on the $F$-distribution. (Moreover, if [11] is unavailable, the next source is the original journal article, in Russian!) The point is that what seems a relatively simple inference problem can sometimes require a complicated, arcane solution. For this reason, an alternate, "quick and dirty" approach based on confidence rectangles is developed in Section 6.3.6.2 for inference about $\lambda/\rho$.

### 6.3.5 Prediction Intervals

Besides intervals for population parameters it is possible to utilize the confidence interval concept to produce an interval that, with a given confidence, will contain *a new observation from the population*. These *prediction intervals* are frequently used in SLR to lend precision to prediction of a future observation at a value $x^*$ of the independent variable between or beyond those values at which observations were collected.

For example, based on the fitted log failure time versus log current density regression line of Figure 3 we may wish to claim that a *particular* chip in the population, perhaps the one sold to Customer A, will have a certain lifetime at the particular current density $x^*$ that Customer A operates the chip.

For this case a $(1 - \alpha)$-level prediction interval for a future observation at $x^*$ is

$$\hat{\beta}_0 + \hat{\beta}_1 x^* \pm t_{\alpha/2, n-2} \sqrt{\frac{SSE}{n-2} \left(1 + \frac{1}{n} + \frac{n(x^* - \bar{x})^2}{n \sum x_i^2 - (\sum x_i)^2}\right)} \quad (6.28)$$

Note that the standard error term in Eq. (6.28) depends not only on $SSE$ but also on the value of $x^*$ along with those of $\bar{x} = 14.17$, $\sum x_i = 1275$, and $\sum x_i^2 = 18{,}069.9$. In particular, a 95% prediction interval for the log failure time at log current density $x^* = 10.80$ is

$$34.5 - (2.14 \cdot 10.80)$$

$$\pm 1.99 \sqrt{\frac{27.97}{88} \left(1 + \frac{1}{90} + \frac{90(10.80 - 14.17)^2}{(90 \cdot 18{,}069.9) - 1275^2}\right)} \quad (6.29)$$

$$= (9.60, 13.18)$$

By applying the exp transformation (just as legitimate for prediction intervals as confidence intervals), we find that (14,800 hrs, 530,000 hrs) is a 95% prediction interval on the lifetime of Customer A's chip.

There is an uncontrolled risk here from extrapolating well beyond the range of observation of the independent variable $x$, since no data are available to judge whether the linear model $Y = \beta_0 + \beta_1 x + \epsilon$ holds in the extrapolation region.

It should also be noted that the prediction interval of Eq. (6.28) is wider than the equivalent confidence interval on the mean log failure time $\beta_0 + \beta_1 x^*$, since both the sampling variability of the new observation at $x^*$ *and* the uncertainty of the parameter estimators $\hat{\beta}_0$ and $\hat{\beta}_1$ must be accounted for.

This prediction interval for Customer A's chip is quite wide, and several publications support a theory of electromigration that holds that the current density exponent $m$ equals 2. Consider now the prediction interval that results if we adopt this theory by *setting* $\beta_1 = -2$ rather than estimating it. By the method of least squares we minimize

$$SSE(\beta_0) = \sum_{i=1}^{n} [Y_i - \beta_0 - (-2)x_i]^2 \quad (6.30)$$

which (again by simple calculus) yields $\hat{\beta}_0 = \bar{Y} + 2\bar{x}$. The resulting prediction interval is

$$\hat{\beta}_0 - 2x^* \pm 1.99 \sqrt{\frac{SSE}{89} \left(1 + \frac{1}{90}\right)} \quad (6.31)$$

The same Monte Carlo data for these revised estimators, $\hat{\beta}_0$ and $SSE(\hat{\beta}_0)$, yields realized values of 32.44 and 28.12. Thus the revised (under the

$\beta_0 = -2$ assumption) 95% prediction interval is $32.44 - (2 \cdot 10.80) \pm 1.99 \sqrt{(28.12/89)(91/90)}$, or (16,600 hrs, 157,000 hrs) after applying the $\exp(x)$ transformation. Note how much narrower (more "precise" in the special confidence interval way) the latter prediction interval is—the benefit of replacing our uncertainty about $\beta_1$ with an *assumption* of $\beta_1 = -2$.

### 6.3.6 Statistical Intervals for Two or More Parameters

When inference is needed on two or more parameters based on the same set of observations, methods of *simultaneous* inference should be used to optimize the precision and control the confidence level of the inference.

#### 6.3.6.1 Confidence Regions.
The concept of a confidence *interval* can be generalized for carrying out simultaneous inference on two or more parameters. The basis is the same—a probability statement that relates statistics (which can be calculated from the observable) to the (unknown) parameters of interest is developed, and then manipulated to describe a *region* in the *parameter space*.

For example, the statistics $\hat{\beta}_0$, $\hat{\beta}_1$, and $SSE$ combine with the parameters of interest $\beta_0$ and $\beta_1$ in the probability statement

$$P\left[\frac{n(\hat{\beta}_0 - \beta_0)^2 + 2(\Sigma x_i)(\hat{\beta}_0 - \beta_0)(\hat{\beta}_1 - \beta_1) + (\Sigma x_i^2)(\hat{\beta}_1 - \beta_1)^2}{2SSE/(n-2)} \right.$$

$$\left. < F_{1-\alpha, 2, n-2} \right] = 1 - \alpha \qquad (6.32)$$

which defines a *random* region in the plane (parameter space) of all possible pairs $\langle \beta_0, \beta_1 \rangle$. Replacing $\hat{\beta}_0$, $\hat{\beta}_1$, and $SSE$ in Eq. (6.32) with the realized values 34.5, 2.14, and 27.97 from the Monte Carlo simulation (along with $\Sigma x_i = 1275$, $\Sigma x_i^2 = 18{,}069.9$, and $F_{.99, 2, 88} = 4.85$) yields the elliptical region

$$141.6(34.5 - \beta_0)^2 + 2550(34.5 - \beta_0)(-2.14 - \beta_1)$$
$$+ (-2.14 - \beta_1)^2 < 4.85 \qquad (6.33)$$

centered at $\langle 34.5, -2.14 \rangle$. Figure 4 shows a plot of this realized 99% elliptical *confidence region* for $\beta_0$ and $\beta_1$ defined by inequality (6.33) along with their point estimates.

Generalizations of Eq. (6.32) for generating joint confidence (ellipsoidal) regions for the (three or more) parameters in linear models beyond SLR are derivable via the theory of quadratic forms for the normal distribution. However, joint confidence regions not based on the normal error model are

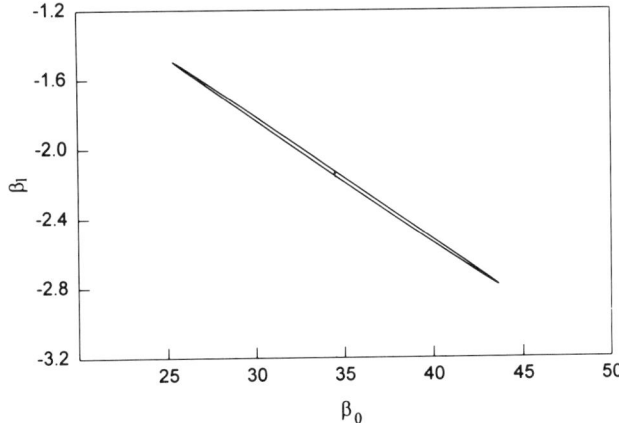

FIG. 4. A 99% confidence ellipse for electromigration linear regression parameters $\beta_0$ and $\beta_1$.

uncommon, due the difficulty of deriving the analog of Eq. (6.32) under such models.

**6.3.6.2 Confidence Rectangles.** Forming a confidence region for two parameters generally requires knowledge of the joint probability distribution of the two estimators (as in Eq. (6.32)), which may be complex if the two estimators are not statistically independent. An alternate, simpler way to carry out simultaneous inference on two parameters is via *joining* individual confidence intervals on the two parameters to form a *confidence rectangle*.

Since each (single-parameter) confidence interval is based on a probability statement about an random event, probability calculus can be employed to derive the confidence rectangle for the pair of parameters. For example, for the grain boundary triple point frequency inference (Section 6.3.1), the event

$$A_1 = \left\{ \frac{\mathcal{L}_1}{13.4} < \lambda < \frac{\mathcal{U}_1}{13.4} \right\} \qquad (6.34)$$

will have $P(A_1) = .90$ based on Eq. (6.21) and (6.22) for $\alpha = .10$. Analogously, the event $A_2 = \{\mathcal{L}_2/11.4 < \rho < \mathcal{U}_2/11.4\}$ will have $P(A_2) = .90$, also. Assuming $A_1$ and $A_2$ are independent events (based on our knowledge that the *annealed* line was processed and examined independent of the *as deposited* line), we have $P(A_1 \cap A_2) = .90 * .90 = .81$. Thus the combined confidence intervals (from Section 6.3.1)

$$\{1.65 < \lambda < 3.37 \quad \text{and} \quad 4.20 < \rho < 6.91\} \tag{6.35}$$

form a rectangular confidence region of confidence level 81% in the plane of all pairs $\langle \lambda, \rho \rangle$.

Moreover, we are now in a position to solve the two-sample inference problem for $\lambda/\rho$ (the relative change in triple point frequency due to annealing) from Section 6.3.4. The rectangular region defined in Statement 6.35 is a *subset* of the region

$$\left\{ \frac{1.65}{6.91} < \frac{\lambda}{\rho} < \frac{3.37}{4.20} \right\} \tag{6.36}$$

obtained by combining the two inequalities. Hence the confidence interval (.24, .80) in Eq. (6.36) has confidence level of at least 81% for $\lambda/\rho$. Therefore we are at least 81% confident that annealing *reduces* the triple point frequency by at least 20% (that is, $\lambda < .80\rho$). Figure 5 compares the confidence rectangle from statement (6.35) (darkly shaded) to the confidence region on the ratio $\lambda/\rho$ from statement (6.36) (lightly shaded).

In many cases independence cannot be assumed for the events that define the formula for the confidence rectangle—either because the samples are not independently obtained or because the individual parameter estimators are correlated (as in Eq. (6.32)). Then the *Bonferroni* method can be applied.

The combination of $k$ (not necessarily independent) confidence intervals

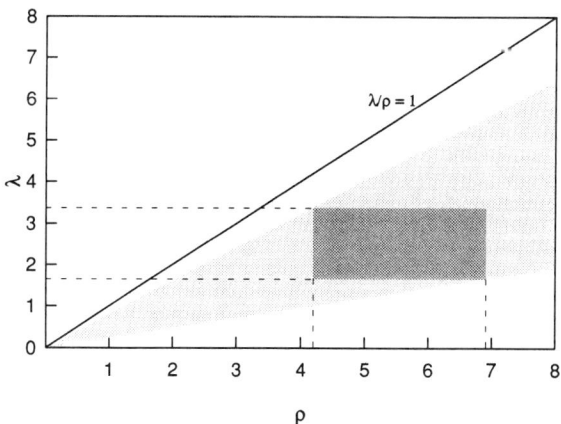

FIG. 5. An 81% confidence rectangle for grain boundary triple point frequencies $\lambda$ and $\rho$.

to form a $k$-dimensional confidence rectangle (analogous to statement (6.35)) has a confidence level governed by the Bonferroni inequality:

$$P\left(\bigcap_{j=1}^{k} A_j\right) \geq 1 - \sum_{j=1}^{k} \alpha_j \qquad (6.37)$$

where $A_j$ is the event that $j$th confidence interval (having confidence coefficient $\alpha_j$) contains the $j$th parameter (analogous to statement (6.34)). Note that the intervals combined need not have equal confidence coefficients. Also, if the confidence intervals are mutually independent (which, however, may be difficult to confirm or simply not worth the effort), then there is equality in statement (6.37). Thus the Bonferroni method is conservative (but only slightly so in our triple point frequencies example—80% confidence by Bonferroni versus 81% confidence by independence).

Finally, the confidence level for a joint confidence interval can be quite low when several intervals are joined. For example, joining $k = 10$ confidence intervals by the Bonferroni method, each with confidence level 95%, yields a joint confidence level bounded by only 50%.

A further generalization of confidence intervals applies to a *function* of the unknown parameters and some other known, continuous variable. One example already seen of such a function is the SLR line $\beta_0 + \beta_1 x$. The *family* of confidence intervals over the continuum of values of the known variable (such as $x$) is called a *confidence band*. A confidence *band* allows efficient inference over *all* values of the known variable *simultaneously*. In contrast, applying the Bonferroni method to more than few values of the function (corresponding to different values of the known variable) would yield a confidence level degraded to a useless level.

In the SLR example, with a confidence band over all values of log current density $x$ the analyst can make a future inference for reliability at any group of values of current density—the values of interest do not have be known at the time the data is analyzed. A $(1 - \alpha)$-level *hyperbolic* confidence *band* for the function $\beta_0 + \beta_1 x$ is

$$\hat{\beta}_0 + \hat{\beta}_1 x \pm \sqrt{\frac{2 \cdot \text{SSE} \cdot F_{\alpha/2,2,n-2}}{n-2} \left(\frac{1}{n} + \frac{n(x - \bar{x})^2}{n \sum x_i^2 - (\sum x_i)^2}\right)} \qquad (6.38)$$

The values realized in the Monte Carlo simulation of electromigration log failure times yield the 90% confidence band graphed in Figure 6.

Confidence bands are also available for parametric inference on a *distribution function*. For example, for the statistical model of Eq. (6.3) through

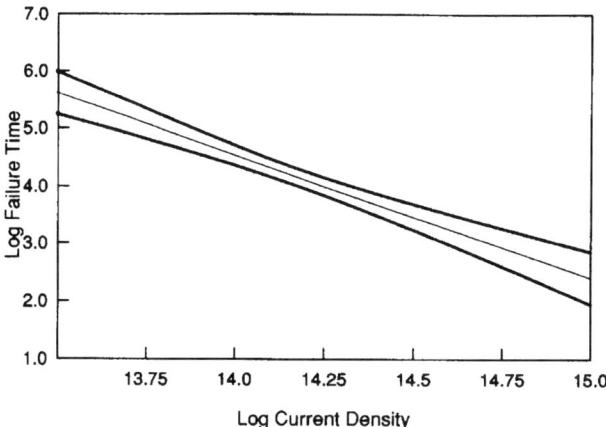

FIG. 6. A 90% confidence band for the electromigration linear regression line.

(6.6) the distribution of log time to failure is $\Phi[(x - \mu)/\sigma]$. Based on the the estimators $\bar{X}$ and $S$, [5] gives a formula (too complex to reproduce here) for a $(1 - \alpha)$-level confidence band for $\Phi[(x - \mu)/\sigma]$. Figure 7 plots this band for the realized values $\bar{x} = 3.97$ and $s = .42$ from the data of Table I, along with the estimated distribution function $\Phi[(x - 3.97)/.42]$.

Note carefully the distinction between these two confidence bands. For the SLR example the band is for the *mean* log failure time as a function

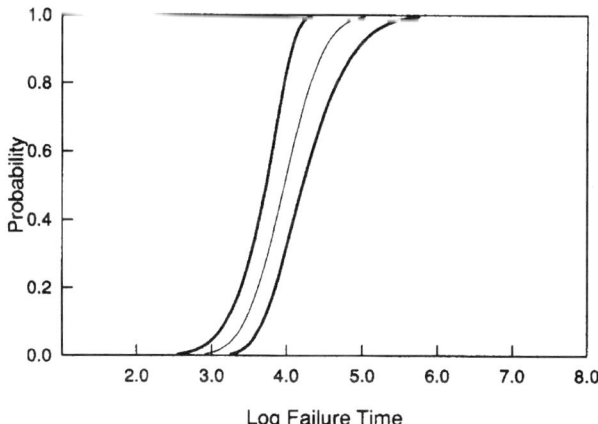

FIG. 7. A 90% confidence band for the normal distribution of log failure time.

the log current density. In the normal distribution model for log failure time example the band is for the *probability of failure* as a function of the log failure time. However, both 90% bands have (conceptually) the same interpretation as a 90% confidence interval—in the long-run over repeated trials for which each band is calculated by the formula given, 90% of the realized bands will completely contain the (unknown) function.

### 6.3.7 Looking Further on Confidence Intervals

Many statistics texts go into much greater detail than here about the generation of confidence intervals, often indexing the development by the type of statistical model (normal, binomial, one sample, two sample, etc.). See, for example, [7].

Typically, an elliptical confidence region (Section 6.3.6.1) has greater precision (is smaller) than the analogous Bonferroni rectangular region (Section 6.3.6.2) of the same confidence level, due partly to the latter's being a (conservatively) approximate region. However, the elliptical region is generally *not* a subset of the Bonferroni region. See Chapter 5 of [16] for a direct comparison of the two and more extensive discussion of confidence regions and the Bonferroni method for SLR.

An extensive guidebook that presents (typically without derivation) graphical and tabular methods for confidence intervals, along with sample size formulas and large-sample approximations, for many common statistical models is [8]. This reference also covers prediction intervals and tolerance intervals[5] for many models.

Though we mention the usefulness of pivots in deriving confidence intervals, pivots are difficult or even impossible to find in many statistical models. Special distributional relationships like the one underlying Eqs. (6.21) and (6.22) are uncommon and often require specialized knowledge to uncover. General methods of obtaining confidence intervals that work for most parametric statistical models include numerically inverting *likelihood ratio tests,* approximate intervals based on the large-sample normality of maximum likelihood estimators, and intervals derived from *profile likelihood functions* of the observable. See Chapter 8. But whatever method is employed to obtain the limits $\mathcal{L}$ and $\mathcal{U}$, the interpretation of the confidence interval ($\mathcal{L} < \theta < \mathcal{U}$) remains the same.

Finally, see [6] for one-sided confidence bands, analogous to the one-sided confidence intervals discussed in Section 6.3.3.

---

[5]Tolerance intervals bracket, with a given confidence, a fixed fraction of the values of a population.

## 6.4 Statistical Tests

As we stated in the Motivation section for this chapter, statistical inference enables scientific inference based on observations subject to statistical uncertainty. In this section we briefly introduce *statistical testing*. Again by analogy to the scientific method, statistical testing enables a process for *statistical falsification* of a (properly constructed) scientific hypothesis, within the limits imposed by the uncertainty in the observations.

### 6.4.1 Statistical Hypotheses and Decision Making

A *statistical hypothesis* is a formal claim about a state of nature structured within the framework of a statistical model. For example, one could claim that the median time to failure from (accelerated) electromigration of the chip population described in Section 6.1.4 is at least 60 hrs, perhaps to address Question I of Table II where 60 hrs represents a reliability requirement.

Within the framework of the statistical model for the chip population failure times (again, see Section 6.1.4), the reliability claim would be stated as

$$H_0: \mu \geq 4.1 \tag{6.39}$$

since $\log(60) \doteq 4.1$ and the log of the median of a lognormal distribution is the mean of the corresponding normal distribution.

The label $H_0$ arises from the term *null hypothesis* or more generally *working hypothesis*. Scientifically speaking, $H_0: \mu \geq 4.1$ is posed and allowed to stand until it can be falsified. The *statistical decision* that $H_0$ is false (and is rejected) must be based on a *decision procedure* that combines some function of the observable in the statistical model with the stipulations of the hypothesis—data meets theory.

More generally for a parameter $\theta$, a working hypothesis can be given as $H_0: \theta \in \Omega_0$, where $\Omega_0$ is a set of real numbers bounded by a $\theta_0$, yielding one of three cases: $\theta = \theta_0$, $\theta \leq \theta_0$, or (like our reliability requirement example above) $\theta \geq \theta_0$.

### 6.4.2 Significance Tests

A *significance test* is a decision procedure for possibly rejecting $H_0: \theta \in \Omega_0$ based on the observable $\mathbf{X}$. Associated with such a test is the *significance level* $\alpha$ given by

$$\alpha \stackrel{\text{def}}{=} P \text{ (reject } H_0 \text{ when } \theta = \theta_0) \geq P \text{ (reject } H_0 \text{ when } H_0 \text{ is true)} \quad (6.40)$$

The significance level $\alpha$ is set to control the risk of falsely rejecting $H_0$, but the actual false-rejection probability depends on the (unknown) value of the parameter $\theta$. Since the maximum (worst-case) probability of false rejection occurs at the boundary value $\theta_0$ of $\Omega_0$ (for reasonable decision procedures), $\alpha$ in statement (6.40) is specified at $\theta = \theta_0$, so as to be an upper bound on the false-rejection probability. The decision procedure is usually constructed so that $\alpha$ is small, say .10 or 0.05, to protect against (make unlikely) an erroneous conclusion, calculated supposing $H_0$ to be true.

In classical inference the significance level is the *a priori* probability that the significance test will lead to an erroneous decision. Similarly to confidence intervals (see Section 6.3.1.1), there is no *a posteriori* error protection imputed by the significance test—the decision based on the realized observations is *either correct or in error.*

### 6.4.2.1 Choosing a Test Statistic and Rejection Region.

The decision procedure for a statistical test requires some function of the observable (data) be compared to the hypothesis (theory). Typically this comparison is through a *test statistic* $T$ (a random variable as in Section 6.2.1) whose probability distribution is completely specified through $H_0$. Often the search for a test statistic begins with a good estimator of the parameter on which $H_0$ is defined.

In the statistical model for the electromigration example $\overline{X}$ is minimum variance and unbiased for $\mu$. The association of $\overline{X}$ to the $\mu_0 = 4.1$ of the hypothesis leads to the test statistic $T = (\overline{X} - \mu_0)/(S/\sqrt{n})$, which has a Student $t$-distribution with $n - 1$ degrees of freedom when the unknown $\mu$ is at the (worst case) boundary value $\mu_0$.

Now needed is a *rejection region* $R$ to specify *a priori* (before the sample is realized) the basis for deciding whether to reject $H_0$ while ensuring that the significance level $\alpha$ satisfies the relation

$$\alpha = P(T \in R \text{ when } \theta = \theta_0) \quad (6.41)$$

Intuitively for our reliability requirement example such a relationship $T \in R$ should be of the form $T < c$, for some *critical value c*. If the sample yields an $\bar{x}$ that is low relative to the minimum reliability requirement (hypothesis boundary) $\mu_0 = 4.1$ (that is, if $\bar{x} - 4.1$ is sufficiently negative), then the hypothesized conformance to the reliability requirement can be reasonably rejected.

If we take "$< -t_{\alpha,n-1}$" as defining the rejection region $R$, then we have constructed a general $\alpha$-level significance test for all hypotheses of the form $H_0$: $\mu \geq \mu_0$ in the statistical model of our reliability requirement example, because $P[(\overline{X} - \mu_0)/(S/\sqrt{n}) < -t_{\alpha,n-1}] = \alpha$. In particular, the rejection criteria for the reliability requirement example for an $\alpha = .05$ test of significance is not met since $(3.97 - 4.1)/(.42/\sqrt{16}) \doteq -1.18$ is *not* less than $-1.75 = t_{.05,15}$. Thus, the hypothesis $H_0$: $\mu \geq 4.1$ is *not* rejected.

**6.4.2.2 Relationship of Confidence Intervals to Significance Tests.** Confidence intervals have a *direct* relationship to significance tests. In general, the decision procedure

$$\textbf{Reject } H_0\text{: } \theta = \theta_0 \textbf{ If } (\mathcal{L}, \mathcal{U}) \text{ does not contain } \theta_0 \quad (6.42)$$

defines an $\alpha$-level significance test for $H_0$ when $(\mathcal{L}, \mathcal{U})$ is a $(1 - \alpha)$-level confidence interval for $\theta$.

For example, the confidence interval constructed in Section 6.3.6.2 for the relative change in triple point frequency due to annealing gives an $\alpha = .19$ significance level test of $H_0$: $\lambda/\rho = 1$ (the hypothesis of no change in triple point frequency) when implemented as a decision procedure of the form in statement (6.42).

**6.4.2.3 Use of *p*-Values.** After the test statistic in a significance test is realized (that is, *a posteriori*) and compared to the rejection region, the test decision is either correct or in error—there is no "probability" of correctness. However, within this limitation in the interpretation of classical statistical testing, *p-values* are a frequently used method for quantifying the "weight of evidence" against the working hypothesis.

The *p*-value *of a realization* of a statistical test is the significance level of the test for which the realized test statistic straddles the boundary between the rejection region and its complement (the "do not reject" region).

In Figure 8 the *p*-value .127 obtained for the reliability requirement example (from the fact that $P[(\overline{X} - 4.1)/(S/\sqrt{16}) < -1.18] = .127$) is plotted relative to the $\alpha = .05$ rejection region. Again, the *p*-value is a function of the realized sample value (based on $\overline{x} = 3.97$ and $s = .42$) but the *a priori* $\alpha = .05$ rejection region is not (and moreover cannot be). In this way *p*-values give flexibility in evaluating the results of the significance test beyond the Accept–Reject decision based on an arbitrary choice of $\alpha = .05$ made before the data are realized.

**6.4.2.4 Statistical versus Practical Significance.** The die probe travel study of Section 6.3.4 reveals the distinction between *statistical* significance and *practical* significance.

A significance test of $H_0$: $\mu_1 - \mu_2 = 0$ can be carried out when the pivot $Z$ of Eq. (6.25) is transformed into a test statistic by setting $\mu_1 -$

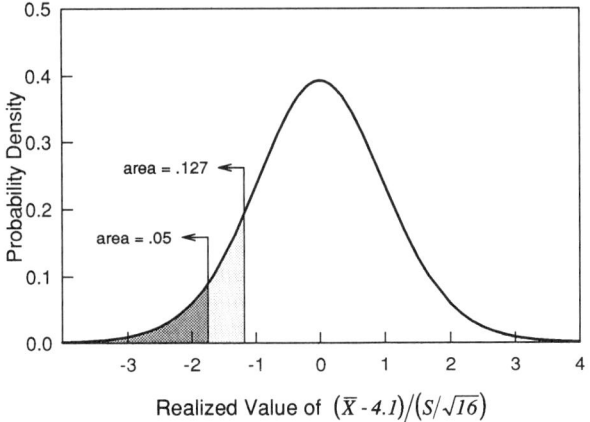

FIG. 8. Rejection region and *p*-value of significance test of $H_0$: $\mu \geq 4.1$.

$\mu_2 = 0$. Substituting the realized statistical values of Table V into Eq. (6.25) yields

$$z = \frac{(0.897 - 0.870) - 0}{\sqrt{\frac{(0.854)^2}{8177} + \frac{(0.610)^2}{8180}}} = 2.33 \tag{6.43}$$

The corresponding *p*-value of the test is <.01, and thus the tested difference is highly statistically significant. However, the observed difference is less than 3 parts in a 100, deemed practically irrelevant by the investigators. The statistical significance of the observed difference is due to the large sample sizes available, but this statistical test result has no *practical* significance. Thus the data are sufficient to establish convincingly that the mean resistances of short and long pin die probes *differ*, but the *size* of the difference is not large enough to be of scientific importance.

### 6.4.3 Looking Further on Statistical Tests

The results of the significance test in the reliability requirement example of Section 6.4.2 present an unanswered question: "What probability of error attaches to the decision *not to reject* $H_0$: $\mu \geq 4.1$ if in fact $H_0$ is *false?*" Such is clearly important because this probability is the *consumer's risk* of receiving a lot of chips that does not meet the reliability standard. Neymann–Pearson *hypothesis testing* in one sense completes significance testing by formalizing the "$H_0$ is false" case with an *alternative* hypothesis, usually of the form $H_a$: $\theta \in \Omega_a$, where $\Omega_0$ and $\Omega_a$ are disjoint. For our reliability requirement example, $H_a$: $\mu < 3.9$ might be appropriate (since $\exp(3.9) \doteq 50$ hrs, measurably less than the 60-hr requirement). The *a priori* probability of the error of *not* rejecting $H_0$ when $H_a$ is true (called $\beta$) can be evaluated and controlled. Neymann–Pearson hypothesis testing is presented in most statistics texts. See [7] for an elementary introduction with applications in the physical sciences. For a more mathematical introduction see [3].

In our reliability requirement example, the rejection region was of the form $T < c$ but any region giving the same $\alpha$-level is a candidate. However, regions of the form $T < c$ or $T > c$ are frequently "best" based on formal criteria explained in references on the optimality theory of statistical testing. A standard reference is the classic [14], though [3] is more readable by nonstatisticians.

The *likelihood ratio principle* is a powerful approach to constructing optimal hypothesis tests based on the ratio of the two values of the likelihood function of the statistical model (see Section 6.2.3.3) when it is maximized separately over $\Omega_0$ and $\Omega_a$. See Section 6.4 of [3] for an introduction.

Another important class of statistical tests, especially for analysis of discrete data, is based on Pearson's $\chi^2$ (chi-square). These tests are often employed as "goodness of fit" tests to answer Question V (from Table II): "Does this model fit the data?" The (meta-) statistical model invoked to structure such a goodness-of-fit test is a often a nonparametric one for which the parametric testing approaches introduced here do not apply. See Chapter 8 of [3] or Chapter 14 of [7].

Finally, a standard criticism of classical statistical testing is that there is no formal consideration of the *consequences* of an erroneous decision. We see from the reliability standard example that the consequences of the two types of error might differ greatly in impact. *Decision theoretic* methods generalize statistical testing to include consequence by employing the concept of *utility*. See [1] for an introduction.

## 6.5 Beyond Basic Statistical Inference

The goals of this chapter were to only briefly introduce and illustrate some of the methods available for classical statistical inference. A more comprehensive (and classic) introductory text is [18]. More up to date and focused on the physical sciences is [7].

All of the methods and examples introduced in this chapter are, for simplicity, based on having (or more generally, modeling for) *complete* samples. Had the electromigration testing of Example 1 ended at 75 hrs,

$x_{15}$ and $x_{16}$ would have both been recorded as $\log(75) \doteq 4.32$, and the sample of 16 log failure times would be termed *censored*. Our statistical model would have to incorporate the fixed 75 hr ending point for the accelerated testing since the resulting censoring changes the probability distribution $P_\theta$ of the observable. Chapter 8 considers classical statistical inference for censored samples.

Another complication not addressed in this chapter is that of *outliers*—observations seemingly disparate from the body of the sample and often identified by descriptive statistics and graphical methods applied to the sample before settling on the statistical model. (For example, if $x_1$ in Table I equaled 1.95 instead of 3.29, it would be termed an outlier.) Statistical inference methods for formally identifying outliers (and thus excluding them from the sample with confidence) are found in [2].

A good estimator for a parameter under one statistical model may have inferior properties under another model with a different underlying distribution. When there is uncertainty about the form of the underlying distribution, *robust estimation* provides specially modified estimators (for example, a trimmed sample mean to estimate a population mean) that have reasonably good properties under several parametric statistical models. Refer to [10] for an introduction to robust estimation methods.

Besides confidence intervals (to contain population parameters) and prediction intervals (to contain a future observation from the population) there are *tolerance* intervals to contain a prespecified proportion of the members of the population, with a confidence level attached. For an introduction, formulas, and relevant statistical tables see [8]

Our brief examination of simple linear regression as an application of the least squares method of estimation was meant only to open the door to the whole area of *linear models*—statistical models for which the observable is a linear function of other important observable variables and random error. Statistical inference for linear models under the normal error model is perhaps the most completely developed and frequently exercised application of classical statistical inference. The previously cited [16] gives a readable and comprehensive introduction to this broad area of statistical practice.

The planning of experiments to maximize the efficiency of the inferences and the accuracy of the predictions (typically based in linear models) is called *experimental design*. A basic and popular introduction is [4].

For simplicity of introduction the inference methods in this chapter are all based on a *fixed*, that is, predetermined, sample size $n$, though this constraint is not realistic in many statistical inference problems. *Sequential* methods build and utilize statistical models in which the sample size or the duration of the observation period is itself part of the random observable. For an introduction and examples of application see [20].

In the *parametric* statistical models assumed in this chapter (as formally defined in Eqs. (6.1) through (6.4)) the underlying probability distribution of the observable, **X,** is specified up to a finite vector of parameters $\Theta$. A *nonparametric* statistical model does not require such a specification of the probability distribution. Hence, nonparametric inference methods demand fewer assumptions and are thus more general. They can be especially useful in identifying an accurate parametric model for conducting statistical inference. However, their generality comes at the price of less precision in estimation. See [9] for an introduction to nonparametric statistical models, estimators, and tests for assessing distributional assumptions.

We introduce classical statistical inference as a framework for doing scientific inference under uncertainty in a manner free of expediency, the subjective belief of the experimenter, and the consequences of the decision. While all will agree that expediency has no place in scientific inference, many have criticized classical inference for its inability to incorporate subjective belief or consequence. Chapter 14 introduces *Bayesian inference,* viewed by many as a competitor to classical inference, which does allow the incorporation of the experimenter's subjective prior beliefs into the inference. *Decision theoretic* methods allow the incorporation of the consequences of the decision into the inference through use of the concept of *utility.* For a thorough and balanced comparative introduction to these (now three) schools of statistical inference, see [1].

## References

1. Barnett, V. (1982). *Comparative Statistical Inference,* 2nd ed. John Wiley and Sons, New York.
2. Barnett, V., and Lewis, T. (1984). *Outliers in Statistical Data.* John Wiley and Sons, New York.
3. Bickel, P. J., and Doksum, K. A. (1977). *Mathematical Statistics: Basic Ideas and Selected Topics.* Holden-Day, San Francisco.
4. Box, G. E. P., Hunter, W. G., and Hunter, J. S. (1978). *Statistics for Experimenters.* John Wiley and Sons, New York.
5. Cheng, R. C. H. and Iles, T. C. (1983). "Confidence Bands for Cumulative Distribution Functions of Continuous Random Variables." *Technometrics* **25,** 77–86.
6. Cheng, R. C. H., and Iles, T. C. (1988). "One-Sided Confidence Bands for Cumulative Distribution Functions." *Technometrics* **30,** 155–159.
7. Devore, J. L. (1991). *Probability and Statistics for Engineering and the Sciences,* 2d ed. Brooks/Cole, Pacific Grove, CA.

8. Hahn, G. J., and Meeker, W. Q. (1991). *Statistical Intervals: A Guide for Practitioners.* John Wiley and Sons, New York.
9. Hollander, M., and Wolfe, D. A. (1973). *Nonparametric Statistical Methods.* John Wiley and Sons, New York.
10. Huber, P. J. (1981). *Robust Statistics.* John Wiley and Sons, New York.
11. Johnson, N. L., and Kotz, S. (1969). *Discrete Distributions.* John Wiley and Sons, New York.
12. Johnson, N. L., and Kotz, S. (1970). *Continuous Univariate Distributions,* **1.** John Wiley and Sons, New York.
13. Johnson, N. L., and Kotz, S. (1972). *Continuous Univariate Distributions,* **2.** John Wiley and Sons, New York.
14. Lehmann, E. L. (1959). *Testing Statistical Hypotheses.* John Wiley and Sons, New York.
15. Moore, D. (1985). *Statistics: Concepts and Controversies,* 2nd ed. W. H. Freeman, San Francisco.
16. Neter, J., and Wasserman, W. (1974). *Applied Linear Statistical Models.* Richard D. Irwin, Homewood, IL.
17. Sachs, L. (1984). *Applied Statistics: A Handbook of Techniques.* Springer-Verlag, New York.
18. Snedecor, W. S., and Cochran, W. G. (1967). *Statistical Methods,* 6th ed. Iowa State University Press, Ames.
19. Tukey, J. (1977). *Exploratory Data Analysis.* Addison-Wesley, Reading, MA.
20. Wetherill, G. B. (1966). *Sequential Methods in Statistics.* Methuen, MA, London.

# 7. METHODS FOR ASSESSING DISTRIBUTIONAL ASSUMPTIONS IN ONE- AND TWO-SAMPLE PROBLEMS

Vijayan N. Nair

University of Michigan, Ann Arbor, Michigan

Anne E. Freeny

AT&T Bell Laboratories, Murray Hill, New Jersey

## 7.1 Introduction

Readers have already been introduced to statistical models and methods of estimation in Chapter 6. Efficient methods of estimation for different parametric models will be discussed in later chapters. The use of these methods, however, requires knowing which parametric model is appropriate for a given situation. Sometimes, prior knowledge or subject matter expertise can be used to select the right model. More often, however, such information is unavailable or incomplete, and empirical methods are needed for assessing goodness of fit and identifying appropriate models. This is the topic of the present chapter.

### 7.1.1 Scope

The focus in this chapter is on assessing distributional assumptions in one- and two-sample problems. In a one-sample situation, we have data from a population of interest, and the goal is to model the underlying cumulative distribution function (*cdf*) $F(x)$. The simplest goodness-of-fit problem in this situation is assessing the null hypothesis:

$$F(x) = F_0(x) \quad \text{for all } x \tag{7.1}$$

where $F_0(x)$ is a completely specified *cdf*. This is called a *simple hypothesis* because $F_0(x)$ is completely known. A more common problem is assessing whether

$$F(x) = F_0(x; \theta) \quad \text{for some } \theta \tag{7.2}$$

where $F_0(x; \theta)$ is a specified parametric family, such as the normal, Weibull or gamma. This is called a *composite hypothesis* because the *cdf* $F(x)$ is not completely specified under the null hypothesis. Of particular interest are parametric location–scale families

$$F(x) = F_0\left(\frac{x - \mu}{\sigma}\right) \qquad (7.3)$$

where $\mu$ is a location parameter and $\sigma$ is a scale parameter. Examples are the normal location–scale family and the exponential scale family.

In a two-sample situation, we have data from two populations characterized by the two *cdf*s $F(x)$ and $G(x)$. We will consider testing the simple hypothesis

$$F(x) = G(x) \qquad \text{for all } x \qquad (7.4)$$

as well as methods for assessing the common shape of the two *cdf*s.

Two types of techniques, graphical methods and formal goodness-of-fit tests, will be discussed. Graphical methods are attractive because they allow one to visually examine the data and look for systematic patterns as well as aberrant behavior. If the specified model does not fit the data, plots will often suggest what alternative models to consider. However, they are informal in nature and hence the interpretations are often subjective. Goodness-of-fit tests, on the other hand, provide formal criteria for accepting or rejecting a hypothesis, but if a hypothesis is rejected, they provide little information on what alternative models might be suitable. We will show how the advantages of both methods can be combined by using graphical methods with confidence bands.

### 7.1.2 Organization

The well-known chi-squared tests are discussed in Section 7.2. These tests are, however, best suited for discrete distributions or grouped data whereas the emphasis of the chapter is on modeling data from continuous distributions. Sections 7.3–7.5 deal with the one-sample problem. The quantile–quantile (Q–Q) plotting technique and associated simultaneous confidence bands are developed in Section 7.3. Formal test procedures for testing the simple hypothesis (7.1) and the composite hypothesis (7.2) are discussed in Section 7.4. Extensions to censored data are given in Section 7.5, and the two-sample problem is considered in Section 7.6.

## 7.2 Chi-Squared Tests

The simplest, and perhaps most versatile, goodness-of-fit procedure is the chi-squared test introduced by Pearson in 1900. The general form of the test statistic is

$$X^2 = \sum_{\{\text{all cells}\}} \frac{(O - E)^2}{E} \qquad (7.5)$$

where the $O$ is the actual number of observations and $E$ is the "expected" number of observations in a given cell. To see what these are in concrete terms, let us consider specific testing problems.

### 7.2.1 Simple Hypothesis

We discuss first the testing problem (7.1). Suppose the actual observations $X_1, \ldots, X_n$, which are independent and identically distributed from $F(x)$, have been grouped or "binned" into $J$ cells $(T_{j-1}, T_j], j = 1, \ldots, J$. Then $O_j$ is the number of $X_i$ that fall in the $j$th cell, $j = 1, \ldots, J$. The "expected" number of observations in the $j$th cell, $E_j$, is given by $E_j = np_{0j}$ where $p_{0j}$ is the probability content of the $j$th cell under the null hypothesis, $F_0(T_j) - F_0(T_{j-1})$.

The test is carried out by computing the value of $X^2$ and comparing it to the appropriate percentile of the distribution of $X^2$ under the null hypothesis. The hypothesis (7.1) is rejected at level $\alpha$ if $X^2$ exceeds the $100(1 - \alpha)$th percentile; otherwise it is accepted. The exact distribution of $X^2$ under the null hypothesis is complicated. In large samples, however, this distribution can be well approximated by a $\chi^2$ distribution with $(J - 1)$ degrees of freedom. Since percentiles of $\chi^2$ distributions are widely available, it is easy to implement this test.

### 7.2.2 Composite Hypothesis

Next consider the composite hypothesis (7.2) that $F(x) = F_0(x; \boldsymbol{\theta})$, where $F_0(\cdot)$ is a specified parametric family and $\boldsymbol{\theta}$ is some unknown $p$-dimensional parameter. $O_j$ is again defined as the number of observations falling in the $j$th cell. To compute $E_j$, however, we now have to estimate the unknown $\boldsymbol{\theta}$. Let $\hat{\boldsymbol{\theta}}$ be an estimate, and $\hat{p}_{0j}$ be the probability content of the $j$th cell under $F_0(x; \hat{\boldsymbol{\theta}})$. Then $E_j = n\hat{p}_{0j}$. By substituting these values of $O_j$ and $E_j$ in (7.5), we get the chi-squared statistic for testing the composite hypothesis (7.2). The limiting distribution of this test statistic depends on the method of estimation used to obtain $\hat{\boldsymbol{\theta}}$. Fisher [1] showed that if maximum likelihood estimation based on the grouped data is used, the large-sample distribution is again $\chi^2$ but now with $(J - p - 1)$ degrees of freedom, to account for the $p$ unknown parameters that have been estimated. The result holds true if other asymptotically equivalent methods of estimation are used. Use of maximum likelihood estimators based on the original values of $X_i$ (see Chapter 8) leads to a more complicated distribution for the test statistic, however.

**Example** Consider the following data from a well-known experiment by Rutherford and Geiger [2] on the emission of α-particles from a radioactive source. The data in Table I are $n_j$ = the number of intervals (of a given duration) in which $j$ α-particles were observed, $j = 0, 1, \ldots, 11$.

We now illustrate how the chi-squared test can be used to assess the null hypothesis that the emission of α-particles has a constant rate of decay. Under this null hypothesis, it is well-known that the the data in Table I come from a Poisson distribution. If the data have a Poisson distribution with mean $\lambda$, the probability content of the $j$th cell is $p_{0j}(\lambda) = \lambda^j e^{-\lambda}/j!$. We have to estimate the unknown parameter $\lambda$ in order to get the values of $E_j$. For this data set, the maximum likelihood estimate of $\lambda$ based on the grouped data is very close to the sample mean $\bar{X} = 3.88$. Using this value as $\hat{\lambda}$, the chi-squared statistic is computed to be 11.39. The $P$-value from a $\chi^2$ distribution with $J - p - 1 = 10$ degrees of freedom is $P\{\chi^2_{10} > 11.39\} = 0.33$, so there is no evidence to reject the null hypothesis of a Poisson model.

Before proceeding with chi-squared tests for other situations, we digress slightly to describe a graphical technique for assessing the fit of a Poison model. Under the Poisson assumption, the values of $O_j$ should be close to $n\lambda^j e^{-\lambda}/j!$. If we let $Y_j = \log(O_j) + \log(j!) - \log(n)$, then $Y_j$ should be close to $-\lambda + \log(\lambda)j$. Thus, a plot of $Y_j$ vs. $j$ should be approximately linear with slope $\log(\lambda)$ and intercept $-\lambda$. A nonlinear configuration would suggest that the assumption of a Poisson model does not hold. This plot was suggested by Hoaglin [3] who called it the *Poissonness* plot. See Hoaglin and Tukey [4] for graphical techniques for checking the shapes of other discrete distributions.

**Example** Figure 1 shows the Poissonness plot for the emission data. The plot is remarkably linear, indicating that the Poisson model is reasonable for this data set. Note that we did not have to estimate the unknown value of $\lambda$ to assess the fit of the Poisson model. In fact, one can estimate $\lambda$ from Figure 1 by fitting a line to the linear configuration of points.

TABLE I. Data on Emission of α-Particles
(number of intervals in which $j$ α-particles are observed)

| $j$ | 0 | 1 | 2 | 3 | 4 | 5 | 6 | 7 | 8 | 9 | 10 | 11 |
|---|---|---|---|---|---|---|---|---|---|---|---|---|
| $O_j$ | 57 | 203 | 383 | 525 | 532 | 408 | 273 | 139 | 49 | 27 | 10 | 6 |

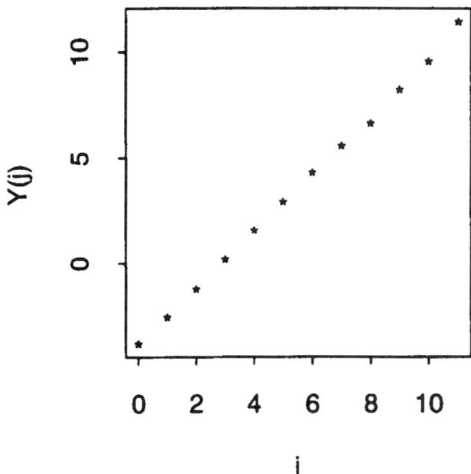

FIG. 1. Poissonness plot for data on emission of α-particles.

### 7.2.3 Two-Sample Tests

The chi-squared test can also be used for two- and $K$-sample problems. We will describe it for the two-sample hypothesis (7.4). Extension to $K > 2$ is straightforward.

Let $X_i$, $i = 1, \ldots, n$ and $Y_k$, $k = 1, \ldots, m$ be independent and identically distributed from $F(\cdot)$ and $G(\cdot)$, respectively. We assume that the two data sets are mutually independent. Then $O_{1j}$ is the number of $X_i$ in cell $j$ and $O_{2j}$ is the number of $Y_k$ in cell $j$, $j = 1, \ldots, J$. The corresponding "expected" number of observations under the null hypothesis (7.4) are $E_{1j} = n\hat{p}_j$ and $E_{2j} = m\hat{p}_j$, where $\hat{p}_j = (O_{1j} + O_{2j}) / (n + m)$. The chi-squared statistic for the two-sample problem is then obtained from (7.5) by summing over all the cells involving both samples. If $n$ and $m$ are both large, the distribution of the $X^2$ statistic under the null hypothesis for the two-sample problem can also be well approximated by a $\chi^2$ distribution with $(J - 1)$ degrees of freedom. For the general $K$-sample case, the test statistic has a limiting $\chi^2$ distribution with $(K - 1)(J - 1)$ degrees of freedom.

### 7.2.4 Discussion

The $\chi^2$ approximations to the distributions of the test statistics work well provided the expected number of observations in each cell is not too small. A conservative guideline is that the expected number in each cell is at least five. There are also guidelines available for selecting the number of cells and cell sizes. It is generally recommended that the cells are chosen to have equal probability under the null hypothesis. Moore [5] suggests a rough rule of thumb for the number of cells as $J \approx 2n^{2/5}$.

The chi-squared tests are based on grouped data, and so they are ideally suited for situations where data are available only in grouped form or for testing hypotheses about discrete distributions. They should not be used in other more general situations, however, as there will be considerable loss of information in grouping the data into cells. The choice of cells or breakpoints for binning the data can also affect the results. In subsequent sections of this chapter, we present alternative methods that are more appropriate for general situations.

## 7.3 Quantile–Quantile (Q–Q) Plots

### 7.3.1 Assessing the Fit of a Completely Specified Distribution

Suppose we have a set of data $X_1, \ldots, X_n$ independent and identically distributed from $F(x)$, and we want to determine if $F(x)$ can be reasonably approximated by a given distribution $F_0(x)$. Based on the data, we can estimate $F(x)$ by the empirical cumulative distribution function (*ecdf*):

$$\hat{F}(x) = n^{-1} \sum_{i=1}^{n} I[X_i \le x] \tag{7.6}$$

where $I[X_i \le x] = 1$ if $X_i \le x$ and equals 0 otherwise. Let $X_{(1)} \le \ldots \le X_{(n)}$ be the order statistics of the sample data; i.e., the values of $X_i$ arranged in increasing order. Then, it can be seen that the *ecdf* is a step function that jumps a height of $1/n$ at the value of each order statistic $X_{(i)}$.

**Example** We use a data set on failure times of mechanical devices to illustrate the techniques in this section. The data in Table II are failure times (in millions of operations) of 37 devices. (We have given here only part of the data from the study. See Nair [6] for the complete data set.) There were two possible modes of failure—two springs—for these devices indicated by the codes A and B. For now, however, we will focus on just the failure times of the devices.

There are 37 observations, so the *ecdf*, given in Figure 2, jumps by 1/37 at each of the order statistics and is constant in between the values of the order statistics. The estimated cumulative probabilities and quantiles can be read off from this plot. In addition, it is possible to identify features peculiar to the data set, such as the striking gap immediately preceding 2 million operations, indicating a relatively long time between successive failures at that point.

TABLE II. Data on Failure Times of a Mechanical Device
(in millions of operations)

| Time | Mode | Time | Mode |
|---|---|---|---|
| 1.151 | B | 2.116 | B |
| 1.170 | B | 2.119 | B |
| 1.248 | B | 2.135 | A |
| 1.331 | B | 2.197 | A |
| 1.381 | B | 2.199 | B |
| 1.499 | A | 2.227 | A |
| 1.508 | B | 2.250 | B |
| 1.534 | B | 2.254 | A |
| 1.577 | B | 2.261 | B |
| 1.584 | B | 2.349 | B |
| 1.667 | A | 2.369 | A |
| 1.695 | A | 2.547 | A |
| 1.710 | A | 2.548 | A |
| 1.955 | B | 2.738 | B |
| 1.965 | A | 2.794 | A |
| 2.012 | B | 2.910 | A |
| 2.051 | B | 3.015 | A |
| 2.076 | B | 3.017 | A |
| 2.109 | A | | |

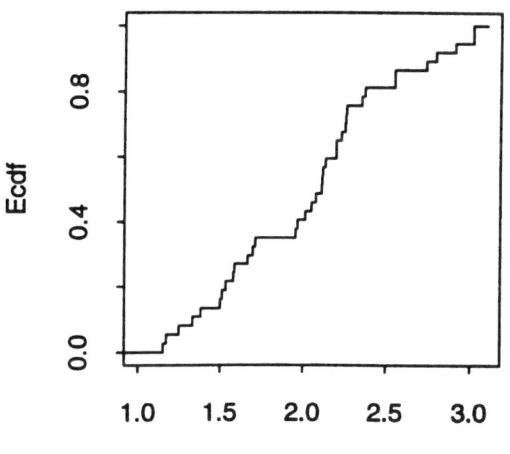

FIG. 2. Plot of the ecdf for the mechanical device failure time data.

To assess the null hypothesis in (7.1), it is natural to plot the *ecdf* $\hat{F}(x)$ and the hypothesized *cdf* $F_0(x)$ on the same figure and visually compare the fit. Under the null hypothesis, $\hat{F}(x)$ is an estimate of $F_0(x)$, so we can expect them to be close to each other. Large, systematic differences would indicate that the null hypothesis is not true. This "comparative" plot has several drawbacks. First, it is hard to visually assess differences between two curves. Second, if the hypothesis is not true, it is not easy to determine from this plot what other distributions might be reasonable. It is much easier to assess and interpret deviations from a straight line. We discuss a well-known method based on this idea called *quantile–quantile* (Q–Q) plots. See Wilk and Gnanadesikan [7] and Gan, Koehler, and Thompson [8] for more details and other plotting techniques.

The Q–Q plot, as the name suggests, is a plot of the quantiles from the hypothesized *cdf* against the quantiles from the *ecdf*. More specifically, let $F_0^{-1}(p)$ be the *p*th quantile ($=$ the 100*p*th percentile) of the hypothesized *cdf*. Then, the Q–Q plot is a plot of

$$F_0^{-1}\left(\frac{i - .5}{n}\right) \text{ vs. } X_{(i)} \qquad (7.7)$$

for $i = 1, \ldots, n$, where the values of $X_{(i)}$ denote the order statistics. There has been a discussion of other plotting positions besides $(i - .5)/n$ in the literature (see, e.g., Chambers et al. [9]), but we will restrict attention here to this choice.

**Example:** Consider the data set on failure times of a mechanical device in Table II. Suppose we want to assess the simple null hypothesis that the distribution of failure times of the devices is a standard exponential distribution with parameter $\lambda = 1$; i.e., $F_0(x) = 1 - \exp(-x)$. To obtain the exponential Q–Q plot, we just substitute in (7.7) the quantiles of the exponential distribution with parameter $\lambda = 1$, $F_0^{-1}(p) = -\log(1 - p)$. Figure 3(a) shows this plot with the order statistics plotted on the *x*-axis and the standard exponential quantiles on the *y*-axis. The probabilities are also shown on the righthand side of the *y*-axis. If the data are from the standard exponential distribution, the plot should follow the straight line with slope 1 and intercept 0. It is quite clear from this plot, however, that the null hypothesis does not hold. In fact, as we shall see later, none of the exponential distributions fit this data set well.

The Q–Q plot is essentially a plot of $F_0^{-1}[\hat{F}(x)]$; i.e., a plot of the *ecdf* $\hat{F}(x)$ with the *y*-axis transformed. It still retains all of the original information in the plot of $\hat{F}(x)$. Since the *y*-axis is also labeled in the probability scale,

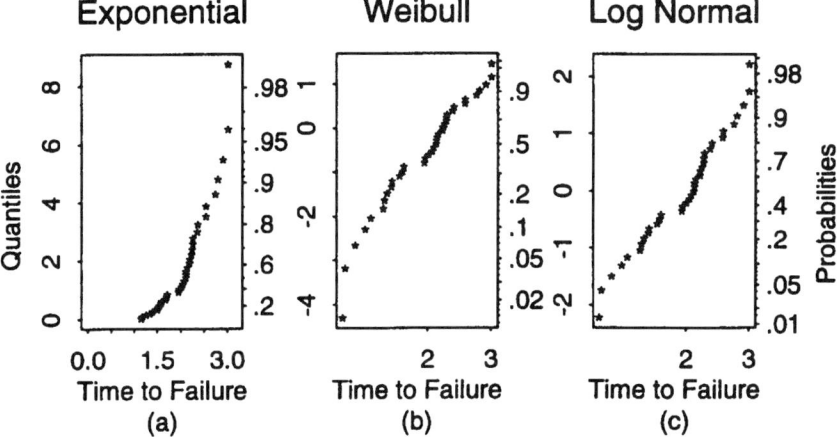

FIG. 3. Some Q–Q plots for the mechanical device failure time data. The quantile axis is plotted on the left of the plot, the probability axis on the right of the plot for the distribution given in the plot heading. A logarithmic scale is used on the x-axis for the Weibull and log-normal plots.

the *cdf* can still be visually estimated from this plot. Unlike the *ecdf* plot, however, it is now easier to interpret the deviations from linearity and identify possible alternative distributions. To develop guidelines on how to interpret the deviations, let us consider the possible shapes of the Q–Q plot

$$\tau(x) = F_0^{-1}[\hat{F}(x)] \tag{7.8}$$

- If $F(x) = F_0([x - \mu]/\sigma)$, i.e., the location-scale model (7.3) holds, it is easy to see that $\tau(x)$ will be approximately equal to $[x - \mu]/\sigma$. Thus, if the Q–Q plot does not follow the straight line with slope 1 and intercept 0 but still appears linear, this would suggest that a location-scale model is reasonable. This will be discussed in more detail in the next subsection.
- If $\tau(x)$ has an *S* or inverted *S* shape, then $F(x)$ is *symmetric with respect to $F_0(x)$*. Thus, if $F_0(x)$ itself is symmetric, an *S* or inverted *S* shaped Q–Q plot would suggest that $F(x)$ is also symmetric. An *S* shape implies that $F(x)$ is heavier tailed than $F_0(x)$ while an inverted *S* shape implies that it has lighter tails.
- If $\tau(x)$ is piecewise linear, that suggests that the data could be from two different populations so a mixture model might be appropriate.

Fowlkes [10] provides an extensive catalog of the shapes of the theoretical Q–Q plot for different distributions. This catalog is very useful in interpreting the shapes and identifying specific parametric alternatives.

## 7.3.2 Assessing the Fit of Parametric Families

We discuss first the special case of location–scale families, which include the normal and exponential families. The method can also be applied to check the fit of models that can be transformed to location–scale families. For example, the lognormal and Weibull models can be transformed to location–scale families by applying a logarithmic transformation to the data.

As we have already noted, under the location–scale model $F_0([x - \mu]/\sigma)$, the Q–Q plot is approximately linear with slope $1/\sigma$ and intercept $-\mu/\sigma$. So we can assess the location–scale hypothesis by checking if the Q–Q plot follows some linear configuration without specifying the slope and intercept of the line. Thus, we can use the same Q–Q plot, without any modification, to check the fit of an entire location–scale family of $cdf$s. Note also that we do not have to estimate the unknown parameters $\mu$ and $\sigma$. These properties make the use of the Q–Q plot particularly attractive. Let us consider some important special cases of location–scale families.

The one-parameter exponential model with $cdf$ $1 - \exp(-\lambda x)$ is a scale family. It arises naturally in reliability and other applications where interest centers on the failure rate (or rate of decay). The exponential distribution is characterized by the fact that it has a constant failure rate (or rate of decay). Constructing an exponential Q–Q plot was already discussed in the last section. Under exponentiality, the plot should be approximately linear with slope $\lambda$ and intercept 0.

If the exponential model does not fit the data, a natural next step is to consider the more general Weibull model with $cdf$ $1 - \exp(-[x/\alpha]^\beta)$ for some $\alpha > 0$ and $\beta > 0$. This family of distributions includes the exponential model as a special case when $\beta = 1$. It is a flexible family that can be used to model a variety of problems. The Weibull family is not a location–scale family but can be made into one by applying the log transformation. Specifically, if $Y$ has a Weibull distribution, then $X = \log(Y)$ has a smallest extreme-value (SEV) distribution which is a location–scale family with location parameter $\mu = \log(\alpha)$ and scale parameter $\sigma = \beta^{-1}$. Thus, the Q–Q plot with $X_{(i)} = \log(Y_{(i)})$ and $F_0^{-1}(p_i) = \log[-\log(1 - p_i)]$ can be used to assess the fit of Weibull models. Under a Weibull model, the plot will be linear with slope $1/\sigma = \beta$ and intercept $-\mu/\sigma = \beta \log(\alpha)$. In particular, if the exponential model holds, the line will have slope 1. The guidelines discussed in the previous subsection can be used to interpret deviations from linearity.

The normal distribution plays a central role in statistics and assessing whether a given distribution belongs to a normal family is a problem that occurs frequently. In addition, the lognormal distribution has been found useful in modeling data with heavy tails, such as oil and gas exploration data and environmental data on air quality. The normal Q–Q plot is obtained by taking $F_0(x)$ in (7.7) to be the standard normal $cdf$ $\Phi(x)$. Under a normal model, the plot will be approximately linear with slope $1/\sigma$ and intercept $-\mu/\sigma$. The Q–Q plot for assessing lognormality is obtained by first taking logarithms of the data and then using the normal Q–Q plot.

**Example:** As we have already seen, Figure 3(a) is the exponential Q–Q plot for the mechanical device failure data. Since no line through the origin fits the data well, we can conclude that the exponential model is not reasonable for this data set. The plot appears to be nonlinear with a convex shape, suggesting a Weibull model as a possible alternative. The Weibull Q–Q plot in Figure 3(b) is approximately linear, suggesting that the Weibull model is plausible. The lognormal Q–Q plot is shown in Figure 3(c) and suggests that the lognormal model might be a slightly better fit.

To extend the Q–Q plotting technique to more general parametric families such as beta or gamma, we have to first estimate the unknown parameters $\theta$. The maximum likelihood method of estimation, to be discussed in Chapter 8, can be used to estimate these parameters. Once an estimate of $\theta$ is available, the Q–Q plot for assessing the composite hypothesis (7.2) is obtained by replacing $F_0^{-1}[(i - .5)/n]$ in (7.7) with $F_0^{-1}[(i - .5)/n; \hat{\theta}]$, where $F_0^{-1}(p; \hat{\theta})$ is the $p$th quantile of $F_0(x; \hat{\theta})$. Under the null hypothesis, the plot should be approximately linear with slope 1 and intercept 0. Deviations from linearity can be interpreted using the guidelines discussed in the last section.

### 7.3.3 Plots with Simultaneous Confidence Bands

Since the interpretation of Q–Q plots is subjective and can be difficult, we develop confidence bands that allow us to account for the uncertainty in the Q–Q plots and provide more objective assessment. We first discuss confidence bands for the true $cdf$ $F(x)$ and use these to get bands associated with the Q–Q plots.

Interval estimation was discussed in the previous chapter. For our purposes we need simultaneous confidence bands; i.e., confidence intervals that are valid simultaneously for a range of parameter values. Throughout this chapter, we denote a confidence band of the form $(a - b, a + b)$ by the notation $a \pm b$. Further, confidence bands for the $cdf$ $F(x)$ can always be restricted to the interval [0, 1]. Although we will not state it explicitly, we assume throughout that this is always done.

The simplest and most well-known confidence band for $F(x)$ is the Kolmogorov [11] band (or $K$-band):

$$\hat{F}(x) \pm n^{-1/2}c_\alpha \qquad (7.9)$$

where $c_\alpha$ is the $100(1 - \alpha)$th percentile of the distribution of a corresponding test statistic (see next section for discussion of test statistics). This confidence band is interpreted as follows. The probability that the confidence band will contain the true *cdf* $F(x)$ *simultaneously for all values of x* is at least $1 - \alpha$.

Note that the width of the *K*-band is constant for all values of $x$. An alternate band that has width proportional to the estimated standard deviation of the *ecdf* is the *S*-band, given by

$$\hat{F}(x) \pm n^{-1/2}d_\alpha[\hat{F}(x)(1 - \hat{F}(x))]^{1/2} \qquad (7.10)$$

where $d_\alpha$ is the appropriate percentile value. This confidence band was discussed in the context of two-sample comparisons by Doksum and Sievers [12]. The band is valid simultaneously for $F(x)$ for all values of $x$ in the range $\{x$ such that $a \leq \hat{F}(x) \leq b\}$. The restriction to the interval $[a \leq \hat{F}(x) \leq b]$ is necessary because the percentile value $d_\alpha \to \infty$ as either $a \to 0$ or $b \to 1$. In our experience, the choice $a = 1 - b = .05$ provides a reasonable compromise between the size of the percentile value (and hence the width of the band) and coverage.

Table III in the next section provides large-sample approximations to the percentile values for the *K*- and *S*-bands for some commonly used $\alpha$-levels. For small to moderate sample sizes, the asymptotic percentile values for the *K*-band can be used with the correction $c_\alpha/(n^{1/2} + .12 + .11/n^{1/2})$ instead of $c_\alpha/n^{1/2}$ [see (7.9)]. A similar correction is not available for the *S*-band, and readers are referred to Niederhausen [13] for exact percentile values.

**Example:** Figure 4 shows the *ecdf* given in Figure 2 together with the 95% *K*- and *S*-bands for all of the mechanical device failure data. For the *S*-band, $a = 1 - b = .05$ so the band is valid for all $x$ such that $.05 \leq \hat{F}(x) \leq .95$. Notice that the *K*-band is narrower in the middle of the distribution while the *S*-band is narrower in the tails. This property can be shown to hold in general. For detecting alternative distributions, most of the information is typically in the tails, and thus the *S*-band is to be preferred in this regard.

The simultaneous confidence bands for $F(x)$ can be used to formally test the simple null hypothesis (7.1). The null hypothesis is accepted at level $\alpha$ if the corresponding $(1 - \alpha)$-level band contains the *cdf* $F_0(x)$; otherwise it is rejected. This test can be implemented visually by plotting the simultaneous band and the *cdf* $F_0(x)$ on the same plot and checking whether the band contains $F_0(x)$. However, as we have already seen, visual comparisons are

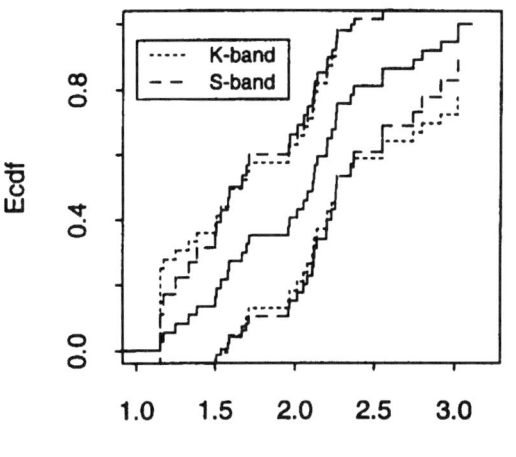

FIG. 4. Comparison of 95% K- and S-bands for the mechanical device failure time data.

much easier when the plots are displayed on the quantile scale. Thus, we want bands associated with the Q–Q plots.

Recall that the Q–Q plot is just a plot of $\hat{F}(x)$ with the y-axis transformed by $F_0^{-1}(\cdot)$. By applying the same transformation to the simultaneous bands we get bands associated with Q–Q plots. Specifically, let $L(x)$ and $U(x)$ denote the lower and upper envelopes of a particular simultaneous confidence band for $F(x)$. (As noted before, we implicitly assume that they are restricted to [0, 1].) Then, the confidence band associated with the Q–Q plot is just $[F_0^{-1}(L(x)), F_0^{-1}(U(x))]$.

This band is a simultaneous $(1 - \alpha)$-level confidence band for the true (theoretical) Q–Q plot. The simple null hypothesis that $F(x) = F_0(x)$ can be tested at level $\alpha$ by checking if the band contains the straight line with slope 1 and intercept 0. The band also provides a conservative test of the composite hypothesis of a location–scale model (7.3). This is done by checking if the band contains some straight line (with any slope and intercept). This test is conservative in the sense that the level of the test is less than $\alpha$. Another way of saying this is that the band is wider than necessary. Discussion of alternative procedures for testing for the location–scale model is beyond the scope of the present chapter.

**Example:** Figure 5 shows the exponential, Weibull and lognormal Q–Q plots (same as Figure 2) with a 95% S-band. Let us first consider Figure

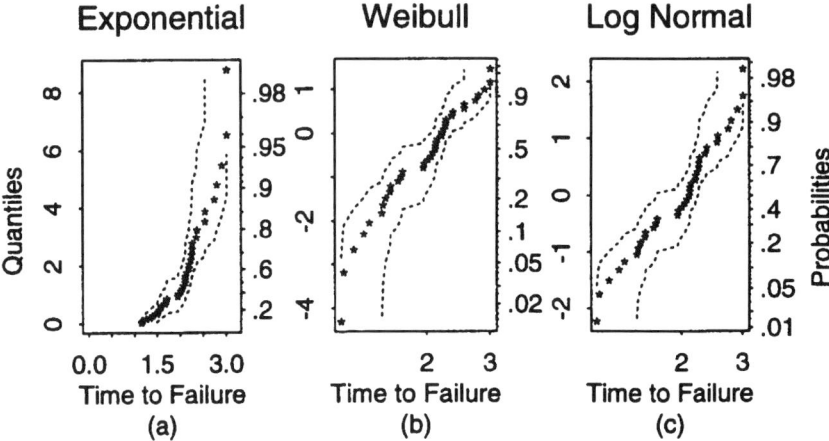

FIG. 5. A 95% S-band associated with the Q–Q plots for the mechanical device failure time data.

5(a). Since the straight line with slope 1 and intercept 0 does not fit into the band, we can reject, at level .05, the simple null hypothesis that $F(x)$ is given by the standard exponential distribution with $\lambda = 1$. To determine if any exponential distribution will fit the data, we have to determine if a line with intercept equal to 0 and an arbitrary slope $\lambda$ will fit into the band. Since this is not the case, we can reject, at level .05, the more general hypothesis of exponentiality for this data set. Figures 5(b) and 5(c) can be used similarly to assess the fit of Weibull and lognormal models, respectively. For example, suppose we want to test the hypothesis that $F(x)$ is a lognormal distribution with $\mu = 1.0$ and $\sigma = 1.0$. Under this null hypothesis, the lognormal Q–Q plot in Figure 5(c) will be linear with slope $1/\sigma = 1$ and intercept $-\mu/\sigma = 0$. Since this line does not fit into the S-band, we can reject this hypothesis at level .05. We can, however, fit some straight line into the band, so there is no evidence to reject the composite hypothesis of a general lognormal model.

In addition to being a formal hypothesis test, the band provides us with an indication of the amount of variability in the plot at different $x$-values and thus allows us to account for the uncertainty while making visual judgments. It retains all the advantages of the original graphical method. In particular, the righthand side of the $y$-axis is labeled in probability scale, so one can still use it to obtain simultaneous interval estimates for $F(x)$ and its quantiles.

## 7.4 Formal Test Procedures

### 7.4.1 Assessing the Fit of a Completely Specified cdf

We have already seen that the simultaneous confidence bands in Section 7.3 yield formal tests of the simple hypothesis (7.1). In fact, there is a one-to-one correspondence between the $K$-band in (7.9) and the well-known Kolmogorov [11] test statistic:

$$K = \max_x n^{1/2}|\hat{F}(x) - F_0(x)| \tag{7.11}$$

This statistic has the desirable property that, if the null hypothesis (7.1) holds and if $F_0(x)$ is continuous and strictly increasing, then its distribution does not depend on what $F_0(x)$ is. This property is often referred to as being distribution free, and it makes the computation of the percentiles of the distribution of the statistic especially easy. Table III provides values of some large-sample percentiles.

However, we have already seen in the context of confidence bands that the $K$-band is not as desirable as the $S$-band. In fact, although the $K$-test is well-known and simple to use, it does not perform as well as other possible tests and is not recommended. We now discuss some alternatives.

The $S$-test that corresponds to the $S$-band is given by

$$S = \max_x n^{1/2} \frac{|\hat{F}(x) - F_0(x)|}{[\hat{F}(x)(1 - \hat{F}(x))]^{1/2}} \tag{7.12}$$

The $S$-test is also distribution-free under the null hypothesis. See Table III for some approximate percentile values.

The $S$- and $K$-tests are both based on *maximal* weighted distances between the *ecdf* $F(x)$ and the hypothesized *cdf* $F_0(x)$. There are also corresponding tests based on *average* distances. We describe two well-known tests of this type. The first is the Cramer–von Mises test:

TABLE III. Approximate Percentile Values for $K$, $S$, $CM$, and $AD$ Tests

| Band | $\alpha = .01$ | $\alpha = .05$ | $\alpha = .10$ |
|---|---|---|---|
| $K$ | 1.63 | 1.36 | 1.22 |
| $S: a = 1 - b = .01$ | 3.81 | 3.31 | 3.08 |
| $S: a = 1 - b = .05$ | 3.68 | 3.16 | 2.91 |
| $S: a = 1 - b = .10$ | 3.59 | 3.06 | 2.79 |
| $CM$ | 0.74 | 0.46 | 0.35 |
| $AD$ | 3.86 | 2.49 | 1.93 |

$$\text{CM} = n \int [\hat{F}(x) - F_0(x)]^2 \, dF_0(x) \tag{7.13}$$

This is simply the average unweighted squared distance between the *ecdf* and the hypothesized *cdf*. A weighted average distance test is the Anderson–Darling test given by

$$\text{AD} = n \int \frac{[\hat{F}(x) - F_0(x)]^2}{[F_0(x)(1 - F_0(x))]} \, dF_0(x) \tag{7.14}$$

The motivation for this weighting scheme is similar to the one used with the *S*-test. Like the maximal distance tests, the average distance tests are also distribution free. Again, Table III provides approximate percentile values for several commonly used $\alpha$-levels. See Stephens [14] and references therein for additional tables of critical values.

**Example:** Suppose we want to formally test the hypothesis that the distribution $F(x)$ of the mechanical device failure data is lognormal with $\mu = 1.0$ and $\sigma = 1.0$. The computed statistics are $K = 2.78$, $S = 11.27$, $CM = 1.98$, and $AD = 9.46$. All of these test statistics are significant at the .01 level, and so this null hypothesis must be rejected. This is consistent with our conclusion in the last section based on the 95% *S*-band.

We turn now to a comparison of the performances of the various tests. As noted already, most of the information for discriminating between parametric models is in the tails, so procedures that give more weight to the tails will perform better. We have already seen this in the context of confidence bands where the *S*-band did better than the *K*-band. Similarly, the *AD* test, which gives greater weight to the tails, is better than the *CM* test in this regard. Overall, the *K*-test is dominated by the other tests discussed in this section and is not recommended. The others are all competitive. The advantage of the *S*-test is that it yields a confidence band that can be displayed graphically. However, it has to be restricted to a range $0 < a \leq \hat{F}(x) \leq b < 1$. The *AD* test, on the other hand, does not suffer from this problem.

There are many other tests in the literature for testing the simple hypothesis (7.1). Interested readers should refer to D'Agostino and Stephens [15], Chapters 4 and 8.

### 7.4.2 Tests for Parametric Families

Consider now the more general problem of testing the composite hypothesis in (7.2). For example, we may want to assess the fit of a location–scale family such as the normal. Or we may be interested in the fit of a gamma model.

The test statistics discussed in the previous subsection can be extended to this more general situation by replacing $F_0(x)$ with $F_0(x; \hat{\boldsymbol{\theta}})$, where $\hat{\boldsymbol{\theta}}$ is a suitable estimate of the unknown parameter $\boldsymbol{\theta}$. Unfortunately, however, the test statistics are no longer distribution free, and it is more difficult to get appropriate percentile values for the test statistics. Tables of approximate percentile values are available in the literature for selected tests and parametric families, especially location–scale families. See Stephens [14] and references therein.

### 7.4.3 Special Tests for Normality

There is a large variety of tests in the literature for assessing the assumption of normality. We have already considered distance based tests for testing normality. We consider briefly some other important tests here. See D'Agostino [16] for additional details and references to other procedures.

Perhaps the test with best overall performance for assessing normality is the Shapiro–Wilk [17] test. It is simply a ratio of two estimators of the variance $\sigma^2$,

$$SW = \hat{\sigma}_1^2 / \hat{\sigma}_2^2 \tag{7.15}$$

The term $\hat{\sigma}_1$ in the numerator is the best linear unbiased estimator of $\sigma$ under the assumption of normality while the term $\hat{\sigma}_2$ in the denominator is the usual sample standard deviation $S$. If the data are normal, both will estimate $\sigma$, and hence the ratio will be close to 1. If normality does not hold, $\hat{\sigma}_1$ does not estimate $\sigma$, and hence the ratio will be quite different from 1. The exact computation of $\hat{\sigma}_1$ involves calculating the expected values, variances, and covariances of the order statistics of a sample of size $n$ from a standard normal distribution. This is quite involved, although tables of these values are available in the literature. Many approximations to the Shapiro–Wilk test have been suggested to ease the computational problem. In particular, Weisberg and Bingham [18] show that the numerator, $\hat{\sigma}_1^2$, can be approximated well by

$$\hat{\sigma}_1^2 = \frac{1}{n-1} \frac{[\sum_{i=1}^{n} \Phi^{-1}(p_i) X_{(i)}]^2}{\sum_{i=1}^{n} [\Phi^{-1}(p_i)]^2} \tag{7.16}$$

where $p_i = [i - 3/8] / [n + 1/4])$. Note that the $p_i$ are close to the plotting positions used in the normal Q–Q plot. In fact, the Shapiro–Wilk test is directly related to the measure of linearity in the normal Q–Q plot. Stephens [14] provides an extensive table of critical values for the statistic $n(1 - SW)$ with $SW$ computed using Weisberg and Bingham's approximation in (7.16). The hypothesis of normality is rejected at level $\alpha$ if the computed value of $n(1 - SW)$ exceeds the critical value.

The procedures we have discussed thus far are omnibus tests in the sense that they have not been designed with any particular alternatives in mind. Among the various omnibus tests for normality, the Shapiro–Wilk and its modifications have the best performance over a wide variety of alternatives. With the exception of the $K$-test, the distance based tests also perform reasonably well. The $S$-test yields a simultaneous band that can be used with Q–Q plots and hence is attractive.

There are also directional tests for detecting particular types of deviations. For detecting asymmetry, moment tests based on the sample skewness have been proposed. Similarly, for detecting heavier or lighter tailed distributions, moment tests based on the sample kurtosis can be used. For details about these tests, readers should consult Bowman and Shenton [19]. When certain types of deviation from normality are suspected a priori, directional tests specifically geared for such an alternative, such as the moment tests, should be used.

## 7.5 Extensions to Censored Data

In this section, we briefly discuss extensions of the methods discussed thus far to situations with *censored* data. We will consider only a specific type of censored data here. See Chapter 8 for discussion of other situations.

Consider the mechanical switch data in Table II. As we have already noted, there were two possible modes of failure for these mechanical devices. The codes A and B in Table II indicate the particular spring that failed in each case. Note that when a device failed due to mode A (B), we can no longer observe the failure of the corresponding mode B (A). In this case, one says that the failure time of mode B (A) was *randomly right censored*. This data set is a typical example of a competing-risks problem, where a system fails due to one or more competing causes, and one observes only the time to failure of the system and the corresponding failure mode (component). Because the two springs were subjected to different stress levels, the life distributions of the two failure modes are likely to be different. Thus it is of interest in this problem to make inferences about the failure time distributions of the individual springs in addition to the failure distribution of the device itself.

If the failure times due to the two different modes are independent, it is possible to estimate the underlying distributions of both failure modes without making any parametric assumptions. To be specific, suppose we want to estimate the failure distribution of mode A. Let $X_i$, $i = 1, \cdots ,$ 37, denote the failure times of the 37 mechanical devices in Table II. Let $R_i$ denote the rank of $X_i$ among the 37 observations, let $\delta_i = 1$ if the failure was due to mode A, and let $\delta_i = 0$ otherwise. Then, the Kaplan–Meier estimator of the failure distribution for mode A is given by

$$\hat{F}_{KM}(x) = 1 - \prod_{\{i: X_i \leq x\}} [(n - R_i)/(n - R_i + 1)]^{\delta_i}, \qquad x \leq X_{(n)} \quad (7.17)$$

where $X_{(n)}$ is the largest order statistic of the $X_i$. It can be seen that this estimator is a step function that jumps only at the times where failure is due to mode A.

**Example:** Figure 6 is a plot of the Kaplan–Meier estimates of the *cdf*s of both failure modes A and B for the mechanical device data. It is in fact quite interesting that one can estimate both the individual distributions from such data. The Kaplan–Meier estimate of the *cdf* for mode B goes only up to about 0.7. It lies above the estimate for mode A for up to about 2.5 million operations indicating that the system is more likely to fail because of mode B than mode A in the early stages.

The Q–Q plot for randomly censored data can be based on the Kaplan–Meier estimators. Again, we use the specific case for mode A to illustrate the ideas. Let $r_1 \leq \ldots \leq r_K$ denote the ranks of the $X_i$ that correspond to mode A failures; i.e., the ranks of those $X_i$ for which $\delta_i = 1$. Recall that $X_{(i)}$ is the $i$th order statistic, and let $p_i = [\hat{F}_{KM}(X_{(r_i)}) +$

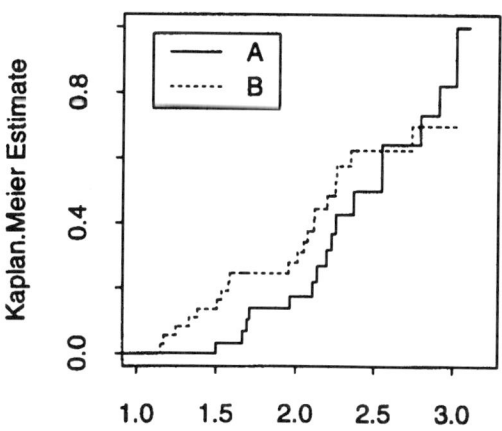

FIG. 6. Plot of the Kaplan–Meier estimates of the cdf's for the two failure modes for the mechanical device data.

$\hat{F}_{KM}(X_{(r_{(i-1)})})]/2$. Then, the Q–Q plot for randomly censored data is just a plot of

$$F_0^{-1}(p_i) \quad \text{vs.} \quad X_{(r_i)} \qquad (7.18)$$

In the absence of censoring, this reduces to the Q–Q plot we have discussed previously.

**Example:** Figures 7(a)–(c) show the exponential, Weibull and lognormal Q–Q plots based on the Kaplan–Meier estimates of the *cdf*s for both failure modes A and B. Again, the exponential model does not fit the data well. The Weibull model seems reasonable for both failure modes, but the slopes of the two plots appear to be different. This suggests that a simple Weibull model with the same shape parameter for both distributions may not be appropriate. Figure 7(c) indicates that the lognormal model is also plausible.

Confidence bands associated with the *cdf* $F(x)$ and with Q–Q plots as well as formal goodness-of-fit tests for randomly censored data are available in the literature. A complete discussion of these is beyond the scope of the present chapter. Interested readers can refer, for example, to Nair [6] and Stephens [14].

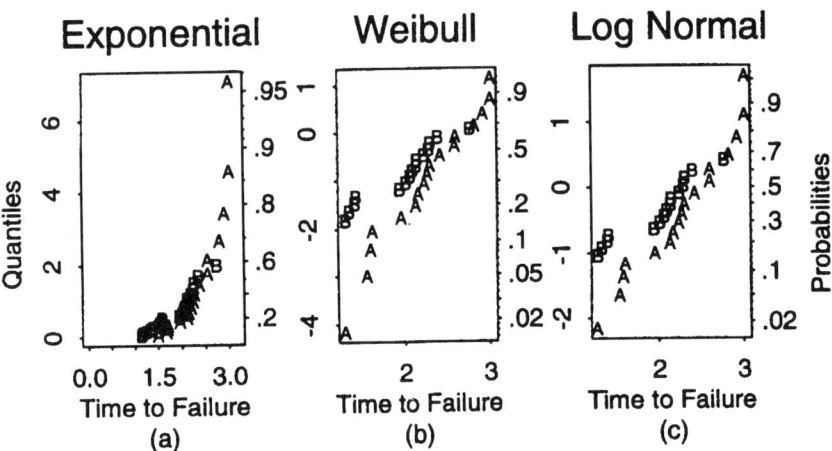

FIG. 7. Q–Q plots for the distributions of the two failure modes for the mechanical device data.

## 7.6 Two-Sample Comparisons

In this section, we consider the situation where we have two *cdfs* $F(x)$ and $G(x)$ corresponding to two populations of interest. We have to compare the two *cdfs* based on two mutually independent sets of data from the two populations. We consider methods for testing the simple hypothesis (7.4) as well as methods for assessing the common shape of the two *cdfs*.

### 7.6.1 Two-Sample Q–Q Plots

The Q–Q plotting technique has been extended to two-sample problems (see Wilk and Gnanadesikan [7]). For simplicity, we will consider only the case where the sample sizes for both populations are the same. Then, the two-sample Q–Q plot is just a plot of

$$Y_{(i)} \text{ vs. } X_{(i)}, \quad i = 1, \ldots, n \qquad (7.19)$$

where the $X_{(i)}$ and $Y_{(i)}$ are the order statistics from the two samples.

If the two distributions are the same, the Q–Q plot will be linear with slope 1 and intercept 0. If the distributions differ only in location and scale, i.e., $F(x) = G([x - \mu]/\sigma)$, then the plot will be linear with slope $1/\sigma$ and intercept $-\mu/\sigma$. Thus, the same Q–Q plot can be used to check the composite hypothesis that the two distributions belong to the same location–scale family. Unlike the one-sample situation, however, we are not specifying here the shape of the underlying location–scale family. If the Q–Q plot appears to be nonlinear, the guidelines suggested in Section 7.3.1 can be used to interpret the deviations, such as whether $F$ is symmetric or heavier/lighter-tailed compared to $G$.

**Example:** The data in Table IV are rainfall in acre–feet from 52 clouds, of which 26 were chosen at random to be seeded with silver oxide. This data set comes from Simpson, Olsen, and Eden [20]. Figure 8 shows the two-sample Q–Q plot of rainfall from the seeded clouds plotted against rainfall from the control clouds. The solid line is drawn at $y = x$. Clearly there is more rainfall from the seeded clouds. In fact the line $y = 2.5x$ (chosen by eye), which passes through the data very well, tells us that the seeded clouds have produced about 2.5 times as much rain as the unseeded clouds in this experiment. For a further analysis of this data, as well as a discussion of how to handle the situation of unequal sample size, see Chambers et al. [9].

### 7.6.2 Formal Tests

Two-sample versions of the formal test procedures in Section 7.4 can be obtained. We discuss here only the simple Kolmogorov–Smirnov test.

TABLE IV.  Data on Rainfall from Cloud Seeding
(in acre–feet)

| Control Clouds | | Seeded Clouds | |
|---|---|---|---|
| 1202.6 | 41.1 | 2745.6 | 200.7 |
| 830.1  | 36.6 | 1697.8 | 198.6 |
| 372.4  | 29.0 | 1656.0 | 129.6 |
| 345.5  | 28.6 | 978.0  | 119.0 |
| 321.2  | 26.3 | 703.4  | 118.3 |
| 244.3  | 26.1 | 489.1  | 115.3 |
| 163.0  | 24.4 | 430.0  | 92.4  |
| 147.8  | 21.7 | 334.1  | 40.6  |
| 95.0   | 17.3 | 302.8  | 32.7  |
| 87.0   | 11.5 | 274.7  | 31.4  |
| 81.2   | 4.9  | 274.7  | 17.5  |
| 68.5   | 4.9  | 255.0  | 7.7   |
| 47.3   | 1.0  | 242.5  | 4.1   |

Let $m$ and $n$ denote the (possibly unequal) sizes of the two samples, and let $N = mn/(m + n)$. Let $\hat{F}(x)$ and $\hat{G}(x)$ be the *ecdf*s of $F(x)$ and $G(x)$, respectively. To test the null hypothesis (7.4), we can use the Kolmogorov–Smirnov test statistic

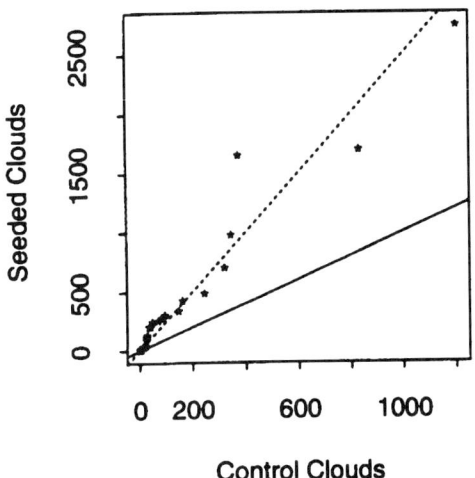

FIG. 8. Two-sample Q–Q plot of rainfall from seeded clouds vs. rainfall from control clouds, with the solid line $y = x$ and dashed line $y = 2.5x$ drawn.

$$\text{KS} = \max_x N^{1/2}|\hat{F}(x) - \hat{G}(x)| \tag{7.20}$$

the maximum distance between the two *ecdf*s. Under the null hypothesis, the distribution of this test statistic does not depend on the (unknown) common distribution. Its finite-sample distributions have been tabulated extensively and percentile values are readily available. The large-sample distribution is the same as that of the one sample Kolmogorov test statistic, and hence the percentile values in Table III can be used with large samples.

**Example:** The value of the KS test statistic for the data of Table IV is 5563, indicating that the two samples do not have the same distribution. This result could have been anticipated from the two-sample Q–Q plot.

The other tests and bands discussed in Sections 7.3 and 7.4 can also be extended to the two-sample situation. Readers can refer to Doksum and Sievers [12] and references therein for details.

## References

1. Fisher, R. A. (1924). "The Conditions Under Which $\chi^2$ Measures the Discrepancy Between Observation and Hypothesis." *Journal of the Royal Statistical Society* **87,** 442–450.
2. Rutherford, E., and Geiger, H. (1910). "The Probability Variations in the Distribution of $\alpha$-Particles." *Phil. Mag. S6,* **20,** 698.
3. Hoaglin, D. C. (1980). "A Poissonness Plot." *The American Statistician* **34,** 146–149.
4. Hoaglin, D. C., and Tukey, J. W. (1985). "Checking the Shape of Discrete Distributions," Chapter 9 in *Exploring Data Tables, Trends, and Shapes,* D. C. Hoaglin, F. Mosteller and J. W. Tukey (eds.). John Wiley and Sons, New York.
5. Moore, D. S. (1986). "Tests of Chi-Squared Type," Chapter 3 in *Goodness-of-Fit Techniques,* R. B. D'Agostino and M. A. Stephens (eds.). Marcel Dekker, New York.
6. Nair, V. N. (1984). "Confidence Bands for Survival Functions with Censored Data: A Comparative Study." *Technometrics* **26,** 265–275.
7. Wilk, M. B, and Gnanadesikan, R. (1968). "Probability Plotting Methods for the Analysis of Data." *Biometrika* **55,** 1–19.
8. Gan, F. F., Koehler, K. J., and Thompson, J. C. (1991). "Probability Plots and Distribution Curves for Assessing the Fit of Probability Models." *The American Statistician* **45,** 14–21.
9. Chambers, J. M., Cleveland, W. S., Kleiner, B., and Tukey, P. A. (1983). *Graphical Methods for Data Analysis.* Wadsworth International Group, Belmont, CA.

10. Fowlkes, E. B. (1987). *A Folio of Distributions: A Collection of Theoretical Quantile–Quantile Plots.* Marcel Dekker, New York.
11. Kolmogorov, A. N. (1933). "Sulla Determinazione Empirica di una Legge di Distribuziane." *Giorna. Inst. Attuaari.* **4,** 83–91.
12. Doksum, K. A., and Sievers, G. L. (1976). "Plotting with Confidence: Graphical Comparisons of Two Populations." *Biometrika* **63,** 421–434.
13. Niederhausen, H. (1981). "Tables and Significant Points for the Variance Weighted Kolmogorov–Smirnov Statistic." Technical Report 298, Stanford University, Dept. of Statistics.
14. Stephens, M. A. (1986). "Tests Based on EDF Statistics," Chapter 4 in *Goodness-of-Fit Techniques,* R. B. D'Agostino and M. A. Stephens (eds.). Marcel Dekker, New York.
15. D'Agostino, R. B., and Stephens, M. A. (1986). *Goodness-of-Fit Techniques.* Marcel Dekker, New York.
16. D'Agostino, R. B. (1986). "Tests for the Normal Distribution," Chapter 9 in *Goodness-of-Fit Techniques,* R. B. D'Agostino and M. A. Stephens (eds.). Marcel Dekker, New York.
17. Shapiro, S. S., and Wilk, M. B. (1965). "An Analysis of Variance Test for Normality (Complete Samples)." *Biometrika* **52,** 591–611.
18. Weisberg, S., and Bingham, C. (1975). "An Approximate Analysis of Variance Test for Normality Suitable for Machine Calculations." *Technometrics* **17,** 133–134.
19. Bowman, K. O., and Shenton, L. R. (1986). "Moment ($\sqrt{b_1}$, $b_2$) Techniques," Chapter 7 in *Goodness-of-Fit Techniques,* R. B. D'Agostino and M. A. Stephens (eds.). Marcel Dekker, New York.
20. Simpson, J., Olsen, A., and Eden, J. C. (1975). "A Bayesian Analysis of a Multiplicative Treatment Effect in Weather Modification." *Technometrics* **17,** 161–166.

# 8. MAXIMUM LIKELIHOOD METHODS FOR FITTING PARAMETRIC STATISTICAL MODELS

William Q. Meeker

Statistics Department
Iowa State University, Ames

Luis A. Escobar

Department of Experimental Statistics
Louisiana State University, Baton Rouge

## 8.1 Introduction

The method of maximum likelihood (ML) is perhaps the most versatile method for fitting statistical models to data. The method dates back to early work by Fisher [1] and has been used as a method for constructing estimators ever since. In typical applications, the goal is to use a parametric statistical model to describe a set of data or a process that generated a set of data. The appeal of ML stems from the fact that it can be applied to a large variety of statistical models and kinds of data (e.g., continuous, discrete, categorical, censored, truncated, etc.), where other popular methods, like least squares, do not, in general, provide a satisfactory method of estimation. Statistical theory (e.g., Kulldorff [2], Kempthorne and Folks [3], Rao [4], and Cox and Hinkley [5]) shows that, under standard regularity conditions, ML estimators are optimal in large samples.

Modern computing hardware and software have tremendously expanded the feasible areas of application for ML methods. There have been a few reported problems with some applications of the ML method. These are typically due to the use of inappropriate approximations or inadequate numerical methods used in calculation. In this chapter we will illustrate the method of ML on a range of problems. We suggest methods to avoid potential pitfalls and to effectively quantify the information (including limitations) in one's data.

### 8.1.1 Modeling Variability with a Parametric Distribution

Chapter 1 shows how to use probability models to describe the variability in data generated from physical processes. Chapter 2 gives examples of some probability distributions that are commonly used in statistical models. For the most part, in this chapter, we will discuss problems with an underlying continuous model, although much of what we say also holds for discrete models.

The natural model for a continuous random variable, say $Y$, is the cumulative distribution function (cdf). Specific examples given in Chapters 1 and 2 are of the form

$$\Pr(Y \leq y) = F_Y(y; \boldsymbol{\theta})$$

where $\boldsymbol{\theta}$ is a vector of parameters and the range of $Y$ is specified. One simple example that we will use later in this chapter is the exponential distribution for which

$$\Pr(Y \leq y) = F_Y(y; \theta) = 1 - \exp\left(-\frac{y}{\theta}\right) \quad (8.1)$$

where $\theta$ is the single parameter of the distribution (equal to the first moment or mean, in this example). The most commonly used parametric probability distributions have between one and four parameters, although there are distributions with more than four parameters. More complicated models involving mixtures or other combinations of distributions or models that include explanatory variables could contain many more parameters.

### 8.1.2 Basic Ideas of Statistical Inference

Chapter 6 provides some of the basic ideas of statistical inference. In this chapter we will focus on the problem of estimating the model parameters $\boldsymbol{\theta}$ and important functions of $\boldsymbol{\theta}$. The choice of $\boldsymbol{\theta}$, a set of parameters (the values of which are often assumed to be unknown) to describe a particular model, is somewhat arbitrary and may depend on tradition, on physical interpretation, or on having parameters with desirable numerical properties. For example, the traditional definition for the parameters of a normal distribution are $\theta_1 = \mu$ and $\theta_2 = \sigma^2 > 0$, the mean and variance, respectively. An alternative unrestricted parameterization would be $\theta_1 = \mu$ and $\theta_2 = \log(\sigma)$. Another parameterization, which has better numerical properties for some data sets is $\theta_1 = \mu - 2\sigma$ and $\theta_2 = \log(\sigma)$.

Interest often centers on quantities that are functions of $\boldsymbol{\theta}$ like

- The probability $p = F_Y(y; \boldsymbol{\theta})$ for a specified $y$. For example, if $Y$ is the tensile strength of a particular unit, then $p$ is the probability that the unit's tensile strength $Y$ is less than $y$.

# INTRODUCTION

- The $p$ quantile of the distribution of $Y$ is the value of $Y_p$ such that $F_Y(Y_p; \boldsymbol{\theta}) = p$. We will express this quantity as $Y_p = F_Y^{-1}(p; \boldsymbol{\theta})$. For the tensile strength example, $Y_p$ is the strength below which we will find $100p\%$ of the units in the product population.
- The mean of $Y$

$$E(Y) = \int_{-\infty}^{\infty} y f_Y(y) dy$$

which is sometimes, but not always, one of the model parameters.

In this chapter we will emphasize both *point estimation* and methods of obtaining *confidence intervals* (for scalars) and *confidence regions* (for simultaneous inference on a vector of two or more quantities). Confidence intervals and regions quantify the uncertainty in parameter estimates arising from the fact that we only have a finite number of observations from the process or population of interest. Importantly, confidence intervals and regions do *not* account for possible model misspecification.

## 8.1.3 Basic Ideas of Modeling and Inference with the Likelihood Function

The practice of statistical modeling is an iterative process of fitting successive models in search of a model that provides an adequate description without being unnecessarily complicated. Application of ML methods generally starts with a set of data and a tentative statistical model for the data. The tentative model is often suggested by the initial graphical analysis (Chapter 7) or previous experience with similar data or other "expert knowledge."

We can consider the likelihood function to be the *probability of the observed data*, written as a function of the model's parameters. For a set of $n$ independent observations the likelihood function can be written as the following joint probability:

$$L(\boldsymbol{\theta}) = L(\boldsymbol{\theta}; \text{DATA}) = \prod_{i=1}^{n} L_i(\boldsymbol{\theta}; \text{DATA}_i) \qquad (8.2)$$

where $L_i(\boldsymbol{\theta}) = L_i(\boldsymbol{\theta}; \text{DATA}_i)$, the interval probability for the $i$th case, is computed as shown in Section 8.2. The dependency of the likelihood on the data will be understood and is usually suppressed in our notation. For a given set of data, values of $\boldsymbol{\theta}$ for which $L(\boldsymbol{\theta})$ is relatively large are more plausible than values of $\boldsymbol{\theta}$ for which the probability of the data is relatively small. There may or may not be a unique value of $\boldsymbol{\theta}$ that maximizes $L(\boldsymbol{\theta})$. Regions in the space of $\boldsymbol{\theta}$ with large $L(\boldsymbol{\theta})$ can be used to define confidence regions for $\theta$. We can also use ML estimation to make inferences on *functions* of $\theta$. In the rest of this chapter, we will show how to make these concepts operational, and we provide examples for illustration.

## 8.2 Data from Continuous Models

Many scientific observations are modeled on a continuous scale. Because of inherent limitations in measurement precision, however, actual data are *always* discrete. Moreover, other limitations in our ability or in our measuring instrument's ability to observe can cause our data to be censored or truncated. We will illustrate several different types of observations, give examples, and show how to compute the "probability of the data" as a function of $F_Y(y; \theta)$. Typical data sets, illustrated by our examples, often contain a mixture of these different types of observations.

### 8.2.1 Interval-Censored Observations

In this section we describe "interval-censored" data that arise due to round-off or binning or when inspections are done at discrete time points in a life test.

Meeker and LuValle [6] give data on failure times of printed circuit boards. The failures were caused by the growth of conductive anodic filaments. The data consist of the number of failures that were observed in a series of 4-hr and 12-hr long intervals over the entire life test. This experiment resulted in what is known as "interval-censored" data because the "exact" failure time is censored, and we know only the interval in which each failure occurred. If the $i$th unit failed between times $y_i^L$ and $y_i^U$ the probability of this event is

$$L_i(\theta) = F_Y(y_i^U; \theta) - F_Y(y_i^L; \theta) \tag{8.3}$$

In this example all test units had the same inspection times, but this is not always the case.

Berkson [7] investigates the randomness of $\alpha$-particle emissions of Americium-241 (which has a half-life of about 458 yrs). Based on physical theory one expects that, over a short period of time, the interarrival times of observed particles would be approximately distributed according to an exponential distribution

$$F_Y(y; \theta) = 1 - \exp\left(-\frac{y}{\theta}\right) \tag{8.4}$$

where $\lambda = 1/\theta$ is the intensity of the corresponding homogeneous Poisson process on the real-time line. For the interarrivals of $\alpha$-particles times, $\lambda$ is proportional to the Americium-241 decay rate.

Berkson's data consisted of 10220 observed interarrival times (time units 1/5000 sec), binned into intervals running from 0 to 4000 time units with interval sizes ranging from 25 to 100 units, with one additional interval for observed times exceeding 4000 time units. To illustrate the effects of sample size on the inferences, we have selected simple random samples (i.e., each interarrival time having equal probability) of size $n = 2000$, 200, and 20 from these interarrival times. To save space, we have used a coarser set of bins; Meeker [8] uses asymptotic variances to show that reducing the number of bins in this way will not seriously affect the precision of ML estimates. These data are shown in Table I. We will return to this example in Section 8.3.

Meeker [9] gives the observed failure times from a life test on 4139 integrated circuits (ICs). These data are shown in Table II. There were 21 failures in the first 593 hrs, but no additional failures had been observed when the test was stopped at 1370 hrs.

TABLE I. Binned $\alpha$-Particle Interarrival Time Data in 1/5000 sec

| Time Interval Endpoint | | Interarrival Times Frequency of Occurrence | | | |
|---|---|---|---|---|---|
| | | All Times | Random Samples of Times | | |
| Lower | Upper | $n = 10220$ | $n = 2000$ | $n = 200$ | $n = 20$ |
| 0 | 100 | 1609 | 292 | 41 | 3 |
| 100 | 300 | 2424 | 494 | 44 | 7 |
| 300 | 500 | 1770 | 332 | 24 | 4 |
| 500 | 700 | 1306 | 236 | 32 | 1 |
| 700 | 1000 | 1213 | 261 | 29 | 3 |
| 1000 | 2000 | 1528 | 308 | 21 | 2 |
| 2000 | 4000 | 354 | 73 | 9 | 0 |
| 4000 | ∞ | 16 | 4 | 0 | 0 |

TABLE II. Integrated Circuit Failure Times in Hours. When the Test Ended at 1370 hrs, 4128 Units Were Still Running

| | | | | | |
|---|---|---|---|---|---|
| .10 | .10 | .15 | .60 | .80 | .80 |
| 1.20 | 2.5 | 3.0 | 4.0 | 4.0 | 6.0 |
| 10.0 | 10.0 | 12.5 | 20. | 20. | 43. |
| 43. | 48. | 48. | 54. | 74. | 84. |
| 94. | 168. | 263. | 593. | | |

216   MAXIMUM METHODS FOR FITTING PARAMETRIC STATISTICAL MODELS

Although time is a continuous variable, these data (as with *all* data) are actually discrete. For the ICs that failed, failure times were recorded to the nearest .05 hr for times before 2 hrs, to the nearest .5 hr for times between 2 and 20 hrs, and to the nearest hour for times beyond 20 hrs. (Actually, we inferred that this was the scheme, based on the observed ties in the data.) Thus the "correct" likelihood is one based on interval-censored data (Eq. (8.3)). For example, the probability of the failure recorded as .15 hrs is

$$L_3(\theta) = F_Y(.175; \theta) - F_Y(.125; \theta)$$

and for the failure recorded at 593 hrs, we have

$$L_{28}(\theta) = F_Y(593.5; \theta) - F_Y(592.5; \theta)$$

### 8.2.2 The "Density Approximation" for "Exact" Observations

Let $\Delta_i$ represent the difference between the $i$th recorded "exact" observation and the upper and lower endpoints of the actual discrete observation. For most statistical models, the contribution of the exact observations to the likelihood (i.e., probability of the data) can, for small $\Delta_i$, be approximated by

$$[F_Y(y_i + \Delta_i; \theta) - F_Y(y_i - \Delta_i; \theta)] \approx 2f_Y(y_i; \theta)\Delta_i \qquad (8.5)$$

It is this approximation that leads to the traditional form of the likelihood for the $i$th failure time $y_i$, which would be written as

$$L_i(\theta) = f_Y(y_i; \theta) \qquad (8.6)$$

where $f_Y(y_i; \theta) = dF_Y(y_i; \theta)/dy$ is the assumed probability density function (pdf) for the random variable $Y$. The density approximation in equation (8.6) is convenient, easy to use, and in simple cases, allows the derivation of closed form equations to compute ML estimates. Because the righthand sides of Eqs. (8.5) and (8.6) differ by a factor of $2\Delta_i$, when the density approximation is used, the approximate likelihood in (8.6) differs from the probability in Eq. (8.5) by a constant scale factor. As long as the approximation in (8.5) is adequate and $\Delta_i$ does not depend on $\theta$, however, this general character (i.e., the shape and the location of the maximum) of the likelihood is not affected.

### 8.2.3 Potential Problems with Using the "Density Approximation" for "Exact" Observations

Although the density approximation in Eq. (8.6) provides an adequate approximation for most commonly used statistical models, there are a number of special cases where it fails. Generally is sufficient that the approximation

$$\prod_{i=1}^{r} [F_Y(y_i + \Delta_i; \boldsymbol{\theta}) - F_Y(y_i - \Delta_i; \boldsymbol{\theta})] \approx 2 \prod_{i=1}^{r} f_Y(y_i; \boldsymbol{\theta}) \Delta_i \quad (8.7)$$

hold as $\Delta_i \to 0$ over *all* values of $\boldsymbol{\theta}$ inside $\Theta$ and especially as one approaches *the boundary* of $\Theta$. Here $\Theta$ the "parameter space," which consists of the possible values of $\boldsymbol{\theta}$.

The breakdown of this approximation is the primary source of cited failures of the ML method. Consider the following problem of looking for signals of known periodicity in a point process (see, for example, Protheroe [10] and Chapter 12 of this volume). Assume that events are observed in time. Events caused by background noise occur according to a Poisson process with a constant intensity. Events from the "signal" occur according to a process with an intensity that is periodic with a known period $\mathcal{P}$. If we take the event times $T_1, T_2, \ldots$ and transform them according to

$$Y_i = \mathrm{mod}(T_i, \mathcal{P})/\mathcal{P}$$

the periodicity is mapped to a circle with a unit circumference. Then the probability density of $Y$ might be described by

$$f_Y(y; \boldsymbol{\theta}) = p + (1 - p) f_N(y; \mu, \sigma)$$

where $0 \leq y \leq 1$, $0 < p < 1$, $0 \leq \mu \leq 1$, $\sigma > 0$,

$$f_N(y; \mu, \sigma) = \frac{1}{\sigma} \sum_{i=-\infty}^{\infty} \phi_{\mathrm{nor}}\left(\frac{y + i - \mu}{\sigma}\right)$$

$$\approx \frac{1}{\sigma} \sum_{i=-K}^{K} \phi_{\mathrm{nor}}\left(\frac{y + i - \mu}{\sigma}\right)$$

$\phi_{\mathrm{nor}}(\cdot)$ is the standardized normal density and $p$ is the proportion of events caused by noise. Here $f_N$ is the density for a "wrapped normal distribution" and $K$ must be chosen large enough so that the neglected terms are negligible (for $\sigma < .5$, we choose $K = 8$). The corresponding cdf for $Y$ is

$$F_Y(y; \boldsymbol{\theta}) = py + (1 - p) F_N(y; \mu, \sigma), \quad 0 \leq y \leq 1$$

where

$$F_N(y; \mu, \sigma) = \sum_{i=-\infty}^{\infty} \left[\Phi_{\mathrm{nor}}\left(\frac{y + i - \mu}{\sigma}\right) - \Phi_{\mathrm{nor}}\left(\frac{i - \mu}{\sigma}\right)\right]$$

$$\approx \sum_{i=-K}^{K} \left[\Phi_{\mathrm{nor}}\left(\frac{y + i - \mu}{\sigma}\right) - \Phi_{\mathrm{nor}}\left(\frac{i - \mu}{\sigma}\right)\right]$$

and $\Phi_{\text{nor}}(\cdot)$ is the standardized normal cumulative distribution function. The wrapped normal is approximately the same as the von Mises distribution used in Protheroe [10]. In either case, if one uses the density approximation on the righthand side of (8.7) for the likelihood, the approximation breaks down. To see this, fix $p$ to any value between 0 and 1, let $\mu$ equal any observed $Y$ and as $\sigma \to 0$, the product of the densities approaches $\infty$. With the correct likelihood on the lefthand side of (8.7), however, the likelihood is bounded and there are no problems.

For other examples, see Le Cam [11] and Friedman and Gertsbakh [12], who describe models and likelihoods that have paths in the parameter space along which $L(\theta)$ approaches $\infty$. Also see Section 8.6.1. The potential problem of encountering an unbounded likelihood can be avoided by using, instead, the "correct likelihood" of the data, which, because it is a probability, is bounded by 0 and 1 for *all* values of $\theta$ in the "parameter space" $\Theta$. The advantages of using the density approximation, *if* the approximation is adequate, are that one does not need to specify the mapping (often unknown) from $y_i$ to $(y_i^L, y_i^U)$ and the amount of computer time needed is typically reduced by about 50%. If, however, the approximation is inadequate, serious numerical and statistical problems are likely to occur.

### 8.2.4 Left-Censored Observations

Left-censored observations arise when we have an upper bound on a response. Such observations are common when an instrument lacks the sensitivity needed to measure observations below a threshold, the value of which is known. Table III gives observed values of fractional variation of X-ray intensity for 53 active galaxies, first reported in Tennant and Mushotzky [13]. The data were also analyzed by Feigelson and Nelson [14]. The numbers marked with a < are left censored (upper bounds on the actual values) due to limitations in observation sensitivity.

TABLE III. Observed Fractional Variation of X-Ray Intensity for 53 Active Galaxies

| | | | | | |
|---|---|---|---|---|---|
| <1.6 | 3.0  | <0.8 | <0.8 | 1.9  | 1.1  |
| <0.8 | 2.0  | 1.6  | <0.7 | 0.9  | <0.7 |
| 0.6  | 0.8  | <0.6 | <0.6 | <0.7 | 0.8  |
| <1.3 | 1.3  | 1.4  | 1.0  | <0.6 | <0.5 |
| <0.6 | <0.7 | <0.8 | <0.9 | 1.6  | <0.9 |
| 1.2  | 1.5  | 5.7  | <2.3 | <1.8 | 1.5  |
| <0.8 | <0.8 | 8.4  | <0.8 | 1.3  | 1.5  |
| 0.7  | 1.6  | <0.8 | <1.2 | <0.7 | 1.5  |
| <0.9 | <1.0 | <1.0 | <0.9 | <0.8 |      |

Left-censored observations also occur in life-test applications when a unit has failed at the time of its first inspection; all we know is that the unit failed before the inspection time (e.g., the first row of Table I).

In general, if we have an upper bound $y_i^U$ for the $i$th observation, the probability of the observation is

$$L_i(\boldsymbol{\theta}) = F_Y(y_i^U; \boldsymbol{\theta}) - F_Y(-\infty; \boldsymbol{\theta}) = F_Y(y_i^U; \boldsymbol{\theta}) \tag{8.8}$$

If the lower limit of support for $F_Y(y; \boldsymbol{\theta})$ is 0 (e.g., when $Y$ is concentration level, tensile strength, or waiting time), we replace $-\infty$ with 0 in Eq. (8.8).

### 8.2.5 Right-Censored Observations

At the end of the IC life test there were 4128 units that ran until 1370 hrs (the end of the test) without failure. These are "right-censored" observations because all we know is that the units' failure times are beyond 1370 hrs. They can also be thought of as intervals running from 1370 hrs to $\infty$. In general, if we have a lower bound $y_i^L$ for the $i$th unit, the probability of the observation is

$$L_i(\boldsymbol{\theta}) = F_Y(\infty; \boldsymbol{\theta}) - F_Y(y_i^L; \boldsymbol{\theta}) = 1 - F_Y(y_i^L; \boldsymbol{\theta})$$

### 8.2.6 Left Truncation

Although left truncation is similar to left censoring, the concepts differ in important ways. Consider monitoring a channel to measure the amplitude of randomly arriving signals. There are two possibilities:

- We know when a signal is present but cannot measure its amplitude when its value lies below a specified threshold, say $\tau^L$. Then we know the number of signals that were not measured exactly and the data are left censored.
- A signal is not detected at all when its amplitude lies below the threshold $\tau^L$. Then we observed only the signals that are greater than $\tau_L$ in amplitude; we do not know the number of signals below $\tau_L$ and this leads to left truncation.

For most problems the additional information provided by the number of missed signals would importantly improve estimation precision. Left truncation also arises in life-test applications when failures that occur before $\tau^L$ are not recorded and we have no information, even on the number of units that failed before this time.

If all units below $\tau^L$ in a population or process are screened out before observation, the remaining data are from a "left-truncated" distribution.

Depending on the application, interest could center on either the original untruncated distribution or on the distribution after truncation.

In general, if a random variable $Y_i$ is truncated when it falls below $\tau_i^L$, then the probability of an observation is the conditional probability

$$L_i(\theta) = \Pr(y_i^L < Y_i \leq y_i^U \mid Y_i > \tau_i^L) = \frac{F_Y(y_i^U; \theta) - F_Y(y_i^L; \theta)}{1 - F_Y(\tau_i^L; \theta)}$$

where $y_i^L \geq \tau_i^L$. It is possible to have either right or left censoring when sampling from a left-truncated distribution. To obtain $L_i(\theta)$, one simply replaces the numerator by the appropriate probability. For "exact" observations, the numerator can, for some distributions, be replaced by the density approximation $f_Y(y; \theta)$.

### 8.2.7 Right Truncation

Right truncation is similar to left truncation and occurs when the upper tail of the values of the distribution are removed. If the random variable $Y_i$ is truncated when it lies above $\tau_i^U$ then the probability of an observation is

$$L_i(\theta) = \Pr(y_i^L < Y_i \leq y_i^U \mid Y_i \leq \tau_i^U) = \frac{F_Y(y_i^U; \theta) - F_Y(y_i^L; \theta)}{F_Y(\tau_i^U; \theta)}$$

where $y_i^U \leq \tau_i^U$. As with left truncation it is possible to have either left or right censoring when sampling from the right-truncated distribution.

### 8.2.8 Regression and Explanatory Variables

It is often useful or important to have a model in which one or more of the elements of $\theta$ are allowed to be functions of explanatory variables. This is a generalization of statistical "regression analysis" in which the most common models have the mean of the normal distribution depending linearly on a vector **x** of explanatory variables. For example, if $x_i$ is a scalar explanatory variable for the $i$th observation,

$$\mu_i = \beta_0 + \beta_1 x_i$$

In this case we think of $x_i$ as being a fixed auxiliary part of $\text{DATA}_i$ (when the variables $x_i$ are themselves random, the common statistical methods and models provide inferences that are conditional on the fixed, observed values of these explanatory variables) and the unknown regression model coefficients ($\beta_0$ and $\beta_1$) replace the model parameter $\mu_i$ in $\theta$. Chapter 9 of this volume describes statistical methods for regression models (that are linear and nonlinear in the parameters) and discusses standard regression assumptions, including the standard assumption that the explanatory variables have negligible measurement error. Lawless [15], Cox and Oakes [16], and Nelson [17] describe applications of such models with censored data.

In some situations, there may be more than one model parameter that will depend on one or more explanatory variables. Nelson [18] gives an example in which both $\mu$ and $\sigma$ (parameters of the lognormal distribution) depend on a stress variable. Meeker and LuValle [6] describe a regression model in which two different kinetic rate constants depend on temperature.

Given appropriate data, all of the methods in this chapter can be applied to problems with explanatory variables. To do this, the basic ideas in Chapter 9 can be applied with ML estimation using the alternative distributions and different kinds of data used in this section.

## 8.3 General Method and Application to the Exponential Distribution (a One-Parameter Model)

### 8.3.1 Data and Model

This section describes the general method of ML and illustrates its use by fitting the exponential distribution to Berkson's data in Table I. We compare the results that one obtains with samples of size $n = 10220, 2000$, 200 and 20. Figure 1 shows an exponential probability plot of the $n = 200$

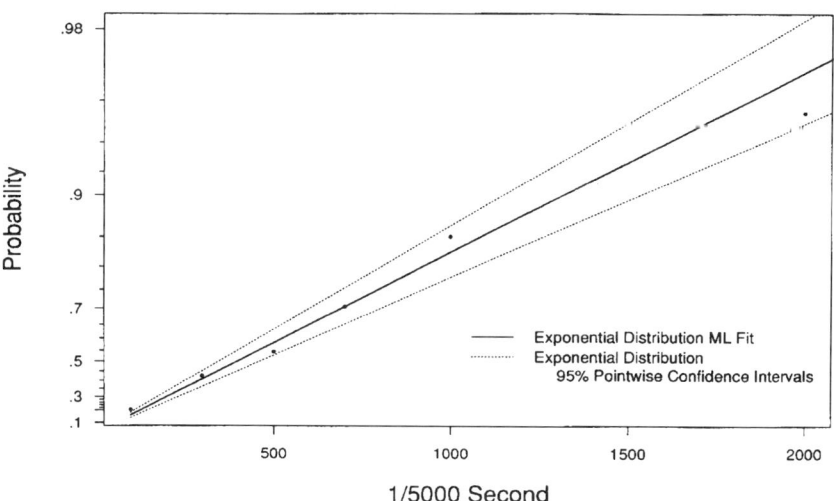

FIG. 1. Exponential probability plot, ML estimate and 95% confidence intervals for $F_Y(y; \theta)$ based on the $n = 200$ sample of $\alpha$-particle interarrival time data.

**222**   MAXIMUM METHODS FOR FITTING PARAMETRIC STATISTICAL MODELS

sample. The solid line on the graph is the ML estimate of the exponential. The dotted lines are drawn through a set of 95% pointwise normal-theory confidence intervals; these intervals will be explained in Section 8.3.6. The approximately linearity of the points on the plot indicates that the exponential distribution provides a good fit to these data.

### 8.3.2 The Likelihood Function and Its Maximum

If we have a sample of $n$ independent observations, denoted generically by $DATA_i$, $i = 1, \ldots, n$, and a specified model, then the total likelihood $L(\theta)$ for the sample is given by Eq. (8.2). The ML estimate of the parameter vector $\theta$ is found by maximizing $L(\theta)$. When there is a unique maximum we use $\hat{\theta}$ to denote the value of $\theta$ that maximizes $L(\theta)$. Note that, in general, the maximum may not be unique. The surface $L(\theta)$ can have flat spots or, more commonly, ridges along which $L(\theta)$ changes slowly, if at all. These may or may not be at the maximum value of $L(\theta)$. Such behaviors generally indicate "identifiability problems" for a parameter or parameters that could be the result of (a) overparameterization, (b) inadequate data, or (c) an inappropriate model. The shape and magnitude of $L(\theta)$ relative to $L(\hat{\theta})$ over $\theta \in \Theta$ describe the information on $\theta$ that is contained in $DATA_i$, $i = 1, \ldots, n$.

Operationally, for some purposes, it is good practice to use the log likelihood $\mathcal{L}_i(\theta) = \log[L_i(\theta)]$. For all practical problems $\mathcal{L}_i(\theta)$ will be representable in computer memory without special scaling (which is not so for $L(\hat{\theta})$ because of possible extreme exponent values), and some theory for ML is developed more naturally in terms of sums like

$$\mathcal{L}(\theta) = \log[L(\theta)] = \sum_{i=1}^{n} \mathcal{L}_i(\theta)$$

than in terms of the products in Eq. (8.2).

When $F_Y(y; \mu, \sigma)$ is a normal distribution with known variance $\sigma^2$ and when there is no censoring or truncation, $\mathcal{L}(\theta)$ will be quadratic in $\mu$ (which implies that $L(\mu)$ will have the shape of a normal or Gaussian density). This will be approximately so for other models under regularity conditions (see Section 8.7.2) and large samples. Otherwise, and particularly in unfamiliar data or model situations, it is important to investigate $L(\theta)$ graphically. This is simple when $\theta$ has length 1 or 2. When the length of $\theta$ is 3 or more we will view 1 (and 2) dimensional "profiles" of $L(\theta)$.

**Example:** Substituting Eq. (8.4) into (8.3), and (8.3) into (8.2) gives the exponential distribution likelihood function for Berkson's data (Table I) as

$$L(\theta) = \prod_{j=1}^{8} \left[ \exp\left(-\frac{y_j^L}{\theta}\right) - \exp\left(-\frac{y_j^U}{\theta}\right) \right]^{r_j} \qquad (8.9)$$

where $r_j$ is the number of interarrival times that had lengths between $y_j^L$ and $y_j^U$.

In this case, the unknown parameter $\theta$ is a scalar, and this makes the analysis particularly simple and provides a useful first example to illustrate basic concepts. In subsequent sections we will generalize with other examples. Table IV summarizes the results of fitting exponential distributions to the four different samples in Table I. We provide results specifically for $\theta$, the mean (which is also the .63 quantile) of the exponential distribution. In Section 8.3.4 we show how to obtain similar results for $\lambda = 1/\theta$, the arrival intensity rate (per unit of time).

Figure 2 shows the *relative likelihood* functions $R(\theta) = L(\theta)/L(\hat{\theta})$ for the sample of sizes $n$ = 2000, 200 and 20 and a vertical line at the ML estimate for $n$ = 10220. These functions allow one to judge the probability of the data for a value of $\theta$ *relative* to the probability at the ML. For example, $R(\theta) = .2$ implies that the probability is 5 times larger at $\hat{\theta}$ than at $\theta$. In the next section we explain how to use $R(\theta)$ to compute confidence intervals for $\theta$.

We see that the spread of the likelihood function tends to decrease with increasing sample size. The likelihood function for the larger samples are

TABLE IV. Summary and Comparison of ML Estimates and Approximate 95% Confidence Intervals from Fitting the Exponential Distribution to Samples of the α-Particle Interarrival Data

|  | All Times $n$ = 10220 | Sample of Times | | |
|---|---|---|---|---|
|  |  | $n$ = 2000 | $n$ = 200 | $n$ = 20 |
| ML estimate $\hat{\theta}$ | 596 | 613 | 572 | 440 |
| Standard error $s_{\hat{\theta}}$ | 6.1 | 14.1 | 42.7 | 101 |
| 95% confidence intervals for $\theta$ based on |  |  |  |  |
|   Likelihood | [585, 608] | [586, 641] | [498, 662] | [289, 713] |
|   $\log(\hat{\theta}) \sim$ normal | [585, 608] | [586, 641] | [496, 660] | [281, 690] |
|   $\hat{\theta} \sim$ normal | [585, 608] | [585, 640] | [490, 653] | [242, 638] |
| ML estimate $\hat{\lambda} \times 10^5$ | 168 | 163 | 175 | 227 |
| Standard error $s_{\hat{\lambda} \times 10^5}$ | 1.7 | 3.8 | 13 | 52 |
|   Likelihood | [164, 171] | [156, 171] | [151, 201] | [140, 346] |
|   $\log(\hat{\lambda}) \sim$ normal | [164, 171] | [156, 171] | [152, 202] | [145, 356] |
|   $\hat{\lambda} \sim$ normal | [164, 171] | [156, 171] | [149, 200] | [125, 329] |

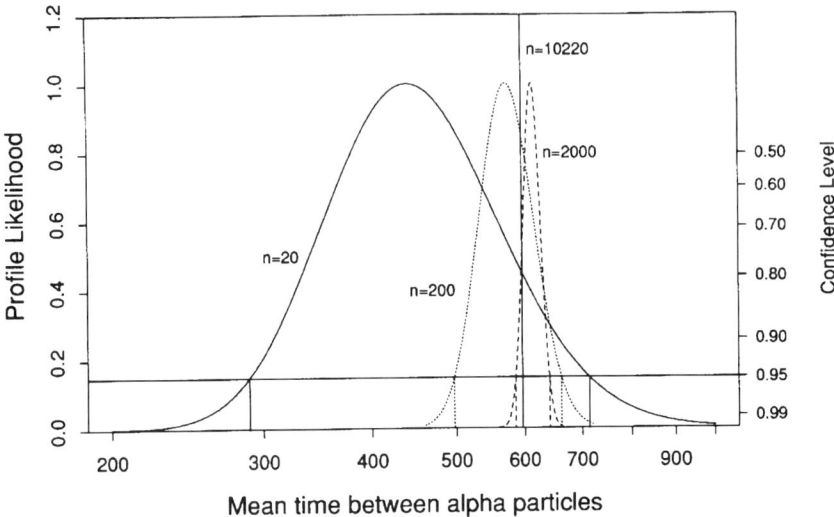

FIG. 2. Relative likelihood functions $R(\theta) = L(\hat{\theta})$ for $n = 2000$, 200, and 20 samples and ML estimate for the $n = 10220$ sample of the $\alpha$-particle interarrival data.

much "tighter" than those for the smaller samples, indicating that the larger samples contain more information about $\theta$. The $\theta$ values at which the different likelihood functions are maximized is random and depends on the particular times chosen for the samples. As expected, the four values of $\hat{\theta}$ differ, but are consistent with the overlapping of the $R(\theta)$ functions in Figure 2. When focusing on just one sample, we will use the sample of $n = 200$ interarrival times.

### 8.3.3 Likelihood Ratios, Profile Likelihoods, and Approximate Likelihood-Based Confidence Intervals on All or Part of $\theta$

The likelihood function provides a versatile method for assessing the information that one's data contains on parameters or functions of parameters. Specifically, use of the likelihood function provides a generally useful method for finding approximate confidence regions or intervals for parameters and functions of parameters.

For the general problem, assume that we want to make inferences on $\theta_1$, from the partition $\theta = (\theta_1', \theta_2')'$, where $'$ denotes vector transpose. Let $k_1$ denote the length of $\theta_1$. The profile likelihood for $\theta_1$ is

$$R(\boldsymbol{\theta}_1) = \max_{\boldsymbol{\theta}_2} \left[ \frac{L(\boldsymbol{\theta}_1, \boldsymbol{\theta}_2)}{L(\hat{\boldsymbol{\theta}})} \right]$$

When the length of $\boldsymbol{\theta}_2$ is 0 (as in our exponential example), this is a relative likelihood for $\boldsymbol{\theta} = \boldsymbol{\theta}_1$. Otherwise we have a "maximized relative likelihood" for $\boldsymbol{\theta}_1$. In either case, $R(\boldsymbol{\theta}_1)$ is commonly known as a *profile likelihood* because it provides a view of the profile of $L(\boldsymbol{\theta})$. When

- $\boldsymbol{\theta}_1$ is of length 1, $R(\boldsymbol{\theta}_1)$ is a curve projected onto a plane.
- $\boldsymbol{\theta}_1$ is of length 2 or more, $R(\boldsymbol{\theta}_1)$ is a surface projected onto a hyperplane.

In either case the projection is in a direction perpendicular to the coordinate axes for $\boldsymbol{\theta}_1$. When $\boldsymbol{\theta}_1$ is of length 1 or 2, it is useful to display $R(\boldsymbol{\theta}_1)$ graphically.

An approximate $100(1 - \alpha)\%$ likelihood-based confidence region for $\boldsymbol{\theta}_1$ is the set of all values of $\boldsymbol{\theta}_1$ such that $-2 \log[R(\boldsymbol{\theta}_1)] < \chi^2_{(1-\alpha;k_1)}$, or $R(\boldsymbol{\theta}_1) > \exp(-\chi^2_{(1-\alpha;k_1)}/2)$, where $k_1$ is the length of $\boldsymbol{\theta}_1$. Simulation studies for different applications and models (e.g., Ostrouchov and Meeker [19], Meeker [9], and Vander Wiel and Meeker [20]) have shown that the likelihood-based intervals have important advantages over the standard normal-theory intervals (discussed in Section 8.3.6), especially when one has only a small sample. Specifically, in repeated sampling, normal-theory intervals tend to have actual confidence levels that are smaller than the nominal levels. Likelihood-based intervals tend to have confidence levels that are much closer to the nominal. These intervals will be compared later, in our numerical examples.

**Example:** For the particle arrival data, Table IV gives approximate 95% confidence intervals for $\theta$ based on the likelihood method and the normal-theory method (explained in Section 8.3.6). Figure 2 illustrates the application of the likelihood method. The horizontal line is at $\exp(-\chi^2_{(.95,1)}/2) = .147$, corresponding to 95% confidence intervals. The vertical lines dropping from the respective curves give the confidence intervals for the different samples and can be compared with the values in Table IV. This figure shows that increasing sample size decreases confidence interval length. Asymptotic (large-sample) theory shows that confidence interval length under standard regularity conditions is approximately proportional to $\sqrt{n^{-1}}$.

### 8.3.4 Approximate Likelihood-Based Confidence Regions and Intervals for Functions of θ

The parameters $\boldsymbol{\theta}$ in a model are often chosen for convenience, by tradition so that the parameters have scientific meaning, or for numerical reasons.

Interest, however, often centers on functions of these parameters like probabilities $p = F_Y(y; \boldsymbol{\theta})$ or quantiles like $Y_p = F_Y^{-1}(p; \boldsymbol{\theta})$. In general, the ML estimator of a transformation $\mathbf{g}(\boldsymbol{\theta})$ is $\mathbf{g}(\hat{\boldsymbol{\theta}})$. Due to this *invariance* property of ML estimators, likelihood-based methods can, in principle, be applied, as described previously, to make inferences about such transformations. For each function of interest, this can be done by defining a one-to-one transformation (or reparameterization), $\mathbf{g}(\boldsymbol{\theta})$, that contains the function of interest among its elements. Some of the new parameters may be identical to the old ones. To compute confidence intervals for the elements of $\mathbf{g}(\boldsymbol{\theta})$ we require that the first partial derivations of $\mathbf{g}(\boldsymbol{\theta})$ be continuous.

Then ML procedures can be carried out for this new parameterization in a manner that is the same as that described previously for $\boldsymbol{\theta}$ to find ML estimates and confidence intervals for any scalar or vector function of $\boldsymbol{\theta}$. If one can readily compute $\mathbf{g}(\boldsymbol{\theta})$ and its inverse, this method is straightforward. Otherwise, iterative numerical methods are needed and computing time could become an issue.

**Example:** For one-parameter distributions, like the exponential in Eq. (8.4) used in the α-particle interarrival data, one can translate estimates and confidence intervals for θ directly into estimates and confidence intervals for monotone functions of θ. For example, interest might center on the arrival rate $\lambda = 1/\theta$. Using the ML estimate and the likelihood-based confidence interval for the $n = 200$ sample in Table IV, we have

$$\hat{\lambda} = 1/\hat{\theta} = .00175$$
$$[\underset{\sim}{\lambda}, \tilde{\lambda}] = [1/\tilde{\theta}, 1/\underset{\sim}{\theta}] = [.00151, .00201]$$

where $\underset{\sim}{\theta}$ and $\tilde{\theta}$ are lower and upper endpoints of the confidence interval for θ.

In this case, because λ is a *decreasing* function of θ, we substituted the upper limit for θ to get a lower limit for λ and vice versa. Section 8.3.6 describes the "normal-theory" approximate confidence intervals in Table IV. Sections 8.4.3 and 8.5.4 illustrate confidence intervals for the more complicated situation in which $\boldsymbol{\theta}$ is a vector.

### 8.3.5 Asymptotic Distribution of ML Estimators, Fisher Information, and Variances

Under standard regularity conditions on the assumed model and form of the resulting data (see Section 8.7.2), ML estimators have certain desirable statistical properties. In particular, in large samples, the random vector $\hat{\boldsymbol{\theta}}$ approximately follows a multivariate normal distribution with mean $\boldsymbol{\theta}$ and covariance matrix $\Sigma_{\hat{\boldsymbol{\theta}}} = I_{\boldsymbol{\theta}}^{-1}$, where

$$I_{\boldsymbol{\theta}} = \mathrm{E}\left[-\frac{\partial^2 \mathcal{L}(\boldsymbol{\theta})}{\partial \boldsymbol{\theta} \partial \boldsymbol{\theta}'}\right] = \sum_{i=1}^{n} \mathrm{E}\left[-\frac{\partial^2 \mathcal{L}_i(\boldsymbol{\theta})}{\partial \boldsymbol{\theta} \partial \boldsymbol{\theta}'}\right]$$

where $E[\cdot]$ is the expectation operator. This matrix $I_{\boldsymbol{\theta}}$ is often known as the Fisher information or "expected information" matrix for $\boldsymbol{\theta}$. For large samples we can say that $I_{\boldsymbol{\theta}}$ approximately quantifies the amount of information that we "expect" to get from our future data. Generally, more information implies smaller variance. Asymptotic (large-sample) statistical theory shows that the elements of $\Sigma_{\hat{\boldsymbol{\theta}}}$ are of the order of $n^{-1}$.

For an assumed model if there is to be no censoring or truncation, and if the density approximation (Eq. (8.6)) is used for $L_i(\boldsymbol{\theta})$, then $\Sigma_{\hat{\boldsymbol{\theta}}}$ is a function of the sample size $n$, the unknown parameters $\boldsymbol{\theta}$, and the levels of the explanatory variables (if any). Otherwise, $I_{\boldsymbol{\theta}}$ also depends on the type of censoring, truncation, rounding, etc. that will be encountered in the data. If any of these measurement or observation "limitations" are random, then $I_{\boldsymbol{\theta}}$ depends on the distribution(s) of these limitations. Generally, the effect of round-off or binning on the "correct likelihood" is not large (e.g., Meeker [8]). Because $\Sigma_{\hat{\boldsymbol{\theta}}}$ does not depend on data, if one has "planning values" for $\boldsymbol{\theta}$, it is generally straightforward to evaluate $\Sigma_{\hat{\boldsymbol{\theta}}}$ numerically to compute the variances of $\hat{\boldsymbol{\theta}}$ and of smooth functions of $\hat{\boldsymbol{\theta}}$ (see the details later), and these variances are useful for planning experiments; see, for example, Escobar and Meeker [21] and Nelson [17], Chapter 6.

When data are available, one can get a "local" estimate $\hat{\Sigma}_{\hat{\boldsymbol{\theta}}}$ of $\Sigma_{\boldsymbol{\theta}}$ as the inverse of the "observed" information matrix

$$\hat{I}_{\boldsymbol{\theta}} = \left[-\frac{\partial^2 \mathcal{L}(\boldsymbol{\theta})}{\partial \boldsymbol{\theta} \partial \boldsymbol{\theta}'}\right] = \sum_{i=1}^{n}\left[-\frac{\partial^2 \mathcal{L}_i(\boldsymbol{\theta})}{\partial \boldsymbol{\theta} \partial \boldsymbol{\theta}'}\right]$$

where the derivatives are evaluated at $\hat{\boldsymbol{\theta}}$. Under certain regularity conditions (see Section 8.7.2), $\hat{\Sigma}_{\hat{\boldsymbol{\theta}}}$ is a consistent estimator of $\Sigma_{\hat{\boldsymbol{\theta}}}$.

In general, one is interested in inferences on functions of $\boldsymbol{\theta}$. For example, consider a vector function $\mathbf{g}(\boldsymbol{\theta})$ of the parameters such that the function is one to one and all the first derivatives with respect to the elements of $\boldsymbol{\theta}$ exist and are continuous. The ML estimator of $\mathbf{g}(\boldsymbol{\theta})$ is $\hat{\mathbf{g}} = \mathbf{g}(\hat{\boldsymbol{\theta}})$. In large samples, $\mathbf{g}(\hat{\boldsymbol{\theta}})$ approximately follows a normal distribution with mean $\mathbf{g}(\boldsymbol{\theta})$ and covariance matrix

$$\Sigma_{\hat{g}} = \left[\frac{\partial \mathbf{g}(\boldsymbol{\theta})}{\partial \boldsymbol{\theta}}\right]' \Sigma_{\hat{\boldsymbol{\theta}}} \left[\frac{\partial \mathbf{g}(\boldsymbol{\theta})}{\partial \boldsymbol{\theta}}\right] \qquad (8.10)$$

The "local" estimates of $\Sigma_{\hat{g}}$ can be obtained by substituting $\hat{\Sigma}_{\hat{\theta}}$ for $\Sigma_{\hat{\theta}}$ and evaluating the derivatives at $\hat{\boldsymbol{\theta}}$. This Taylor series approximation is sometimes known as *propagation of error* in the engineering literature and the *delta method* in the statistical literature.

For scalar $g$ and $\theta$ the formula simplifies to

$$\text{Var}[g(\hat{\theta})] = \left[\frac{\partial g(\theta)}{\partial \theta}\right]^2 \Sigma_{\hat{\theta}} = \left[\frac{\partial g(\theta)}{\partial \theta}\right]^2 \text{Var}(\hat{\theta})$$

where $\text{Var}[\cdot]$ is the variance function. For example, if $\theta$ is positive and $g(\theta)$ is the logarithmic function, the asymptotic variance of $\log(\hat{\theta})$ is $\text{Var}[\log(\hat{\theta})] = \text{Var}(\hat{\theta})/\theta^2$.

### 8.3.6 Approximate Confidence Regions and Intervals Based on Asymptotic Normality of ML Estimators

The large sample normal approximation for the distribution of ML estimators can be used to compute approximate confidence intervals (regions) for scalar (vector) functions of $\boldsymbol{\theta}$. In particular, an approximate $100(1 - \alpha)\%$ confidence region for $\boldsymbol{\theta}$ is the set of all values of $\boldsymbol{\theta}$ in the ellipsoid

$$(\hat{\boldsymbol{\theta}} - \boldsymbol{\theta})' \left[\hat{\Sigma}_{\hat{\boldsymbol{\theta}}}\right]^{-1} (\hat{\boldsymbol{\theta}} - \boldsymbol{\theta}) \leq \chi^2_{(1-\alpha;k)} \qquad (8.11)$$

where $k$ is the length of $\boldsymbol{\theta}$. This is sometimes known as *Wald's method*, but we will refer to it as the *normal-theory method*. Normal-theory intervals are easy to compute and are used in most commercial statistical packages. Their main shortcomings are (a) they have actual coverage probabilities that can be importantly different from the nominal specification, unless sample sizes are large and (b) unlike the likelihood-based intervals, they depend on the transformation used for the parameter, as illustrated in the following example. With moderate to large samples they are useful for initial data analyses, where rapid interactive analysis is important.

Normal-theory confidence regions are based on a *quadratic approximation* to the log likelihood and will be adequate when the log likelihood is nearly quadratic over the range of the confidence region. This region will be large for small samples and small for larger samples (see Figure 2). In "large samples," under the usual regularity conditions (Section 8.7.2) the log likelihood is approximately quadratic, and thus the two regions will be in close agreement. The sample size required to have an adequate approximation is not easy to characterize because it depends on the model, the amount of censoring and truncation, and on the desired inference. In some extreme examples, $n$ on the order of thousands is not sufficient for a good approximation. When the quadratic approximation is poor, likelihood-based intervals (Section 8.3.3) should be used instead, especially when reporting final results.

For the general problem, the $100(1 - \alpha)\%$ normal-theory confidence region for a $k_1$ dimensional subset $\mathbf{g}_1 = \mathbf{g}_1(\boldsymbol{\theta})$, from the partition $\mathbf{g}(\boldsymbol{\theta}) = (\mathbf{g}_1(\boldsymbol{\theta}), \mathbf{g}_2(\boldsymbol{\theta}))'$ is the set of all $\mathbf{g}_1$ in the ellipsoid

$$[\hat{\mathbf{g}}_1 - \mathbf{g}_1]' \left[ \hat{\Sigma}_{\hat{\mathbf{g}}_1} \right]^{-1} [\hat{\mathbf{g}}_1 - \mathbf{g}_1] \leq \chi^2_{(1-\alpha;k_1)}$$

where $\hat{\mathbf{g}}_1 = \mathbf{g}_1(\hat{\boldsymbol{\theta}})$ is the ML estimator of $\mathbf{g}_1(\boldsymbol{\theta})$ and $\hat{\Sigma}_{\hat{\mathbf{g}}_1}$ is the local estimate of the covariance matrix of $\hat{\mathbf{g}}_1$, which can be obtained from the local estimate of $\hat{\Sigma}_{\hat{g}}$ in equation (8.10). As shown in Escobar and Meeker [22], this normal-theory confidence region can be viewed as a quadratic approximation for the log profile likelihood of $\mathbf{g}_1(\boldsymbol{\theta})$ at $\hat{\mathbf{g}}_1$.

When $k_1 = 1$, $g_1 = g_1(\boldsymbol{\theta})$ is a scalar function of $\boldsymbol{\theta}$, the $100(1 - \alpha)\%$ normal-theory interval is obtained from the familiar formula

$$[\underset{\sim}{g}_1, \tilde{g}_1] = \hat{g}_1 \pm z_{(1-\alpha/2)} s_{\hat{g}_1}$$

where $s_{\hat{g}_1} = \sqrt{\widehat{\text{Var}}[\hat{g}_1(\boldsymbol{\theta})]}$ is the local estimate for the standard error of $\hat{g}_1$ and $z_{(1-\alpha/2)}$ is the $1 - \alpha/2$ quantile of the standard normal distribution.

When using approximate confidence intervals based on the normal-theory method, the choice of parameterization or parameter transformation can be extremely important to the accuracy of the confidence level approximation, especially with small samples. There are some general rules that tend to improve the approximation. In particular, one should try to find a transformation that will improve the approximation to the likelihood-based method by, for example, making $R(\boldsymbol{\theta})$ nearly symmetric. Sometimes this can be done by transforming bounded parameters so that they are unrestricted. For positive parameters (e.g., a scale parameter $\sigma > 0$), using the approximate distribution of $\log(\sigma)$, which is unrestricted, will often (but not always) improve the approximation. Similarly, for estimating $p = F_Y(y; \boldsymbol{\theta})$ for a specified $y$ or some other model probability $p$, the logistic transformation $\text{logit}(p) = \log[p/(1-p)]$ can improve the approximation. For further discussion, see Lawless [15], pages 108 and 403, and Nelson [23], page 362. Of course, direct use of the likelihood renders the transformation question moot, as the likelihood-based confidence regions and intervals are invariant to the model parameterization.

**Example:** For the particle arrival data, Table IV compares 95% approximate confidence intervals for $\theta$ and $\lambda$ based on (a) the likelihood method, (b) the large sample approximate normal distribution for $\hat{\theta}$ and $\hat{\lambda}$, and (c) the large sample approximate normal distribution for $\log(\hat{\theta})$ and $\log(\hat{\lambda})$. As expected, there is close agreement for the larger sample sizes. For the sample with $n = 20$, however, there are large differences among the methods. The comparison shows that, in this case, the normal theory method with the log transformation provides a better approximation to the likelihood-based method.

The dotted lines in Figure 1 are drawn through a set of normal-theory 95% pointwise confidence intervals for the exponential $F_Y(y; \theta)$, based on the logit transformation. Because $\theta$ is the only unknown parameter for this model, these "bands" will contain the unknown exponential $F_Y(y; \theta)$ if and only if the confidence interval for $\theta$ contains the unknown $\theta$. Thus this set of intervals can also be interpreted as simultaneous confidence bands for the entire exponential $F_Y(y; \theta)$. In subsequent examples, with more than one parameter, we will have to view such a collection of intervals differently because the confidence level applies only to the process of constructing an interval for a single point on the distribution. To make a "simultaneous" statement would require either a wider set of bands or a lower level of confidence.

## 8.4 Fitting the Weibull with Left-Censored Observations (a Two-Parameter Model)

### 8.4.1 Example and Data

In this section we return to the data on fractional variation of X-ray intensity from active galaxies, introduced and described in Section 8.2.4. Figure 3 is a Weibull probability plot of these data, constructed using methods similar to those described in Chapter 7. The first point is plotted at a probability of about .43, due to the large amount of left censoring in the lower tail of the distribution. The plotted points in the upper tail of the distribution seem to depart somewhat from linearity, indicating a small departure from the Weibull model. Even if the distribution of galaxy variability could be described by a Weibull distribution, it is possible that such a departure could arise from the particular sample of galaxies that were chosen for observation (assuming that the sample was, in fact "random" from a list of galaxies that could have been observed).

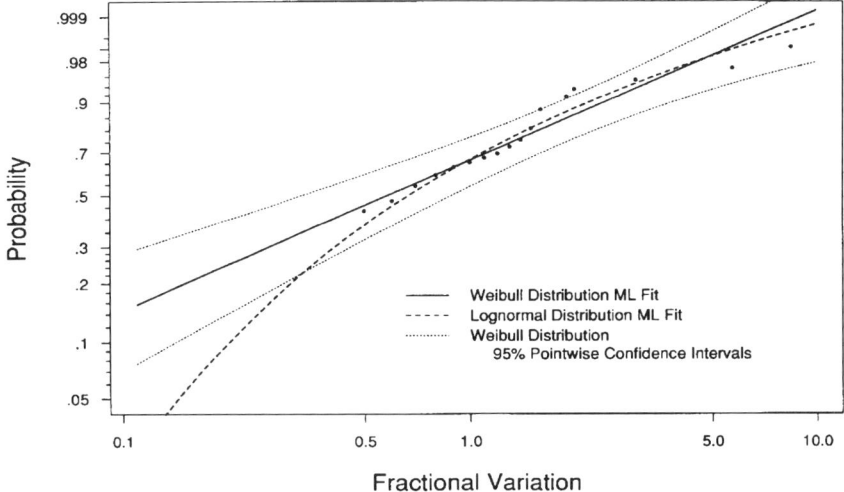

FIG. 3. Weibull probability plot of X-ray variability data with maximum likelihood estimates and 95% pointwise confidence intervals for $F_Y(y; \mu, \sigma)$.

### 8.4.2 The Likelihood Function and Its Maximum

The departure from linearity in Figure 3 was not too pronounced. Therefore we fit the two-parameter Weibull distribution to the data using ML. This is equivalent to fitting a straight line through the data on the Weibull probability paper to estimate the cdf, using the ML criterion to choose the line. The Weibull cdf is commonly written as

$$F_Y(y; \alpha, \beta) = 1 - \exp\left[-\left(\frac{y}{\alpha}\right)^\beta\right] \qquad (8.12)$$

where $\beta$ is a shape parameter and $\alpha$ is a scale parameter. The Weibull cdf also can be written as

$$F_Y(y; \mu, \sigma) = \Phi_{\text{sev}}\left[\frac{\log(y) - \mu}{\sigma}\right] \qquad (8.13)$$

where $\Phi_{\text{sev}}(z) = 1 - \exp[-\exp(z)]$ is the standardized smallest extreme value distribution. This is so because natural logarithms of Weibull random variables follow the simpler smallest extreme value distribution with location and scale parameters $\mu = \log(\alpha)$ and $\sigma = 1/\beta$; see Nelson [23], page 43. The second expression is convenient because we can immediately switch to other distributions by changing the definition of $\Phi$. For example, when we replace $\Phi_{sev}(z)$ with $\Phi_{nor}(z)$, the standard normal distribution, $Y$ follows a lognormal distribution.

The Weibull likelihood function for these data is

$$L(\mu, \sigma) = \prod_{i=1}^{n} \left\{ \Phi_{sev}\left[\frac{\log(y_i) - \mu}{\sigma}\right] \right\}^{1-\delta_i} \\ \times \left\{ \frac{1}{\sigma y_i} \phi_{sev}\left[\frac{\log(y_i) - \mu}{\sigma}\right] \right\}^{\delta_i} \quad (8.14)$$

where $\delta_i = 1$ for an "exact" observation and $\delta_i = 0$ for a left-censored observation and $\phi_{sev}(z) = d\Phi_{sev}(z)/dz = \exp[z - \exp(z)]$ is the standardized smallest extreme value density. In this case, the density approximation to the "correct" likelihood is adequate for the two-parameter Weibull distribution and ML estimates computed with the approximation generally agreed with those computed with the "correct" likelihood to within $\pm 1$ in the third significant digit.

Figure 4 provides a contour plot of the relative likelihood function $R[\mu, \sigma] = L(\mu, \sigma)/L(\hat{\mu}, \hat{\sigma})$ for the Weibull model. The surface is well behaved with a unique maximum defining the ML estimates. Table V gives the ML estimates, standard errors, and confidence intervals for both distributions. The straight line on Figure 3 is the ML estimate of the Weibull

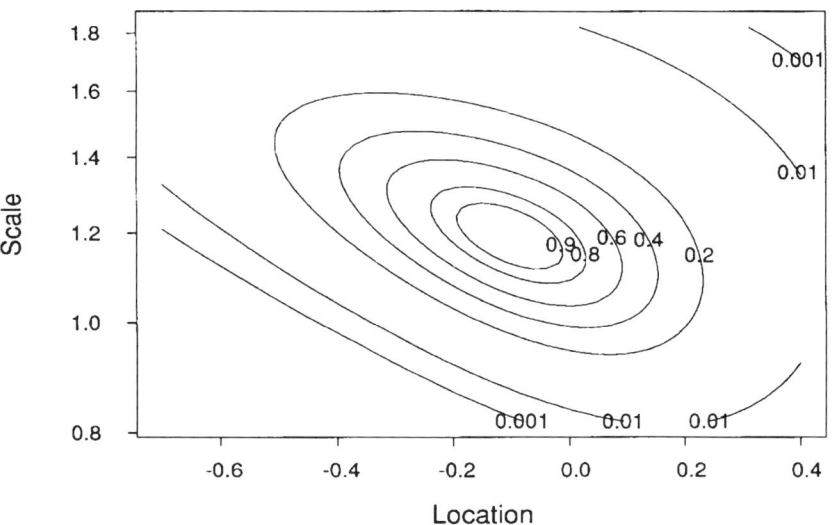

FIG. 4. Relative likelihood for the X-ray variability data and the Weibull model.

TABLE V. Summary and Comparison of Results from Fitting the Weibull and Lognormal Distributions to the X-Ray Variability Data

|  | Distribution | |
| --- | --- | --- |
|  | Weibull | Lognormal |
| ML estimate $\hat{\mu}$ | −.101 | −.400 |
| Standard error $s_{\hat{\mu}}$ | .200 | .171 |
| 95% confidence intervals for $\mu$ |  |  |
| Based on the likelihood | [−.551, .259] | [−.809, −.104) |
| Based on $\hat{\mu} \overset{.}{\sim}$ normal | [−.490, .289] | [−.734, −.065] |
| ML estimate $\hat{\sigma}$ | 1.19 | .93 |
| Standard error $s_{\hat{\sigma}}$ | .17 | .15 |
| 95% confidence intervals for $\sigma$ |  |  |
| Based on the likelihood | [.92, 1.64] | [.70, 1.32] |
| Based on $\log(\hat{\sigma}) \overset{.}{\sim}$ normal | [.89, 1.59] | [.68, 1.28] |
| Based on $\hat{\sigma} \overset{.}{\sim}$ normal | [.85, 1.53] | [.64, 1.22] |
| ML estimate $\hat{Y}_{.5}$ | .584 | .671 |
| Standard error $s_{\hat{Y}_{.5}}$ | .140 | .114 |
| 95% confidence intervals for $Y_{.5}$ |  |  |
| Based on the likelihood | [.332, .882] | [.445, .901] |
| Based on $\log(Y_{.5}) \overset{.}{\sim}$ normal | [.366, .934] | [.480, .937] |
| Based on $Y_{.5} \overset{.}{\sim}$ normal | [.311, .860] | [.446, .895) |

$F_Y(y; \mu, \sigma)$. The curved line going through the points is the corresponding ML estimate of the lognormal $F_Y(y; \mu, \sigma)$. The dotted lines are drawn through a set of pointwise normal-theory confidence intervals for the Weibull $F_Y(y; \mu, \sigma)$; the corresponding set of intervals for the lognormal $F_Y(y; \mu, \sigma)$ (not shown on the figure) were similar in width. As is frequently the case, there is good agreement for inferences from these two models *within the range of the data*. Without more information, however, one cannot make inferences in the lower tail of the distribution because the estimates are importantly different *and* the data do not strongly suggest one model over the other. In general, it is useful and important to fit different models to compare results on questions of interest.

### 8.4.3 Confidence Regions and Intervals for $\mu$ and $\sigma$ and for Functions of $\mu$ and $\sigma$

As described in Section 8.3.3, a contour line on Figure 4 defines an approximate joint confidence region for $\mu$ and $\sigma$ that can be accurately calibrated, even in moderately small samples, by using the large sample $\chi^2$ approximation for the distribution of the likelihood-ratio statistic. For example, the region $R[\mu, \log(\sigma)] > \exp(-\chi^2_{(.90;2)}/2) = .100$ provides an approximate 90% joint confidence region for $\mu$ and $\sigma$.

The numerical values of the likelihood-based confidence intervals and intervals based on the normal-theory approximation are given in Table V. Individual profile likelihood plots for $\mu$ and $\sigma$ (not shown here) were used to obtain the likelihood-based confidence intervals. Note that because $\mu$ and $\sigma$ have different meanings in the two different distributions, they are not comparable. It is, however, interesting to compare estimates of the same quantile using alternative methods.

Profile plots for the $Y_5$, the median of the respective distributions, computed as described in Section 8.3.4, are compared in Figure 5. Likelihood-based confidence intervals for $Y_5$ shown in Table V for both models can also be read directly from the figure. The ML estimates differ somewhat, but not importantly, relative to the width of the confidence intervals. It is interesting to note that (a) the lognormal profile likelihood for $Y_5$ is narrower than the Weibull profile and (b) when using the normal-theory intervals, the log transformation of the positive quantile $Y_5$ provides, in this case, a worse approximation to the likelihood-based confidence interval than doing no transformation at all.

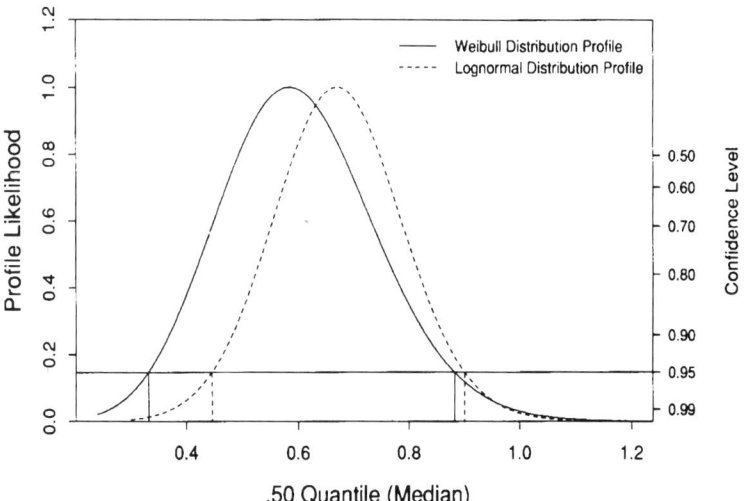

FIG. 5. Comparison profile likelihoods for the median of the X-ray variability distribution using Weibull and lognormal distributions.

## 8.5 Fitting the Limited Failure Population Model (a Three-Parameter Model)

### 8.5.1 Example and Data

We now return to the IC failure time data (Table II) that we introduced in Section 8.2.1. Meeker [9] provides a more complete analysis and more technical details for this example.

Figure 6 is a Weibull probability plot of the right-censored failure data. The last failure occurred at 593 hrs and the test was stopped at 1370 hrs. We see that the points on the probability plot are leveling off at something less than 1% failing. The failures were caused by manufacturing defects that could not be detected without a life test and the reliability engineers responsible for this product wanted to estimate $p$, the proportion of defects being manufactured by their process, in its current state. Moreover, they were interested in making improvements to the process and wondered if informative tests could, in the future, be run without waiting so long.

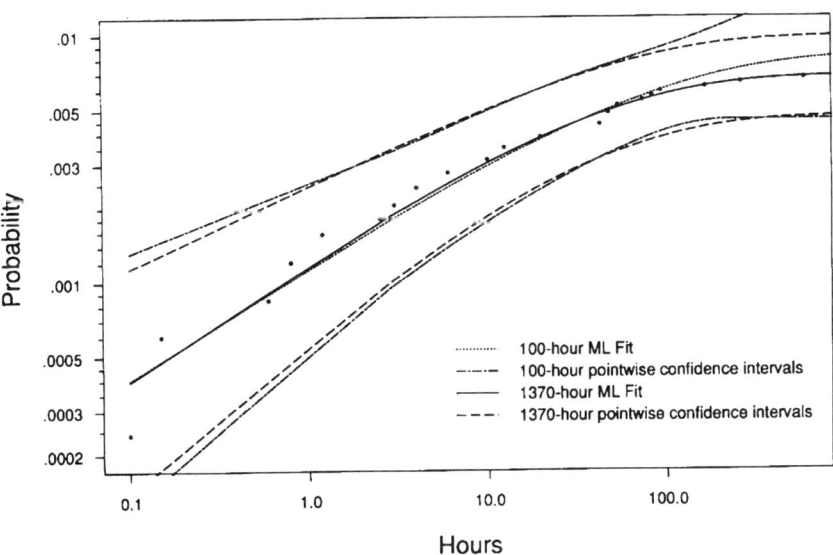

FIG. 6. Weibull probability plot of integrated circuit failure time data with ML estimates of the Weibull/LFP model after 1370 hrs and 100 hrs of testing.

### 8.5.2 The Limited Failure Population Model

Meeker [9] used the limited failure population (LFP) model for the IC data, assuming that a proportion $p$ of units from the manufacturing process is defective and will fail according to a distribution $F_Y(y; \mu, \sigma)$ and that the remaining proportion $(1 - p)$ would never fail. Assuming that $F_Y(y; \mu, \sigma)$ is Weibull leads to

$$\Pr(Y \leq y) = G_Y(y; \mu, \sigma, p) = pF_Y(y; \mu, \sigma)$$
$$= p\Phi_{\text{sev}}\left[\frac{\log(y) - \mu}{\sigma}\right] \quad (8.15)$$

for the entire product population. Note that as $y \to \infty$, $G_Y(y) \to p$.

### 8.5.3 The Likelihood Function and Its Maximum

The likelihood function for the IC data under the Weibull LFP model is

$$L(\mu, \sigma, p) = \prod_{i=1}^{n} \left\{\frac{p}{\sigma y_i} \phi_{\text{sev}}\left[\frac{\log(y_i) - \mu}{\sigma}\right]\right\}^{\delta_i}$$
$$\left\{1 - p\Phi_{\text{sev}}\left[\frac{\log(y_i) - \mu}{\sigma}\right]\right\}^{1-\delta_i} \quad (8.16)$$

where the notation is similar to that used in Section 8.4. Again, for this example, the probability density is approximately proportional to the "correct likelihood" for the "exact" failure times and we will use it here. Table VI summarizes and compares the results of the analyses for the data that were available at 1370 hrs and the data that would have been available after only 100 hrs. Figure 6 shows the ML estimates and 95% pointwise normal-theory confidence intervals for $G_Y(y)$. As might be expected, there is close agreement until approximately 100 hrs, when the estimates of $G_Y(y)$ begin to differ importantly. The upper bounds of the pointwise normal-theory confidence intervals for $G_Y(y)$ are larger for $y > 100$. As suggested later, however, similar confidence intervals computed with the likelihood-based method would be *much* wider.

### 8.5.4 Profile Likelihood Functions and Likelihood-Based Confidence Intervals for $\mu$, $\sigma$, and $p$

Table VI gives numerical values for the likelihood-based confidence intervals for $\mu$, $\sigma$, and $p$ based on these and other profiles (not shown here). The table also gives confidence intervals based on the normal-

TABLE VI. Summary and Comparison of Results from Fitting the Weibull Limited Failure Population Model to the Integrated Circuit Failure Data After 1370 Hrs and 100 Hrs

|  | Analysis with Test Run Until | |
| --- | --- | --- |
|  | 1370 Hours | 100 Hours |
| ML estimate $\hat{\mu}$ | 3.34 | 4.05 |
| Standard error $s_{\hat{\mu}}$ | .41 | 1.70 |
| 95% confidence intervals for $\mu$ based on | | |
|   The likelihood | [2.50, 4.20] | [2.43, 24.99] |
|   $\hat{\mu} \sim$ normal | [2.55, 4.12] | [.72, 7.38] |
| ML estimate $\hat{\sigma}$ | 2.02 | 2.12 |
| Standard error $s_{\hat{\sigma}}$ | .31 | .55 |
| 95% confidence intervals for $\sigma$ based on | | |
|   The likelihood | [1.53, 2.82] | [1.40, 3.96] |
|   $\log(\hat{\sigma}) \sim$ normal | [1.50, 2.71] | [1.28, 3.51] |
|   $\hat{\sigma} \sim$ normal | [1.42, 2.62] | [1.05, 3.19] |
| ML estimate $\hat{p}$ | .00674 | .00827 |
| Standard error $s_{\hat{p}}$ | .00127 | .00380 |
| 95% confidence intervals for $p$ based on | | |
|   The likelihood | [.00455, .00955] | [.00463, 1.000] |
|   Logit($\hat{p}$) $\sim$ normal | [.00466, .00975] | [.0033, .0203] |
|   $\hat{p} \sim$ normal | [.00426, .00923] | [.00081, .0157] |

theory approximation. Figure 7 is a two-dimensional likelihood profile for $p$ and $\sigma$ for the 100-hr data. Figure 8 provides a comparison of the one-dimensional profiles for $p$ for the 1370- and the 100-hr data. The results of this comparison show that, for the 1370-hr data, the log likelihood is approximately quadratic and the different methods of computing confidence intervals give similar results. For the 100-hr data, however, the story is quite different. In particular, the 100-hr profile likelihood tells us that the data available after 100 hrs could reasonably have come from a population with $p = 1$. That is, the 100-hr data do not allow us to clearly distinguish between a situation where there are many defectives failing slowly and a situation with just a few defectives failing rapidly. The 1370-hr data, however, allow us to say with a high degree of confidence, that $p$ is small.

For the 100-hr data, the likelihood and normal-theory confidence intervals for $p$ are vastly different. This is because the log likelihood is not well approximated by a quadratic function over the range of the confidence interval. Meeker [9] used Monte Carlo simulation to show that the likelihood-based confidence intervals provide a much better approximation to the nominal confidence levels over a wide range of parameter values for the LFP model.

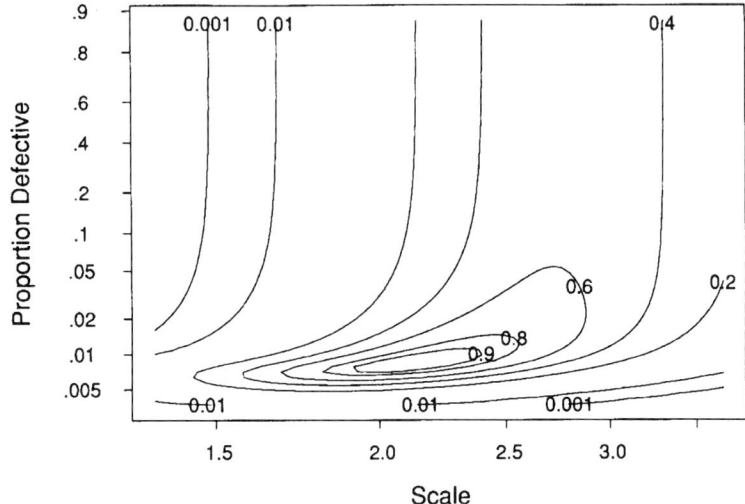

FIG. 7. Two-dimensional profile likelihood for LFP parameters $p$ and $\sigma$ after 1370 hrs of testing.

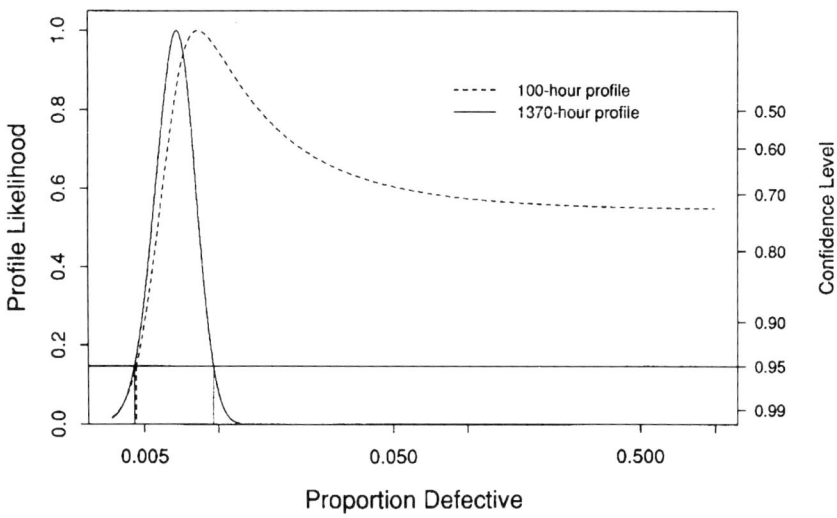

FIG. 8. Comparison of profile likelihoods for the LFP proportion defective after 1370 and 100 hrs of testing.

## 8.6 Some Other Applications

### 8.6.1 Distributions with Threshold Parameters

In many areas of application, analysts wish to fit models with a "threshold parameter" $\gamma$ that shifts the distribution of a positive random variable to the right by an amount $\gamma$. For example, the three-parameter Weibull distribution can be written as

$$F_Y(y; \mu, \sigma, \gamma) = \Phi_{\text{sev}}\left[\frac{\log(y - \gamma) - \mu}{\sigma}\right]$$

for $y > \gamma$. This is the classic example for which the "density approximation" in Eq. (8.6) can cause serious numerical and statistical problems in the application of ML estimation; see Giesbrecht and Kempthorne [24]. Also the asymptotic theory for "exact observations" is complicated (see, e.g., Smith [25] and, arguably, inappropriate for data with finite precision (see Section 8.7.2). Clearly, it is important to use the "correct" likelihood for these models. Griffiths [26] and Smith and Naylor [27] describe likelihood-based inferences for the three-parameter lognormal and Weibull distributions, respectively.

### 8.6.2 The Generalized Gamma Distribution

The generalized gamma distribution includes the gamma, Weibull, exponential, and lognormal distributions as special cases. As such, it provides a flexible distribution structure for modeling data. Farewell and Prentice [28] show that a judicious choice of parameters for this model can make an important difference in one's ability to apply ML methods. Lawless ([15], Chapter 5) shows how to use likelihood-based methods with the generalized gamma distribution to assess and compare results from the special case distributions.

### 8.6.3 Random Truncation

Morrell and Johnson [29] used ML estimation and a model with random left truncation to estimate the energies of neutrinos from Supernova 1987A. Under their random left-truncation model, the probability of observing a response depended on the value of the response. They used likelihood-based statistical methods to estimate and make confidence statements about the parameters of the energy distribution.

### 8.6.4 Point Process Data

Data from point processes occur frequently in science and engineering. The likelihood-based methods in this chapter can be applied directly to these problems. Snyder [30] describes many such applications, appropriate point processes, and likelihood-based inference methods for the models. Cox and Lewis [31] and Lawless [15], Chapter 10, also describe ML methods for point process models.

## 8.7 Other Topics and Sources of Additional Information

### 8.7.1 Computer Software, Numerical Methods, and Other Concerns for Fitting Statistical Models

Nelson [17], pages 237–239, provides a detailed summary of the features and capabilities of various software packages that can analyze such data.

Some (but not all) parts of the examples in this chapter could have been done with commercial or other publicly available software. In particular, because commercial software lacks the necessary capabilities, we had to use specially written software to fit the LFP model in Section 8.5 and to compute the profile likelihood plots and confidence intervals in *all* of the examples. We did our computations with a combination of Fortran and S (Becker, Chambers, and Wilks [32]) programming. (To get a copy of the S-Plus functions used in our examples, send electronic mail to wqmeeker@iastate.edu.)

As previously stated in Section 8.1.2, model parameterizations are often chosen on the basis of tradition, because certain parameters are of particular interest in an analysis, or for numerical reasons, in an effort to have a "well-behaved" likelihood function that would help avoid numerical difficulties. In computer software, the model parameterization that users see in software output typically is chosen on the basis of tradition or for the convenience of the user. Internally, however, a different set of "stable parameters" should be used for optimization and other numerical work, but their use should, by default, be hidden from the users. Choosing stable parameters, in general, is a challenging task and is discussed in some detail by Ross [33], who also discusses other important aspects of nonlinear estimation. Griffiths [26] suggests the use of quantiles as stable parameters for two parameters of the 3-parameter lognormal distribution. Griffiths [26] also suggests using $\log(y_{(1)} - \gamma)$ in place of the threshold parameter $\gamma$, where $y_{(1)}$ is the smallest observed response. Farewell and Prentice demonstrate the importance of parameterization for the three-parameter generalized gamma distribution. Also see Nelson [23], pages 386–395.

The choice of starting values for ML estimation often makes the difference between success and failure. Again, it is difficult to give general prescriptions on how to find good starting values. Typically, for specific models, and especially if one has already defined a set of "stable parameters" it is easy to find "ball park" starting values that will be satisfactory. Analytical implementation of simple graphical estimation methods based, for example, on the distribution-free estimates covered in Chapter 7, generally works well. Chapters 13 and 15 in Seber and Wild [34] contain a more complete discussion of these issues.

## 8.7.2 Regularity Conditions and Nonregular Models

In general, the ML estimators obtained from the "correct" likelihood behave well in the sense that when the sample size $n$ increases the estimators approach the true parameter values and the estimators are approximately normally distributed with a mean equal to the true parameter and a variance obtained from the Fisher information matrix (see Section 8.3.5). This asymptotic behavior is assured for problems that meet certain conditions that depend on the distribution, $F_Y(y; \theta)$, and the censoring mechanisms that generate the data. When the conditions are not met, ML estimators still generally have good asymptotic properties but details may differ from the standard results (e.g., limiting distributions may change as described by Smith [25] and Boente and Fraiman [35] and one should be aware of these potential differences.

The asymptotic behavior of ML estimators is, in general, complicated and we do not intend to make a full account here. Our purpose is to provide guidelines, references, and some heuristic arguments derived from our analytical and numerical studies of the problem.

First, and without loss of generality, we consider the case when $y$ is nonnegative. One can think of this range as divided into $m$ small time intervals $(0, y_1), \ldots, (y_{m-1}, y_m)$, where $y_1 = \Delta$, $y_{i+1} - y_i = 2\Delta$, $i = 1, 2, \ldots$, and $m$ could be finite or infinite. Then, it is easy to see that the "correct" likelihood can be written in terms of the probabilities of the intervals; i.e., $\pi_i = \pi_i(\theta) = F_Y(y_i; \theta) - F_Y(y_{i-1}; \theta)$, $i = 1, \ldots$, with $y_0 = 0$. An "exact" failure time observation contributes to the "correct" likelihood the probability $\pi_i$ of the interval $(y_{i-1}, y_i)$ into which it failed. An observation right or left censored at $y_k$ contributes $\sum_{i=k+1}^{m} \pi_i$ or $\sum_{i=1}^{k} \pi_i$, respectively. Similarly, an observation censored between $y_j$ and $y_k$ ($y_j < y_k$) contributes $\sum_{i=j+1}^{k} \pi_i$. Then the "correct" likelihood is essentially a parametric multinomial likelihood with some observations covering more than one cell. This is a generalization of the likelihood for grouped data. With this setup, one can use the standard regularity conditions (see Rao [4], page 359, Cox and Hinkley [5], page 281, or Kulldorff [2]).

In general, the "regularity conditions" for the standard large-sample asymptotic results require that (a) different values of $\theta$ assign different probabilities to the intervals (this is an identifiability condition); (b) the first three derivatives of $\log[\pi_i(\theta)]$ with respect to $\theta$ exist in a neighborhood of the true parameters and the derivatives up to third order are bounded in this neighborhood; (c) the Fisher information matrix for $\theta$ is finite and positive definite in a neighborhood of the true parameter values. When the identifiably condition is not met, ML methods are still useful, but the likelihood may have "flat spots," even at values of $\theta$ that maximize the likelihood.

When the range of the sample space of the underlying random variable $Y$ depends on the parameters $\theta$, the regularity conditions for the "correct" likelihood are similar; see for example Kulldorff [2] and Giesbrecht and Kempthorne [24].

### 8.7.3 Simulation-Based Confidence Intervals

Although approximate normal-theory confidence intervals are commonly used in practice, it is possible, particularly with small samples, that these intervals could give misleading results. In Section 8.3.3, we gave references to simulation studies showing that likelihood-based confidence intervals generally have coverage properties that are closer to nominal, even with moderate to small sample sizes. Computational complexity is the price that one pays for the better asymptotic approximation.

Another alternative that can lead to better confidence intervals, also at the cost of additional computational effort, is to replace asymptotic approximations of sampling distributions with sampling distributions obtained by using Monte Carlo simulation. A particular class of processes for generating such distributions and for inverting them to obtain confidence intervals is known as the "parametric bootstrap." These simulation-based methods and the closely related "nonparametric bootstrap" methods have been discussed in a large number of published papers since the pioneering work of Efron [36]. Efron and Tibshirani [37] and Hall [38] describes general theory and methods for the bootstrap method. Also see Chapter 17 of this volume.

## Acknowledgments

We would like to thank Gerry Hahn, John Hauptman, Molly Isbell, Mark Kaiser, Dave Lewis, Chuck Lerch, Joseph Lu, Chris Novak, George Seber,

Juergen Shroeder, Scott Vander Wiel, and Chris Wild for helpful suggestions on an earlier version of this chapter. We would like to express special appreciation for the encouragement and helpful suggestions John Stanford and Steve Vardeman provided to us while we worked on this project.

## References

1. Fisher, R. A. (1925). "Theory of Statistical Estimation." *Proceedings of the Cambridge Philosophical Society* **22,** 700–725.
2. Kulldorff, G. (1961). *Contributions to the Theory of Estimation from Grouped and Partially Grouped Samples.* Almqvist and Wiksell, Stockholm.
3. Kempthorne, O., and Folks, L. (1971). *Probability, Statistics, and Data Analysis.* Iowa State University Press, Ames.
4. Rao, C. R. (1973). *Linear Statistical Inference and Its Applications.* John Wiley and Sons, New York.
5. Cox, D. R., and Hinkley, D. V. (1974). *Theoretical Statistics.* Chapman and Hall, London.
6. Meeker, W. Q., and LuValle, M. J. (1991). "An Accelerated Life Test Model Based on Reliability Kinetics." Department of Statistics, Iowa State University.
7. Berkson, J. (1966). "Examination of Randomness of $\alpha$-Particle Emissions," *Festschrift for J. Neyman, Research Papers in Statistics,* F. N. David (ed.). John Wiley and Sons, New York.
8. Meeker, W. Q. (1986). "Planning Life Tests in Which Units Are Inspected for Failure." *IEEE Transactions on Reliability* **R-35,** 571–578.
9. Meeker, W. Q. (1987). "Limited Failure Population Life Tests: Application to Integrated Circuit Reliability." *Technometrics* **29,** 51–65.
10. Protheroe, R. J. (1985). "A New Statistic for the Analysis of Circular Data with Applications in Ultra-High Energy Gamma-Ray Astronomy." *Astronomy Express* **1(4–6),** 137–142.
11. Le Cam, L. (1990). "Maximum Likelihood: An Introduction." *International Statistical Review* **58,** 153–171.
12. Friedman, L. B., and Gertsbakh, I. (1981). "Maximum Likelihood Estimation in a Minimum-Type Model with Exponential and Weibull Failure Modes." *Journal of the American Statistical Association* **75,** 460–465.
13. Tennant, A. F., and Mushotzky, R. F. (1983). "The Absence of Rapid X-Ray Variability in Active Galaxies." *Astrophysical Journal* **264,** 92–104.
14. Feigelson, E. D., and Nelson, P. I. (1985). "Statistical Methods for Astronomical Data with Upper Limits. I. Univariate Distributions." *Astronomical Journal* **293,** 192–206.
15. Lawless, J. F. (1982). *Statistical Models and Methods for Life Time Data.* John Wiley and Sons, New York.
16. Cox, D. R., and Oakes, D. (1984). *Analysis of Survival Data.* Chapman and Hall, London.
17. Nelson, W. (1990). *Accelerated Testing: Statistical Models, Test Plans, and Data Analyses.* John Wiley and Sons, New York.
18. Nelson, W. (1984). "Fitting of Fatigue Curves with Nonconstant Standard Deviation to Data with Runouts." *Journal of Testing and Evaluation* **12,** 69–77.

19. Ostrouchov, G., and Meeker, W. Q. (1988). "Accuracy of Approximate Confidence Bounds Computed from Interval Censored Weibull and Lognormal Data." *Journal of Statistical Computation and Simulation* **29**, 43–76.
20. Vander Wiel, S. A., and Meeker, W. Q. (1990). "Accuracy of Approximate Confidence Bounds Using Censored Weibull Regression Data from accelerated Life Tests." *IEEE Transactions on Reliability* **R-39**, 346–351.
21. Escobar, L. A., and Meeker, W. Q. (1994). "Fisher Information Matrix for the Extreme Value, Normal, and Logistic Distributions and Censored Data." *Applied Statistics* **43**, No. 3.
22. Escobar, L. A., and Meeker, W. Q. (1993). "Teaching About Approximate Confidence Regions Based on Maximum Likelihood Estimation." Department of Statistics, Iowa State University, Ames.
23. Nelson, W. (1982). *Applied Life Data Analysis.* John Wiley and Sons, New York.
24. Giesbrecht, F., and Kempthorne, O. (1976). "Maximum Likelihood Estimation in the Three-Parameter Lognormal Distribution." *Journal of the Royal Statistical Society* **B38**, 257–264.
25. Smith, R. L. (1985). "Maximum Likelihood Estimation in a Class of Nonregular Cases." *Biometrika* **72**, 67–90.
26. Griffiths, D. A. (1980). "Interval Estimation for the Three-Parameter Lognormal Distribution via the Likelihood Function." *Applied Statistics* **29**, 58–68.
27. Smith, R. L., and Naylor, J. C. (1987). "A Comparison of Maximum Likelihood and Bayesian Estimators for the Three-Parameter Weibull Distribution." *Applied Statistics* **36**, 358–369.
28. Farewell, V. T., and Prentice, R. L. (1977). "A Study of Distributional Shape in Life Testing." *Technometrics* **19**, 69–75.
29. Morrell, C. H., and Johnson, R. A. (1991). "Random Truncation and Neutrinos." *Technometrics* **33**, 429–440.
30. Snyder, D. L. (1975). *Random Point Processes.* John Wiley and Sons, New York.
31. Cox, D. R., and Lewis, P. A. W. (1966). *The Statistical Analysis of a Series of Events.* Chapman and Hall, London.
32. Becker, R. A., Chambers, J. M., and Wilks, A. R. (1988). *The New S Language.* Wadsworth and Brooks/Cole, Pacific Grove, CA.
33. Ross, G. J. S. (1990). *Nonlinear Estimation.* Springer-Verlag, New York.
34. Seber, G. A. F., and Wild, C. J. (1989). *Nonlinear Regression.* John Wiley and Sons, New York.
35. Boente, G., and Fraiman, R. (1988). "On the Asymptotic Behaviour of General Maximum Likelihood Estimates for the Nonregular Case Under Nonstandard Conditions." *Biometrika* **75**, 45–56.
36. Efron, B. (1979). "Bootstrap Methods: Another Look at the Jackknife." *Annals of Statistics* **7**, 1–26.
37. Efron, B., and Tibshirani, R. J. (1993). *An Introduction to the Bootstrap.* Chapman and Hall, New York.
38. Hall, P. (1992). *The Bootstrap and Edgeworth Expansion.* Springer-Verlag, New York.

# 9. LEAST SQUARES

George A. F. Seber and Christopher J. Wild

Department of Statistics
Auckland University, New Zealand

## 9.1 Statistical Modeling

Much of science is about modeling relationships between variables. We shall consider just two variables $x$ and $y$ to begin with. These models are generally constructed one of two ways. The first method uses knowledge of the subject matter to derive a mathematical law relating $x$ and $y$. The emphasis is then on estimating certain constants in the law. The second method is an empirical one, where we try to find a mathematical relationship, again with unknown constants, that fits the data well. If we get a good fit, the model can be then used for prediction purposes, for example. The mathematical relationship need not summarize insights into the underlying physical mechanisms in any way but may simply provide a "black box" for prediction. Because of fluctuations in experimental conditions and measurement errors in the variables, we find that such relationships, whether empirical or theoretical, have two components—a mathematical part describing the relationship, and a statistical or "error" part describing any random variation about the relationship. Typically our model takes the form

$$\text{observation} = \text{model function} + \text{error}$$

or

$$Y = g(x; \boldsymbol{\theta}) + \epsilon$$

Here $g$ is a known function of $x$ and $\boldsymbol{\theta}$, where $\boldsymbol{\theta}$ is a $p$-dimensional vector of constants (technically called *parameters*) that are generally unknown and need to be estimated from the data. The variable $x$ is variously known as the *explanatory, regressor* or *independent variable,* while $Y$ is called the *response* or *dependent variable* and it is random as $\epsilon$ is random. The terms *dependent* and *independent* are a bit misleading and are therefore not recommended in spite of their popularity outside of statistics. The term *error* is also misleading as the difference between the observation and the model will include random fluctuations as well as measurement error. A better term might be *fluctuation* or *deviation;* economists often use the term *disturbance*. We shall continue to use the terms *error* and *error structure* as they seem well entrenched in the literature.

In fitting this model it is usually assumed that the distribution of $\epsilon$ does not vary with $x$; i.e., the error process is "stable." By writing $x$ in lower case we are assuming that it is not random. However, there is a range of models where $x$ is random so that the preceding model function also contains a random component. A simple example of this is the following.

**Example 9.1:** Consider Ohm's law $v = ru$ relating the voltage $v$ across a resistance $r$ with the current $u$ through the resistor. Because of measurement error, we will not observe the true values $u$ and $v$ of the mathematical variables, only their observed random values $X$ and $Y$, respectively. Setting $Y = v + \epsilon$ and $X = u + \delta$ we have, by substitution,

$$Y - \epsilon = r(X - \delta)$$

or

$$Y = rX + \epsilon - r\delta = rX + \eta \qquad (9.1)$$

We therefore have a straight-line relationship between the observed values, plus an error term $\eta = \epsilon - r\delta$.

**Example 9.2:** A well-known growth model take the form $y = \alpha x^\beta$ for the variables $x$ and $y$. This will only be approximately true for many real situations so that there will be an unknown difference $Y - \alpha x^\beta$, which we call, once again, the error, i.e., $Y = \alpha x^\beta + \epsilon$, with $\theta = (\alpha, \beta)^T$, our vector of parameters. Since $Y \approx \alpha x^\beta$, our natural inclination would be to take logarithms giving

$$Y' = \log Y \approx \log \alpha + \beta \log x = \beta_0 + \beta x'$$

which is approximately a straight-line relationship between the transformed variables $Y'$ and $x'$. (This model is further complicated if $x$ is measured with error). Confirmation of the original model would then be provided by a roughly linear plot of the set of observation pairs $(x'_i, y'_i)$. However, when it comes to estimating $\beta_0$ and $\beta$, transforming to achieve linearity is not always the best thing to do. It depends on the nature of the error $\epsilon$. In Example 9.2, where we have assumed an *additive-error* model, it is also often assumed that $\epsilon$ is stable, i.e., does not depend on $x$. Now

$$Y = \alpha x^\beta + \epsilon$$
$$= \alpha x^\beta (1 + \epsilon/\alpha x^\beta)$$

so that taking logarithms leads to

$$Y' = \beta_0 + \beta x' + \log(1 + \epsilon/\alpha x^\beta)$$
$$= \beta_0 + \beta x' + \epsilon_1$$

If $\epsilon$ is stable, then $\epsilon_1$ is not stable as it changes with $x$. Although we have linearized the model, our new error structure is more complicated. On the other hand, if the error was proportional to begin with, so that the bigger the measurement the bigger is the error, then we would have $\epsilon = k\alpha x^\beta$ and $Y = \alpha x^\beta (1 + k)$, where $k$ does not depend on $x$. If we take logarithms once again,

$$Y' = \beta_0 + \beta x' + \log(1 + k) = \beta_0 + \beta x' + \epsilon_2$$

where $\epsilon_2$ is stable (as $k$ is). We see, then, that in the second situation it is legitimate to take logarithms as it leads not only to a linear model but also to a stable error. However, in the first situation, it is not appropriate as it leads to an unstable error. The decision to transform or not depends, therefore, on the nature of the error structure. In general, determining the transformation scale in which to fit a model is a trial and error process involving fitting models and then inspecting estimates of the errors called *residuals*. This point is taken up again in Example 9.3, and when we discuss diagnostics. Also, with better nonlinear programs available, we do not have to "force" linearity just to simplify the computations. The following example illustrates this point.

**Example 9.3:** Carr [1] studied the catalytic isometrization of *n*-pentane to *i*-pentane in the presence of hydrone. One model proposed for the rate of this reaction, based on a single-site mechanism, was the nonlinear model

$$r \approx \frac{\theta_1 \theta_3 (x_2 - x_3/1.632)}{1 + \theta_2 x_1 + \theta_3 x_2 + \theta_4 x_3} \tag{9.2}$$

Here $r$ is the rate of disappearance of *n*-pentane [in g/(g catalyst)-hr]; $x_1$, $x_2$, and $x_3$ are the partial pressures of hydrogen, *n*-pentane, and *i*-pentane, respectively (in psia); $\theta_1$ is a constant depending on the catalyst; and $\theta_2$, $\theta_3$, and $\theta_4$ are equilibrium adsorption constants (in psia$^{-1}$). Carr [1] used the following transformation to linearize the model (9.2):

$$y \approx \frac{x_2 - x_3/1.632}{r}$$
$$= \frac{1}{\theta_1 \theta_3} + \frac{\theta_2}{\theta_1 \theta_3} x_1 + \frac{1}{\theta_1} x_2 + \frac{\theta_4}{\theta_1 \theta_3} x_3$$

or

$$Y = \beta_0 + \beta_1 x_1 + \beta_2 x_2 + \beta_3 x_3 + \epsilon \qquad (9.3)$$

We shall ignore the effects of any errors in the $x_i$. If a standard linear least squares is carried out on this last model using Carr's data, we obtain negative estimates of $\theta_2$, $\theta_3$, and $\theta_4$. This is contrary to the usual physical interpretation of these (positive) parameters. Box and Hill [2] studied this model and argued from some diagnostic plots that $\epsilon$ was not stable but varied with the mean so that a weighted least squares analysis (see later) was more appropriate. The lesson here is that applying ordinary least squares to a linearized model may not be appropriate. Furthermore, such an approach may not lead even to a useful first approximation to the parameter estimates before applying nonlinear least squares directly. For a further discussion of the analysis of this model see Seber and Wild ([3], pp. 77–86).

In some physical situations more than one $y$ observation is made for each $x$ observation. One such nonlinear model in which this occurs follows.

**Example 9.4:** In the compartmental analysis of humans, a radioactive tracer is injected into the body and the proportion $y_j$ present in component $j$ of the body is measured at various times $x$. For two compartments such as blood and tissue, the following pair of models was used by Beauchamp and Cornell [4]:

$$\begin{aligned}
Y_1 &= g_1(x; \boldsymbol{\theta}) + \epsilon_1 \\
&= \theta_1 e^{-\theta_2 x} + (1 - \theta_1) e^{-\theta_3 x} + \epsilon_1 \\
Y_2 &= g_2(x; \boldsymbol{\theta}) + \epsilon_2 \\
&= 1 - (\theta_1 + \theta_4) e^{-\theta_2 x} + (\theta_1 + \theta_4 - 1) e^{-\theta_3 x} + \epsilon_2,
\end{aligned}$$

where $\boldsymbol{\theta}^T = (\theta_1, \theta_2, \theta_3, \theta_4)$. Although we have a single $x$ variable, we now have a vector $(Y_1, Y_2)^T$ of $y$ observations, and the model is described as a *multivariate* nonlinear model. Such multivariate models are complex, and we will not discuss them further. The theory for linear multivariate models is given by Seber [5], and for nonlinear models by Bates and Watts ([6], Chapter 4) and Seber and Wild ([3], Chapter 11). A number of more applied books on multivariate analysis are available.

In the models given above, there is an explicit relationship between $x$ and $y$, so that $y$ is a function of $x$. However, sometimes we cannot express $y$ explicitly in terms of $x$ as we see in the following example.

**Example 9.5:** The freezing point of a solution of two substances is related to the concentration $x$ of component 1 by the equation

$$\log x = \alpha\left(\frac{1}{T_0} - \frac{1}{y}\right) + \beta \log\left(\frac{y}{T_0}\right) \tag{9.4}$$

where $T_0$ is the freezing point of pure component 1. In this model we cannot write $y = g(x; \boldsymbol{\theta})$ but instead we have

$$h(x, y; \boldsymbol{\theta}) = 0 \tag{9.5}$$

Building in the error structure for these so-called implicit models is more difficult, and the reader is referred to Seber and Wild ([3], pp. 501–513) for the rather complex theory behind the fitting of such models. We shall consider only explicit models in this chapter.

In the preceding five examples the model has been determined from physical considerations, and the unknown parameters can be given physical interpretations. We now turn our attention to empirical models, the simplest being the straight line or *simple linear* model $Y = \beta_0 + \beta_1 x + \epsilon$. This model is said to be linear as it is linear in $\beta_0$ and $\beta_1$. By the same token, the quadratic

$$Y = \beta_0 + \beta_1 x + \beta_2 x^2 + \epsilon$$

is also, technically, a linear model as the parameters $\beta_0$, $\beta_1$, and $\beta_2$ enter the model in a linear fashion. However,

$$Y = \beta_0 + \beta_1 x^{\beta_2} + \epsilon$$

is linear in $\beta_0$ and $\beta_1$, but not in $\beta_2$ so that it is said to be a nonlinear model.

A natural extension of the above simple linear model is the linear model

$$Y = \beta_0 + \beta_1 x_1 + \beta_2 x_2 + \epsilon$$

where $Y$ is now related to two variables $x_1$ and $x_2$. We get the quadratic equation as a special case by writing $x_1 = x$ and $x_2 = x^2$. We would still have a linear model if $x_2 = \sin x$, for example.

Another extension is the general quadratic equation in two variables, namely,

$$Y = \beta_0 + \beta_1 x_1 + \beta_2 x_2 + \beta_3 x_3 + \beta_4 x_4 + \beta_5 x_5 + \epsilon$$

where $x_3 = x_1^2$, $x_4 = x_2^2$, and $x_5 = x_1 x_2$. Polynomials in one or more variables are useful for trying to fit a curve or surface empirically to data. In fact linear models, which include the polynomial family, are generally used in this role, where we try and find some model linear in the parameters $\beta_0$, $\beta_1$, etc. that gives a good fit to the data, without necessarily having a physical interpretation for the parameters.

The method of fit that we shall use is least squares, to be discussed in Section 9.3. This method is straightforward to apply to linear models, but nonlinear models can pose a number of computational and practical problems.

**Example 9.6:** The data in Table I, which is taken from Scott et al. [7], concerns the congealing of spray paint particles. There is a single response variable $Y$, measuring mean surface–volume particle size of the product, and three explanatory variables $x_1$, $x_2$, and $x_3$ measuring, respectively, the feed rate per unit of whetted wheel periphery (gm/sec/cm), the peripheral wheel velocity (cm/sec), and the feed viscosity (poise). The basic task is to use $x$-variables to explain or predict the behaviour of $Y$. Our first approach to such data would be to look at two- and three-dimensional scatter plots relating each of the expanatory variables $x_i$ to $Y$ and also to one another to get a feeling for the variables. Although such plots can be misleading if taken in isolation, they can suggest starting points for in-depth analysis.

If there were no theory available about the processes generating the data and the scatterplots did not strongly suggest transformating the data in some way (e.g., taking logs), the model that we would use as a starting point is

$$Y_i = \beta_0 + \beta_1 x_{i1} + \beta_2 x_{i2} + \beta_3 x_{i3} + \epsilon_i$$

We would choose this model because it is the simplest model linking these three $x$-variables to $Y$.

There is no telling at this stage whether this model will be adequate, or even useful, for this particular set of data. In the process of empirical model building, one now fits the model and then looks at diagnostic information from the data itself about the adequacy of the fit of the model. If there are clearly systematic features of the data that the model fails to describe we will have to complicate the model by adding features until we arrive at a model that "passes" the diagnostic tests. In empirical model building we generally adhere to the principle of parsimony in that the models used are no more complicated than they have to be. The final model can then be used to draw conclusions or make predictions.

We may find several models that appear adequate. In this case we would want to compare the conclusions obtained from each of them. We would not want our conclusions to be too heavily influenced by the assumptions of a particular model in cases where the data cannot discriminate between models.

With the preceding data, the model

$$y \approx \alpha x_1^{\beta_1} x_2^{\beta_2} x_3^{\beta_3}$$

has been proposed from theoretical considerations. We have two simple choices for statistical models to use as a starting point. First, we could take the model as given and allow for additive error, namely,

TABLE I. Congealing of Spray Paint Particles

| Run | Feed Rate per Unit Whetted Wheel Periphery (gm/sec/cm) ($X_1$) | Mean Peripheral Wheel Velocity (cm/sec) ($X_2$) | Mean Feed Viscosity (poise) ($X_3$) | Mean Surface–Volume Particle Size of Product ($\mu$) ($Y$) |
|---|---|---|---|---|
| 1 | 0.0174 | 5300 | 0.108 | 25.4 |
| 2 | 0.0630 | 5400 | 0.107 | 31.6 |
| 3 | 0.0622 | 8300 | 0.107 | 25.7 |
| 4 | 0.0118 | 10800 | 0.106 | 17.4 |
| 5 | 0.1040 | 4600 | 0.102 | 38.2 |
| 6 | 0.0118 | 11300 | 0.105 | 18.2 |
| 7 | 0.0122 | 5800 | 0.105 | 26.5 |
| 8 | 0.0122 | 8000 | 0.100 | 19.3 |
| 9 | 0.0408 | 10000 | 0.106 | 22.3 |
| 10 | 0.0408 | 6600 | 0.105 | 26.4 |
| 11 | 0.0630 | 8700 | 0.104 | 25.8 |
| 12 | 0.0408 | 4400 | 0.104 | 32.2 |
| 13 | 0.0415 | 7600 | 0.106 | 25.1 |
| 14 | 0.1010 | 4800 | 0.106 | 39.7 |
| 15 | 0.0170 | 3100 | 0.106 | 35.6 |
| 16 | 0.0412 | 9300 | 0.105 | 23.5 |
| 17 | 0.0170 | 7700 | 0.098 | 22.1 |
| 18 | 0.0170 | 5300 | 0.099 | 26.5 |
| 19 | 0.1010 | 5700 | 0.098 | 39.7 |
| 20 | 0.0622 | 6200 | 0.102 | 31.5 |
| 21 | 0.0622 | 7700 | 0.102 | 26.9 |
| 22 | 0.0170 | 10200 | 0.100 | 18.1 |
| 23 | 0.0118 | 4800 | 0.102 | 28.4 |
| 24 | 0.0408 | 6600 | 0.102 | 27.3 |
| 25 | 0.0622 | 8300 | 0.102 | 25.8 |
| 26 | 0.0170 | 7700 | 0.102 | 23.1 |
| 27 | 0.0408 | 9000 | 0.613 | 23.4 |
| 28 | 0.0170 | 10100 | 0.619 | 18.1 |
| 29 | 0.0408 | 5300 | 0.671 | 30.9 |
| 30 | 0.0622 | 8000 | 0.624 | 25.7 |
| 31 | 0.1010 | 7300 | 0.613 | 29.0 |
| 32 | 0.0118 | 6400 | 0.328 | 22.0 |
| 33 | 0.0170 | 8000 | 0.341 | 18.8 |
| 34 | 0.0118 | 9700 | 1.845 | 17.9 |
| 35 | 0.0408 | 6300 | 1.940 | 28.4 |

Source: From Table II of M. W. Scott, M. J. Robinson, J. F. Pauls, and R. J. Lantz, "Spray Congealing: Particle Size Relationships Using a Centrifugal Wheel Atomizer," *J. Pharmaceutical Sciences* **53**(6), 670–675 (1964). Reproduced with permission from the American Pharmaceutical Association.

$$Y_i = \alpha x_1^{\beta_1} x_2^{\beta_2} x_3^{\beta_3} + \epsilon_i$$

or we could take logarithms to linearize the relationship first before adding the error to give

$$\log Y_i = \beta_0 + \beta_1 \log x_{i1} + \beta_2 \log x_{i2} + \beta_3 \log x_{i3} + \epsilon_i \quad (9.6)$$

In the analyses that follow we use the second formulation because linear models are easier to work with. We stress, however, that this model is still adopted only as a starting point and that the data may force us to modify it.

## 9.2 The Error Process Viewed Statistically

Having talked loosely about the error $\epsilon$ we now wish to discuss the nature of $\epsilon$ from a statistical point of view. To simplify the discussion we assume that $x$ is not random and not subject to error, but is controlled by the experimenter. The only randomness in the model is then contained in $\epsilon$, which is passed on to $Y$ through the model. With careful measurement we can assume that there is no bias in our measurements so that $\epsilon$ is just as likely to be positive as negative. In statistical terms $E(\epsilon) = 0$, where $E$ denotes the expected value (or population mean) of $\epsilon$. If we take several pairs of measurements, say $(x_i, y_i)$ for $i = 1, 2, \ldots, n$, then

$$Y_i = g(x_i; \boldsymbol{\theta}) + \epsilon_i$$

If $\epsilon$ is stable, we would expect each $\epsilon_i$ to have the same variance so that $V(\epsilon_i) = \sigma^2$, where $V(\epsilon_i)$ is the variance of $\epsilon_i$ and $\sigma^2$ is a constant not depending on the value of $x$. Since adding or subtracting a constant does not change a variance, we have

$$V(Y_i) = V[g(x_i; \boldsymbol{\theta}) + \epsilon_i] = V(\epsilon_i) = \sigma^2$$

With controlled experimental conditions we can expect the $n$ repetitions of the experiment to be independent so that the $\epsilon_i$ (and therefore the $Y_i$) are all statistically independent. For many physical measurements, we find that the error $\epsilon$ is made up of a large number of much smaller environmental changes. Since surfaces are approximately flat (i.e., linear) in small neighborhoods, $\epsilon$ can then be approximated by a linear combination of small errors, and by a famous theorem called the *central limit theorem* for independent (but not necessarily identically distributed) random variables, $\epsilon$ has a statistical distribution that is approximately normal (Gaussian). From this rough justification we see that it is often reasonable, as a starting point, to assume that $\epsilon$ is approximately normal for physical models. However, these arguments never constitute a proof that the errors will be normal in any practical situation, and one should always check this assumption from the data after performing the fit.

We now bring together the preceding ideas and say that $\epsilon$ satisfies the following four assumptions:

1. $E(\epsilon_i) = 0$.
2. $V(\epsilon_i) = \sigma^2$
3. The $\epsilon_i$ are statistically independent.
4. The $\epsilon_i$ are normally distributed.

These assumptions can be described in words as follows. (1) The model is correct in that there is no systematic departure of the observation $Y$ from the model $g(x; \theta)$; $Y$ is just as likely to lie above $g(x; \theta)$ as below it so that the average departure is zero. (2) The underlying level of variability is the same for all observations. Therefore if we fit the the model to the data on a graph then the data will lie within a band of roughly constant width centered on the graph. There will be no tendancy for the departure to change with $x$. (3) The observations are independent and usually come from independent physical experiments. A common departure from this assumption is that observations taken close together in time tend to be correlated. What happens at one instant of time effects what happens at the next. This is called *serial autocorrelation*. Similar effects can occur in space (spatial autocorrelation). (4) The $\epsilon_i$ have a roughly bell-shaped distribution (without too heavy tails), which commonly occurs with physical measurements. If the tails are "heavy" then there is a reasonable probability that a "wild" observation can occur.

The four assumptions may be included under one umbrella by saying that the $\epsilon_i$ are independently and identically distributed (i.i.d.) as $N(0, \sigma^2)$. In terms of the $Y_i$ this amounts to assuming that the $Y_i$ are independently distributed as $N[g(x_i; \beta), \sigma^2]$. We emphasize that these assumptions represent an idealization of the real world only, and at best, we can only hope that they are satisfied approximately. Graphical and numerical diagnostics to provide a rough check on these assumptions are described in Section 9.6. If the assumptions are clearly not appropriate then more general models will have to be fitted: transformations may be useful here (see Seber and Wild [3], Section 2.8) as may weighted least squares (Section 9.4.2). If the four assumptions hold, an extensive theory can be developed, and this is outlined in the next section.

In the preceding discusion we have assumed that $x$ is a controlled variable. What happens when $x$ is random? We discuss this question in Section 9.7.

## 9.3 Least Squares Fitting

Having established the model

$$Y_i = g(x_i; \theta) + \epsilon_i$$

we would like to estimate $\theta$ from the data pairs $(x_i, y_i)$. The method of least squares consists of finding the value of $\theta$ that minimizes the sum of squares of the errors:

$$\sum_{i=1}^{n} [y_i - g(x_i; \theta)]^2 \qquad [= S(\theta), \quad \text{say}]$$

The estimate produced is the maximum likelihood estimate if the errors are normally distributed. However, the method is intuitively attractive in any situation where the underlying level of variability in each of the observations is the same. In the case of linear models, $g(x_i; \theta)$ is conventionally expressed as

$$g(x_i; \theta) = \beta_0 + \beta_1 x_{i1} + \beta_2 x_{i2} + \ldots + \beta_{p-1} x_{i,p-1} \qquad (= \mathbf{x}_i^T \boldsymbol{\beta}), \quad \text{say}$$

where we have $p - 1$ regressors with $x_{ij}$ being the $i$th observation on the $j$th regressor, and the role of $\theta$ is now taken over by $\boldsymbol{\beta}$. In vector form we have $\mathbf{Y} = \mathbf{X}\boldsymbol{\beta} + \boldsymbol{\epsilon}$, where $\mathbf{X}$ is the matrix whose $i$th row is $\mathbf{x}_i^T$. To find the least squares estimate, $\hat{\boldsymbol{\beta}}$, say, of $\boldsymbol{\beta}$, we minimize the sum of squared deviations of the observations from the model, namely,

$$S(\boldsymbol{\beta}) = \sum_{i=1}^{n} (y_i - \beta_0 - \beta_1 x_{i1} - \ldots - \beta_{p-1} x_{i,p-1})^2$$

Differentiating this sum of squares with respect to $\boldsymbol{\beta}$ we find that $\hat{\boldsymbol{\beta}}$ satisfies the so-called normal equations $\mathbf{X}'\mathbf{X}\boldsymbol{\beta} = \mathbf{X}'\mathbf{y}$. When $\mathbf{X}$ has rank $p$, these equations have a unique solution

$$\hat{\boldsymbol{\beta}} = (\mathbf{X}'\mathbf{X})^{-1}\mathbf{X}'\mathbf{y} \qquad (9.7)$$

Unfortunately inverting a matrix is a numerically unstable way of solving the normal equations so that (9.7) does not provide a practically reliable way of computing $\hat{\boldsymbol{\beta}}$. Some of the computational issues for the linear model are discussed in Seber [8], Chapter 11. We strongly advise readers not to write their own programs, but to use any of the reputable packages. One reason for this recommendation is that such packages tend to use numerically stable methods. A more important reason is that, as we shall see later, least squares estimates and their standard errors form a small part of the information needed to analyze data successfully, and good packages can provide a wealth of additional information.

For each observed value of $y_i$ we have a fitted (or predicted) value

$$\hat{y}_i = \hat{\beta}_0 + \hat{\beta}_1 x_{i1} + \ldots + \hat{\beta}_{p-1} x_{i,p-1}$$

and the vector of fitted values is related to the observed vector by the equation $\hat{\mathbf{y}} = \mathbf{Py}$, where the matrix $\mathbf{P}\, (= \mathbf{X}(\mathbf{X}^T\mathbf{X})^{-1}\mathbf{X}^T)$ is called a *projection matrix*, as it represents an orthogonal projection of $\mathbf{y}$ onto the plane determined by $\mathbf{X}$. It is also called the *hat* matrix $\mathbf{H}$ (e.g., Wetherill [9]) as it converts $\mathbf{y}$ into $\hat{\mathbf{y}}$, and its diagonal elements play an important role in the diagnostics discussed later. Even more important in diagnostics are the "residuals" $r_i = y_i - \hat{y}_i$, which contain most of the information about the fit of the model. Defining $\mathbf{r}$ to be the vector of residuals, the following equation will be useful later,

$$\mathbf{r} = (\mathbf{I}_n - \mathbf{P})\mathbf{y} = (\mathbf{I}_n - \mathbf{P})\boldsymbol{\epsilon} = \underset{\sim}{y} - \underset{\sim}{\hat{y}} \qquad (9.8)$$

the last step following from $(\mathbf{I}_n - \mathbf{P})\mathbf{X} = \mathbf{0}$. Here $\mathbf{I}_n$ is the $n \times n$ identity matrix.

In the case of nonlinear models we again minimize the squared deviations of the observations from the model. For example, if we have the exponential model $g(x_i; \boldsymbol{\theta}) = \theta_1 \exp \theta_2 x_i$, then we minimize $S(\boldsymbol{\theta}) = \Sigma_{i=1}^n [y_i - g(x_i; \boldsymbol{\theta})]^2$ with respect to $\boldsymbol{\theta}$. Differentiating $S(\boldsymbol{\theta})$ partially with respect to $\theta_1$ and $\theta_2$ and equating to 0 leads to

$$\sum_{i=1}^n e^{\theta_2 x_i}(y_i - \theta_1 e^{\theta_2 x_i}) = 0$$

$$\sum_{i=1}^n x_i \theta_1 e^{\theta_2 x_i}(y_i - \theta_1 e^{\theta_2 x_i}) = 0$$

the so-called nonlinear normal equations. In general such equations cannot be solved analytically for $\hat{\boldsymbol{\theta}}$, the least squares estimate, so that iterative numerical methods are necessary. The basic idea behind the iterative approach is that, over a small range of $\boldsymbol{\theta}$ in the neighborhood of $\hat{\boldsymbol{\theta}}$, any nonlinear function is *locally* approximately linear. We then find that the role of $\mathbf{X}$ in the linear case is taken over by the matrix of derivatives $\mathbf{G}$ with $i, j$th elements $\partial g(x_i; \boldsymbol{\theta})/\partial \theta_j$. For example, instead of (9.8) we now have

$$\mathbf{r} \approx (\mathbf{I}_n - \mathbf{P}_G)\boldsymbol{\epsilon} \qquad (9.9)$$

where $\mathbf{P}_G = \mathbf{G}(\mathbf{G}^T \mathbf{G})^{-1} \mathbf{G}^T$.

The most common method of solving the nonlinear normal equations is the Gauss–Newton method, which forms the basis of a number of algorithms. A note of warning, however. Although the simple Gauss–Newton method provides the "driving engine" for most nonlinear least squares algorithms, many "bells and whistles" have to be added to make the algorithm practically reliable. We advise the reader to use programs in the major statistical packages or numerical subroutine libraries. It is not worthwhile writing your own software unless you are prepared to make a big investment in finding out about and circumventing the practical difficulties that arise. Chapters 13–15 in Seber and Wild [3] provide a useful starting point. A short summary of some of the computational methods is given by Bates and Watts [6], Chapter 3.

## 9.4 Statistical Properties of Least Squares Estimates

### 9.4.1 Ordinary Least Squares

In this section we focus on linear models and some general references are Chatterjee and Price [10], Daniel and Wood [11], Draper and Smith [12], Montgomery and Peck [13], Weisberg [14], Wetherill [9], Gunst and Mason [15], and the more theoretical book of Seber [8]. Nonlinear models will have similar properties to linear models for large samples: we simply replace $\mathbf{X}$ by $\mathbf{G}$. The first question to ask is, "Why do least squares?" The short answer is that least squares estimates have a number of desirable statistical properties. First, under the first three assumptions they provide estimates that are not only unbiased, i.e., $E(\hat{\beta}_r) = \beta_r$, but among certain classes of unbiased estimates they have the smallest variances. Second, under normality (assumption 4), they are also the maximum likelihood estimates, are normally distributed and are most efficient (have smallest variances for all unbiased estimates). Also exact methods of inference are available. Third, under nonnormality, the least squares estimates may still have reasonable asymptotic properties and are asymptotically normal under fairly general conditions. However, in deriving distribution theory we need to work with the vectors $\hat{\boldsymbol{\beta}}$ or $\hat{\boldsymbol{\theta}}$ rather than the single elements using the so-called multivariate normal (MVN) distribution. Fortunately we need not introduce this rather complex distribution as it has the useful property that every element, and every linear combination of the elements, of a MVN vector has the usual (univariate) normal distribution. For example, if the four assumptions on $\boldsymbol{\epsilon}$ hold then we find that $\hat{\boldsymbol{\beta}}$ is MVN and $\hat{\beta}_r$ is normally distributed, being an element of $\hat{\boldsymbol{\beta}}$. The same is true for the estimate $\mathbf{a}^T\hat{\boldsymbol{\beta}}$ of any linear combination $\mathbf{a}^T\boldsymbol{\beta}$; for example, $\beta_1 - \beta_2$. Since these estimates are unbiased and normally distributed, we require only their variances to complete the picture. These follow from

$$\text{Cov}(\hat{\boldsymbol{\beta}}) = \sigma^2(\mathbf{X}^T\mathbf{X})^{-1} \tag{9.10}$$

the so-called variance–covariance (or dispersion) matrix of $\hat{\boldsymbol{\beta}}$. The diagonal elements of this matrix are the variances of $\hat{\beta}_0, \hat{\beta}_1, \ldots,$ and the $(r, s)$th off-diagonal element is the covariance of $\hat{\beta}_r$ and $\hat{\beta}_s$ ($r \neq s$). To estimate these elements we need an estimate of $\sigma^2$, which we now consider.

Since $\epsilon_1, \epsilon_2, \ldots, \epsilon_n$ represent a random sample from $N(0, \sigma^2)$, then $\sum_{i=1}^{n} (\epsilon_i - \bar{\epsilon})^2/(n - 1)$ would be an unbiased estimate of $\sigma^2$. However, we do not know the true deviations $\epsilon_i$ so we would use the estimated deviations or residuals $r_i$. This would give $\sum_{i=1}^{n} (r_i - \bar{r})^2/(n - 1)$, where $\bar{r}$ is 0, as it can be shown that $\sum_{i=1}^{n} r_i = 0$ for models with a constant term $\beta_0$; i.e., the sum of the positive residuals exactly balance the sum of the negative residuals. (We could use this property to try and fit a least squares straight line by eye.) As $r_i$ is only an estimate of $\epsilon_i$ we have to adjust the degrees of freedom $n - 1$. Putting all this together it transpires that

$$s^2 = \sum_{i=1}^{n} r_i^2/(n - p) \quad [= \text{RSS}/(n - p)] \tag{9.11}$$

is unbiased estimate of $\sigma^2$; i.e., $E(s^2) = \sigma^2$. Here $RSS$ ($= \sum_{i=1}^{n} r_i^2$) is called the *residual sum of squares*.

If we now replace $\sigma^2$ by its estimate $s^2$ in (9.10) and take the square roots of the diagonal elements, we get an estimate of the standard deviation of each $\hat{\beta}_r$. This is usually called the *standard error*, and we shall denoted it by $se(\hat{\beta}_r)$. The standard printout from computer packages will include the estimates of the $\beta_r$ and their standard errors. The whole estimated variance–covariance matrix can usually be printed also on request.

When the four assumptions are satisfied, the vector of residuals $\mathbf{r}$ has a MVN distribution with the mean of each $r_i$ zero, and variance–covariance matrix

$$\text{Cov}(\mathbf{r}) = \sigma^2(\mathbf{I}_n - \mathbf{P}) \tag{9.12}$$

Since this is not $\sigma^2 \mathbf{I}_n$, the $r_i$ are not statistically independent as independent random variables have zero covariances. However, if $\mathbf{P}$ is in some sense "small" (and we shall see later that it is often ignored when making plots of the residuals), then (9.8) implies that $\mathbf{r} \approx \boldsymbol{\epsilon}$. Since the residuals now reflect the deviations, they can be used to investigate whether the assumptions for the $\epsilon_i$ hold. For example, any severe nonnormality in the $\epsilon_i$ will show up in the $r_i$. We discuss diagnostic checks in Section 9.6.

### 9.4.2 Robust and Weighted Least Squares

We now consider what happens when there are departures from the underlying assumptions and suggest some procedures for handling such situations. If systematic biases are present in the model so that assumption (1) does not hold, i.e., $E(\epsilon_i) \neq 0$, then all the estimates will be biased and their standard errors inflated. If the variances are not constant so that assumption (2) is not valid, then, although the least squares estimates of the $\beta_r$ remain unbiased under assumption (1), they are now inefficient and the standard errors are wrong. Clearly the observations with smaller variances are more useful and the idea of giving more weight to the more reliable observations comes under the umbrella term of *weighted least squares*. This technique is used, for example, when the assumption $V(Y_i) = V(\epsilon_i) = \sigma^2$ is replaced by $V(Y_i) = \sigma^2/w_i$, where the "weight" $w_i$ is known or can be satisfactorily estimated. Clearly the larger $w_i$ is, the smaller is the variance of $Y_i$ and the more useful the observation $y_i$. These weights can be taken into account in the least squares procedure by minimizing the weighted sum of squares

$$\sum_{i=1}^{n} w_i(y_i - \beta_0 - \beta_1 x_{i1} - \ldots - \beta_{p-1} x_{i,p-1})^2$$

For example, in the case of $Y_i = \beta_1 x_i + \epsilon_i$, a straight line through the origin, differentiating $\sum_{i=1}^{n} w_i(y_i - \beta_1 x_i)^2$ with respect to $\beta_1$ leads to the weighted least squares estimate

$$\hat{\beta}_1 = \frac{\sum_{i=1}^{n} w_i x_i y_i}{\sum_{i=1}^{n} w_i x_i^2}$$

An example where $w_i$ is known is when each observation $y_i$ is really $\bar{y}_i$, the mean of $n_i$ repeated observations, so that $V(\bar{Y}_i) = \sigma^2/n_i$ and $w_i = n_i$. Most computer packages will fit models by weighted least squares with prechosen weights. However, weights often have to be estimated from the data (Carroll and Ruppert [16]). We note that without weights, assumptions (2) and (3) imply that $Cov(\mathbf{y}) = Cov(\boldsymbol{\epsilon}) = \sigma^2 \mathbf{I}_n$, a diagonal matrix with diagonal elements equal to $\sigma^2$. When weights are introduced we still have a diagonal matrix (because of the independence of the observations) but with different diagonal elements.

Similar problems arise when there is a lack of independence in the data and assumption (3) does not hold. The measures of variability such as $s^2$ and the standard errors are now wrong. In this case $Cov(\mathbf{Y}) = \sigma^2 \mathbf{V}$, where $\mathbf{V}$ is often known or can be estimated, and is no longer diagonal. Fortunately, we can extend the principal of weighting, usually under the title of generalized least squares, to handle this situation. The sum of squares to be minimized is now much more general as it involves the elements of the inverse matrix $\mathbf{V}^{-1}$. Algebraic details of the method are described by Seber ([8], Section 3.6) for linear models, and by Seber and Wild ([3], Section 2.1.4) in the context of nonlinear models. Weighted least squares can be regarded as a special case in which $\mathbf{V}^{-1}$ is a diagonal matrix with elements $w_1, w_2, \ldots, w_n$.

Generalized least squares is particularly important when the $\epsilon_i$ are correlated; for example, having serial or spatial autocorrelation. The correlation structure then must be modeled as well as the variances. For example, it may be necessary to model the values of $\epsilon_i$ as coming from a time series such as the AR(1) model (see Chapters 3 and 11). In this case the elements of **V** are functions of the autocorrelation parameter $\rho$, and an estimate of $\rho$ will provide an estimate of **V**. Linear time series regression models are described by Abraham and Ledolter [17], and nonlinear models by Seber and Wild ([3], Chapter 6) and Gallant ([18], Chapter 2).

The effect of departures from the normality assumption (4) depends very much on the nature of the departure and the size of the sample. If all the $\epsilon_i$ have the same (nonnormal) distribution, then all the results still hold for large samples under fairly general conditions. However, for small or moderate samples, the least squares estimators may perform badly. In the less common situation where we know this nonnormal distribution we can use the method of maximum likelihood described elsewhere in this book and obtain the maximum likelihood estimate of $\beta_r$. Such estimates tend to have good properties for moderate to large samples. In fact, as we have already noted, when the underlying distribution of the errors is actually normal, the maximum likelihood estimate of $\beta_r$ is the same as $\hat{\beta}_r$. Sometimes a transformation of the model may lead to an error structure that is more normal looking.

Another type of departure from assumption (4) is when the $\epsilon_i$ are all normal except for a few values, called *outliers,* that have very different distributions. In this case we need a method that gives less weight, or even no weight, to extreme observations. Such methods are called *robust methods.* There is no space to go into details so we refer the reader to Roousseuw and Leroy [19]. Graphical methods for detecting outliers are discussed in Section 9.6.

### 9.4.3 Nonlinear Least Squares

The role of $\boldsymbol{\beta}$ is now replaced by $\boldsymbol{\theta}$ and the "exact" theory of the previous section is replaced by asymptotic theory with the matrix of derivatives **G** instead of **X**. Thus, for large samples, the least squares estimate $\hat{\boldsymbol{\theta}}$ is approximately multivariate normal with $E(\hat{\theta}_i) \approx \theta_i$ and [cf. (9.10)]

$$\text{Cov}(\hat{\boldsymbol{\theta}}) \approx \sigma^2(\mathbf{G}^T\mathbf{G})^{-1}$$

where **G** can be estimated by $\hat{\mathbf{G}}$, its value at $\boldsymbol{\theta} = \hat{\boldsymbol{\theta}}$. However we might ask, "How large is a large sample in this context?" The typical answer of a statistician is that it depends! As already mentioned, the asymptotic theory is based on the idea that a nonlinear function is locally linear close to $\hat{\boldsymbol{\theta}}$. Mathematically, we use a first-order Taylor expansion

$$\mathbf{g}(\boldsymbol{\theta}) \approx \mathbf{g}(\hat{\boldsymbol{\theta}}) + \hat{\mathbf{G}}(\hat{\boldsymbol{\theta}})(\boldsymbol{\theta} - \hat{\boldsymbol{\theta}}) \tag{9.13}$$

where $\mathbf{g}(\boldsymbol{\theta})$ is a vector with $i$th element $g(x_i; \boldsymbol{\theta})$, and this equation approximates the surface $\mathbf{z} = \mathbf{g}(\boldsymbol{\theta})$ near $\hat{\boldsymbol{\theta}}$ by the tangent plane at $\hat{\boldsymbol{\theta}}$. The validity of this approximation depends on the relative "size" of the second term omitted from the expansion. This second term has two components: the first, called the *intrinsic curvature array*, describes how curved the model is at $\hat{\boldsymbol{\theta}}$; and the second, called the *parameter-effects array*, measures how curved the parameter contours are on the surface. For many models, the intrinsic curvature array (which does not depend on the method of parameterization, only on the shape of the surface $\mathbf{g}(\boldsymbol{\theta})$) is often negligible. The parameter-effects array, however, can vary considerably depending on the choice of parameters. For example, consider the nonlinear model

$$g(x; \boldsymbol{\theta}) = \theta_1 x + \theta_1 \theta_2 (1 - x)$$

Since the intrinsic curvature array is independent of the parameters used, we see that if we choose $\phi_1 = \theta_1$ and $\phi_2 = \theta_1 \theta_2$ as our parameters, then the model is linear as far as intrinsic curvature is concerned. This reparameterized model has no curvature arrays, being a plane, so that the intrinsic array is 0. If we use the original parameters then the model is nonlinear in $\theta_1$ and $\theta_2$ and there will be some parameter-effects curvature, even though the intrinsic curvature is 0. However, if we use $\phi_1$ and $\phi_2$ then the parameter-effects array is also 0.

How do these second-order arrays affect the asymptotic theory? First, it turns out that $\hat{\boldsymbol{\theta}}$ is biased with the bias depending on the parameter-effects array. Second,

$$\mathrm{Cov}(\hat{\boldsymbol{\theta}}) \approx \sigma^2[(\mathbf{G}^T\mathbf{G})^{-1} + \mathbf{C}]$$

where $\mathbf{C}$ depends on both the intrinsic and parameter-effects arrays and is often not negligible. We see, therefore, that it may be inappropriate to base inference about $\boldsymbol{\theta}$ on the approximate properties described at the beginning of this subsection without investigating the arrays or some other measures of nonlinearity and assessing their effects on the bias and variance–covariance matrix. For this reason alternative methods of inference will be given in Section 9.5. Details of the preceding theory are given by Seber and Wild [3].

In conclusion we note that similar comments apply to the residual vector $\mathbf{r} = \mathbf{y} - \mathbf{g}(\hat{\boldsymbol{\theta}})$, which, to just a first-order approximation, is given by (9.9). If we add a second-order term then, instead of $E(r_i) \approx 0$, $r_i$ may be biased away from 0 and $V(r_i)$ may be inflated. Fortunately, the culprit both times is only the intrinsic array, which seems to be negligible for most models.

## 9.5 Statistical Inference

Under the assumptions (1) to (4) of Section 9.2 we saw that for the general linear model each $\hat{\beta}_r$ is normally distributed. Applying standard normal theory we know that

$$t(\beta_r) = \frac{\hat{\beta}_r - \beta_r}{se(\hat{\beta}_r)} \sim t_{n-p} \qquad (9.14)$$

where $t_{n-p}$ is the $t$-distribution with $n - p$ degrees of freedom and $se(\hat{\beta}_r)$ is the standard error of $\hat{\beta}_r$. The latter is computed as $s\sqrt{v_r}$, where $s$ is given by (9.11) and $v_r$ is the $(r + 1)$th diagonal element of $(\mathbf{X}^T\mathbf{X})^{-1}$ ($r + 1$ as there is a $\beta_0$). If our linear model is a theoretical one (e.g., (9.3)), then our attention will focus on the estimation of each $\beta_r$ so that confidence intervals rather than hypothesis testing would be more appropriate. A $100(1 - \alpha)\%$ confidence interval for a given $\beta_r$ is given by

$$\hat{\beta}_r \pm t_{n-p}(\alpha/2) \, se(\hat{\beta}_r) \qquad (9.15)$$

If the linear model is empirical, then we would like to know which of the possible regressors should be included in the model. We could use either the confidence interval (9.15) to see if it contained 0, or else we could carry out a test of the null hypothesis $H_0$: $\beta_r = 0$ using (9.14) with $\beta_r = 0$; i.e., using $t_r = t(0)$. These individual $t$-tests are usually part of the standard computer printout for most statistical packages. The magnitude of $t_r$ will give some idea as to the importance of $\beta_r$ in the model. If $t_r$ is large, thus representing a highly significant result, then this suggests that $\beta_r$ needs to be retained in the model. However, care is needed in interpreting sets of hypothesis tests or confidence intervals. Although each confidence interval may have a probability of .95 of covering the true parameter value, the probability that a set of $k$ such intervals all *simultaneously* cover their true parameter values is not .95 but is a little greater than $p = 1 - k(.05)$. When $k = 10$, this is .5, which does not inspire much confidence! To get round this problem we replace .05 by $.05/k$, which will give us $p = .95$. For confidence intervals we would then use a confidence level of $100(1 - .05/k)$ for each interval. If we are considering significance tests, then we use a significance level of $.05/k$ for each test. Such intervals and tests are called *Bonferroni intervals* and *tests,* as they are based on a probability inequality by Bonferroni. Other methods of constructing simultaneous confidence intervals are also available (cf. Seber [8], p. 125).

Another approach for selecting the regressors is to include all the $x$-variables (and perhaps their squares and products as well) in the model and then see how many we can eliminate without affecting too much how well the model fits the data, as measured by the residual sum of squares $RSS\ [= S(\hat{\beta}) = (n - p)s^2$, cf. (9.11)]. Trying to drop variables out of the model is equivalent to testing the hypothesis $H_0$ that several values, say, $q$, of $\beta$ are simultaneously 0. This can be done using the so-called $F$-ratio:

$$F = \frac{\text{RSS}_H - \text{RSS}}{qs^2} \quad (9.16)$$

where $RSS_H$ is the residual sum of squares for the model with $q$ less parameters. Here $F$ is referred to the upper $\alpha$ quantile value of the $F_{q,n-p}$ distribution. The problem of choosing an appropriate set of candidates for this testing procedure is a complex one, and various procedures have been proposed. For an overview see [8], Chapter 12; Gunst and Mason [15], Chapter 8; and Miller [20].

A useful measure of how well the data fits the model is the *coefficient of determination* $R^2$, where $R$ is the correlation coefficient for the pairs $(y_i, \hat{y}_i)$. It is algebraically equal to $1 - RSS/\Sigma_{i=1}^n (y_i - \bar{y})^2$, and since it represents the proportion of the variation explained by the model, it satisfies $0 \leq R^2 \leq 1$, where large values tend to indicate a good fit. It is usually printed out by most computer packages but needs to be interpreted carefully. For example, as more $x$-variables are brought into the model, $R^2$ increases. However, if the increase is slight, this suggests that the extra variables may not be needed. In the form presented here, $R^2$ should be used to compare only models that are fitted on the same transformation scale, contain an intercept, and have a reasonably stable error.

Before considering the nonlinear case we note that we can also test $H_0$: $\beta_r = c$, say, using a likelihood ratio test based on the $F$-statistic

$$\begin{aligned} F(c) &= \frac{\tilde{S}(c) - S(\hat{\beta})}{S(\hat{\beta})/(n - p)} \\ &= [\tilde{S}(c) - S(\hat{\beta})]/s^2 \\ &= t^2(c) \end{aligned} \quad (9.17)$$

where $\tilde{S}(c)$ is obtained by fixing $\beta_r = c$ and minimizing $S(\beta)$ over the remaining parameters $\beta_0, \beta_1, \ldots, \beta_{r-1}, \beta_{r+1}, \ldots, \beta_{p-1}$, and $t(c)$ follows from (9.14). The value where the minimum occurs will be a function of $c$. We note that $F(c)$ has an $F$-distribution with 1 and $n - p$ degrees of freedom, respectively, when $H_0$ is true. Also

$$t(c) = \text{sign}(\hat{\beta}_r - c) \sqrt{F(c)} \qquad (9.18)$$

where $\text{sign}(\hat{\beta}_r - c)$ is the sign of the the expression in brackets. The confidence interval (9.15) for $\beta_r$ is the set of all $\beta_r$ such that $|t(\beta_r)| \leq t_{n-p}(\alpha/2)$.

For nonlinear models, we saw in the previous section that if the appropriate curvature arrays are small then the asymptotic theory can be used. We then have

$$z_r = \frac{\hat{\theta}_r - \theta_r}{\text{se}(\hat{\theta}_r)}$$

is approximately $N(0, 1)$, where $\text{se}(\hat{\theta}_r)$ is the standard error of $\hat{\theta}_r$ and is computed as $s\sqrt{v_r}$, where $v_r$ is the $r$th diagonal element of $(\hat{\mathbf{G}}^T\hat{\mathbf{G}})^{-1}$, $s^2 = S(\hat{\boldsymbol{\theta}})/(n-p)$ and $p$ is the dimension of $\boldsymbol{\theta}$. However, we saw that the asymptotic theory is seriously affected by any parameter-effects curvature, which is frequently present, so that an alternative method is needed. It transpires that, replacing $\boldsymbol{\beta}$ by $\boldsymbol{\theta}$, $F(c)$ is still approximately $F_{1,n-p}$ when $\theta_r = c$ (cf. [3], Equation (5.23) with $p_2 = 1$). Then, the so-called *profile t function*

$$\tau(\theta_r) = \text{sign}(\hat{\theta}_r - \theta_r)\sqrt{[\tilde{S}(\theta_r) - S(\hat{\boldsymbol{\theta}})]}/s \qquad (9.19)$$

is approximately distributed as $t_{n-p}$. This leads us to a confidence interval for $\theta_r$ consisting of all $\theta_r$ such that $|\tau(\theta_r)| \leq t_{n-p}(\alpha/2)$. Such an interval was proposed by Bates and Watts ([6], p. 205; they use $-\tau(\theta_r)$). When the model is linear, $\tau(\theta_r)$ is the same as $t(\theta_r)$ of (9.14) so that a plot of $\tau(\theta_r)$ versus $\theta_r$ is a straight line. For a nonlinear model, these two functions are different so that the same plot will be curved with the amount of curvature giving information about the nonlinearity of the model. Because the value of $S(\boldsymbol{\theta})$ is invariant under one-to-one reparametrizations, the previous profile method is unaffected by any parameter-effects curvature. It can still be affected by any intrinsic curvature, but this is much less likely. Further plotting aids are described by Bates and Watts [6], including some programming details in their appendix A3.

The test procedure described by (9.16) can also be used with nonlinear models for testing whether several of the parameters are 0. The distribution of $F$ is now only approximately distributed as $F_{q,n-p}$ and several other asymptotically equivalent tests are also available (cf. Seber and Wild [3], Section 5.3). When $n$ is large, $qF_{q,n-p}$ is approximately $\chi_q^2$ so that $(RSS_H - RSS)/s^2$ is approximately $\chi_q^2$. This form of the test is promoted in some books. However, the $F$-test is preferred. As with the $t$-test using (9.19), which is based on the same principal but with $q = 1$, the $F$-test is unaffected by any parameter-effects curvature. In the less likely situation where the effect of intrinsic curvature cannot be neglected, a correction factor is available for $F$. However, this correction factor is very complex (cf. Seber and Wild [3], p. 200) and is currently not available in standard computer packages.

Statistical inferences about parameter values make most sense in two situations: (i) when the parameters have physical interpretations and are of intrinsic interest and (ii) when we wish to decide whether an umbrella model can be simplified or whether the current model must be expanded. The latter situation involves testing whether certain parameters are 0 or not. Apart from these situations, the model fitting is most useful as a predictive device. Then two important kinds of inference are confidence intervals for the *mean response* at various specified combinations of settings of the $x$-variables (often called *confidence limits for the mean* in regression programs) and prediction intervals for a new *actual response* at some combination(s) of settings of the $x$-variables (sometimes described as "confidence limits for an individual predicted value"). Formulas for these intervals are given by Seber [8]. Such intervals can be obtained from most reputable packages.

An important consideration in interpreting any of these inferences is, first, whether the $x$-variables are controlled and set by the experimenter and the $y$-variable responds to these changed settings or, second, whether the data (both $x$- and $y$-variables) are measurements taken on a process over which the experimenter has no such control. Only in the first case can the experimenter reasonably conclude that changes in $y$ are *caused* by changes in $x$. In the second case, inferences about mean responses and predictions can apply only to new observations produced by exactly the same mechanism or process as that which generated the original data.

**Example 9.6 (Continued):** In Table II we have the computer output from the regression program PROC REG in the statistical package SAS obtained from fitting (9.6) to the data in Table I. We note that the least squares estimate (LSE) of $\beta_0$ is 8.549532, the LSE of $\beta_1$ is 0.168424, etc. Also provided are the standard errors corresponding to each LSE, the $t$-test ratios for testing each hypothesis of the form $\beta_r = 0$, and the $p$-value for the test, namely,

$$p\text{-value} = P[|T| > t(\beta_r)], \qquad \text{where } T \sim t_{n-p}.$$

We note that the $t$-tests for the parameters $\beta_0$, $\beta_1$, and $\beta_2$ are highly significant. In contrast, the test for $\beta_3$ is nonsignificant and thus the data provides no

TABLE II. Some of the Output from the SAS Regression Program PROC REG Obtained from Fitting Eq. (9.6) to the Data in Table I

Model: MODEL1
Dependent Variable: LOGY

### Analysis of Variance

| Source | DF | Sum of Squares | Mean Square | F Value | Prob > F |
|---|---|---|---|---|---|
| Model | 3 | 1.69570 | 0.56523 | 220.306 | 0.0001 |
| Error | 31 | 0.07954 | 0.00257 | | |
| C total | 34 | 1.77524 | | | |
| | Root MSE | 0.05065 | R-square | 0.9552 | |
| | Dep mean | 3.23975 | Adj R-sq | 0.9509 | |
| | C.V. | 1.56347 | | | |

### Parameter Estimates

| Variable | DF | Parameter Estimate | Standard Error | T for $H_0$: Parameter = 0 | Prob > $|T|$ |
|---|---|---|---|---|---|
| INTERCEP | 1 | 8.549532 | 0.26602390 | 32.138 | 0.0001 |
| LOGX1 | 1 | 0.168424 | 0.01180812 | 14.263 | 0.0001 |
| LOGX2 | 1 | −0.537137 | 0.03009608 | −17.847 | 0.0001 |
| LOGX3 | 1 | −0.014413 | 0.00981705 | −1.468 | 0.1521 |

evidence against $\beta_3 = 0$. It is therefore possible that changes in this variable do not affect the level of the response.

Appearing above the table of estimates is the results of an $F$-test for an overall hypothesis that all of the regression coefficients $\beta_j$ apart from the intercept are 0. The coefficient of determination $R^2$, which here takes a value of 0.9552, indicates that 95.5% of the variation in log $Y$ has been explained by the regression model. A host of other information is available including the predicted values $\hat{y}_i$, 95% confidence intervals for the mean response at values of the $x$'s, and 95% prediction intervals for the value of a new response at values of the $x$'s.

However, all of these statistical inferences ($R^2$ is merely descriptive) are plausible only if the model being fitted is not contradicted by the data. We will find that, in this case, the model is strongly contradicted by the data, as there are systematic features of the data that the model above fails to account for. This leads us naturally into the next section, which describes model checking or model *diagnostics*.

## 9.6 Diagnostics

The previous methods of inference require that the four assumptions underlying our model are satisfied. We now describe some diagnostic plots for checking out whether these assumptions seem reasonable or not for linear models. There are a large number of these plots, which is confusing, and various statistical packages will have their own favorites. However, it is a good idea to use a variety of plots as they all highlight different features of the data. Since the vector of residuals $\mathbf{r}$ has similar properties to $\boldsymbol{\epsilon}$, we use $\mathbf{r}$ and functions of it as a basis for all the methods. When the $\epsilon_i$ are i.i.d. $N(0, \sigma^2)$, then $E(r_i) = 0$, $Cov(\mathbf{r}) = \sigma^2(\mathbf{I}_n - \mathbf{P})$, and each $r_i$ is $N(0, \sigma^2(1 - p_{ii}))$ where $p_{ii}$ is the $i$th diagonal element of $\mathbf{P}$ and satisfies $0 < p_{ii} < 1$. To make the $r_i$ comparable, it is usual to scale them so that they have approximately unit variances; i.e., we use

$$t_i = \frac{r_i}{s(1-p_{ii})^{1/2}}$$

which is approximately $N(0, 1)$. Although the $r_i$ (and $t_i$) are correlated, it seems that the correlation has little effect on the plots that follow. Here $t_i$ is called an *(internally) studentized* or *standardized* residual. The term *(externally) studentized residual* is usually attached to

$$t_i^* = \frac{r_i}{s_{(i)}(1 - p_{ii})^{1/2}}$$

where $s_{(i)}^2$ is $s^2$ for the regression with the $i$th data set $(y_i, x_{i1}, \ldots, x_{i,p-1})$ omitted and a divisor of $n - p - 1$ used instead of $n - p$. From a theoretical point of view, $t_i^*$ is often preferred as it has a $t_{n-p-1}$ distribution (which is approximately $N(0, 1)$ when $n - p - 1$ is large), while $t_i^2/(n - p)$ actually follows a Beta distribution. For simplicity of notation we shall use $t_i$ in the following but the same remarks apply to $t_i^*$ as well. Often the $p_{ii}$ are negligible for plotting purposes so that some packages simply use $r_i/s$. We now consider a number of diagnostic plots.

### 9.6.1 Plot Residual versus Fitted Value

Four common defects may be revealed by plotting $t_i$ versus $\hat{y}_i$.

1. Model inadequacy: If the model is correct then there will be no trend in the plot. The points will lie roughly in a horizontal band centered on $t_i = 0$ as in Fig. 1(a). However, a curved plot like Fig. 1(c) will indicate that $E(r_i) \neq 0$ for some $i$ so that $E(\epsilon_i) \neq 0$. Further regressors, or even squares and products of the current regressors, will then need to be considered for inclusion in the model.

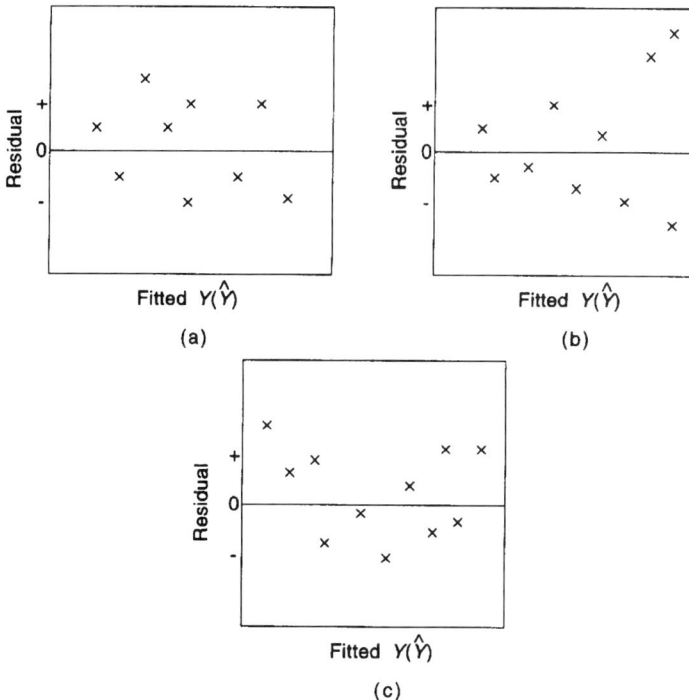

FIG. 1. Plot of residual $r_i$ versus fitted value $\hat{y}_i$: (a) model appears to be correct; (b) increasing variance; (c) model is inadequate.

2. Outliers: A few of $t_i$ may be much larger in absolute magnitude than all the others. Any observation that looks out of place is generally referred to as an *outlier*. These are usually omitted and the model refitted. However, such points need to examined carefully and treated with caution. It has been suggested that an observation should be rejected only if there is strong nonstatistical evidence that it is abnormal. Various studies (see [5], pp. 156–177, for references) suggest that practitioners can usually expect something like 0.1 to 10% of observations (or even more!) to be dubious; that is, wrong measurements, wrong decimal points, wrongly copied, or simply unexplainable. Sometimes a "peculiar" point is more important than the rest of the plot because it may indicate serious shortcomings in the model, e.g., the model is beginning to change its shape at a certain value of $x$.
3. Nonconstant variance: If the plot is funnel-shaped, like Fig. 1(b), there is a strong indication that the variance of $r_i$, and therefore $\epsilon_i$, is increasing

with $i$. We saw in Section 9.1 that models with a multiplicative rather than an additive error can have the variance increasing with the mean. A suitable transformation of the model may get rid of the problem and often helps to make the values of $Y_i$ more normal looking. Various ways of transforming the values of $y$ are available, especially the *power transformations* (the so-called ladder of powers, see Tukey [21]). The $x$-variables can also be transformed. Various transformations of both $y$ and $x$ are described by Draper and Smith [12], Weisberg [14], Atkinson [22] and, for nonlinear models as well, Carroll and Ruppert [16].

4. Correlated observations: Suppose the $\epsilon_i$ are not independent but are correlated, as in a time series. Then if we plot the residuals against time (or time order $i$ if the exact times are not known), we can obtain plots like Fig. 2 where the $r_i$ tend to come in positive or negative runs as in Fig. 2(a) or alternatively switching from positive to negative as in Fig. 2(b). These plots are useful only for small data sets. Correlations are better shown by a plot of the pairs $(t_i, t_{i-1})$, the so-called lagged residual plot, which will show a positive or negative trend. A test for trend in this plot, called the *Durbin–Watson test,* is available (see [8], pp. 167–169, [12], p. 162): most regression books describe this test. This plot is, technically, a test for serial first-order autocorrelation. It is essential to look for such correlations when data have been collected in time order. They can often show breakdowns in the experiment such as drift in meter readings over time.

### 9.6.2 Normal Probability Plot

To investigate the normality assumption for the $\epsilon_i$, the $t_i$ are ranked in order of magnitude and plotted in such a way that observations from a

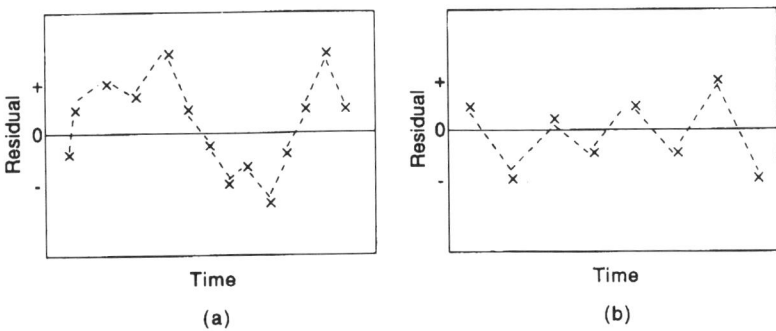

FIG. 2. Residual plots showing correlation between time-consecutive $\epsilon_i$: (a) positive correlation, (b) negative correlation.

normal distribution lie roughly on a straight line. However, some experience is needed in interpreting such plots. Daniel and Wood [11] give examples of plots from the $N(0, 1)$ distribution for samples sizes ranging from 8 to 384, and it is a good idea to look at these. For samples of less than 20 you can get very nonlinear plots with lots of "wobbles," even though the data is truly normal. Also, because of the way such plots are fitted, points at the top and bottom of the plot tend to show more scatter. Therefore such plots of residuals must be regarded as "broad brush" indicators only of any nonnormality. Although they are good at revealing outliers, a particular point needs to be well away from the trend before it is regarded as an outlier. Their use needs to be coupled with a formal test for normality such as the Wilk–Shapiro ($W$) test. A number of such tests are described by Seber ([5], pp. 141–148).

### 9.6.3 Plot the Variables in Pairs

Here all the variables, including the $y$-variable, are plotted in pairs. However, care is need in interpreting the plots of $y_i$ versus $x_{ij}$ for each $j$ because they can be completely misleading, as demonstrated in [11].

### 9.6.4 Plot Residual versus Each Regressor

These plots of $t_i$ versus $x_{ij}$ for $i = 1, 2, \ldots, n$ and each $j$ are useful for detecting a curvilinear relationship with the variable $x_j$. It may be appropriate to include a term $x_{ij}^2$ in the original model (or transform $x_{ij}$ to log $x_{ij}$, for example). They also may show up any marked changes in variance. Another plot for detecting the effect of $x_j$ is the following.

### 9.6.5 Larsen and McCleary Plots

If we define

$$r_{ij}^* = r_i + \hat{\beta}_j x_{ij}$$

then

$$E(r_{ij}^*) = \beta_j x_{ij} \qquad (9.20)$$

This suggests carrying out a plot of $r_{ij}^*$ versus $x_{ij}$ for each $j$. This plot, rediscovered by Larsen and McCleary [23], should give a straight line through the origin. It has the property that if we fit (9.20) by least squares we arrive at $\hat{\beta}_j$ once again and the residuals from this fitting are the correct residuals. It is also called a *partial residual plot* or *residual plus component plot*.

One role of such plots is to assess the importance of $x_j$ in the presence of the other $x$-variables. Another is to assess the presence of any nonlinearity and choose the correct transformation, as we now show. Suppose the true model is our original model but with $\beta_j x_{ij}$ replaced by some nonlinear function of $x_{ij}$, say, $h(x_{ij})$. Then

$$Y_i - \beta_0 - \beta_1 x_{i1} - \ldots - \beta_{p-1} x_{i,p-1} = h(x_{ij}) - \beta_j x_{ij} + \epsilon_i$$

If we estimate the lefthand side by $r_i$ we have, roughly,

$$r_i \approx h(x_{ij}) - \hat{\beta}_j x_{ij} + \epsilon_i$$

or

$$r_{ij}^* \approx h(x_{ij}) + \epsilon_i$$

This algebra is rather crude but it does indicate that a plot of $r_{ij}^*$ versus $x_{ij}$ for a given $j$ may give some idea as to the shape of $h$. A plot similar to this is the *added variable plot* (see Cook and Weisberg [24]). More general plots of this type are described by Chatterjee and Hadi [25, 26] and Atkinson [22].

### 9.6.6 Plots to Detect Influential Observations

An influential observation is one that either individually or collectively with several other observations has a bigger impact on the estimators than most of the other observations. There are two kinds of influential points—outliers and high leverage points. We have already mentioned outliers, and they can brought to our attention by unusual points in a residual plot. It is usually best to carry out the least squares fit with and without an outlier. Frequently the omission of an outlier has little effect so there may be no need to agonize over whether to include it or not. An outlier, therefore, need not be influential. On the other hand, an influential observation need not be an outlier in the sense of having a large residual, as we see in Fig. 3, based on Draper and Smith ([12] p. 169). Here we have four observations, three at $x = a$ and one at $x = b$. The residual for the middle observation, point (1), at $x = a$ is 0. However, it turns out that the residual at $x = b$ is also 0, irrespective of the corresponding $y$-value; i.e., points (2) and (3) both have a 0 residual. Clearly the observation at $x = b$ is extremely influential, and the slopes of the fitted lines can be completely different. What makes this point influential is that its $x$-value ($x = b$) is very different from the $x$-values ($x = a$) for the remaining points. It is clear, therefore,

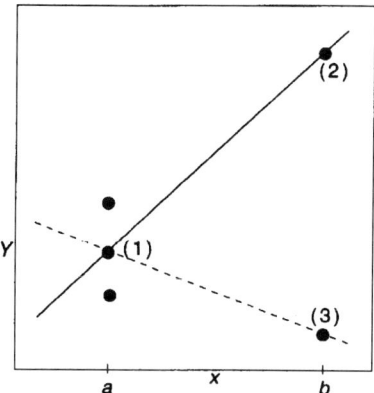

FIG. 3. The point (1) at $x = a$ has a 0 residual. A single point at $x = b$ is very influential but has a 0 residual irrespective of its value of $y$. Here points (2) and (3) represent two such values of $y$.

that graphical methods based on residual alone will may fail to detect these points. This leads us to the concept of leverage.

We recall that

$$\hat{\mathbf{y}} = \mathbf{P}\mathbf{y}$$

or

$$\hat{y}_i = \sum_{j=1}^{n} p_{ij} y_j$$

so that $p_{ii}$, the coefficient of $y_i$ can be thought of as the amount of "leverage" that $y_i$ has on $\hat{y}_i$. Since $p_{ii} = \mathbf{x}_i^T (\mathbf{X}^T\mathbf{X})^{-1} \mathbf{x}_i$, where $\mathbf{x}_i$ is the $i$th row of $\mathbf{X}$; i.e., the $i$th data point in "$\mathbf{X}$-space," we define a high leverage point to be an $\mathbf{x}_i$ with a large $p_{ii}$. Points far removed from the main body of points in the $\mathbf{X}$-space will have high leverage. Since it can be shown that $\sum_{i=1}^{n} p_{ii} = p$, we see that $p/n$ is the average value of the $p_{ii}$. Values greater than $3p/n$ or even $2p/n$ are therefore deemed to have high leverage. As with outliers, a high leverage point need not be influential, but it has the potential to be. In Fig. 4 we have three situations: (a) there is one outlier but it is not influential, (b) there are no outliers but there is one high leverage point that is not influential, and (c) there is one outlier with high leverage so that it has high influence.

There is a wide and somewhat confusing range of measures for detecting influential points, and a good summary of what is available is given

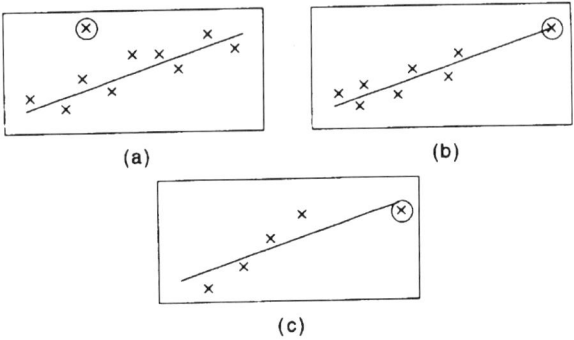

FIG. 4. The effect of extreme points on least squares fitting: (a) there is one outlier but it is not influential; (b) there are no outliers but there is one high leverage point; (c) there is one outlier with high leverage so that it has high influence.

by Chatterjee and Hadi [25] and the ensuing discussion. Some measures highlight problems with $y$ (outliers), others highlight problems with the $x$-variables (high leverage), while some focus on both. As statisticians all have different experiences with data, the advice given on which measure to use is confusing. As Cook [27] emphasizes, the choice of the method depends on the answer to the question, "Influence on what?" One should be most concerned with the influence on the part of the analysis that is most closely related to the *practical* problem under investigation. The choice of measure should therefore be determined by the priorities in the subject matter rather than the purely statistical priorities. However, for the practioner in the physical sciences, what often determines the choice is what is available from the statistical packages being used. We shall therefore begin by looking at some of the measures produced by the regression program PROC REG using the key word INFLUENCE. The first is known as Cook's distance $D_i$, which measures the change that occurs to $\hat{\beta}$ when the $i$th data point is omitted (giving $\hat{\beta}_{(i)}$). It represents a generalized distance between $\hat{\beta}$ and $\hat{\beta}_{(i)}$, namely,

$$\text{Influence on } \hat{\beta} \text{ as a whole: } D_i = (\hat{\beta} - \hat{\beta}_{(i)})^T X^T X (\hat{\beta} - \hat{\beta}_{(i)})/ps^2$$

$$= \frac{t_i^2 p_{ii}}{p(1 - p_{ii})} \quad (9.21)$$

In SAS $p_{ii}$ is denoted by $h_i$. Points of high influence are those for which $D_i$ exceeds the upper $\alpha$ quantile point of the $F_{p,n-p}$ distribution. Equation (9.21) indicates that $D_i$ depends on two quantities: (i) $t_i^2$, which measures the degree to which the $i$th point is an outlier, and (ii) $p_{ii}/(1 - p_{ii})$, which measures leverage.

Another measure is

$$\text{Influence on predicted values: DFFITS}_i = \frac{|\mathbf{x}_i^T(\hat{\boldsymbol{\beta}} - \hat{\boldsymbol{\beta}}_{(i)})|}{s_{(i)}\sqrt{p_{ii}}}$$

$$= |t_i^*|\sqrt{\frac{p_{ii}}{1 - p_{ii}}}$$

which is similar to $\sqrt{D_i}$ except that $s_{(i)}$ is used instead of $s$. Points for which $DFFITS_i > 2\sqrt{p/n}$ are investigated further. While Cook's $D_i$ measures the influence of a single observation on $\hat{\boldsymbol{\beta}}$, $DFFITS_i$ tends to measure its influence on both $\hat{\boldsymbol{\beta}}$ and $s^2$ simultaneously. However, the latter does have some deficiencies in this double role, and the comments by Cook [27] about the relative merits of both measures are helpful.

The preceding measures assume that all the values of $\beta_r$ are of equal importance so that they measure the effect of the observation on all the elements of $\hat{\boldsymbol{\beta}}$. However, an observation can be an outlier or influential only in one dimension or a few dimensions. The following measure

$$\text{Influence on individual coefficients: DFBETAS}_{ij} = \frac{\hat{\beta}_j - \hat{\beta}_{j(i)}}{\text{se}(\hat{\beta}_j)}$$

(with $s_{(i)}$ replacing $s$ in $\text{se}(\hat{\beta}_j)$) measures the *partial influence* of the $i$th observation on the $j$th coefficient and it is to be compared to $2/\sqrt{n}$.

Before leaving these measures we note that Cook [27] recommends the routine use of the so-called likelihood displacement. This method can also be applied to nonlinear models as in Cook [28], though the theory is difficult. The development of techniques for nonlinear models is still in its infancy. However, if the intrinsic curvatures are negligible then the residuals $r_i$ can be used in much the same way as in linear models. Otherwise, so-called projected residuals need to be used (Seber and Wild [3], p. 179). For a further discussion on regression diagnostics for linear models, the reader is referred to Belsey, Kuh, and Welsch, [29], Cook and Weisberg [24, 30], Weisberg [14], and Atkinson [22]. Influence methods are discussed by Escobar and Meeker [31] with applications to censored data. A helpful book on graphical methods in general, as well as regression diagnostics, is given by Chambers et al. [32].

### 9.6.7 ACE Transformation Method

The aim of the ACE (alternating conditional expectation) algorithm is to find suitable transformations of the $Y$ and $x$-variables such that the resulting model is linear; i.e., we look for functions $a$ and $b_i$ ($i = 1, 2, \ldots, p - 1$) such that

$$a(Y_i) = \beta_0 + \beta_1 b_1(x_{i1}) + \ldots + \beta_{p-1} b_{p-1}(x_{i,p-1}) + \epsilon_i$$

The method amounts to choosing the functions so that $R^2$, the square of the multiple correlation coefficient between the transformed variables, is maximized subject to the constraint that the functions are a member of some class of suitable functions such as powers, logs, monotone functions, etc. The package SAS has a procedure called TRANSREG that allows one to choose a particular family of functions. The theory of the method is given by Breiman and Friedman [33] and an application to soil science is described by De Veaux and Steele [34].

### 9.6.8 Multicollinearity

Most statistical computer packages have programs for fitting linear models by least squares. It is important to use a well-known package, as some "home-grown" programs may be based on inaccurate and inefficient algorithms. One problem in particular that can cause computational as well as interpretational problems is that of *ill-conditioning* when the matrix $X^T X$, which has to be inverted, is close to being singular. This happens when one $x$-variable is close to being a linear combination of some of the other $x$-variables so that the columns of $X$ are close to being linearly dependent. A poor program will break down when it meets this kind of problem. From a statistical point of view, a near singular $X^T X$ will have large elements on the diagonals of $(X^T X)^{-1}$ so that variances associated with the $\hat{\beta}_r$ will be large. This means that the least squares estimates will have low precision, and the problem is referred to as the *multicollinearity problem*. Another symptom is that regression coefficients change drastically when a variable is kept in or left out of an analysis. Some tools for detecting this problem are the diagonal elements of $(X^T X)^{-1}$ and the eigenvalues of $X^T X$. For further references see Montgomery and Peck [13], Chapter 8, Belsey et al. [29] and particularly Stewart [35], who gives several measures. Although multicollinearity has large effects on the precision of the estimates $\hat{\beta}_r$ (and thus on the explanatory power of the model), it may not badly affect the predictions made by the model. The effect of multicollinearity can be reduced a little by standardizing the variables so that their sample means are 0 and their sample variances are unity. The linear regression is then carried out on these transformed variables. The theory for this approach is given by Seber ([8], Section 11.7).

Even greater care is needed in using nonlinear packages as there are a wide range of possible problems, including lack of convergence of the least squares algorithm used and difficulties associated with interpreting confidence regions and intervals. Ill-conditioning in nonlinear models is a problem that is closely related to multicollinearity. With some models, very different parameter values can produce almost identical predictions. Bates and Watts ([6], Chapter 3) give a helpful summary of some of the practical issues involved. Many of the problems encountered are discussed by Seber and Wild ([3], Chapter 3) and they include a three chapter summary of the various computational methods available. Gallant [18] shows how to use the SAS package to analyze nonlinear models but avoids virtually all mention of the curvature arrays and their effects. Ross [36] and Ratkowsky [37, 38] demonstrate how a careful choice of parameters can ease many of the problems and provide a helpful practical approach to nonlinear modeling.

**Example 9.6 (Continued):** As a first step we will try to get the systematic part of the model right before worrying too much about the random part. We stressed in the previous section that the model first fitted is just a starting point for the modeling process. Having fitted it, we now look at the series of plots and other diagnostics we have just described in this section. The residuals versus predicted values plot revealed an increasing trend that looked reasonably linear. This is an indication that there are systematic effects not accounted for by the model. The residuals versus $x_1$ plot revealed a curved dish-shaped trend. There was also some evidence of curved trend in residuals versus $x_3$. Each residual plot showed a negative outlier. The curves seen in the residual plots and the Larsen and McClearly plots were fairly gentle so it looked as though they could be modeled by the inclusion of quadratic terms in the variable. When adding additional terms to the model (9.6) we follow the parsimony principle and only include those terms that are necessary.

We next added quadratic terms in $x_1$ and $x_3$ to the model, both separately and together at the same time, to fit

$$\log Y_i = \beta_0 + \beta_1 \log x_{i1} + \beta_2 \log x_{i2} + \beta_3 \log x_{i3} \\ + \beta_4 (\log x_{i1})^2 + \beta_5 (\log x_{i3})^2 + \epsilon_i \quad (9.22)$$

The main panel of output is given in Table III. Both quadratic terms result in coefficients that are weakly significantly different from 0 ($p$-values for the test statistics are on the order of 0.05) indicating that these terms should probably be included in the model. An overall $F$-test for omitting both the quadratic terms (i.e., $\beta_4 = 0$ and $\beta_5 = 0$) is significant, giving a $p$-value of 0.025. Thus we decided that these terms really should be retained.

While exploring the data in this vein, we also checked for the inclusion of cross-product terms *(interactions)* by checking for the inclusion of each

TABLE III. Some of the Output from the SAS Regression Program PROC REG Obtained from Fitting Eq. (9.22)

Model: MODEL1
Dependent Variable: LOGY

| | | Analysis of Variance | | | |
|---|---|---|---|---|---|
| Source | DF | Sum of Squares | Mean Square | F Value | Prob > F |
| Model | 5 | 1.71248 | 0.34250 | 158.278 | 0.0001 |
| Error | 29 | 0.06275 | 0.00216 | | |
| C Total | 34 | 1.77524 | | | |
| | Root MSE | 0.04652 | R-square | 0.9647 | |
| | Dep Mean | 3.23975 | Adj R-sq | 0.9586 | |
| | C.V. | 1.43584 | | | |

| | | Parameter Estimates | | | |
|---|---|---|---|---|---|
| Variable | DF | Parameter Estimate | Standard Error | T for $H_0$: Parameter = 0 | Prob > $\|T\|$ |
| INTERCEP | 1 | 8.906937 | 0.30971636 | 28.758 | 0.0001 |
| LOGX1 | 1 | 0.427631 | 0.13030862 | 3.282 | 0.0027 |
| LOGX2 | 1 | −0.529049 | 0.02780776 | −19.025 | 0.0001 |
| LOGX3 | 1 | 0.041415 | 0.02942214 | 1.408 | 0.1699 |
| LOGX1SQ | 1 | 0.037408 | 0.01881724 | 1.988 | 0.0563 |
| LOGX3SQ | 1 | 0.025065 | 0.01270459 | 1.973 | 0.0581 |

cross-product (e.g., $x_1x_2$ or, in this case, $\log x_1 \log x_2$) one at a time. Since none of the cross-products had coefficients anywhere near significant we decided not to include this level of complexity in the model.

We now have a new candidate model for the data; namely, that given earlier containing the two quadratic terms. Again it is a trial model only. We have to cycle back and use the data to look for discrepancies from the model. In other words, we have to go through all the diagnostics again. This time the residual plots showed no trends, outliers, or other bad features. There was no evidence of the existence of observations that had undue influence on the fitted model or of ill-conditioning except for the reasonably high correlation between a variable and its square, which we do not expect to cause any problems in interpretation. The lagged residual plot and the Durbin–Watson test statistic gave no evidence of serial correlation between residuals. The normal probability plot looked linear and a formal test for the normality of the residuals (Shapiro–Wilk) showed no signs of nonnormality. In short there appears to be no evidence of lack of fit for this model so that we could go ahead and use it to answer any of the substantive questions that the study was intended to explore.

There are several other interesting issues raised by the data. For example, the joint test for $\beta_4 = 0$ and $\beta_5 = 0$ gives us very strong evidence that the theoretically based initial model is inadequate. However, the value of $R^2$ (proportion of variation explained) only increases from 95.5% to 96.5% when we add the quadratic terms. This indicates that the actual point predictions from the earlier model are very little different from those made by the more complicated model. Thus the earlier model probably makes reasonably good approximate point predictions. So why is the discrepancy between the models important? The reason is that when the *statistical* model is inadequate, the *statistical* inferences (e.g., confidence levels for intervals and significance levels of tests) are completely unreliable.

When we first started thinking about how to model this data we stated that, in the absence of a theoretical model, we would probably have started modeling the data using

$$Y_i = \beta_0 + \beta_1 x_{i1} + \beta_2 x_{i2} + \beta_{i3} x_{i3} + \epsilon_i$$

In other words, we would have started with the original variables, not their logarithms. The diagnostic phase of this analysis again showed systematic departures. Similar reasoning led us once again to try adding quadratic and cross-product terms to overcome the discrepancies. This led finally to the model

$$Y_i = \beta_0 + \beta_1 x_{i1} + \beta_2 x_{i2} + \beta_3 x_{i3} + \beta_4 x_{i2}^2 + \beta_5 x_{i1} x_{i3} + \beta_6 x_{i2} x_{i3} + \epsilon_i$$

The diagnostics reveal that this model also fits the data quite well.

## 9.7 Errors in the Regressors

The theory developed in this chapter is based on the assumption that the $x$-variables are not stochastic. As stochastic regressors are the more common situation in physical models, we now have a relook at the least squares theory in this case. In physical models randomness occurs two ways, through natural variation in the material used for an experiment and measurement error. Usually $\epsilon$ contains both components. As all measurements are made with error, an $x$-variable will always have a random component, a fact usually overlooked by users of regression models. The whole subject of errors-in-variables models is not easy so we will try and indicate when the problem can be ignored.

In assessing the effect of random regressors, we again focus on the two kinds of model; namely, theoretical and empirical models. The model given by (9.1) is a typical theoretical model, and we extend it slightly to a straight line not necessarily through the origin; i.e.,

$$v_i = \beta_0 + \beta_1 u_i \tag{9.23}$$

with $Y_i = v_i + \epsilon_i$ and $X_i = u_i + \delta_i$, where the $u_i$ and the $v_i$ are fixed values. The pairs $(\delta_i, \epsilon_i)$ are generally assumed to be a random sample from a bivariate (two-dimensional multivariate) normal distribution. The first question one might ask is, "What happens if we ignore the errors $\delta_i$ and carry out the usual least squares on the model $Y_i = \beta_0 + \beta_1 X_i + \eta_i$ on the assumption that $X_i$ is fixed and not random?" We find, for the usual situation of independent $\epsilon_i$ and $\delta_i$ with variances $\sigma_\epsilon^2$ and $\sigma_\delta^2$, that $\hat{\beta}_0$ and $\hat{\beta}_1$ are biased estimates of $\beta_0$ and $\beta_1$ (see Seber [8], p. 211, Richardson and Wu [39]). However the biases can be ignored if $\sigma_\delta^2$ is much less than $s_x^2 = \sum_{i=1}^n (x_i - \bar{x})^2/(n-1)$; that is, when the error in each $X_i$ is much smaller than the variation in $X_i$. Since $\sigma_\delta^2$ is unknown, we need a rough idea of its magnitude, and this will be available if we know how accurate the device is for measuring $X_i$. Similar comments apply when there are several $x$-variables. Biases can be ignored if each regressor has this property.

The next question to consider is, "What effect do random regressors have on variance estimates?" It transpires that, although $s^2$ overestimates $\sigma^2$, $s^2(\mathbf{X}^T\mathbf{X})^{-1}$ is close to being an unbiased estimate of $Cov(\hat{\boldsymbol{\beta}})$, provided $n$ is moderate and the $x$-variables have small errors (Hodges and Moore [40], p. 189). Therefore if the bias in $\hat{\boldsymbol{\beta}}$ is small, the usual fixed-regressor theory described in Section 9.4 can still be used.

This model (9.23) in which $u_i$ is fixed is called the *functional* model ( $f$ for fixed). Sometimes it is more appropriate to regard $u_i$ as random so that we now have

$$V_i = \beta_0 + \beta_1 U_i \tag{9.24}$$

with $X_i = U_i + \delta_i$ as before. Here $U_i$ and $V_i$ are both random and the model is called a *structural* model ($s$ for stochastic). For example, suppose we take pairs of measurements on $n$ random soil samples and the true measurements satisfy the linear relationship (9.24). Since each sample is randomly chosen, the true measurement $U_i$ can be regarded as a random variable. Fortunately, we do not generally have to worry about all of this as biases can be ignored under the same conditions as for the functional model. The model in Example 9.6 is essentially a structural model. However, the measurement error in each $x$-variable could reasonably be assumed to be a lot less than the variation in its values so that the model can be analyzed as though the regressors were fixed. In fact what we are effectively doing is carrying out an ordinary least squares *conditional* on the observed values of the regressors even though they are random.

What do we do if the bias is not negligible and cannot be estimated? Estimation may now be carried out using the method of maximum likelihood. For the functional model we need more information such as knowledge of the regressor variances. If this is available then maximizing the likelihood provides estimates but they are unsatisfactory in that they do not tend to their true values as $n \to \infty$; i.e., they are said to be inconsistent (see Fuller [41], p. 104). The reason for this is that, in the model (9.23), $u_1, u_2, \ldots, u_n$ are unknown. This breakdown of the method of maximum likelihood is a common occurrence when the number of unknown quantities increases with $n$. For the structural model we assume that the $U_i$ are normally distributed so that instead of having $n$ unknowns we now have just the mean and variance of the normal distribution to be estimated. However, once again the maximum likelihood cannot be used without some knowledge of the variances as there is an "identifiability" problem associated with the parameters. For (9.24) a common assumption made is that $\sigma_\delta^2$ or the ratio $\sigma_\epsilon^2/\sigma_\delta^2$ is known. Given such an assumption, maximum likelihood estimates are now consistent and asymptotically normal. A good summary of both functional and structural models given by Wetherill ([9], Chapter 13). Stewart [35] applies several measures used for detecting multicollinearity to the problem of random regressors.

There is one important application of a structural model in which the observed regressors are fixed and not random. This occurs when the observed value of the $x$-variable is held at some predetermined value. For instance, in Example 9.1 we could adjust the current so that ammeter readings are fixed values like 1 amp, 2 amp, etc. Then, because of measurement error, the true current is unknown so that we have the structural relationship $V = rU$ with $Y = V + \epsilon$ and $x = U + \delta$. Combining these equations we end up with $Y = rx + \eta$ instead of (9.1). We can now analyze this as an ordinary fixed regressor model. The same comments apply to any linear model with targeted $x$-variables.

The preceding discussion applies to both theoretical and empirical models. However, in the latter case, since we are dealing with "black box" models, we can often assume the linear model $\mathbf{Y} = \mathbf{X}\boldsymbol{\beta} + \boldsymbol{\epsilon}$ to hold, where $\mathbf{X}$ is fixed or random and all the errors are contained in $\boldsymbol{\epsilon}$; i.e., we assume away the difficulties and simply carry out a standard least squares *conditional* on $\mathbf{X}$!

What about nonlinear models? As $\mathbf{X}$ is now effectively replaced by $\mathbf{G}$, the situation becomes unclear and further research is needed on the problem. Some of the difficulties are indicated in Seber and Wild ([3], Sections 1.4 and 1.5).

The theory of errors in variables is complex and for the general theory the reader is referred to the rather theoretical treatments of Fuller [41] and, for the nonlinear case, Seber and Wild [3], Chapter 10. Methods of assessing the effects of errors of measurements are also considered by Chatterjee and Hadi [26], Chapter 7.

## References

1. Carr, N. L. (1960). "Kinetics of Catalytic Isomerization of $n$-Pentane." *Indust. Eng. Chem.* **52,** 391–396.
2. Box, G. E. P., and Hill, W. J. (1974). "Correcting Inhomogeneity of Variance with Power Transformation Weighting." *Technometrics* **16,** 385–389.
3. Seber, G. A. F., and Wild, C. J. (1989). *Nonlinear Regression.* John Wiley and Sons, New York.
4. Beauchamp, J. J., and Cornell, R. G. (1966). "Simultaneous Nonlinear Estimation." *Technometrics* **8,** 319–326.
5. Seber, G. A. F. (1984). *Multivariate Observations.* John Wiley and Sons, New York.
6. Bates, D. M., and Watts, D. G. (1988). *Nonlinear Regression Analysis and Its Applications.* John Wiley and Sons, New York.
7. Scott, M. W., Robinson, M. J., Pauls, J. F., and Lantz, R. J. (1964). "Spray Congealing: Particle Size Relationships Using a Centrifugal Wheel Atomizer." *J. Pharmaceutical Sciences* **53,** 670–675.
8. Seber, G. A. F. (1977). *Linear Regression Analysis.* John Wiley and Sons, New York.
9. Wetherill, G. B. (1986). *Regression Analysis with Applications.* Chapman and Hall, London.
10. Chatterjee, S., and Price, B. (1977). *Regression Analysis by Example.* John Wiley and Sons, New York.
11. Daniel, C., and Wood, F. S. (1980). *Fitting Equations to Data: Computer Analysis of Multifactor Data,* 2d ed. John Wiley and Sons, New York.
12. Draper, N. R., and Smith, H. (1981). *Applied Regression Analysis,* 2d ed. John Wiley and Sons, New York.
13. Montgomery, D. C., and Peck, E. A. (1982). *Introduction to Linear Regression Analysis.* John Wiley and Sons, New York.
14. Weisberg, S. (1985). *Applied Linear Regression,* 2d ed. John Wiley and Sons, New York.
15. Gunst, R. F., and Mason, R. L. (1980). *Regression Analysis and Its Application.* Marcel Dekker, New York.
16. Carroll, R. J., and Ruppert, D. (1988). *Transformation and Weighting in Regression.* Chapman and Hall, London.
17. Abraham, B., and Ledolter, J. (1983). *Statistical Methods for Forecasting.* John Wiley and Sons, New York.
18. Gallant, R. A. (1987). *Nonlinear Statistical Models.* John Wiley and Sons, New York.
19. Rousseeuw, P. J., and Leroy, A. M. (1987). *Robust Regression and Outlier Detection.* John Wiley and Sons, New York.
20. Miller, A. J. (1990). *Subset Selection in Regression.* Chapman and Hall, London.

21. Tukey, J. W. (1977). *Exploratory Data Analysis.* Addison-Wesley, Reading, MA.
22. Atkinson, A. C. (1985). *Plots, Transformations, and Regression: An Introduction to Graphical Methods of Diagnostic Regression Analysis.* Clarendon Press, Oxford.
23. Larsen, W. A., and McCleary, S. J. (1972). "The Use of Partial Residual Plots in Regression Analysis." *Technometrics* **14,** 781–790.
24. Cook, R. D., and Weisberg, S. (1983). "Diagnostics for Heterogeneity in Regression." *Biometrika* **70,** 1–10.
25. Chatterjee, S., and Hadi, A. S. (1986). "Influential Observations, High Leverage Points, and Outliers in Linear Regression." *Statistical Science* **1,** 379–416.
26. Chatterjee, S., and Hadi, A. S. (1988). *Sensitivity Analysis in Linear Regression.* John Wiley and Sons, New York.
27. Cook, R. D. (1986). Comment on "Influential Observations, High Leverage points, and Outliers in Linear Regression." *Statistical Science* **1,** 393–397.
28. Cook, R. D. (1986). "Assessment of Local Influence (with Discussion)." *J. Roy. Stat. Soc.* **48,** 133–169.
29. Belsey, D. A., Kuh, E., and Welsch, R. E. (1980). *Regression Diagnostics: Identifying Influential Data and Sources of Collinearity.* John Wiley and Sons, New York.
30. Cook, R. D., and Weisberg, S. (1982). *Residuals and Influence in Regression.* Chapman and Hall, London.
31. Escobar, L. A., and Meeker, W. Q., Jr. (1992). "Assessing Influence in Regression Analysis with Censored Data." *Biometrics* **48,** 507–528.
32. Chambers, J. M., Cleveland, W. S., Kleiner, B., and Tukey, P. A. (1983). *Graphical Methods for Data Analysis.* Wadsworth, Belmont, CA.
33. Breiman, L., and Friedman, J. (1985). "Estimating Optimal Transformations for Multiple Regression and Correlation." *J. Amer. Stat. Assoc.* **80,** 580–597.
34. De Veaux, R. D., and Steele, J. M. (1989). "ACE Guided-Transformation Method for Estimation of the Coefficient of Soil–Water Density." *Technometrics* **31,** 91–98.
35. Stewart, G. W. (1987). "Collinearity and Least Squares Regression." *Statistical Science* **2,** 68–84.
36. Ross, G. J. S. (1990). *Nonlinear Estimation.* Springer-Verlag, New York.
37. Ratkowsky, D. A. (1983). *Nonlinear Regression Modelling.* Marcel Dekker, New York.
38. Ratkowsky, D. A. (1988). *Handbook of Nonlinear Regression Models.* Marcel Dekker, New York.
39. Richardson, D. H., and Wu, De-Min. (1970). "Alternative Estimators in the Error in Variables Model." *J. Amer. Stat. Assoc.* **65,** 724–748.
40. Hodges, S. D., and Moore, P. G. (1972). "Data Uncertainties and Least Squares Regression." *Appl. Stat.* **21,** 185–195.
41. Fuller, W. A. (1987). *Measurement Error Models.* John Wiley and Sons, New York.
42. Davies, R. B., and Hutton, B. (1975). "The Effect of Errors in the Independent Variables in Linear Regression." *Biometrika* **62,** 383–391.

# 10. FILTERING AND DATA PREPROCESSING FOR TIME SERIES ANALYSIS

William J. Randel

National Center for Atmospheric Research
Boulder, Colorado

## 10.1 Filtering Time Series

### 10.1.1 Introduction

The analysis of time series in physical sciences usually involves procedures such as differentiating, integrating, smoothing, extrapolating, or removal of noise. These all involve linear transformations of the original data, and the application of a linear transformation to a time series may be viewed as some sort of digital filter applied to that time series. In this section some general aspects of applying filters to time series are discussed. The behavior of a filter is often characterized in terms of its *frequency response function;* the motivation behind this concept is discussed, and the frequency response function is calculated for several examples. Next, the technique of designing a digital filter for a desired frequency response is discussed, including a simple FORTRAN subroutine for calculation of digital filter weights. Finally, the method of filtering time series by direct Fourier analysis–resynthesis is discussed. The fine details of filtering time series can be a rather complex subject, and several textbooks can be found that discuss these details (the book *Digital Filters* by R. W. Hamming [1] is a very readable and useful reference). The objective of this chapter is to give a brief introduction to some of the more practical aspects of using digital filters.

In practice, a digital filter is applied to a time series $u(t)$ by forming weighted linear combinations of successive subsets of the time series; letting $c(k)$ denote the weights, this produces a new "filtered" time series $y(t)$:

$$y(t) = \sum_{k=-M}^{M} c(k) \cdot u(t - k) \tag{10.1}$$

This process is termed a *convolution* of the data $u(t)$ with the filter coefficients $c(k)$. The total number of filter coefficients here is $(2M + 1)$. Note that the new or filtered time series $y(t)$ is shorter on each end than the original time series by $M$ points.

A familiar example of this procedure is the smoothing of data by application of a "running 3-point average" or a "1-1-1 moving average" filter. The smoothed or filtered data is given by an average of three successive input values:

$$y(t) = [u(t+1) + u(t) + u(t-1)]/3$$

This is represented by (10.1) with $M = 1$ and $c(k) = 1/3$. This is the simplest type of filter to consider, called a *nonrecursive* filter, because it uses only the original time series as input data. Filters may also be considered that use prior calculated values of the output, e.g.,

$$y(t) = \sum_{k=-M}^{M} c(k) \cdot u(t-k) + \sum_{l=1}^{L} d(l) \cdot y(t-l)$$

These are termed *recursive* filters. For a given filter length $M$, recursive filters have better frequency response characteristics than nonrecursive filters, but they require more computational expense. Furthermore, their analysis is somewhat more complex, and the focus here is on nonrecursive filters.

### 10.1.2 Frequency Response of a Filter

Digital filters represent linear transformations of time series. In order to understand specifically what occurs in the transformation, it is useful to consider the process in the frequency domain. The frequency response of a filter indicates the transformation that occurs for each frequency component of the input and output time series. To transform to the frequency domain, begin by considering the (complex) finite Fourier series expansion of the time series $y(t)$ ($t = 0, 1, \ldots, N-1$):

$$\tilde{y}(\omega_j) = \frac{1}{N} \sum_{t=0}^{N-1} y(t) \cdot e^{-i\omega_j t} \quad (10.2)$$

The discrete angular frequencies $\omega_j$ are given by $\omega_j = (2\pi/N) \cdot j$, with $j = -N/2 + 1, \ldots, -1, 0, 1, \ldots, N/2$. For a real function $y(t)$, the positive and negative frequency coefficients are complex conjugates, i.e., $\tilde{y}(\omega_j) = \tilde{y}*(\omega_{-j})$. The inverse transform is

$$y(t) = \sum_{j=-N/2+1}^{N/2} \tilde{y}(\omega_j) \cdot e^{i\omega_j t} \quad (10.3)$$

The total variance of the time series $y(t)$ may be equated to the sum of the squared harmonic coefficients according to

$$\text{Variance } [y(t)] = \frac{1}{N} \sum_{t=0}^{N-1} [y(t) - \bar{y}]^2$$

$$= \sum_{j=1}^{N/2} 2 \cdot |\tilde{y}(\omega_j)|^2 \quad (10.4)$$

where $\bar{y}$ is the average value of $y(t)$. A plot of $2 \cdot |\tilde{y}(\omega_j)|^2$ versus $\omega_j$ is termed a *periodogram;* this quantity measures the contribution of oscillations with frequency near $\omega_j$ to the overall variance of the time series $y(t)$. Such a diagram is similar to the *sample spectral power density,* although the quantity $2 \cdot |\tilde{y}(\omega_j)|^2$ is not a good estimate of the "true" underlying spectral density, because it is biased and its uncertainty does not decrease as the sample size increases. These factors are discussed in more detail in Chapter 11, along with techniques for the proper and consistent estimation of the spectral power density.

Now consider the frequency transform of (10.1):

$$\tilde{y}(\omega_j) = \sum_{t=0}^{N-1} y(t) \cdot e^{-i\omega_j t}$$

$$= \sum_{t=0}^{N-1} \left[ \sum_{k=-M}^{M} c(k) \cdot u(t-k) \right] e^{-i\omega_j t} \quad (10.5)$$

$$= \sum_{k=-M}^{M} c(k) \cdot e^{-i\omega_j k} \cdot \sum_{t=0}^{N-1} u(t-k) \cdot e^{-i\omega_j (t-k)}$$

$$\equiv \tilde{c}(\omega_j) \cdot \tilde{u}(\omega_j)$$

This result shows that the frequency coefficients of the filtered data $\tilde{y}(\omega_j)$ are equal to the frequency coefficients of the original time series $\tilde{u}(\omega_j)$, multiplied by the frequency transform of the filter coefficients $\tilde{c}(\omega_j)$. The frequency transform $\tilde{c}(\omega_j)$ thus measures what the filter does to each frequency coefficient and is termed the *frequency response* of the filter. Because $\tilde{c}(\omega_j)$ is in general a complex quantity, it is composed of both amplitude and phase components. In terms of spectral power densities (10.4),

$$|\tilde{y}(\omega_j)|^2 = |\tilde{c}(\omega_j)|^2 \cdot |\tilde{u}(\omega_j)|^2 \quad (10.6)$$

i.e., the spectral power density of the filtered data at each frequency is $|\tilde{c}(\omega_j)|^2$ times the power density of the input time series.

Examples of the frequency response of several filters are included here to get a feeling for its use and meaning. The frequency response is calculated directly from the discrete transform equation:

$$\tilde{c}(\omega_j) = \sum_{k=-M}^{M} c(k) \cdot e^{-i\omega_j k} \qquad (10.7)$$

For simple filters, the summations can be done directly; more complicated filters require computer calculations.

**Example 1:** We examine the frequency response for a 1-1-1 moving average filter. This is a filter with coefficients

$$c(-1) = 1/3$$
$$c(0) = 1/3$$
$$c(1) = 1/3$$

Using (10.7) the frequency response is evaluated as

$$\begin{aligned}\tilde{c}(\omega_j) &= \frac{1}{3}\left(e^{-i\omega_j} + 1 + e^{-i\omega_j}\right) \\ &= \frac{1}{3}(1 + 2\cos\omega_j)\end{aligned} \qquad (10.8)$$

This response is shown in Fig. 1 as a function of frequency and wavelength (measured in terms of grid spaces or sampling intervals). In this and the following figures the physical frequency ($f = \omega/2\pi$) is used as abscissa, with units of (1/unit time) or (1/grid spacing). This response function shows the fraction of wave amplitude at each frequency that is passed through this filter. The response is near 1.0 at very low frequencies, denoting that

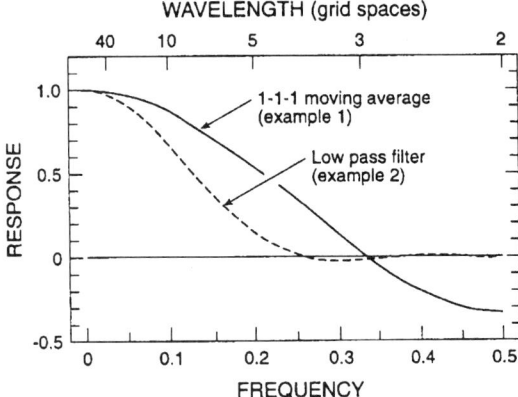

FIG. 1. Frequency response function calculated from (10.7) for the 1-1-1 moving average filter (Example 1) and low-pass filter (Example 2).

low-frequency oscillations (long wavelengths) are nearly unaffected by this filter. Conversely, higher frequency oscillations (shorter wavelengths) are selectively damped out; fluctuations near frequency 0.3 (3-grid space wavelength) are almost completely eliminated. This type of filter is termed a *low-pass filter*, because low frequencies are passed through while high frequencies are removed. This agrees with the intuitive knowledge that application of a 1-1-1 moving average removes high-frequency "noise" from a time series.

Note that the response function for this filter (10.8) is real valued, and this denotes that the phase of each frequency component remains unchanged by this filter. This is a general result for *symmetric* filters; i.e., ones with $c(-k) = c(k)$, and it is often a desirable property to be used in designing a filter. The negative response at high frequencies in Fig. 1 simply means that the input and output frequency coefficients are oppositely signed (i.e., that these frequency components in the filtered data will have opposite sign to those in the original time series).

To clearly show this transformation in the frequency domain, the sample spectral density function of a time series is examined before and after application of this filter. Figure 2 shows a time series of east–west winds near an altitude of 15 km, measured over Canton Island (near the equator in the central Pacific ocean) during 1960–1962. (This is a subset of a time series which was used to first discover the presence of a "40–50 day oscillation" in the winds, temperatures, and pressures over the Pacific Ocean [2]; this oscillation is now recognized as an important mode of atmospheric variability in the tropics.) Figure 3(a) shows a smoothed version of the spectral power density calculated from this time series, using the standard techniques discussed in Chapter 11. The spectral power density shows a strong peak in the frequency band .020–.025 days$^{-1}$ (wave periods of 40–50

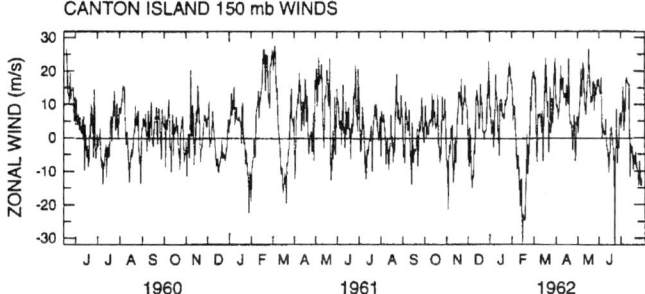

FIG. 2. Time series of east–west wind speed (m/s) near an altitude of 15 km, measured over Canton Island (in the equatorial Pacific Ocean) during 1960–1962.

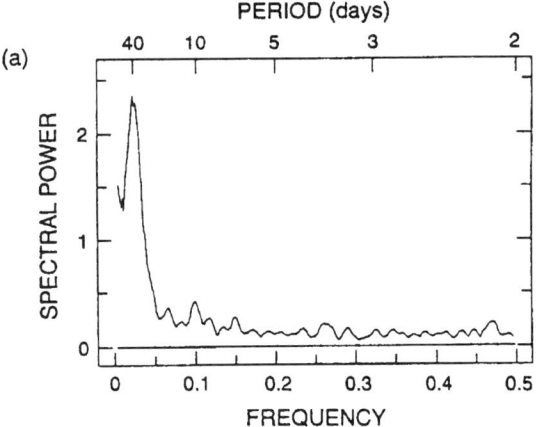

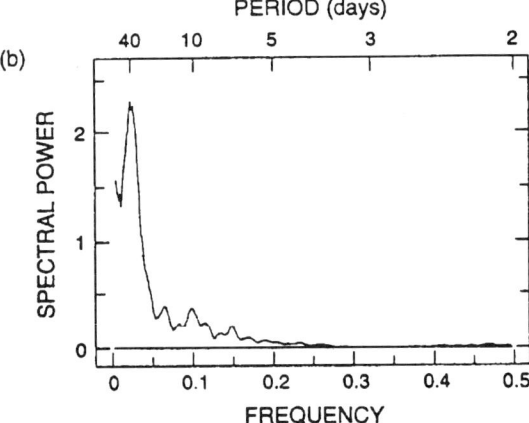

FIG. 3. Spectral power density calculated from the Canton Island time series in Fig. 2, (a) before and (b) after applications of a 1-1-1 moving average filter.

days), and near constant power at frequencies greater than $f = .1-.2$. Figure 3(b) shows the power spectral density for this time series after application of a 1-1-1 moving average filter. The spectral signature shows that the high frequency components ($f > 0.2$) have been selectively damped out. Note from (10.6) that the power spectral density for the filtered data is equal to that for the original data, multiplied by the square of the frequency response function (the square of the curve shown in Fig. 1). There is a small residual of power between $f = 0.4$–$0.5$ in the filtered data in Fig. 3(b), due to the nonzero filter response seen for this frequency range in Fig. 1.

**Example 2:** We now look at a "better" low-pass filter. The nonzero frequency response of the 1-1-1 moving average filter near frequency $f = 0.5$ (see Fig. 1) means that some high-frequency components still remain in the filtered data. This may not be important for the time series in Fig. 1 because there is relatively little power at high frequencies; however, in general a more ideal low-pass filter would have a response function that approaches 0 at all frequencies above some high-frequency cutoff limit $f_{high}$. An example of such a filter is given by the filter coefficients

$$c(0) = 0.260000$$
$$c(1) = 0.217069$$
$$c(2) = 0.120215$$
$$c(3) = 0.034124$$
$$c(4) = 0.002333$$

and again the filter weights are symmetric $c(-k) = c(k)$. These filter weights were calculated by the subroutine included in Section 10.1.3 (see discussion later). The frequency response function for this filter is shown as the dashed curve in Fig. 1, as calculated by numerical summation of (10.7). The response is near 1.0 at low frequencies and near 0.0 at high frequencies, with a transition region near $f = 0.1$–$0.2$ (i.e., this filters most of the oscillations with periods shorter than 5–10 days). The width of the transition region (where the response goes from 1.0 to 0.0) is directly related to the number of coefficients chosen for this filter (in this case $M = 4$). The use of more filter coefficients will result in a sharper transition region, but at the cost of losing more data at the beginning and end of the filtered time series. In practice, the user must balance the choice of sharper frequency cutoff versus longer filter length for the specific time series at hand. In this example relatively few coefficients have been used, with the result that the transition region is broad. The result of this filter applied to the Canton Island time series is shown in Fig. 4, clearly showing the smoothing effect of a low-pass filter; note that four data points have been lost from the beginning and end of the filtered data.

The simple computer subroutine used to generate these low-pass filter coefficients will also generate coefficients for *high-pass filters* (which pass high frequencies and remove low-frequency components) and *bandpass filters* (which pass frequencies only over a specified frequency band). An example of the latter is included later (Example 4).

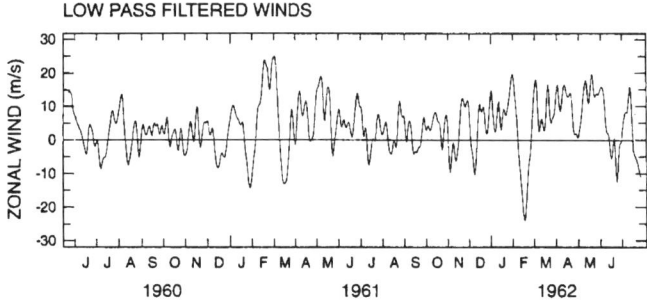

FIG. 4. Time series of Canton Island winds (Fig. 2), after application of the low-pass filter of Example 2.

**Example 3:** Finite difference approximations to derivatives are common calculations made on time series. The finite difference operators used to approximate derivatives are digital filters; the associated frequency response functions indicate how accurate the finite approximations are. Consider first the common centered difference operator

$$y(t) = [u(t + 1) - u(t - 1)]/2$$

This is a filter with coefficients

$$c(-1) = -1/2$$
$$c(0) = 0$$
$$c(1) = 1/2$$

The frequency response for this filter is given by (10.7):

$$\tilde{c}(\omega_j) = \frac{1}{2}(-e^{-i\omega_j} + e^{i\omega_j})$$
$$= i \sin\omega_j$$

Now for each spectral component $y(t) = e^{i\omega_j t}$, the "true" derivative is given by $y'(t) = i\omega_j y(t)$. The ratio of the calculated estimate of the continuous derivative to the true value is thus

$$\frac{\text{calculated}}{\text{true}} = \frac{\sin \omega_j}{\omega_j} \qquad (10.9)$$

This calculated–true ratio was not explicitly considered in Examples 1 and 2, because there the ratio of the output to the input is the natural comparison. However, for derivatives (or integrals) the calculated result should be compared to the ideal result; in both cases the ratio is viewed as a transfer or frequency response function.

The response curve corresponding to (10.9) is shown in Fig. 5. It shows that the derivative estimated from the $[u(t + 1) - u(t - 1)]/2$ formula is very accurate (response near 1.0) for low-frequency waves, but that the estimate is rather poor for very high-frequency oscillations (wavelengths shorter than 5 grid spaces).

A better finite difference approximation to a continuous derivative can be devised by using more filter weights. Again, the disadvantage is that more data are lost at the beginning and end of the filtered time series. Using $M = 3$, the following formula can be obtained (Section 7.2 of [3]):

$$y(t) = [u(t + 3) - 9 \cdot u(t + 2) + 45 \cdot u(t + 1) - 45 \cdot u(t - 1) + 9 \cdot u(t - 2) - u(t - 3)]/60$$

Here the ratio of the calculated–true response is given by

$$\frac{\text{calculated}}{\text{true}} = \frac{45 \cdot \sin \omega_j - 9 \cdot \sin(2\omega_j) + \sin(3\omega_j)}{30 \cdot \omega_j}$$

This response is shown as the dashed line in Fig. 5. Note the substantial increase in accuracy for wavelengths of 3–5 grid spaces compared to the simpler formula.

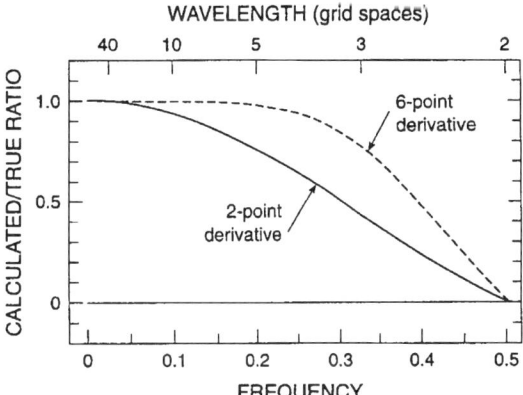

FIG. 5. Calculated vs. true ratio of derivatives estimated using the 2-point and 6-point centered difference operators (Example 3).

### 10.1.3 Designing a Digital Filter

In this section the technique of designing a specific nonrecursive digital filter is reviewed. This discussion includes low-pass, high-pass, and band-pass filters that have symmetric filter coefficients ($c(-k) = c(k)$), and hence no phase distortion between input and output time series. In the last section, the filter frequency response function was calculated as the finite Fourier transform of the filter coefficients (10.7). In order to derive filter weights $c(k)$ for a desired frequency response function $\tilde{c}(\omega_j)$, the inverse transform is used (written here in integral form):

$$c(k) = \frac{1}{2\pi} \int_{-\pi}^{\pi} \tilde{c}(\omega) \cdot e^{i\omega k} \, d\omega \tag{10.10}$$

Now since $c(-k) = c(k)$, only the cosine terms are included in the expansion:

$$c(k) = \frac{1}{\pi} \int_{0}^{\pi} \tilde{c}(\omega) \cdot \cos(\omega k) d\omega \tag{10.11}$$

Ideally, one would need an infinite number of coefficients $c(k)$ to exactly match an arbitrary frequency response function $\tilde{c}(\omega_j)$. In practice, the number of terms must be truncated to a finite length $M$. This truncation produces a rippling effect in the frequency response function called the *Gibbs phenomenon*, and such ripples are a generally undesirable feature for digital filters (the nonzero frequency response of the 1-1-1 moving average filter near $f = 0.4$–$0.5$ (Fig. 1) are an example of such ripples). These ripples in the response function can be reduced by applying a set of weights $w(k)$ to the filter coefficients $c(k)$ to produce a new set of coefficients $c'(k)$; i.e.,

$$c'(k) = c(k) \cdot w(k)$$

These filter weights $w(k)$ are sometimes termed a *window* through which one "sees" the filter weights $c(k)$. The process of truncation of the infinite series $c(k)$ to finite length $M$ can be viewed as using a rectangular window ($w(k) = 1$ for $k = 0$–$M$, and $w(k) = 0$ otherwise). The practical choice of weighting coefficients or "window shape" depends on several factors and is discussed in detail in Chapter 5 of [1].

One commonly used window or set of weight coefficients with the nice property that it removes a large fraction of the ripple effect is termed the Lanczos window:

$$\begin{aligned} w(k) &= \frac{\sin(\pi k / M)}{(\pi k / M)} \quad && k < M \\ &= 0 \quad && k \geq M \end{aligned} \tag{10.12}$$

This window is used in the calculations that follow. Other choices of windows may be more suited to particular problems, such as the Kaiser window ([1], Chapter 9), which produces sharper transition regions than the Lanczos window, but also results in larger frequency response ripples.

A simple FORTRAN subroutine is included here for generation of windowed filter weights for low-pass, high-pass and bandpass filters. These filters approximate idealized rectangular cutoff frequency response functions with bounds at $f_{low}$ and $f_{high}$; i.e.,

$$\tilde{c}(\omega) = 1 \quad f_{low} \leq \frac{\omega}{2\pi} \leq f_{high} \quad (10.13)$$
$$= 0 \quad \text{otherwise}$$

Note that for low-pass filters $f_{low} = 0.0$, and for high-pass filters $f_{high} = 0.5$. The untruncated Fourier expansion for $c(k)$ (10.11) can be evaluated using (10.13):

$$c(k) = \frac{1}{\pi} \int_{2\pi f_{low}}^{2\pi f_{high}} \cos(\omega k) d\omega$$

The unweighted coefficients are then expressed as

$$c(0) = 2(f_{high} - f_{low})$$
$$c(k) = \frac{1}{\pi k} [\sin(2\pi k f_{high}) - \sin(2\pi k f_{low})] \quad (10.14)$$

Finally, these coefficients (10.14) are weighted by the appropriate Lanczos window weights (10.12) to give the final filter coefficients $c'(k)$.

```
      subroutine makefilt (flow, fhigh, nterms, coeffs)
c
c     design nonrecursive bandpass digital filter using Lanczos
c        window
c     → output a symmetric set of filter weights
c
c
c     input:
c              flow - low frequency cutoff (0.0 for a low pass
c                     filter)
c              fhigh - high frequency cutoff (0.5 for high pass
c                      filter)
c                     (frequency units are inverse time or
c                      grid space intervals)
```

```
c                 nterms - resulting digital filter length is
c                          (2*nterms+1)
c
c                → note that the Lanczos weight factor for k= +/-
c                  nterms is zero, so that the filter truncation
c                  point is actually (nterms-1)
c       output:
c            coeffs - digital filter coeffecients
c
c
c       * the maximum (nterms) here is 100, but this can be easily
c         modified
c
      dimension coeffs(-100:100),wt(-100:100)
      do 101 k=-100,100
      coeffs (k)=0.
      wt (k)=0.
  101 continue
c
c     calculate unweighted coefficients
c
      coeffs (0)=2.*(fhigh-flow)
      do 201 k=1,nterms
      tpkfh=2.*3.1415926*k*fhigh
      tpkfl=2.*3.1415926*k*flow
      coeffs(k) = (sin(tpkfh)-sin(tpkfl))/(k*3.1415926)
  201 continue
c
c     now calculate Lanczos weights (sigma factors)
c        (or can substitute a different window here
c         such as the von Hann or Kaiser windows)
c
      wt(0)=1.0
      do 301 k=1,nterms-1
      phi=3.1415926*float(k)/float(nterms)
      wt(k)=sin(phi)/phi
  301 continue
c
c     weighted filter coefficients
c
      do 401 k=0,nterms
      coeffs(k)=coeffs(k)*wt(k)
```

```
  401 continue
c
c       symmetric filter weights
c
        do 403 k=1,nterms
        coeffs(-k)=coeffs(k)
  403 continue
        return
        end
```

This subroutine generated the set of coefficients used in Example 2 by setting $f_{low} = 0.0$, $f_{high} = 0.13$, and $M = 5$ (the Lanczos weights are 0 for the $k = M$ term, so that the filter is effectively truncated at $k = M - 1$).

**Example 4:** As one further example, a bandpass filter is designed to specifically isolate the 40–50 day oscillations in the Canton Island wind time series. The idealized rectangular frequency response function is chosen such that $f_{low} = 0.013$ and $f_{high} = 0.031$ (ideally retaining periods in the range 32–77 days). This is a very narrow spectral band and requires a large number of filter coefficients for an accurate approximation; here $M = 60$ is chosen (so that 60 days of data are lost from each end of the time series). The frequency response of this filter is shown in Fig. 6 (note that the

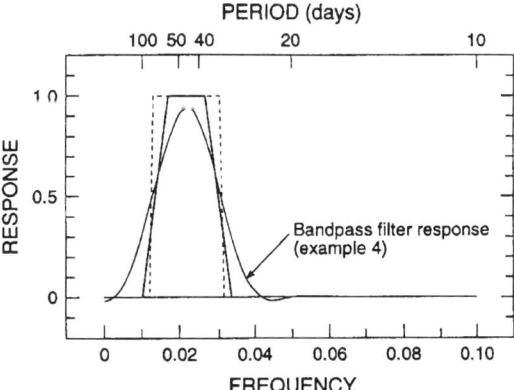

FIG. 6. Smooth continuous line shows frequency response function for bandpass digital filter of Example 4; dashed lines show idealized rectangular response with frequency limits $f_{low} = 0.013$ and $f_{high} = 0.031$. The trapezoid-shaped curve shows the frequency response function used in the direct Fourier analysis–resynthesis calculation of Section 10.1.4. Note the abscissa in this figure only covers frequencies 0.0 to 0.10.

frequency range plotted in Fig. 6 is only $f = 0.0$–$0.1$). The resulting filtered time series is shown as a solid line in Fig. 7, isolating the 40–50 day period oscillations. Note the amplitude modulation of the 40–50 day oscillation in time, with maximum amplitudes over January–February for both years.

### 10.1.4 Filtering via Direct Fourier Analysis–Resynthesis

An alternative method for filtering time series is to use a direct Fourier analysis to obtain frequency coefficients, truncate the coefficients in such a manner so that only a specified frequency band is retained, and then resynthesize the time series using the truncated coefficients. This method is easily implemented by the use of direct and inverse fast Fourier transforms (FFTs), which are now popular and readily available. As an example, this procedure was used to bandpass filter the wind time series over a frequency range similar to that chosen in Example 4. First, an FFT of the entire data is used to generate frequency coefficients. Second, the coefficients are weighted with the trapezoid-shaped frequency response function shown as the heavy curve in Fig. 6; i.e., frequency coefficients outside this band are set to 0. A trapezoid shape is chosen for the frequency window, as opposed to a rectangular shape, because too sharp of a frequency cutoff will result in a rippling effect in the filtered time series, an effect called *ringing*. Finally, the windowed coefficients are used to synthesize the filtered time series. The result is included as a dashed line in Fig. 7. Note the similarity to the digital filtered data; the FFT and digital filter results would be *exactly* the

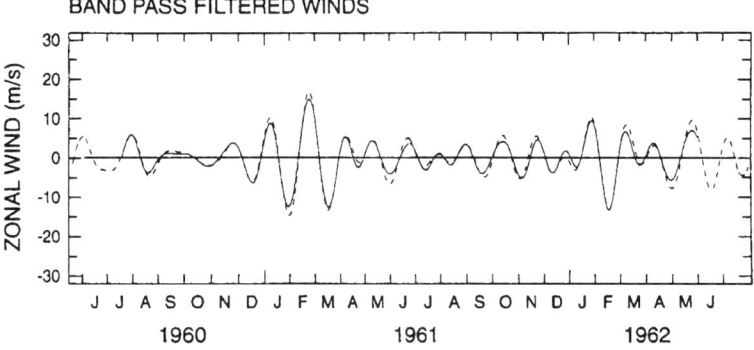

FIG. 7. Time series of Canton Island winds (Fig. 2) after application of 40–50 day bandpass digital filter (Example 4, solid line) and Fourier analysis–resynthesis (Section 10.1.4, dashed line).

same if the FFT window (Fig. 6) had the same shape as the bandpass filter response function.

At first inspection it appears that the FFT filtering method is superior to the digital filtering in that data are not lost at the beginning and end of the time series (see Fig. 7). However, the endpoint data generated by the FFT method are not to be believed (and should not even be displayed in plots of filtered data). The reason for this is that the FFT analysis implicitly assumes that the data are perfectly periodic with repeat distance $N$ (the length of the time series), so that the beginning and end of the time series are implicitly tied together (see the example later in Fig. 11). Hence the FFT bandpass filtered data near the beginning and end of the data are contaminated from the "other" end. This effect is clearly seen upon careful inspection of the beginning and end sections of the original (Fig. 2) versus bandpass filtered data (Fig. 7). The original data show maxima of opposite signs near the ends, whereas the FFT bandpassed data show a "turning over" of the curves near the ends, clearly not tracking the original data.

## 10.2 Data Preprocessing for Spectral Analysis

Spectral analysis is the name given to estimating the power spectral density function (10.4) (or cross-spectral density) from time series of observed data. Spectral analysis is useful in analyzing time series because it allows for a rearranging of the data according to frequency rather than time sequence. This is often useful in physical sciences because many phenomena are naturally separated by their frequency characteristics (such as high-frequency day-to-day weather variations versus low-frequency seasonal changes). Considerations regarding the proper calculation and significance of power and cross-spectral quantities are discussed in Chapter 11. In this section the discussion focuses on two topics that are of practical concern for spectral analysis: data windowing (tapering) and the removal of background trends prior to spectral analysis. As a note, trend removal should be done first and tapering second, prior to spectral analysis; they are discussed in opposite order here because the concepts of data windowing and leakage are central for understanding why trend removal is important.

### 10.2.1 Data Windowing (Tapering)

One fundamental problem in the proper estimation of spectral quantities from real data occurs because the time series analyzed are of finite length. Hence estimates of the "true" frequency spectrum of some variable are made based on a finite length sample of that variable (a measured time

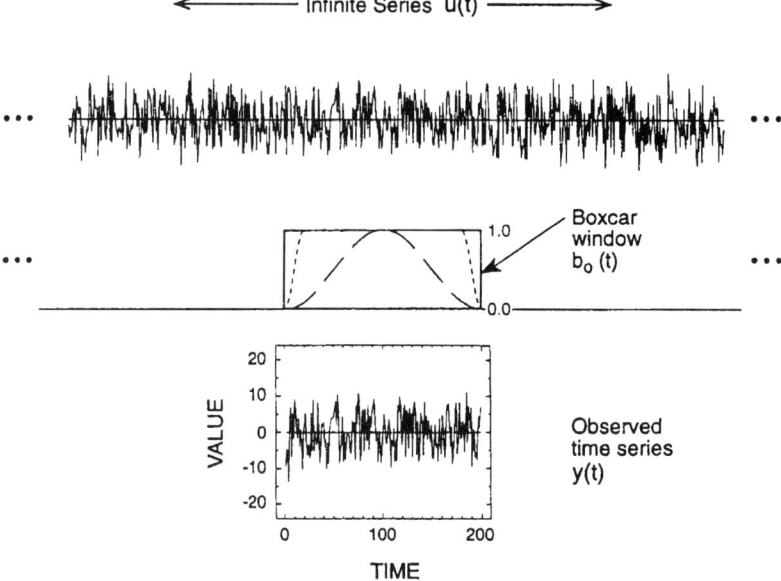

FIG. 8. Schematic representation of a finite length time series (bottom) as an infinite time series (top), sampled with a rectangular "boxcar" window (heavy curve in middle). Also shown in the middle are the Hanning and cosine-taper windows (long and short dashes, respectively.).

series). This situation is shown graphically in Fig. 8. The observed time series $y(t)$ can be viewed mathematically as an infinite time series $u(t)$ multiplied by a "boxcar" function $b_0(t)$ (where $b_0(t) = 1$ for $t = -M, \ldots,$ $M$ and $b_0(t) = 0$ otherwise; the data length $N$ is thus $2M + 1$):

$$y(t) = b_0(t) \cdot u(t)$$

Now the discrete Fourier transform (10.2) of the observed series $y(t)$ may be written as

$$\begin{aligned}
\tilde{y}(\omega_j) &= \sum_{t=-M}^{M} y(t) \cdot e^{-i\omega_j t} \\
&= \sum_{t=-\infty}^{\infty} b_0(t) \cdot u(t) \cdot e^{-i\omega_j t} \\
&= \sum_{t=-\infty}^{\infty} b_0(t) \cdot e^{-i\omega_j t} \sum_{l=-\infty}^{\infty} \tilde{u}(\omega_l) \cdot e^{i\omega_l t} \quad (10.15)
\end{aligned}$$

$$= \sum_{l=-\infty}^{\infty} \tilde{u}(\omega_l) \cdot \sum_{t=-\infty}^{\infty} b_0(t) \cdot e^{-i(\omega_j - \omega_l)t}$$

$$= \sum_{l=-\infty}^{\infty} \tilde{u}(\omega_l) \cdot \tilde{b}_0(\omega_j - \omega_l)$$

This shows that the frequency coefficients calculated from the observed time series are equivalent to the "true" coefficients from an infinite time series $\tilde{u}(\omega_l)$, convoluted with the transform of the boxcar function $\tilde{b}_0(\omega_j - \omega_l)$. In other words, each calculated spectral estimate $\tilde{y}(\omega_j)$ is a sum of the "true" spectral estimates near $\omega_j$, weighted with this function $\tilde{b}_0(\omega_j - \omega_l)$. From the definition of $b_0(t)$, $\tilde{b}_0(\omega_j - \omega_l)$ is easily evaluated:

$$\tilde{b}_0(\omega_j - \omega_l) = \sum_{t=-\infty}^{\infty} b_0(t) \cdot e^{-i(\omega_j - \omega_l)t}$$

$$= \sum_{t=-M}^{M} e^{-i(\omega_j - \omega_l)t} \qquad (10.16)$$

$$= \frac{2 \cdot \sin(\omega_j - \omega_l) \cdot M}{(\omega_j - \omega_l)}$$

This function is shown in Fig. 9. The finite width of the central maximum in $\tilde{b}_0(\omega_j - \omega_l)$ (called the *bandwidth* of the spectral analysis, because it sets

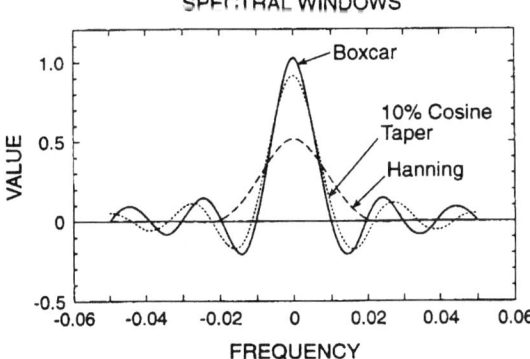

FIG. 9. Frequency transforms of the boxcar, Hanning, and cosine-taper windows (shown in the middle of Fig. 8), calculated as in (10.16), using $M = 100$ (i.e., for a time series of length 200).

a limit on the frequency resolution), and the presence of ripples, results in smearing and mixing up the "true" frequency components. This general problem is termed *leakage,* in that the true spectral coefficients are transferred or "leak" between adjacent and distant frequency bands. Note that, as the length $N$ of the time series increases, the transform function $\tilde{b}_0(\omega_j - \omega_l)$ (10.16) gets narrower, and hence better estimates of the "true" spectrum are made for longer time series. However, this problem remains to some degree for all finite length time series.

Several methods are used in practice in an attempt to minimize the leakage due to the "boxcar" sampling and associated transform $\tilde{b}_0(\omega_j - \omega_l)$. These methods are all similar in that they choose a different "window" than the boxcar function, with the aim of reducing the ripples in the transform $\tilde{b}_0(\omega_j - \omega_l)$. These windows have smooth transitions to 0 at the beginning and end of the data record, and this general process is termed *tapering* the data. Tapering is applied to a time series by simply multiplying by the chosen window function. One common choice is to use a cosine-shaped window of the form

$$b_1(t) = \frac{1}{2}\left[1 + \cos\left(\frac{\pi t}{M}\right)\right] = \cos^2\left(\frac{\pi t}{2M}\right) \qquad (10.17)$$

This is called a Hanning window and is shown for comparison to the boxcar window in Fig. 8. The transform of this function, $\tilde{b}_1(\omega_j - \omega_l)$, is shown in Fig. 9. Note that there is a reduction in the ripples compared to the boxcar transform (and hence a reduction in leakage from "distant" frequencies), but also that the bandwidth for the Hanning window transform is approximately twice as wide (i.e., the resulting spectrum will have less frequency resolution). Also, the Hanning window uses only the central half or so of the data record and will give a bit different results if the signals are not statistically stationary (i.e., the same throughout the record). One other frequently used window is a rectangular or boxcar function, modified so that the beginning and end sections have a smooth cosine-tapered structure. For example, the first and last 10% of the data may be tapered according to

$$b_2(t) = \frac{1}{2}\left[1 + \cos\left(\frac{5\pi t}{M}\right)\right] \qquad -M \leq t \leq -\frac{4}{5}M$$

$$= 1 \qquad -\frac{4}{5}M < t < \frac{4}{5}M$$

$$= \frac{1}{2}\left[1 + \cos\left(\frac{5\pi t}{M}\right)\right] \qquad \frac{4}{5}M \leq t \leq M$$

This window and its transform are shown in Figs. 8 and 9, respectively. This "10% cosine taper" window provides only a slight reduction in leakage compared to the boxcar window.

The inherent problem of leakage is the reason why harmonic amplitude estimates for individual frequency coefficients are biased and possess large uncertainties, even for very long time series; this is why the periodogram (10.4) is *not* a good estimator for the spectral power density. As discussed in more detail in Chapter 11, one method to reduce both the bias and variance of spectral power estimates is to average the periodogram over the individual frequency bands. This can be done in one of two ways: (1) averaging several different realizations of the spectra, i.e., average the power spectra from several time series, or (2) averaging a single spectrum over several adjacent frequency bands (i.e., applying a smoothing filter to the periodogram). For the latter case, the character of the resulting spectrum depends on the degree of smoothing: averaging over few frequency bands (narrow bandwidth) will result in many peaks in the spectrum, some of which may be spurious, while a wider bandwidth may smooth the spectrum so much that no peaks are distinguishable. In practice, the type of smoothing needs to be determined for each problem at hand, usually by testing several variations; further examples may be found in [4], [5], and [6].

An example of spectral smoothing is shown in Fig. 10, based on analysis of a thousand point sample of the synthetic time series shown in Fig. 8. This time series was generated by choosing two sharp spectral peaks at frequencies $f = 0.05$ and $f = 0.35$ and adding a component of random number generated noise to the time series (the latter adds variance to the spectrum at all frequencies, so-called white noise). Figure 10(a) shows the sample power spectrum obtained by applying a Hanning window (10.17) to the data, calculating the harmonic coefficients using a fast Fourier transform, and calculating the spectrum from (10.4). Note the relatively narrow bandwidth associated with this analysis and the large variability between adjacent frequency bands. Figure 10(b) shows the spectrum obtained using no data tapering (i.e., the boxcar window), but with the periodogram power estimates smoothed in frequency using a moving Gaussian-shaped filter (whose width if indicated as $BW$ in Fig. 10(b)). Note the much smoother character of this spectrum compared to that in Fig. 10(a), although in this example the spectral peaks at $f = 0.05$ and $0.35$ are clearly evident in both calculations. One note regarding the calculations in Fig. 10: because the Hanning window tapers the ends of the data, the time series variance (and spectral density estimates) are reduced compared to that for the full time series. For direct comparisons here, the Hanning window spectral power estimates have been multiplied by a constant factor, so that the sums over frequency of the two spectral estimates in Fig. 10 are equal.

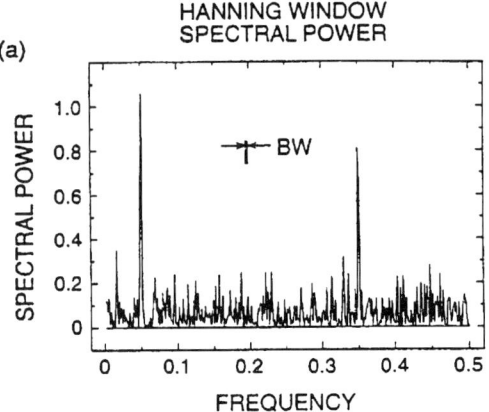

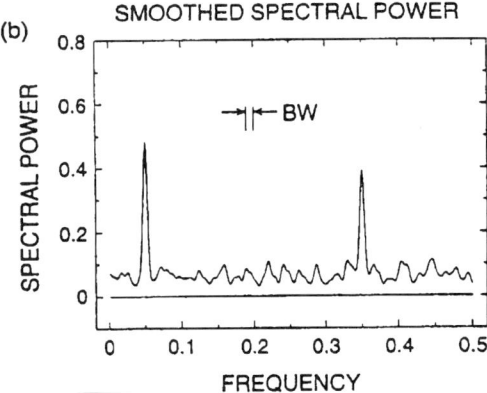

FIG. 10. Sample spectral power densities calculated from 1000-point records of the time series shown in Fig. 8. Spectra in (a) are the periodogram of the data using a Hanning data window, while the estimates in (b) result from frequency smoothing the periodogram derived from untapered data. BW denotes the bandwidth of the analyses.

### 10.2.2 Removal of Background Trends

Another potential problem in spectral analysis of finite length time series due to the presence of strong background signals or trends over the length of the data record. The fundamental problem is that the background signal contributes to additional background frequency components, which in turn get mixed with the "true" coefficients via the leakage arguments discussed previously. One way to visualize the problem is to note that a finite length Fourier analysis of time series $u(t)$ ($t$ = 0, 1, 2, . . . , $N - 1$) implicitly assumes that the data are perfectly periodic with a repeat distance $N$. This is shown schematically in Fig. 11, using for an example a 30-yr time series of atmospheric carbon dioxide ($CO_2$) measurements taken over Hawaii, obtained from [7]. There is a clear increase in time (positive trend) in these data that is a result of the systematic build-up of carbon dioxide in the atmosphere due to burning fossil fuels; there is also a strong annual cycle related to seasonal plant growth in the Northern Hemisphere. The result of the implicit periodic nature of the finite Fourier transform is that the time series effectively has the overall shape of a "sawtooth" curve, with strong discontinuities every 30 yrs due to the trend in the data. Power spectra calculated from this time series will have strong amplitudes over a broad frequency range introduced solely as a result of this trend (and mixed into the "true" spectrum by the leakage arguments discussed earlier). Figure 12 compares power spectrum estimates calculated from this time series, with and without the background trend removed. The trend was

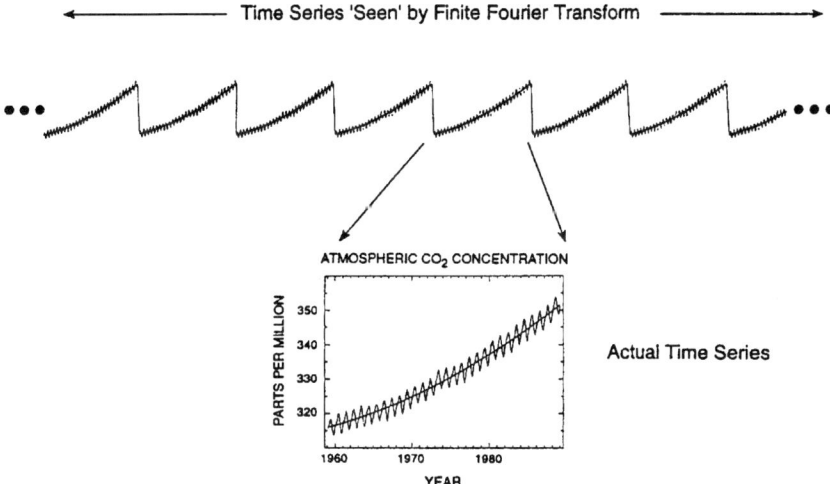

FIG. 11. Top curve shows a schematic representation of the infinitely periodic time series "seen" by a finite Fourier transform of the 30-yr record below. The time series is atmospheric carbon dioxide concentration (in parts per million by volume) measured at Launa Loa, Hawaii, over 1959–1988 [7]. Note that the strong trend (denoted by the smoothed curve in the lower figure) introduces a "sawtooth" shape to the periodic time series above.

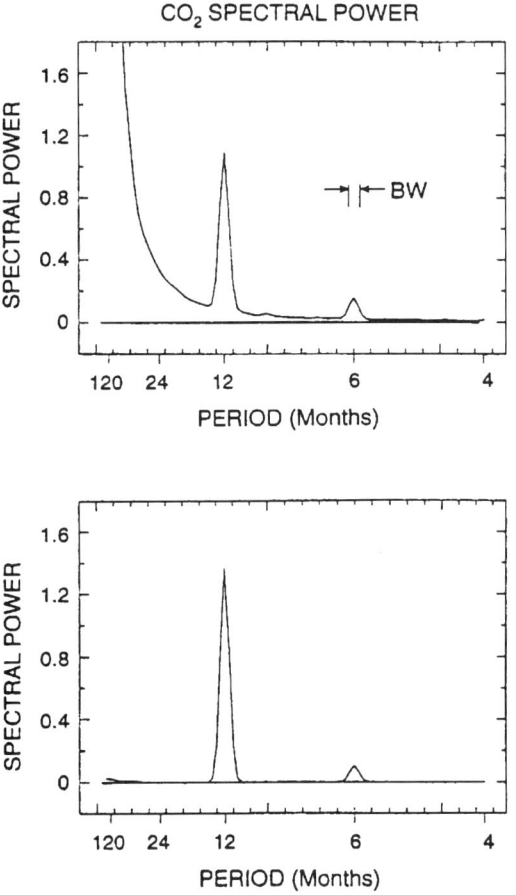

FIG. 12. Spectral power density for the carbon dioxide time series in Fig. 11, calculated before (top) and after (bottom) removal of the background trend (the trend is indicated by the smooth curve in the lower panel of Fig. 11).

estimated with a least squares fit to a quadratic function over the 30 yrs; i.e.,

$$y(t) = a_0 + a_1 t + a_2 t^2$$

with the coefficients $a_0$, $a_1$, and $a_2$ calculated according to the formulas given in Section 8.3 of [8]. This calculated background curve is indicated in the center of Fig. 11. (The choice of using a quadratic function for the background was somewhat arbitrary; other functional forms could have been chosen, or the data could be high-pass filtered to remove the slow background trend.) Comparison of the power spectra before and after trend removal (Fig. 12) clearly shows the spurious power introduced by the trend in this example: note also the changed magnitude of the annual harmonic between these two calculations and the fact that the detrended data spectrum shows a *larger* peak for the annual cycle. The smaller annual cycle peak in the original data spectrum results from the negative sidelobes associated with the boxcar convolution function in Fig. 9, so that the annual harmonic amplitudes (and resulting spectral power densities) were *decreased* by leakage in this example.

Trend removal is the simplest form of *prewhitening* of data prior to spectral analysis. In general, *prewhitening* refers to some linear transformation of the data in order to get an overall smoother spectrum in frequency space (constant power versus frequency is termed a *white* power spectrum, hence the name *prewhitening*). This preprocessing step can be important to minimize the effects of leakage discussed previously.

## 10.3 Imperfectly Sampled Time Series

The preceding section discussed biases that occur in spectral analysis because of finite data length and the presence of background trends. One other subject discussed here is how to handle time series with missing data values or time series that have unequally spaced observations. A simple technique is shown here for the estimation of power spectra from such data. There is one further general problem that occurs as a result of undersampling the true variability in a time series; undersampling in this context means not sampling frequently enough. This undersampling results in a misrepresentation of the true frequency dependence of the spectrum, an effect known as *aliasing*. An example of aliasing is included to illustrate this effect.

### 10.3.1 Calculating Power Spectra for Time Series with Missing Data

Time series may have missing data points for several reasons. Observations may simply be unavailable for certain time periods; this is a frequent problem in historical records of meteorological data, for example. Data values that are clearly spurious may also be present in a data record due to some detector malfunction. Removal of these wild points (or outliers) is important prior to subsequent spectral analyses, or else they will contaminate the entire spectrum. There are several ways to identify such outliers; often a simple plot of the time series will reveal obviously bad data points. Statistical methods may also be used to identify outliers. For example, the standard deviation of the entire data record can be calculated, and data that are more than (say) three standard deviations from the mean can be omitted. In any case, we wish to consider how to calculate spectra from time series with such missing values.

There are three straightforward methods to calculate spectra from such time series. First, the missing data can be interpolated in some manner (using linear interpolation or some more complicated scheme) and then spectral analysis performed on the resulting "complete" time series. This is the easiest solution for a relatively small amount of missing data. The problem with this approach in general is that structure is built into the resulting power spectrum by the type of interpolation used (any interpolation scheme can be thought of as a filter, with a specified frequency response, see [1], Section 3.7; this frequency response is then partially mirrored in the resulting power spectrum).

Second, discrete Fourier transform coefficients can be estimated from the time series by the least-squares fitting of the data to the individual harmonics, one at a time. Equations for these calculations can be found in Section 2.2 of [9].

A third technique is based on calculating the power spectrum using the lag-correlation technique (e.g., Section 7.4 of [4]). This calculation is based on the fact that the power spectrum is equivalent to the finite Fourier transform of the lag autocovariance function of a time series. This technique was a commonly used method of spectral analysis prior to the introduction of fast Fourier transforms. Briefly the calculation is as follows:

a. Calculate the lag autocovariance function $C(\tau)$ up to some finite maximum lag $T$:

$$C(\tau) = \frac{1}{N} \sum_{t=0}^{N-1-\tau} u(t) \cdot u(t + \tau) \quad (10.18)$$

where the time mean values (and trends) have been removed from the series $u(t)$. Note that some authors choose a normalization factor of $1/(N - \tau)$ instead of the $1/N$ in front of the definition of $C(\tau)$, e.g., Bath ([5], Section 3.3.3). Jenkins and Watts ([10], Section 5.3.3) discuss these two options, choosing the $1/N$ expression because it has a smaller mean squared error.

b. Calculate the (unsmoothed) power spectrum by the finite Fourier transform of $C(\tau)$. Because $C(\tau)$ is symmetric in $\tau$, only the cosine terms survive:

$$Y(\omega_j) = \frac{1}{T}\left[C(0) + 2 \sum_{\tau=1}^{T-1} C(\tau) \cdot \cos(\omega_j \tau) + C(T)(-1)^T\right] \quad (10.19)$$

c. The resulting spectrum is smoothed in some manner; a common choice is to use a running .25–.50–.25 smoothing in frequency (see Section 7.4 of [4]).

The maximum number of lags ($T$) is chosen based on the length of the time series and the desired spectral resolution; it is inversely proportional to the bandwidth of the analysis (larger $T$ results in higher spectral resolution but less statistical stability). In practice the user needs to choose $T$ and the method of smoothing the spectrum that balances resolution versus stability.

This calculational procedure is directly applicable to time series with missing data by simply ignoring the missing data in step (a); i.e., only the available data are used in calculating $C(\tau)$:

$$C(\tau) = \frac{1}{L} \sum_{t=0}^{L-1-\tau} u(t) \cdot u(t+\tau) \qquad (10.20)$$

with $L$ being the amount of data pairs that are available at each respective time lag. A recent example of application of this method to the spectral analysis of satellite ozone data with missing observations is found in [11].

We test this analysis technique here by analyzing the same synthetic time series (of length 1000) used in Section 10.2.1. The power spectrum calculated from this time series by the lag correlation technique with maximum lag $T = 500$ is shown in Fig. 13. Note that this spectrum is nearly identical to that produced by the smoothed periodogram estimate shown in Fig. 10(b); this

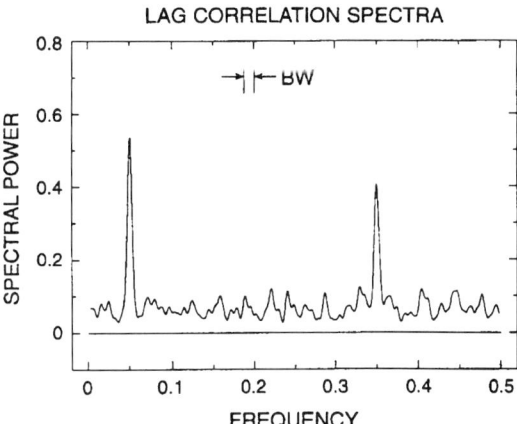

FIG. 13. Power spectral density calculated from 1000 day time sample of the time series shown in Fig. 8, calculated using the lag-correlation analysis technique (Section 10.3.1).

similarity is due to choosing the maximum lag (in 10.19) to be $T = 500$, resulting in a bandwidth similar to that used in the periodogram smoothing. The effect of missing data on these spectral estimates is tested by removing some percentage of the data from the time series in a random manner and recalculating the power spectrum. Figure 14 shows power spectra calculated from the data with 10%, 30%, and 50% of the data randomly removed (compare this to the "original" spectrum in Fig. 13). For 10% and 30% of the data missing the spectra are quite similar to the original, with a slight reduction in intensity of the peaks and increase in the background "noise" level. The peaks

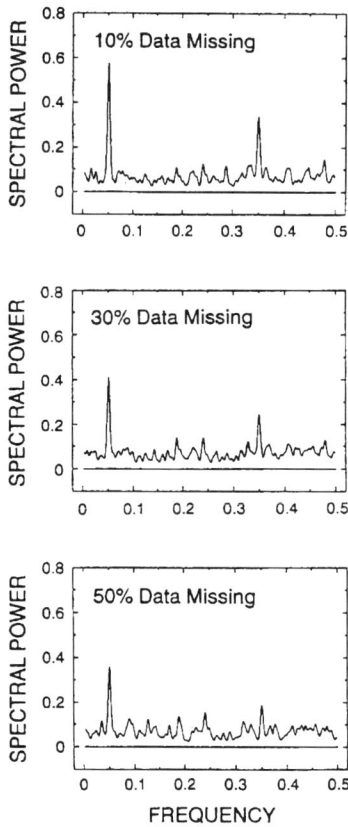

FIG. 14. Power spectral density estimates for the same time series analyzed in Fig. 13, but with successively larger amounts of the data removed prior to calculation of (10.20). Shown are spectra for 10% of the data missing (top), 30% and 50% (bottom). The spectrum for the full time series (no data missing) is shown in Fig. 13.

are reduced and the background increased further for the case with 50% of the data missing, so that the peak at $f = 0.35$ cannot be distinguished from other spurious maxima. If much more than 50% of the data is removed in this example, the larger peak at $f = 0.05$ also becomes indistinguishable, showing that this is near the limit where useful spectra can be calculated in this case. Note that the sharp, well-defined spectral peaks chosen for this example (Fig. 13) make it a highly idealized case and that if the peaks were smaller they would not stand out against the background for the 50% missing data case. In general the amount of missing data that can be tolerated depends on the character of the data being analyzed. For analyses of real time series, it is recommended that tests similar to this one be done to determine the effects of variable missing data.

### 10.3.2 Aliasing

One additional important problem in spectral analysis arises due to the presence of variability with frequencies higher than those that can be resolved by the sampling rate. For a given data spacing $\Delta t$, the highest frequency oscillation that can be resolved is one with period $2\Delta t$, or frequency $f = 1/(2\Delta t)$. This high-frequency limit is called the *Nyquist frequency* $f_N$. In the case where there is substantial variability at frequencies above $f_N$, the effect of sampling with spacing $\Delta t$ is that power at frequencies above $f_N$ will appear as power at frequencies lower than $f_N$. This effect is called *aliasing*. Specifically, power at frequencies $f$, $2f_N \pm f$, $4f_N \pm f$, . . . are all aliased into (appear as) power at frequency $f$, due solely to the sampling rate $\Delta t$.

An example is shown here based on the synthetic time series analyzed previously, whose spectrum is shown in Figs. 10 and 13 (based on sampling at every time step, $\Delta t = 1$). This same data is sampled at every second point instead of at every point, so that $\Delta t = 2$ versus $\Delta t = 1$ previously. The Nyquist frequency for the $2\Delta t$ sampled data is $f_N = 1/(2 * 2\Delta t) = 0.25$. The original power spectrum (Fig. 13) showed that there is a spectral peak in the original data at $f = 0.35$, above this new Nyquist limit, so that this power will be aliased in the new sampling at $\Delta t = 2$. The frequency where this aliased power will occur is at $f_a = 2 f_N - f = 2 (0.25) - 0.35 = 0.15$. A power spectrum calculated from the newly sampled data is shown in Fig. 15, and a clear peak is indeed found at $f = 0.15$, due to aliasing. (One way to visualize the effect of this aliasing is that the power in the "true" spectrum (Fig. 13) has been "folded" back about the Nyquist frequency $f_N = 0.25$). Note that if the higher frequency peak in the "original" spectrum (Fig. 13) had been located near $f = 0.45$, it would have aliased onto the preexisting spectral peak at $f = 0.05$, so that the power for that peak would have been severely overestimated.

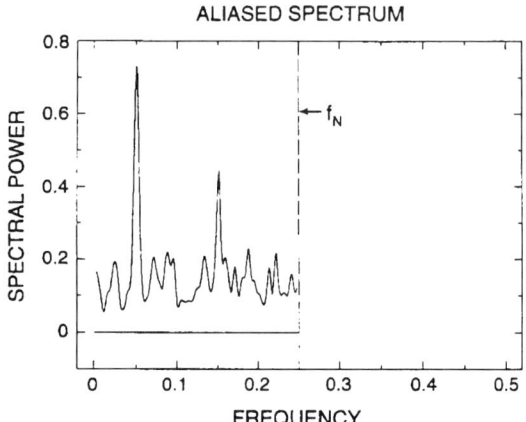

FIG. 15. Power spectral density calculated from a sample of the time series in Fig. 8, with data sampled at every second point. The Nyquist frequency is $f_N = 0.25$. The spectral peak near $f = 0.15$ results from aliasing of the power in the "true" spectrum near $f = 0.35$ (shown in Fig. 13).

Note that once the data have been sampled at $2\Delta t$, there is nothing to distinguish the peak at $f = 0.15$ as a "false" peak resulting from aliasing; nothing can be done once the data are sampled. The only way to avoid aliasing is to choose a high enough sampling rate (small $\Delta t$). This is not always an option in practice, however, and the user should be aware of potential aliasing problems in analysis of any time series.

## Acknowledgments

The author thanks Dennis Shea and Roland Madden of NCAR for helpful discussions and Christopher Chatfield, Don Percival, John Stanford, and Dale Zimmerman for constructive comments on the manuscript. This work was partially supported under NASA grant W-16215. NCAR is sponsored by the National Science Foundation.

## References

1. Hamming, R. W. (1989). *Digital Filters*, 3d ed. Prentice-Hall, Englewood Cliffs, NJ.
2. Madden, R. A., and Julian, P. R. (1971). "Detection of a 40–50 Day Oscillation in the Zonal Wind in the Tropical Pacific." *J. Atmos. Sci.* **28,** 702–708.

3. Maron, M. J. (1982). *Numerical Analysis: A Practical Approach.* Collier Macmillan, New York.
4. Chatfield, C. (1989). *The Analysis of Time Series: An Introduction,* 4th ed. Chapman and Hall, London.
5. Bath, M. (1974). *Spectral Analysis in Geophysics.* Elsevier, Amsterdam.
6. Priestly, M. B. (1981). *Spectral Analysis and Time Series.* Academic Press, London.
7. Boden, T. A., Karciruk, P., and Farrel, M. P. (1990). "Trends '90. A Compendium of Data on Global Change." Carbon Dioxide Information Analysis Center, Oak Ridge, TN.
8. Chatfield, C. (1983). *Statistics for Technology,* 3d ed. Chapman and Hall, London.
9. Bloomfield, P. (1976). *Fourier Analysis of Time Series: An Introduction.* John Wiley and Sons, New York.
10. Jenkins, G. M., and Watts, D. G. (1969). *Spectral Analysis and Its Applications.* Holden-Day, San Francisco.
11. Randel, W. J., and Gille, J. C. (1991). "Kelvin Wave Variability in the Upper Stratosphere Observed in Satellite Ozone Data." *J. Atmos. Sci.* **48,** 2336–2349.

# 11. SPECTRAL ANALYSIS OF UNIVARIATE AND BIVARIATE TIME SERIES

## Donald B. Percival

Applied Physics Laboratory
University of Washington, Seattle

## 11.1 Introduction

The spectral analysis of time series is one of the most commonly used data analysis techniques in the physical sciences. The basis for this analysis is a representation for a time series in terms of a linear combination of sinusoids with different frequencies and amplitudes. This type of representation is called a *Fourier representation.* If the time series is sampled at instances in time spaced $\Delta t$ units apart and if the series is a realization of one portion of a real-valued stationary process $\{X_t\}$ with zero mean, then we have the representation (by Cramér [1])

$$X_t = \int_{-f_{(N)}}^{f_{(N)}} e^{i 2\pi f t \Delta t} \, dZ(f), \qquad t = 0, \pm 1, \pm 2, \ldots \qquad (11.1)$$

where $f_{(N)} = 1/(2\,\Delta t)$ is the Nyquist frequency (if the units of $\Delta t$ are measured in, say, seconds, then $f_{(N)}$ is measured in Hertz (Hz), i.e., cycles per second); $i = \sqrt{-1}$; and $\{Z(f)\}$ is an orthogonal process (a complex-valued stochastic process with quite special properties). This representation is rather formidable at first glance, but the main idea is simple: since, by definition, $e^{i 2\pi f t \Delta t} = \cos(2\pi f t \Delta t) + i \sin(2\pi f t \Delta t)$, Equation (11.1) says that we can express $X_t$ as a linear combination of sinusoids at different frequencies $f$, with the sinusoids at frequency $f$ receiving a random amplitude generated by the increment $dZ(f) = Z(f + df) - Z(f)$ (here $df$ is a small positive increment in frequency). The expected value of the squared magnitude of this random amplitude defines the *spectral density function* $S_X(\cdot)$ for the stationary process $\{X_t\}$ in the following way:

$$E[|dZ(f)|^2] = S_X(f)\, df, \qquad -f_{(N)} \leq f \leq f_{(N)}$$

(the notation $E(X)$ refers to the expected value (mean) of the random variable (rv) $X$). Because $|dZ(f)|^2$ is a nonnegative rv, its expectation must be nonnegative, and hence the sdf $S_X(\cdot)$ is a nonnegative function of frequency. Large values of the sdf tell us which frequencies in Eq. (11.1) contribute the most in constructing the process $\{X_t\}$. (We have glossed over many details here, including the fact that a "proper" sdf does not exist for some stationary processes unless we allow use of the Dirac delta function. See Koopmans [2] or Priestley [3] for a precise statement and proof of Cramér's spectral representation theorem or Section 4.1 of Percival and Walden [4] for a heuristic development.)

Because $\{X_t\}$ is a real-valued process, the sdf is an even function; i.e., $S_X(-f) = S_X(f)$. Our definition for the sdf is "two-sided" because it uses both positive and negative frequencies, the latter being a nonphysical—but mathematically convenient—concept. Some branches of the physical sciences routinely use a "one-sided" sdf that, in terms of our definition, is equal to $2S_X(f)$ over the interval $[0, f_{(N)}]$.

Let us denote the $\tau$th component of the *autocovariance sequence* (acvs) for $\{X_t\}$ as $C_{\tau, X}$; i.e.,

$$C_{\tau, X} = \text{Cov}(X_t, X_{t+\tau}) = E(X_t X_{t+\tau})$$

(the notation $\text{Cov}(X, Y)$ refers to the covariance between the rv's $X$ and $Y$). The spectral representation in Eq. (11.1) can be used to derive the important relationship

$$C_{\tau, X} = \int_{-f_{(N)}}^{f_{(N)}} S_X(f) e^{i 2\pi f \tau \Delta t} \, df \qquad (11.2)$$

(for details, see [4], Section 4.1). In words, $S_X(f)$ is the (nonrandom) amplitude associated with the frequency $f$ in the preceding Fourier representation for the acvs $\{C_{\tau, X}\}$. If we recall that $C_{0, X}$ is just the process variance, we obtain (by setting $\tau = 0$ in the preceding equation)

$$V(X_t) = C_{0, X} = \int_{-f_{(N)}}^{f_{(N)}} S_X(f) \, df$$

(the notation $V(X)$ refers the variance of the rv $X$). The sdf thus represents a decomposition of the process variance into components attributable to different frequencies. In particular, if we were to run the process $\{X_t\}$ through a narrow-band filter with bandwidth $df$ centered at the frequencies $\pm f$, the variance of the process coming out of the filter would be approximately given by $2S_X(f)df$ (the factor of 2 arises because $S_X(\cdot)$ is a two-sided sdf). Spectral analysis is an analysis of variance technique in which we portion out contributions to $V(X_t)$ across different frequencies. Because variance is closely related to the concept of power, $S_X(\cdot)$ is sometimes referred to as a *power spectral density function*.

# INTRODUCTION

In this chapter we discuss estimation of the sdf $S_X(\cdot)$ based upon a time series that can be regarded as a realization of a portion $X_1, \ldots, X_n$ of a stationary process. The problem of estimating $S_X(\cdot)$ in general is quite complicated, due both to the wide variety of sdfs that arise in physical applications and also to the large number of specific uses for spectral analysis. To focus our discussion, we use as examples two time series that are fairly representative of many in the physical sciences; however, there are some important issues that these series do not address and others that we must gloss over due to space (for a more detailed exposition of spectral analysis with a physical science orientation, see [4]). The two series are shown in Figure 1 and are a record of the height of ocean waves as a function of time as measured by two instruments of quite different design. Both instruments were mounted 6 m apart on the same platform off Cape Henry near Virginia Beach, Virginia. One instrument was a wire wave gauge, while the other was an infrared wave gauge. The sampling frequency for both instruments was 30 Hz (30 samples per second) so the sampling

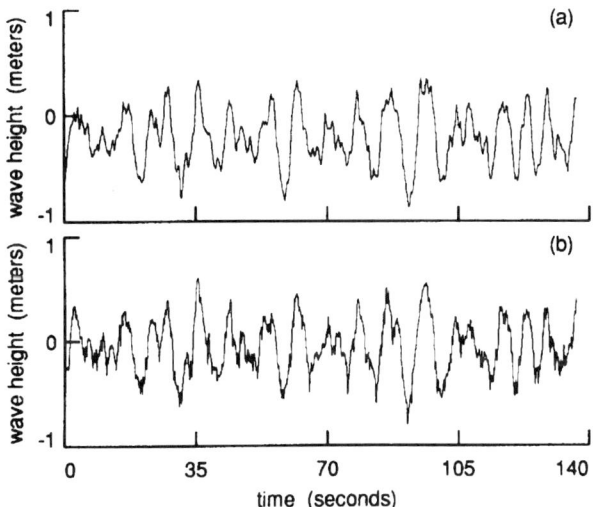

FIG. 1. Plot of height of ocean waves versus time as measured by a wire wave gauge (a) and an infrared wave gauge (b). Both series were collected at a rate of 30 samples per second. There are $n = 4096$ data values in each series. (These series were supplied through courtesy of A. T. Jessup, Applied Physics Laboratory, University of Washington. As of 1993, they could be obtained via electronic mail by sending a message with the single line "send saubts from datasets" to the Internet address statlib@lib.stat.cmu.edu—this is the address for Stat-Lib, a statistical archive maintained by Carnegie Mellon University.)

period is $\Delta t = 1/30$ sec and the Nyquist frequency is $f_{(N)} = 15$ Hz. The series were collected mainly to study the sdf of ocean waves for frequencies from 0.4 to 4 Hz. The frequency responses of the instruments are similar only over certain frequency ranges. As we shall see, the infrared wave gauge inadvertently increases the power in the measured spectra by an order of magnitude at frequencies 0.8 to 4 Hz. The power spectra for the time series have a relatively large dynamic range (greater than 50 dB), as is often true in the physical sciences. Because the two instruments were 6 m apart and because of the prevalent direction of the ocean waves, there is a lead/lag relationship between the two series. (For more details, see Jessup, Melville, and Keller [5] and references therein.)

## 11.2 Univariate Time Series

### 11.2.1 The Periodogram

Suppose we have a time series of length $n$ that is a realization of a portion $X_1, X_2, \ldots, X_n$ of a zero mean real-valued stationary process with sdf $S_X(\cdot)$ and acvs $\{C_{\tau,X}\}$ (note that, if $E(X_t)$ is unknown and hence cannot be assumed to be 0, the common practice is to replace $X_t$ with $X_t - \bar{X}$ prior to all other computations, where $\bar{X} = 1/n \sum_{t=1}^{n} X_t$ is the sample mean). Under a mild regularity condition (such as $S_X(\cdot)$ having a finite derivative at all frequencies), we can then write

$$S_X(f) = \Delta t \sum_{\tau=-\infty}^{\infty} C_{\tau,X} e^{-i2\pi f \tau \Delta t} \quad (11.3)$$

Our task is to estimate the sdf $S_X(\cdot)$ based upon $X_1, \ldots, X_n$. Equation (11.3) suggests the following "natural" estimator. Suppose that, for $|\tau| \leq n - 1$, we estimate $C_{\tau,X}$ via

$$\hat{C}_{\tau,X}^{(p)} = \frac{1}{n} \sum_{t=1}^{n-|\tau|} X_t X_{t+|\tau|}$$

(the rationale for the superscript $(p)$ is explained later). The estimator $\hat{C}_{\tau,X}^{(p)}$ is known in the literature as the *biased estimator* of $C_{\tau,T}$ since its expected value is

$$E(\hat{C}_{\tau,X}^{(p)}) = \frac{1}{n} \sum_{t=1}^{n-|\tau|} E(X_t X_{t+|\tau|}) = \left(1 - \frac{|\tau|}{n}\right) C_{\tau,X} \quad (11.4)$$

and hence $E(\hat{C}_{\tau,X}^{(p)}) \neq C_{\tau,X}$ in general. If we now decree that $\hat{C}_{\tau,X}^{(p)} = 0$ for $|\tau| \geq n$ and substitute the $\hat{C}_{\tau,X}^{(p)}$ for $C_{\tau,X}$ in Eq. (11.3), we obtain the spectral estimator

## UNIVARIATE TIME SERIES

$$\hat{S}_X^{(p)}(f) = \Delta t \sum_{\tau=-(n-1)}^{n-1} \hat{C}_{\tau,X}^{(p)} e^{-i2\pi f \tau \Delta t} \tag{11.5}$$

This estimator is known in the literature as the *periodogram*—hence the superscript $(p)$—even though it is more natural to regard it as a function of frequency $f$ than of period $1/f$. By substituting the definition for $\hat{C}_{\tau,X}^{(p)}$ into the preceding equation and making a change of variables, we find also that

$$\hat{S}_X^{(p)}(f) = \frac{\Delta t}{n} \left| \sum_{t=1}^{n} X_t e^{-i2\pi f t \Delta t} \right|^2 \tag{11.6}$$

Hence we can interpret the periodogram in two ways: it is the Fourier transform of the biased estimator of the acvs (with $\hat{C}_{\tau,X}^{(p)}$ defined to be 0 for $|\tau| \geq n$), and it is—to within a scaling factor—the squared modulus of the Fourier transform of $X_1, \ldots, X_n$.

Let us now consider the statistical properties of the periodogram. Ideally, we might like the following to be true:

1. $E[\hat{S}_X^{(p)}(f)] \approx S_X(f)$ (approximately unbiased);
2. $V[\hat{S}_X^{(p)}(f)] \to 0$ as $n \to \infty$ (consistent);
3. $\mathrm{Cov}[\hat{S}_X^{(p)}(f), \hat{S}_X^{(p)}(f')] \approx 0$ for $f \neq f'$ (approximately uncorrelated).

The "tragedy of the periodogram" is that in fact
1. $\hat{S}_X^{(p)}(f)$ can be a badly biased estimator of $S_X(f)$ even for large sample sizes (Thomson [6] reports an example in which the periodogram is severely biased for $n = 1.2$ *million* data points).
2. $V[\hat{S}_X^{(p)}(f)]$ does *not* decrease to 0 as $n \to \infty$ (unless $S_X(f) = 0$, a case of little practical interest).

As a consolation, however, we do have that $\hat{S}_X^{(p)}(f)$ and $\hat{S}_X^{(p)}(f')$ are approximately uncorrelated under certain conditions (see later).

We can gain considerable insight into the nature of the bias in the periodogram by studying the following expression for its expected value:

$$E[\hat{S}_X^{(p)}(f)] = \int_{-f_{(N)}}^{f_{(N)}} \mathcal{F}(f - f') S_X(f') df',$$

$$\text{with } \mathcal{F}(f) = \frac{\Delta t \sin^2(n\pi f \Delta t)}{n \sin^2(\pi f \Delta t)} \tag{11.7}$$

(for details, see [4], Section 6.3). The function $\mathcal{F}(\cdot)$ is known as *Fejér's kernel*. We also call it the *spectral window* for the periodogram. Figure 2(a) shows $\mathcal{F}(f)$ versus $f$ with $-f_{(N)} \leq f \leq f_{(N)}$ for the case $n = 32$ with $\Delta t = 1$ so that $f_{(N)} = 1/2$ (note that $\mathcal{F}(-f) = \mathcal{F}(f)$; i.e., Fejér's kernel is an even

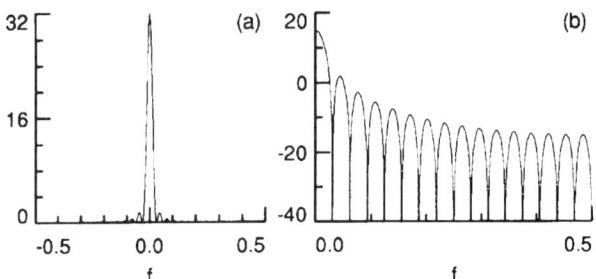

FIG. 2. Fejér's kernel for sample size $n = 16$ with $f_{(N)} = 1/2$.

function). Figure 2(b) plots $10 \cdot \log_{10}[\mathcal{F}(f)]$ versus $f$ for $0 \le f \le 1/2$ (i.e., $\mathcal{F}(\cdot)$ on a decibel scale). The numerator of Eq. (11.7) tells us that $\mathcal{F}(f) = 0$ when the product $nf \Delta t$ is equal to a nonzero integer—there are 16 of these nulls evident in Figure 2(b). The nulls closest to 0 frequency occur at $f = \pm 1/(n\Delta t) = \pm 1/32$. Figure 2(a) indicates that $\mathcal{F}(\cdot)$ is concentrated mainly in the interval of frequencies between these two nulls, the region of the "central lobe" of Fejér's kernel. A convenient measure of this concentration is the ratio $\int_{-1/(n\Delta t)}^{1/(n\Delta t)} \mathcal{F}(f)df / \int_{-f_{(N)}}^{f_{(N)}} \mathcal{F}(f)df$. An easy exercise shows that the denominator is unity for all $n$, while—to two decimal places—the numerator is equal to 0.90 for all $n \ge 13$. As $n \to \infty$, the length of the interval over which 90% of $\mathcal{F}(\cdot)$ is concentrated shrinks to 0, so in the limit Fejér's kernel acts like a Dirac delta function. If $S_X(\cdot)$ is continuous at $f$, Eq. (11.7) tells us that $\lim_{n\to\infty} E[\hat{S}_X^{(p)}(f)] = S_X(f)$ i.e., the periodogram is asymptotically unbiased.

While this asymptotic result is of some interest, for practical applications we are much more concerned about possible biases in the periodogram for finite sample sizes $n$. Equation (11.7) tells us that the expected value of the periodogram is given by the convolution of the true sdf with Fejér's kernel. Convolution is often regarded as a smoothing operation. From this viewpoint, $E[\hat{S}_X^{(p)}(\cdot)]$ should be a smoothed version of $S_X(\cdot)$—hence, if $S_X(\cdot)$ is itself sufficiently smooth, $E[\hat{S}_X^{(p)}(\cdot)]$ should closely approximate $S_X(\cdot)$. An extreme example of a process with a smooth sdf is white noise. Its sdf is constant over all frequencies, and in fact $E[\hat{S}_X^{(p)}(f)] = S_X(f)$ for a white noise process.

For sdfs with more structure than white noise, we can identify two sources of bias in the periodogram. The first source, often called a *loss of resolution*, is because the central lobe of Fejér's kernel will tend to smooth out spectral features with widths less than $1/(n\Delta t)$. Unfortunately, unless a priori information is available (or we are willing to make a modeling assumption), the cure for this bias is to increase the sample size $n$, i.e., to collect a longer time series, the prospect of which might be costly or—in the case of certain geophysical time series spanning thousands of years—impossible within our lifetimes.

The second source of bias is called *leakage* and is attributable to the sidelobes in Fejér's kernel. These sidelobes are prominently displayed in Figure 2(b). Figure 3 illustrates how these sidelobes can induce bias in the periodogram. The thick curve in Figure 3(a) shows an sdf plotted on a decibel scale from $f = -f_{(N)}$ to $f = f_{(N)}$ with $f_{(N)} = 1/2$ (recall that the sdf is symmetric about 0 so that $S_x(-f) = S_x(f)$). The thin bumpy curve is Fejér's kernel for $n = 32$, shifted so that its central lobe is at $f = 0.2$. The

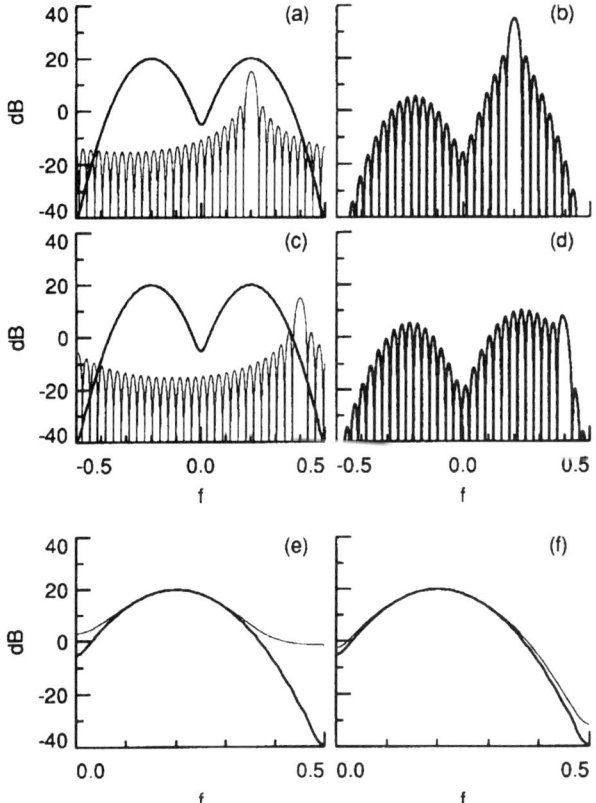

FIG. 3. Illustration of leakage (plots (a) to (e)). The Nyquist frequency $f_{(N)}$ here is taken to be 1/2. Plot (f) shows the alleviation of leakage via tapering and is discussed in Section 11.2.2.

product of this shifted kernel and the sdf is shown in Figure 3(b) (again on a decibel scale). Equation (11.7) says that $E[\hat{S}_X^{(p)}(0.2)]$ is the integral of this product. The plot shows this integral to be determined mainly by values close to $f = 0.2$; i.e., $E[\hat{S}_X^{(p)}(0.2)]$ is due largely to the sdf at values close to this frequency, a result that is quite reasonable. Figures 3(c) and 3(d) show the corresponding plots for $f = 0.4$. Note that $E[\hat{S}_X^{(p)}(0.4)]$ is substantially influenced by values of $S_X(\cdot)$ away from $f = 0.4$. The problem is that the sidelobes of Fejér's kernel are interacting with portions of the sdf that are the dominant contributors to the variance of the process so that $E[\hat{S}_X^{(p)}(0.4)]$ is biased upwards. Figure 3(e) shows a plot of $E[\hat{S}_X^{(p)}(f)]$ versus $f$ (the thin curve), along with the true sdf $S_X(\cdot)$ (the thick curve). While the periodogram is essentially unbiased for frequencies satisfying $0.1 \leq |f| \leq 0.35$, there is substantial bias due to leakage at frequencies close to $f = 0$ and $f = \pm 1/2$ (in the latter case, the bias is almost 40 dB, i.e., four orders of magnitude).

While it is important to know that the periodogram can be severely biased for certain processes, it is also true that, if the true sdf is sufficiently lacking in structure (i.e., "close to white noise"), then $S_X(\cdot)$ and $E[\hat{S}_X^{(p)}(\cdot)]$ can be close enough to each other so that the periodogram is essentially bias free. Furthermore, even if leakage is present, it might not be of importance in certain practical applications. If, for example, we were performing a spectral analysis to determine the height and structure of the sdf in Figure 3 near $f = 0.2$, then the bias due to leakage at other frequencies is of little concern.

If the portions of the sdf affected by leakage are in fact of interest or if we are carrying out a spectral analysis on a time series for which little is known a priori about its sdf, we need to find ways to recognize when leakage is a problem and, if it is present, to minimize it. As is the case for loss of resolution, we can decrease leakage by increasing the sample size (more data can solve many problems!), but unfortunately a rather substantial increase might be required to obtain a periodogram that is essentially free of leakage. Consider again the sdf used as an example in Figure 3. Even with a 32-fold increase in the sample size from $n = 32$ to $1024$, there is still more than a 20 dB difference between $E[\hat{S}_X^{(p)}(0.5)]$ and $S_X(0.5)$.

If we regard the sample size as fixed, there are two well-known ways of decreasing leakage, namely, *data tapering* and *prewhitening*. Both of these techniques have a simple interpretation in terms of the integral in Eq. (11.7). On the one hand, tapering essentially replaces Fejér's kernel $\mathcal{F}(\cdot)$ by a function with substantially reduced sidelobes; on the other hand, prewhitening effectively replaces the sdf $S_X(\cdot)$ with one that is closer to white noise. Both techniques are discussed in the next subsection.

### 11.2.2 Correcting for Bias

**Tapering.** For a given time series $X_1, X_2, \ldots, X_n$, a *data taper* is a finite sequence $h_1, h_2, \ldots, h_n$ of real-valued numbers. The product of this sequence and the time series, namely, $h_1X_1, h_2X_2, \ldots, h_nX_n$, is used to create a *direct spectral estimator* of $S_X(f)$, defined as

$$\hat{S}_X^{(d)}(f) = \Delta t \left| \sum_{t=1}^{n} h_t X_t \, e^{-i 2\pi f t \, \Delta t} \right|^2 \tag{11.8}$$

Note that, if we let $h_t = 1/\sqrt{n}$ for all $t$ (the so-called rectangular data taper), a comparison of Eqs. (11.8) and (11.6) tells us that $\hat{S}_X^{(d)}(\cdot)$ reduces to the periodogram. The acvs estimator corresponding to $\hat{S}_X^{(d)}(\cdot)$ is just

$$\hat{C}_{\tau,X}^{(d)} = \sum_{t=1}^{n-|\tau|} h_t X_t \, h_{t+|\tau|} X_{t+|\tau|},$$

and so $\quad E(\hat{C}_{\tau,X}^{(d)}) = C_{\tau,X} \sum_{t=1}^{n-|\tau|} h_t h_{t+|\tau|}$ \hfill (11.9)

If we insist that $\hat{C}_{0,X}^{(d)}$ be an unbiased estimator of the process variance $C_{0,X}$, then we obtain the normalization $\sum_{t=1}^{n} h_t^2 = 1$ (note that the rectangular data taper satisfies this constraint).

The rationale for tapering is to obtain a spectral estimator whose expected value is close to $S_X(\cdot)$. In analogy to Eq. (11.7), we can express this expectation as

$$E[\hat{S}_X^{(d)}(f)] = \int_{-f_{(N)}}^{f_{(N)}} \mathcal{H}(f - f') \, S_X(f') \, df' \tag{11.10}$$

where $\mathcal{H}(\cdot)$ is proportional to the squared modulus of the Fourier transform of $h_t$ and is called the *spectral window* for the direct spectral estimator $\hat{S}_X^{(d)}(\cdot)$ (just as we called Fejér's kernel the spectral window for the periodogram). The claim is that, with a proper choice of $h_t$, we can produce a spectral window that offers better protection against leakage than Fejér's kernel. Figure 4 supports this claim. The lefthand column of plots shows four data tapers for sample size $n = 32$, while the righthand plots show the corresponding spectral windows. For the sake of comparison, the data taper in Figure 4(a) is just the rectangular data taper, so the corresponding spectral window is Fejér's kernel (see Figure 2(b)). The data taper in Figure 4(c) is the well-known *Hanning data taper*, which we define to be $h_t = \sqrt{2/3(n+1)} \, [1 + \cos(2\pi t/(n+1))]$ for $t = 1, \ldots, n$, (there are other, slightly different, definitions

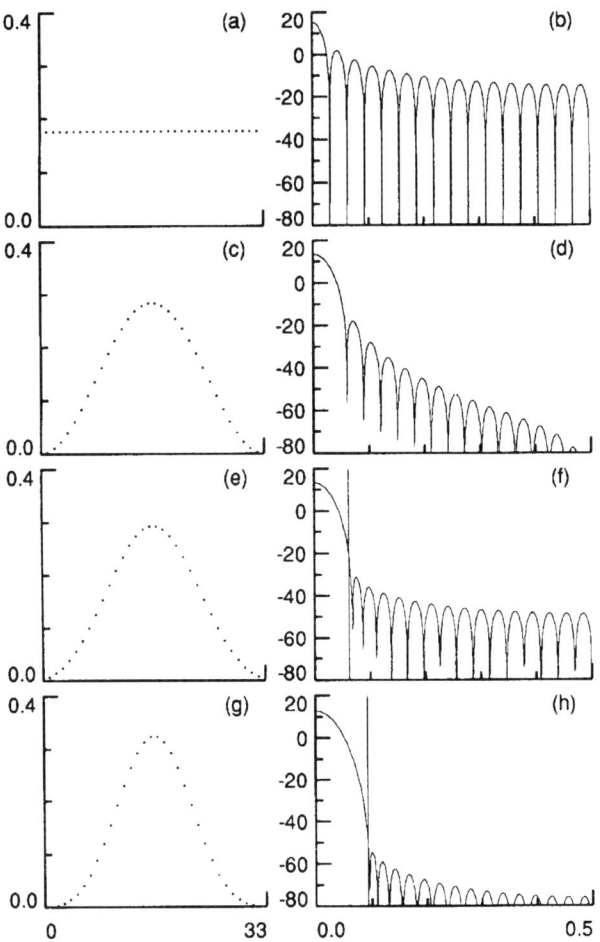

FIG. 4. Four data tapers (lefthand column of plots) and their spectral windows (righthand column) for a sample size of $n = 32$. From top to bottom, the tapers are the rectangular taper, the Hanning taper, the dpss data taper with $nW = 2/\Delta t$ and the dpss data taper with $nW = 3/\Delta t$ (here $\Delta t = 1$).

in the literature). Note carefully the shape of the corresponding spectral window in Figure 4(d): its sidelobes are considerably suppressed in comparison to those of Fejér's kernel, but the width of its central lobe is markedly larger. The convolutional representation for $E[\hat{S}_X^{(d)}(f)]$ given in Eq. (11.10) tells us that an increase in central lobe width can result in a loss of resolution when the true sdf has spectral features with widths smaller than the central lobe width. This illustrates one of the trade-offs in using a data taper, namely, that tapering typically decreases leakage at the expense of a potential loss of resolution.

A convenient family of data tapers that facilitates the trade-off between sidelobe suppression and the width of the central lobe is the *discrete prolate spheroidal sequence* (dpss) tapers. These tapers arise as the solution to the following "concentration" problem. Suppose we pick a number $W$ that, roughly speaking, we think of as half the desired width of the central lobe of the resulting spectral window (typically $1/(n\Delta t) \leq W \leq 4/(n\Delta t)$, although larger values for $W$ are sometimes useful). Under the requirement that the taper must satisfy the normalization $\sum_{t=1}^{n} h_t^2 = 1$, the dpss taper is, by definition, the taper whose corresponding spectral window is as concentrated as possible in the frequency interval $[-W, W]$ in the sense that the ratio $\int_{-W}^{W} \mathcal{H}(f)df / \int_{-f_{(N)}}^{f_{(N)}} \mathcal{H}(f)df$ is as large as possible (note that, if $\mathcal{H}(\cdot)$ were a Dirac delta function, this ratio would be unity). The quantity $2W$ is sometimes called the *resolution bandwidth*. For a fixed $W$, the dpss tapers have sidelobes that are suppressed as much as possible as measured by the concentration ratio. To a good approximation (Walden [7]), the dpss tapers can be calculated as $h_t = C \times I_0(\tilde{W}\sqrt{1 - (1 - g_t)^2})/I_0(\tilde{W})$ for $t = 1, \ldots, n$, where $C$ is a scaling constant used to force the normalization $\sum h_t^2 = 1$; $\tilde{W} = \pi W(n - 1)\Delta t$; $g_t = (2t - 1)/n$; and $I_0(\cdot)$ is the modified Bessel function of the first kind and zeroth order (this can be computed using the Fortran function `bessj0` in Section 6.5 of Press et al. [8]).

Figures 4(e) and (g) show dpss tapers for $n = 32$ and with $W$ set such that $nW = 2/\Delta t$ and $nW = 3/\Delta t$, while plots (f) and (h) show the corresponding spectral windows $\mathcal{H}(\cdot)$ (the quantity $2nW$ is known as the *duration–bandwidth product*). The vertical lines in the latter two plots mark the locations of $W$. Note that in both cases the central lobe of $\mathcal{H}(\cdot)$ is approximately contained between $[-W, W]$. As expected, increasing $W$ suppresses the sidelobes and hence offers increasing protection against leakage. A comparison of the spectral windows for the Hanning and $nW = 2/\Delta t$ dpss tapers (Figures 4(d) and 4(f)) shows that, whereas their central lobes are comparable, their sidelobe structures are quite different. While specific examples can be constructed in which one of the tapers offers better protection against leakage than the other, generally the two tapers are quite comparable in practical applications. The advantage of the dpss tapers is that, if, say, use of an $nW = 2/\Delta t$ dpss taper produces a direct spectral estimator that still suffers from leakage, we can easily obtain a greater degree of protection against leakage by merely increasing $nW$ beyond $2/\Delta t$. The choice $nW = 1/\Delta t$ yields a spectral window with a central lobe closely resembling that of Fejér's kernel but with sidelobes about 10 dB smaller.

Let us now return to the example of Figure 3. Recall that the thin curve in Figure 3(e) shows $E[\hat{S}_X^{(p)}(f)]$ versus $f$ for a process with an sdf given by the thick curve. In Figure 3(f) the thin curve now shows $E[\hat{S}_X^{(d)}(\cdot)]$ for a direct spectral estimator employing an $nW = 2/\Delta t$ dpss taper. Note that tapering has produced a spectral estimator that is overall much closer in expectation to $S_x(\cdot)$ than the periodogram is; however, mainly due to the small sample $n = 32$, $\hat{S}_X^{(d)}(\cdot)$ still suffers from leakage at some frequencies (about 10 dB at $f = 0.5$).

In practical situations, we can determine if leakage is present in the periodogram by carefully comparing it with a direct spectral estimate constructed using a dpss data taper with a fairly large value of $W$. As an example, Figure 5(a) shows the periodogram for the wire wave gauge time series show in Figure 1. Since these data were collected mainly to investigate the roll-off rate of the sdf from 0.8 to 4 Hz, we have plotted only the low-frequency portion of the periodogram. Figure 5(b)

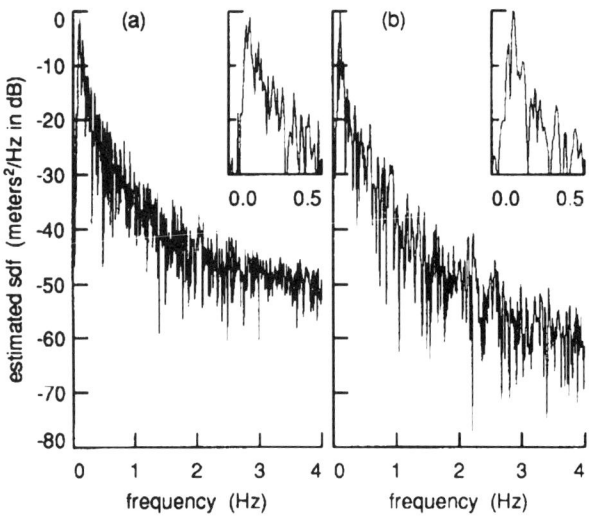

FIG. 5. Periodogram (plot a) and a direct spectral estimate (plot b) using an $nW = 4/\Delta t$ dpss data taper for the wire gauge time series of Figure 1(a). Both estimates are plotted on a decibel scale. The small subplots in the upper righthand corner of each plot give an expanded view of the estimators at $0 \leq f \leq 0.5$ Hz (the vertical scales of the subplots are the same as those of the main plots).

shows a direct spectral estimate for which we have used a dpss taper with $nW = 4/\Delta t$. Note that the direct spectral estimate is markedly lower than the periodogram at frequencies with a relatively small contribution to the overall variance—for example, the former is about 10 dB below the latter at frequencies close to 4 Hz. This pattern is consistent with what we would expect to see when there is leakage in the periodogram. Increasing $nW$ beyond $4/\Delta t$ to, say, $6/\Delta t$ and comparing the resulting direct spectral estimate with that of Figure 5(b) indicates that the $nW = 4/\Delta t$ estimate is essentially leakage free; on the other hand, an examination of an $nW = 1/\Delta t$ direct spectral estimate indicates that it is also essentially leakage free for the wave gauge series. In general, we can determine an appropriate degree of tapering by carefully comparing the periodogram and direct spectral estimates corresponding to dpss tapers with different values of $nW$. If the periodogram proves to suffer from leakage (as it often does in the physical sciences), we then seek a leakage-free direct spectral estimate formed using as small a value of $W$, and hence $nW$, as possible. A small $W$ is desirable from two viewpoints: first, resolution typically decreases as $W$ increases, and second, the distance in frequency between approximately uncorrelated spectral estimates increases as $W$ increases (this causes a loss in degrees of freedom when we subsequently smooth across frequencies; see Section 11.2.3 for details).

In checking for the presence of leakage, comparison between different spectral estimates is best done graphically in the following ways:

1. by comparing spectral estimates side by side (as in Figure 5) or on top of each other using different colors;
2. by plotting the ratio of two spectral estimates on a decibel scale versus frequency and searching for frequency bands over which an average of this ratio is nonzero (presumably a statistical test could be devised here to assess the significance of departures from zero);
3. by constructing scatter plots of, say, $\hat{S}_X^{(p)}(f)$ versus $\hat{S}_X^{(d)}(f)$ for values of $f$ in a selected band of frequencies and looking for clusters of points consistently above a line with unit slope and zero intercept.

The use of interactive graphical displays would clearly be helpful in determining the proper degree of tapering (Percival and Kerr [9]).

**Prewhitening.** The loss of resolution and of degrees of freedom inherent in tapering can be alleviated considerably if we can *prewhiten* our time series. To explain prewhitening, we need the following result from the theory of linear time-invariant filters. Given any set of $p + 1$ real-valued numbers $a_0, a_1, \ldots, a_p$, we can filter $X_1, \ldots, X_n$ to obtain

$$W_t = \sum_{k=0}^{p} a_k X_{t-k}, \qquad t = p + 1, \ldots, n \qquad (11.11)$$

The filtered process $\{W_t\}$ is a zero mean real-valued stationary process with an sdf $S_W(\cdot)$ related to $S_X(\cdot)$ via

$$S_W(f) = |A(f)|^2 S_X(f), \qquad \text{where } A(f) = \sum_{k=0}^{p} a_k e^{-i2\pi f k \Delta t} \qquad (11.12)$$

$A(\cdot)$ is the transfer function for the filter $\{a_k\}$. The idea behind prewhitening is to find a set of $a_k$ such that the sdf $S_W(\cdot)$ has substantially less structure than $S_X(\cdot)$ and ideally is as close to a white noise sdf as possible. If such a filter can be found, we can easily produce a leakage-free direct spectral estimate $\hat{S}_W^{(d)}(\cdot)$ of $S_W(\cdot)$ requiring little or no tapering. Equation (11.12) tells us that we can then estimate $S_X(\cdot)$ using $\hat{S}_X^{(pc)}(f) = \hat{S}_W^{(d)}(f)/|A(f)|^2$, where the superscript (pc) stands for *postcolored*.

The reader might note an apparent logical contradiction here: in order to pick a reasonable set of $a_k$ we must know the shape of $S_X(\cdot)$, the very function we are trying to estimate! There are two ways around this inherent difficulty. In many cases, there is enough prior knowledge concerning $S_X(\cdot)$ from, say, previous experiments so that, even though we do not know the exact shape of the sdf, we can still design an effective prewhitening filter. In other cases we can create a prewhitening filter based upon the data itself. A convenient way of doing this is by fitting *autoregressive models* of order $p$ (hereafter AR($p$)). In the present context, we postulate that, for some reasonably small $p$, we can model $X_t$ as a linear combination of the $p$ prior values $X_{t-1}, \ldots, X_{t-p}$ plus an error term; i.e,

$$X_t = \sum_{k=1}^{p} \phi_k X_{t-k} + W_t, \qquad t = p + 1, \ldots, n \qquad (11.13)$$

where $\{W_t\}$ is an "error" process with an sdf $S_W(\cdot)$ that has less structure than $S_X(\cdot)$. The stipulation that $p$ be small is desirable because the prewhitened series $\{W_t\}$ will be shortened to length $n - p$. The preceding equation is a special case of Eq. (11.11), as can be seen by letting $a_0 = 1$ and $a_k = -\phi_k$ for $k > 0$.

For a given $p$, we can obtain estimates of the $\phi_k$ from our time series using a number of different methods (see [4], Chapter 9). One method that generally works well is Burg's algorithm, a computationally efficient way of producing estimates of the $\phi_k$ that are guaranteed to correspond to a stationary process. The FORTRAN subroutine `memcof` in [8], Section 13.6, implements Burg's algorithm (the reader should note that the time series used as input to memcof is assumed to be already adjusted for any nonzero mean value).

Burg's algorithm (or any method that estimates the $\phi_k$ in Eq. (11.13)) is sometimes used by itself to produce what is known as an *autoregressive spectral estimate* (this is one form of parametric spectral estimation). If we let $\bar{\phi}_k$ represent Burg's estimate of $\phi_k$, then the corresponding autoregressive spectral estimate is given by

$$\bar{S}_X^{(ar)}(f) = \frac{\bar{\sigma}_p^2 \, \Delta t}{|1 - \sum_{k=1}^{p} \bar{\phi}_k e^{-i2\pi f k \Delta t}|^2} \tag{11.14}$$

where $\bar{\sigma}_p^2$ is an estimate of the variance of $\{W_t\}$. The key difference between an autoregressive spectral estimate and prewhitening is in the use of Eq. (11.13): the former assumes the process $\{W_t\}$ is *exactly* white noise, whereas the latter merely postulates $\{W_t\}$ to have an sdf with less structure than $\{X_t\}$. Autoregressive spectral estimation works well for many time series, but it depends heavily on a proper choice of the order $p$; moreover, simple approximations to the statistical properties of $\bar{S}_X^{(ar)}(\cdot)$ are currently lacking. (Autoregressive spectral estimation is sometimes misleadingly called *maximum entropy* spectral estimation; see [4] for a discussion about this misnomer.)

Figure 6 illustrates the construction of a prewhitening filter for the wire gauge series. The dots in plot (a) depict the same direct spectral

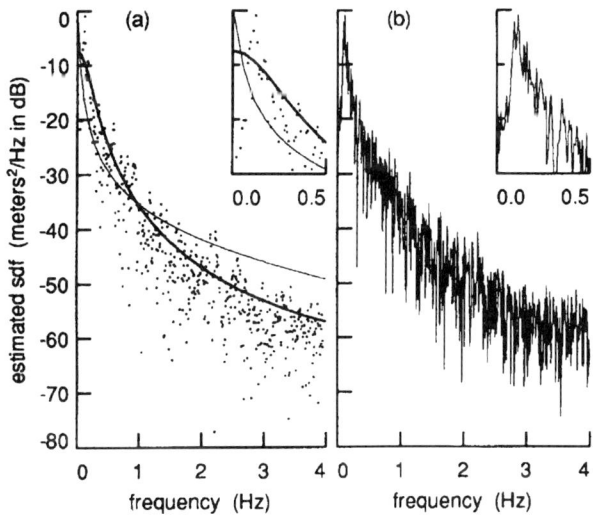

FIG. 6. Determination of a prewhitening filter.

estimate as shown in Figure 5(b). This estimate is essentially leakage free and hence serves as a "pilot estimate" for comparing different prewhitening filters. Using Burg's algorithm, we computed autoregressive spectral estimates for various orders $p$ via Eq. (11.14). The thin and thick solid curves in plot (a) show, respectively, the AR(2) and AR(4) estimates. The AR(4) estimate is the estimate with lowest order $p$ that reasonably captures the structure of the pilot estimate over the frequency range of main interest (namely, 0.4 to 4.0 Hz); moreover, use of these values of $\overline{\phi}_k$ to form a prewhitening filter yields a prewhitened series $W_t$, $t = 5, \ldots, 4096$, with a sdf that has no regions of large power outside of 0.4 to 4.0 Hz. The periodogram of $W_t$ appears to be leakage free for $f$ between 0.4 and 4.0 Hz. The spectral estimate in plot (b) is $\hat{S}_X^{(pc)}(\cdot)$, the result of postcoloring the periodogram $\hat{S}_W^{(p)}(\cdot)$ (i.e., dividing $\hat{S}_W^{(p)}(f)$ by $|1 - \sum_{k=1}^{4} \phi_k e^{-i2\pi f k \Delta t}|^2$). Note that $\hat{S}_X^{(pc)}(\cdot)$ generally agrees well with the leakage-free pilot estimate in plot (a).

We close this section with three comments. First, the prewhitening procedure we have outlined might require a fairly large-order $p$ to work properly and can fail if $S_X(\cdot)$ contains narrow-band features (i.e., line components). Low-order AR processes cannot capture narrow-band features adequately—their use can yield a prewhitened process with a more complicated sdf than that of the original process. Additionally, Burg's algorithm is known to be problematic for certain narrow-band processes for which it will sometimes incorrectly represent a line component as a double peak (this is called *spontaneous line splitting*). The proper way to handle narrow-band processes is to separate out the narrow-band components and then deal with the resulting "background" process. One attractive method for handling such processes uses multitapering (for details, see [6], Section XIII, or [4], Sections 10.11 and 10.13).

Second, in our example of prewhitening we actually *introduced* leakage into the very low-frequency portion of the sdf. To see this fact, note that the spectral level of the prewhitened estimate of Figure 6(b) near $f = 0$ is about 10 dB *higher* than those of the two direct spectral estimates in Figure 5. This happened because we were interested mainly in creating a prewhitening filter to help estimate the sdf from 0.4 to 4.0 Hz. An AR(4) filter accomplishes this task, but it fails to represent the very low-frequency portion of the sdf at all (see Figure 6(a)). As a result, the prewhitened series $\{W_t\}$ has an sdf that is evidently deficient in power near 0 frequency and hence suffers from leakage at very low frequencies. Use of a much higher order AR filter (say, $p = 256$) can correct this problem.

Finally, we note that, if prewhitening is not used, we define $W_t$ to be equal to $X_t$ and set $p$ equal to 0 in what follows.

### 11.2.3 Variance Reduction

A cursory examination of the spectral estimates for the wire gauge series in Figures 5 and 6 reveals substantial variability across frequencies, so much so that it is difficult to discern the overall structure in the spectral estimates without a fair amount of study. All direct spectral estimators suffer from this inherent choppiness, which can be explained by considering the distributional properties of $\hat{S}_W^{(d)}(f)$. First, if $f$ is not too close to 0 or $f_{(N)}$ and if $S_W(\cdot)$ satisfies a mild regularity condition, then $2\hat{S}_W^{(d)}(f)/S_W(f) \stackrel{d}{=} \chi_2^2$; i.e., the rv $2\hat{S}_W^{(d)}(f)/S_W(f)$ is approximately equal in distribution to a chi-square rv with 2 degrees of freedom. If tapering is not used, $f$ is considered "not too close" to 0 or $f_{(N)}$ if $1/(n-p)\Delta t < f < f_{(N)} - 1/(n-p)\Delta t$; if tapering is used, we must replace $1/(n-p)\Delta t$ by a larger term, reflecting the increased width of the central lobe of the spectral window (for example, the term for the Hanning data taper is approximately $2/(n-p)\Delta t$ so $f$ is "not too close" if $2/(n-p)\Delta t < f < f_{(N)} - 2/(n-p)\Delta t$).

Since a chi-square rv $\chi_\nu^2$ with $\nu$ degrees of freedom has a variance of $2\nu$, we have the approximation $V[\hat{S}_W^{(d)}(f)] = S_W^2(f)$. This result is independent of the number of $W_t$ we have: unlike statistics such as the sample mean of independent and identically distributed Gaussian rvs, the variance of $\hat{S}_W^{(d)}(f)$ does not decrease to 0 as the sample size $n - p$ gets larger (except in the uninteresting case $S_W(f) = 0$). This result explains the choppiness of the direct spectral estimates shown in Figures 5 and 6. In statistical terminology, $\hat{S}_W^{(d)}(f)$ is an inconsistent estimator of $S_W(f)$.

We now outline three approaches for obtaining a consistent estimator of $S_W(f)$. Each approach is based upon combining rvs that, under suitable assumptions, can be considered as approximately pairwise uncorrelated estimators of $S_W(f)$. Briefly, the three approaches are to

1. smooth $\hat{S}_W^{(d)}(f)$ across frequencies, yielding what is known as a *lag window spectral estimator;*
2. break $\{X_t\}$ (or $\{W_t\}$) into a number of segments (some of which can overlap), compute a direct spectral estimate for each segment, and then average these estimates together, yielding Welch's *overlapped segment averaging* (WOSA) *spectral estimator;*
3. compute a series of direct spectral estimates for $\{W_t\}$ using a set of orthogonal data tapers and then average these estimates together, yielding Thomson's *multitaper spectral estimator.*

**Lag Window Spectral Estimators** A lag window spectral estimator of $S_W(\cdot)$ takes the form

$$\hat{S}_W^{(lw)}(f) = \int_{-f_{(N)}}^{f_{(N)}} W_m(f-f')\,\hat{S}_W^{(d)}(f')\,df' \qquad (11.15)$$

where $W_m(\cdot)$ is a *smoothing window* whose smoothing properties are controlled by the smoothing parameter $m$. In words, the estimator $\hat{S}_W^{(lw)}(\cdot)$ is obtained by convolving a smoothing window with the direct spectral estimator $\hat{S}_W^{(d)}(\cdot)$. A typical smoothing window has much the same appearance as a spectral window. There is a central lobe with a width that can be adjusted by the smoothing parameter $m$: the wider this central lobe is, the smoother $\hat{S}_W^{(lw)}(\cdot)$ will be. There can also be a set of annoying sidelobes that cause *smoothing window leakage*. The presence of smoothing window leakage is easily detected by overlaying plots of $\hat{S}_W^{(lw)}(\cdot)$ and $\hat{S}_W^{(d)}(\cdot)$ and looking for ranges of frequencies where the former does not appear to be a smoothed version of the latter.

If we have made use of an AR prewhitening filter, we can then postcolor $\hat{S}_W^{(lw)}(\cdot)$ to obtain an estimator of $S_X(\cdot)$, namely,

$$\hat{S}_X^{(pc)}(f) = \frac{\hat{S}_W^{(lw)}(f)}{|1 - \sum_{k=1}^{p} \overline{\phi}_k e^{-i2\pi f k \Delta t}|^2}$$

The statistical properties of $\hat{S}_W^{(lw)}(\cdot)$ are tractable because of the following large sample result. If $\hat{S}_W^{(d)}(\cdot)$ is in fact the periodogram (i.e., we have not tapered the values of $W_t$), the set of rvs $\hat{S}_W^{(d)}(j/(n-p)\Delta t)$, $j = 1, 2, \ldots, J$, are approximately pairwise uncorrelated, with each rv being proportional to a $\chi_2^2$ rv (here $J$ is the largest integer such that $J/(n-p) < 1/2$). If we have used tapering to form $\hat{S}_W^{(d)}(\cdot)$, a similar statement is true over a smaller set of rvs defined on a coarser grid of equally spaced frequencies—as the degree of tapering increases, the number of approximately uncorrelated rvs decreases. Under the assumptions that the sdf $S_W(\cdot)$ is slowly varying across frequencies (prewhitening helps to make this true) and that the central lobe of the smoothing window is sufficiently small compared to the variations in $S_W(\cdot)$, it follows that $\hat{S}_W^{(d)}(f)$ in Eq. (11.15) can be approximated by a linear combination of uncorrelated $\chi_2^2$ rvs. A standard "equivalent degrees of freedom" argument can be then used to approximate the distribution of $\hat{S}_W^{(lw)}(f)$. (see Eq. (11.17) later).

There are two practical ways of computing $\hat{S}_W^{(lw)}(\cdot)$. The first way is to discretize Eq. (11.15), yielding an estimator proportional to a convolution of the form $\sum_k W_m(f - f'_k) \hat{S}_W^{(d)}(f'_k)$, where the values of $f'_k$ are some set of equally spaced frequencies. The second way is to recall that "convolution in one Fourier domain is equivalent to multiplication in the other" to rewrite Eq. (11.15) as

$$\hat{S}_W^{(lw)}(f) = \sum_{\tau=-(n-p-1)}^{n-p-1} w_{\tau,m} \, \hat{C}_{\tau,W}^{(d)} \, e^{-i2\pi f \tau \Delta t} \qquad (11.16)$$

where $\hat{C}_{\tau,W}^{(d)}$ is the acvs estimator given in Eq. (11.9) corresponding to $\hat{S}_W^{(d)}(\cdot)$, and $\{w_{\tau,m}\}$ is a *lag window* (this can be regarded as the inverse Fourier transform of the smoothing window $W_m(\cdot)$). In fact, because $\hat{S}_W^{(d)}(\cdot)$ is a trigonometric polynomial, all discrete convolutions of the form $\Sigma_k W_m(f - f'_k)\hat{S}_W^{(d)}(f'_k)$ can also be computed via Eq. (11.16) with an appropriate choice of $w_{\tau,m}$ values (for details, see [4], Section 6.7). Our two practical ways of computing $\hat{S}_W^{(lw)}(\cdot)$ thus yield equivalent estimators. Unless the discrete convolution is sufficiently short, Eq. (11.16) is computationally faster to use.

Statistical theory suggests that, under reasonable assumptions,

$$\frac{\nu \hat{S}_W^{(lw)}(f)}{S_W(f)} \stackrel{d}{=} \chi_\nu^2 \qquad (11.17)$$

to a good approximation, where $\nu$ is called the *equivalent degrees of freedom* for $\hat{S}_W^{(lw)}(f)$ and is given by $\nu = 2(n - p)B_W \Delta t/C_h$. Here $B_W$ is a measure of the bandwidth of the smoothing window $W_m(\cdot)$ and can be computed via $B_W = 1/\Delta t \sum_{\tau=-(n-p-1)}^{n-p-1} w_{\tau,m}^2$; on the other hand, $C_h$ depends only on the taper applied to the values of $W_t$ and can be computed via $C_h = (n - p) \sum_{t=p+1}^{n} h_t^4$. Note that, if we do not explicitly taper, then $h_t = 1/\sqrt{n - p}$ and hence $C_h = 1$; for a typical data taper, the Cauchy inequality tells us that $C_h > 1$ (for example, $C_h \approx 1.94$ for the Hanning data taper). The equivalent degrees of freedom for $\hat{S}_W^{(lw)}(f)$ thus increase as we increase the smoothing window bandwidth and decrease as we increase the degree of tapering. Equation (11.17) tells us that $E[\hat{S}_W^{(lw)}(f)] \approx S_W(f)$ and that $V[\hat{S}_W^{(lw)}(f)] \approx S_W^2(f)/\nu$, so increasing $\nu$ decreases $V[\hat{S}_W^{(lw)}(f)]$.

The approximation in Eq. (11.17) can be used to construct a confidence interval for $S_W(f)$ in the following manner. Let $\eta_\nu(\alpha)$ denote the $\alpha \times 100\%$ percentage point of the $\chi_\nu^2$ distribution; i.e., $P[\chi_\nu^2 \le \eta_\nu(\alpha)] = \alpha$. A $100(1 - 2\alpha)\%$ confidence interval for $S_W(f)$ is approximately given by

$$\left[\frac{\nu \hat{S}_W^{(lw)}(f)}{\eta_\nu(1 - \alpha)}, \frac{\nu \hat{S}_W^{(lw)}(f)}{\eta_\nu(\alpha)}\right] \qquad (11.18)$$

The percentage points $\eta_\nu(\alpha)$ are tabulated in numerous textbooks or can be computed using an algorithm given by Best and Roberts [10].

The confidence interval of (11.18) is inconvenient in that its length is proportional to $\hat{S}_W^{(lw)}(f)$. On the other hand, the corresponding confidence interval for $10 \cdot \log_{10}(S_W(f))$ (i.e., $S_W(f)$ on a decibel scale) is just

$$\left[10 \cdot \log_{10}(\nu/\eta_\nu(1 - \alpha)) + 10 \cdot \log_{10}\left(\hat{S}_W^{(lw)}(f)\right), 10 \cdot \log_{10}(\nu/\eta_\nu(\alpha)) + 10 \cdot \log_{10}\left(\hat{S}_W^{(lw)}(f)\right)\right]$$

which has a width that is independent of $\hat{S}_W^{(lw)}(f)$. This is the rationale for plotting sdf estimates on a decibel (or logarithmic) scale.

A bewildering number of different lag windows has been discussed in the literature (see [3]). Here we give only one example, the well-known *Parzen lag window* (Parzen [11]):

$$w_{\tau,m} = \begin{cases} 1 - 6\tilde{\tau}^2 + 6|\tilde{\tau}|^3, & |\tau| \leq m/2 \\ 2(1 - \tilde{\tau})^3, & m/2 < |\tau| \leq m \\ 0, & |\tau| > m \end{cases}$$

where $m$ is taken to be a positive integer and $\tilde{\tau} = \tau/m$. This lag window is easy to compute and has sidelobes whose envelope decays as $f^{-4}$ so that smoothing window leakage is rarely a problem. To a good approximation, the smoothing window bandwidth for the Parzen lag window is given by $B_W = 1.85/(m\Delta t)$. As $m$ increases, the smoothing window bandwidth decreases, and the resulting lag window estimator becomes less smooth in appearance. The associated equivalent degrees of freedom are given approximately by $\nu = 3.71(n - p)/(mC_h)$. The Parzen lag window for $m = 32$ and its associated smoothing window are shown in Figure 7.

As an example, Figure 8(a) shows a postcolored lag window estimator for the wire wave gauge data (the solid curve), along with the corresponding postcolored direct spectral estimator (the dots, these depict the same estimate as shown in Figure 6(b)). The Parzen lag window was used here with a value of $m = 237$ for the smoothing window parameter (the corresponding equivalent degrees of freedom $\nu$ is 64). This value was chosen after some experimentation and seems to produce a lag window estimator that captures all of the important spectral features indicated by the direct spectral estimator for frequencies between 0.4

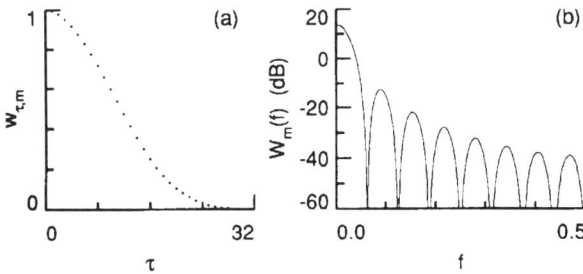

FIG. 7. Parzen lag window (a) and the corresponding smoothing window (b) for $m = 32$. The smoothing window bandwidth is $B_W = 0.058$.

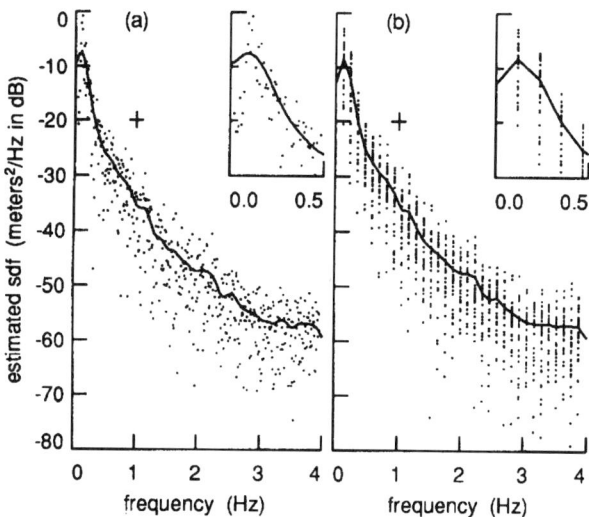

FIG. 8. Postcolored Parzen lag window spectral estimate—solid curve on plot (a)—and WOSA spectral estimate—solid curve on (b)—for wire wave gauge time series. The smoothing window parameter for the Parzen lag window was $m = 237$, yielding $v = 64$ equivalent degrees of freedom. The WOSA spectral estimate was formed using a Hanning data taper on blocks with 256 data points, with adjacent blocks overlapping by 50%. The equivalent degrees of freedom for this estimate is $v = 59$.

and 4.0 Hz (note, however, that this estimator smears out the peak between 0.0 and 0.4 Hz rather badly). We have also plotted a crisscross whose vertical height represents the length of a 95% confidence interval for $10 \cdot \log_{10}(S_x(f))$ (based upon the postcolored lag window estimator) and whose horizontal width represents the smoothing window bandwidth $B_w$.

**WOSA Spectral Estimators.** Let us now consider the second common approach to variance reduction, namely, Welch's overlapped segment averaging (Welch [12]; Carter [13] and references therein). The basic idea is to break a time series into a number of blocks (i.e., segments), compute a direct spectral estimate for each block, and then produce the WOSA spectral estimate by averaging these spectral estimates together. In general, the blocks are allowed to overlap, with the degree of overlap being determined by the degree of tapering—the heavier the degree of tapering, the more the blocks should be overlapped (Thomson [14]). Thus, except at the very beginning and end of the time series, data values that are heavily tapered in one block are lightly tapered in another block, so intuitively we are recapturing "information" lost due to tapering in one block from blocks overlapping it. Because it can be implemented in a computationally efficient fashion (using the fast Fourier transform algorithm) and because it can handle very long time series (or time series with a time varying spectrum), the WOSA estimation scheme is the basis for many of the commercial spectrum analyzers on the market.

To define the WOSA spectral estimator, let $n_S$ represent a block size, and let $h_1, \ldots, h_{n_S}$ be a data taper. We define the direct spectral estimator of $S_X(f)$ for the block of $n_S$ contiguous data values starting at index $l$ as

$$\hat{S}^{(d)}_{l,X}(f) = \Delta t \left| \sum_{t=1}^{n_S} h_t X_{t+l-1} e^{-i2\pi f t \Delta t} \right|^2, \qquad 1 \leq l \leq n + 1 - n_S$$

(there is no reason why we cannot use a prewhitened series $\{W_t\}$ here rather than $X_t$, but prewhitening is rarely used in conjunction with WOSA, perhaps because block overlapping is regarded as an efficient way of compensating for the degrees of freedom lost due to tapering). The WOSA spectral estimator of $S_X(f)$ is defined to be

$$\hat{S}^{(wosa)}_X(f) = \frac{1}{n_B} \sum_{j=0}^{n_B-1} \hat{S}^{(d)}_{js+1,X}(f) \qquad (11.19)$$

where $n_B$ is the total number of blocks and $s$ is an integer shift factor satisfying $0 < s \leq n_S$ and $s(n_B - 1) = n - n_S$ (note that the block for $j = 0$ uses data values $X_1, \ldots, X_{n_S}$, while the block for $j = n_B - 1$ uses $X_{n-n_S+1}, \ldots, X_n$).

The large sample statistical properties of $\hat{S}^{(wosa)}_X(f)$ closely resemble those of lag window estimators. In particular, we have the approximation that $\nu \hat{S}^{(wosa)}_X(f)/S_X(f) \stackrel{d}{=} \chi^2_\nu$, where the equivalent degrees of freedom $\nu$ are given by

$$\nu = \frac{2n_B}{1 + 2\sum_{m=1}^{n_B-1} \left(1 - \frac{m}{n_B}\right) |\sum_{t=1}^{n_S} h_t h_{t+ms}|^2}$$

(here $h_t = 0$ by definition for all $t > n_S$). If we specialize to the case of 50% block overlap (i.e., $s = n_S/2$) with a Hanning data taper (a common recommendation in the engineering literature), this can be approximated by the simple formula $\nu \approx 36n_B^2/(19n_B - 1)$. Thus, as the number of blocks $n_B$ increases, the equivalent degrees of freedom increase also, yielding a spectral estimator with reduced variance. Unless $S_X(\cdot)$ has a relatively featureless sdf, we cannot, however, make $n_B$ arbitrarily small without incurring severe bias in the individual direct spectral estimators mainly due to loss of resolution. (For details on the above results, see [4], Section 6.17.)

Figure 8(b) shows a WOSA spectral estimator for the wire wave gauge data (the solid curve). This series has $n = 4096$ data values. Some experimentation indicated that a block size of $n_S = 256$ and the Hanning data taper are reasonable choices for estimating the sdf between 0.4 and 4.0 Hz using WOSA. With a 50% block overlap, the shift factor is $s = n_S/2 = 128$; the total number of blocks is $n_B = \frac{1}{s}(n - n_S) + 1 = 31$; and $\nu$, the equivalent degrees of freedom, is approximately 59. The 31 individual direct spectral estimates that were averaged together to form the WOSA estimate are shown as the dots in Figure 8(b).

We have also plotted a "bandwidth/confidence interval" crisscross similar to that on Figure 8(a), but now the "bandwidth" (i.e., the horizontal width) is the distance in frequency between approximately uncorrelated spectral estimates. This measure of bandwidth is a function of the block size $n_S$ and the data taper used in WOSA. For the Hanning taper, the bandwidth is approximately $1.94/(n_S \Delta t)$. The crisscrosses in Figures 8(a) and 8(b) are quite similar, indicating that the statistical properties of the postcolored Parzen lag window and WOSA spectral estimates are comparable: indeed, the actual estimates agree closely, with the WOSA estimate being slightly smoother in appearance.

**Multitaper Spectral Estimators.** An interesting alternative to either lag window or WOSA spectral estimation is the multitaper approach of Thomson [6]. Multitaper spectral estimation can be regarded as a way of producing a direct spectral estimator with more than just two equivalent degrees of freedom (typical values are 4 to 16). As such, the multitaper method is different in spirit from the other two estimators in that it does not seek to produce highly smoothed spectra. An increase in degrees of freedom from 2 to just 10 is enough, however, to shrink the width of a 95% confidence interval for the sdf by more than an order of magnitude and hence to reduce the variability in the spectral estimate to the point where the human eye can readily discern the overall structure. Detailed discussions on the multitaper approach are given in [6] and Chapter 7 of [4]. Here we merely sketch the main ideas.

Multitaper spectral estimation is based upon the use of a set of $K$ data tapers $\{h_{t,k}: t = 1, \ldots, n\}$, where $k$ ranges from 0 to $K - 1$. We assume that these tapers are orthonormal (i.e., $\Sigma_{t=1}^{n} h_{t,j} h_{t,k} = 1$ if $j = k$ and 0 if $j \neq k$). The simplest multitaper estimator is defined by

$$\hat{S}_X^{(mt)}(f) = \frac{1}{K} \sum_{k=0}^{K-1} \hat{S}_{k,X}^{(mt)}(f) \quad \text{with } \hat{S}_{k,X}^{(mt)}(f) = \Delta t \left| \sum_{t=1}^{n} h_{t,k} X_t e^{-i2\pi ft \Delta t} \right|^2$$

(Thomson [6] advocates adaptively weighting the $\hat{S}_{k,X}^{(mt)}(f)$ rather than simply averaging them together). A comparison of this definition for $\hat{S}_{k,X}^{(mt)}(\cdot)$ with Eq. (11.8) shows that $\hat{S}_{k,X}^{(mt)}(\cdot)$ is in fact just a direct spectral estimator, so the multitaper estimator is just an average of direct spectral estimators employing an orthonormal set of tapers. Under certain mild conditions, the orthonormality of the tapers translates into the frequency domain as approximate independence of each individual $\hat{S}_{k,X}^{(mt)}(f)$; i.e., $\hat{S}_{j,X}^{(mt)}(f)$ and $\hat{S}_{k,X}^{(mt)}(f)$ are approximately independent for $j \ne k$. Approximate independence in turn implies that $2K\,\hat{S}_{X}^{(mt)}(f)/S_X(f) \stackrel{d}{=} \chi_{2K}^2$ approximately, so that the equivalent degrees of freedom for $\hat{S}_{X}^{(mt)}(f)$ is equal to twice the number of data tapers employed.

The key trick then is to find a set of $K$ orthonormal sequences, each one of which does a proper job of tapering. One appealing approach is to return to the concentration problem that gave us the dpss taper for a fixed resolution bandwidth $2W$. If we now refer to this taper as the *zeroth-order* dpss taper and denote it by $\{h_{t,0}\}$, we can recursively construct the remaining $K - 1$ "higher order" dpss tapers $\{h_{t,k}\}$ as follows. For $k = 1, \ldots, K - 1$, we define the $k$th-order dpss taper as the set of $n$ numbers $\{h_{t,k}: t = 1, \ldots, n\}$ such that

1. $\{h_{t,k}\}$ is orthogonal to each of the $k$ sequences $\{h_{t,0}\}, \ldots, \{h_{t,k-1}\}$ (i.e., $\sum_{t=1}^{n} h_{t,j}\, h_{t,k} = 0$ for $j = 0, \ldots, k - 1$);
2. $\{h_{t,k}\}$ is normalized such that $\sum_{t=1}^{n} h_{t,k}^2 = 1$;
3. subject to conditions 1 and 2, the spectral window $\mathcal{H}_k(\cdot)$ corresponding to $\{h_{t,k}\}$ maximizes the concentration ratio

$$\int_{-W}^{W} \mathcal{H}_k(f)\, df \Big/ \int_{-f_{(N)}}^{f_{(N)}} \mathcal{H}_k(f)\, df = \lambda_k(n, W)$$

In words, subject to the constraint of being orthogonal to all lower order dpss tapers, the $k$th-order dpss taper is "optimal" in the restricted sense that the sidelobes of its spectral window are suppressed as much as possible as measured by the concentration ratio. Methods for calculating the dpss data tapers are discussed in [4], Chapter 8.

In a series of papers, Slepian [15] (and references therein) has extensively studied the nature of dpss. One important fact he discusses is that the concentration ratio $\lambda_k(n, W)$ strictly decreases as $k$ increases in a manner such that $\lambda_k(n, W)$ is close to unity for $k < 2nW\,\Delta t$, after which it rapidly approaches 0 with increasing $k$ (the value $2nW\,\Delta t$ is sometimes called the *Shannon number*). Since $\lambda_k(n, W)$ must be close to unity for $\{h_{t,k}\}$ to be a decent data taper, multitaper spectral estimation is restricted to the use of at most—and, in practice, usually less than—$2nW\,\Delta t$ orthonormal dpss tapers.

An example of multitaper spectral estimation is shown in Figure 9. The lefthand column of plots shows the $k$th-order dpss data tapers for $n = 4096$, $nW = 4/\Delta t$, and $k$ ranging from 0 (top plot) to $K - 1 = 5$ (bottom plot). The thin horizontal lines in each of these plots indicate the zero level, so, whereas the zeroth-order dpss is strictly positive everywhere (but quite close to 0 near $t = 1$ and $t = n$), the higher order tapers assume both positive and negative values. Note also that the zeroth-order taper heavily downweights values of the time series close to $t = 1$ and $t = n$, but that these values are given successively more weight by the higher order tapers (one interpretation of multitapering is that the higher order tapers are recapturing information "lost" when but a single data taper is used). The solid curve in Figure 9(b) shows a multitaper spectral estimate $\hat{S}_X^{(mt)}(\cdot)$ for the wire wave gauge data based upon these 6 dpss tapers, whereas the dots show the six individual direct spectral estimates $\hat{S}_{k,X}^{(mt)}(\cdot)$. Note that the number of tapers that we have used is below the Shannon number $2nW \Delta t = 8$ and that $\nu$, the equivalent degrees of freedom, is here $2K = 12$. The multitaper spectral estimate is much choppier in appearance than either the lag window spectral estimate of Figure 8(a) or the WOSA estimate of Figure 8(b), both of which have a markedly higher number of equivalent degrees of freedom ($\nu = 64$ and $\nu = 59$, respectively). Nonetheless, the variability in the multitaper spectral estimate is small enough so that the eye can readily

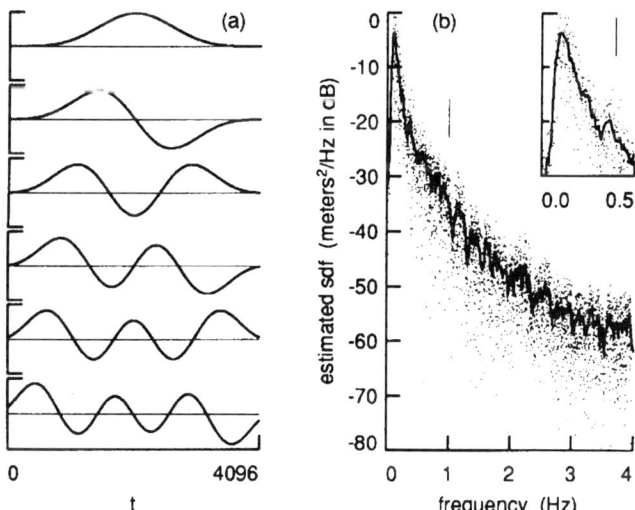

FIG. 9. Multitaper spectral estimation.

detect the overall structure (cf. $\hat{S}_X^{(mt)}(\cdot)$ with the two spectral estimates in Figure 5), and because it is not highly smoothed, the multitaper estimate does markedly better at capturing the spectral structure near $f = 0$.

Based upon performance bounds, Bronez [16] argues that the multitaper spectral estimator has statistical properties that are superior to WOSA for sdfs with very high dynamic ranges (more research is required, however, to verify that these bounds translate into an actual advantage in practice). In comparison to prewhitening, multitapering is useful in situations where leakage is a concern but it is not practical to carefully design prewhitening filters (this occurs in, for example, exploration geophysics due to the enormous volume of time series routinely collected). Finally, we note that Thomson and Chave [17] describe an appealing scheme in which multitapering is used in conjunction with WOSA.

### 11.2.4 Evaluating the Significance of Spectral Peaks

A common use for spectral analysis is the detection of a periodic signal in the presence of noise. In the simpliest case, we assume that the periodic signal is a sinusoid and that the noise is additive zero mean white noise $\{W_t\}$; i.e., we assume that our time series can be modeled as

$$X_t = D \cos(2\pi f_l t \, \Delta t + \phi) + W_t, \qquad t = 1, \ldots, n$$

where $D$, $f_l$, and $\phi$ are constants. Suppose first that the frequency $f_l$ of the sinusoid is known a priori to be equal to the Fourier frequency $l/(n\Delta t)$, where $l$ is an integer such that $0 < l/n < 1/2$. We wish to test the null hypothesis that $D$ is 0 (i.e., that our time series in fact does *not* contain a significant sinusoidal component at the proposed frequency $f_l$). Let $J$ be the largest integer such that $J/n < 1/2$, and let $f_j = j/(n\Delta t)$ for $j = 1, \ldots, J$ be the set of all nonzero Fourier frequencies less than $f_{(N)}$. If the null hypothesis is true and if $\{W_t\}$ is a Gaussian process, then the set of periodogram ordinates $\hat{S}^{(p)}(f_1), \hat{S}^{(p)}(f_2), \ldots, \hat{S}^{(p)}(f_J)$ constitute a set of independent and identically distributed $\chi_2^2$ rvs multiplied by the constant $V(W_t)/2$. The ratio

$$(J - 1)\hat{S}^{(p)}(f_l) \Big/ \sum_{\substack{j=1,\ldots,J \\ j \neq l}} \hat{S}^{(p)}(f_j)$$

thus follows an $F_{2,2(J-1)}$ distribution under the null hypothesis. Under the alternative hypothesis $D \neq 0$, the periodogram ordinate $\hat{S}^{(p)}(f_l)$ will tend to be large, and hence the preceding ratio will also tend to be large. We thus reject the null hypothesis at level of significance $\alpha$ if the ratio exceeds the upper $(1 - \alpha) \times 100\%$ percentage point of the $F_{2,2(J-1)}$ distribution. In general, this percentage point for an $F_{2,k}$ distribution is given by $k(1 - \alpha^{2/k})/(2\alpha^{2/k})$.

This test statistic can be easily modified to handle a periodic *nonsinusoidal* signal with known period $1/f_l$. In general, such periodic signals can be written as a linear combination of sinusoids with frequencies $f_l, 2f_l, 3f_l, \ldots,$ where $f_l$ is known as the *fundamental frequency* of the signal and $(k + 1)f_l$ is called the $k$th *harmonic frequency*. For example, suppose that the signal can be assumed *a priori* to be represented by a fundamental frequency $f_l$ and its first harmonic $2f_l$ with $2f_l < 1/(2\ \Delta t)$; i.e., because $2f_l = f_{2l}$, our assumed model is

$$X_t = D_1 \cos(2\pi f_l t\ \Delta t + \phi_1) + D_2 \cos(2\pi f_{2l} t\ \Delta t + \phi_2) + W_t$$

We can then test the null hypothesis $D_1 = D_2 = 0$ using the ratio

$$(J - 2)\ [\hat{S}^{(p)}(f_l) + \hat{S}^{(p)}(f_{2l})]\ /2 \sum_{\substack{j=1,\ldots,J \\ j \ne l, j \ne 2l}} \hat{S}^{(p)}(f_j)$$

Under the null hypothesis, this ratio follows an $F_{4,2(J-2)}$ distribution. We thus reject the null hypothesis with level of significance $\alpha$ if the preceding ratio exceeds the upper $(1 - \alpha) \times 100\%$ percentage point of this distribution.

Suppose now that we do not know a priori the period of the potential signal in our time series, but that the periodogram for $X_1, X_2, \ldots, X_n$ has apparent large values at one or more Fourier frequencies. To assess whether these large values are ascribable to just random fluctuations, we can use *Fisher's g statistic,* defined by

$$g = \max_{1 \le j \le J} \hat{S}^{(p)}(f_j)\ /\ \sum_{j=1}^{J} \hat{S}^{(p)}(f_j)$$

Under the null hypothesis that our time series is a portion of a realization of a Gaussian white noise process, Fisher [18] derived the exact distribution for $g$. In practice, a simple approximation to this distribution tells us to reject the null hypothesis at the (approximate) level of significance $\alpha$ if $g$ exceeds the value $1 - (\alpha/J)^{1/(J-1)}$. Siegel [19] proposed a test statistic similar to Fisher's test, with the key modification that it considers *all* large values of the periodogram and hence is more powerful against an alternative hypothesis of multiple sinusoidal components plus Gaussian white noise (for details, including some simple approximations to the percentage points for Siegel's test, see [4], Section 10.9).

The three statistical tests we have briefly discussed so far can all give misleading results when the additive noise $\{W_t\}$ cannot reasonably be assumed to be white noise. An appealing approach in this case is to use Thomson's multitaper $F$-test, which is based the notion of frequency domain regression analysis. Details concerning—and examples of the use of—this test are given in [6] and Sections 10.11 and 10.13 of [4].

## 11.3 Bivariate Time Series

### 11.3.1 Basic Concepts

Let us now consider the bivariate real-valued stationary process $\{X_t, Y_t\}$ (see Chapter 3 of this book). We assume that the univariate real-valued stationary processes $\{X_t\}$ and $\{Y_t\}$ are both zero mean processes with sdfs given by, respectively, $S_X(\cdot)$ and $S_Y(\cdot)$. Under these conditions, the cross-spectral properties of $\{X_t, Y_t\}$ are given by the *cross spectrum* (sometimes called the *cross spectral density function*), which can be written as

$$S_{XY}(f) = \Delta t \sum_{\tau=-\infty}^{\infty} C_{\tau,XY} e^{-i2\pi f\tau \Delta t}, \quad -f_{(N)} \leq f \leq f_{(N)} \quad (11.20)$$

(see Eq. (11.3), where $\{C_{\tau,XY}\}$ is the *cross covariance sequence* (ccvs) defined by

$$C_{\tau,XY} = \text{Cov}(X_t, Y_{t+\tau}) = E(X_t Y_{t+\tau})$$

To learn what the cross spectrum can tell us, let us consider a specific example appropriate for the bivariate wave height time series described in Section 11.1 and shown in Figure 1. Let $\{O_t\}$ be a zero mean stationary process (with sdf $S_O(\cdot)$) that represents the true height of the ocean waves at the wire wave gauge. Because of instrumentation noise, what the wire wave gauge actually records at index $t$ is $X_t = O_t + U_t$, where we assume that $\{U_t\}$ is a zero mean stationary process (with sdf $S_U(\cdot)$) that represents the instrumentation noise and is pairwise uncorrelated with $\{O_t\}$ (i.e., $C_{\tau,OU} = E(O_t U_{t+\tau}) = 0$ for all $\tau$). At the infrared wave gauge, we assume that we measure $\{Y_t\}$, which is the sum of a displaced version of $\{O_t\}$ and a separate instrumentation noise process $\{V_t\}$; i.e., $Y_t = O_{t+l} + V_t$, where $l$ is a fixed integer, and $\{V_t\}$ is a zero mean stationary process (with sdf $S_V(\cdot)$) that is pairwise uncorrelated with both $\{O_t\}$ and the other instrumentation noise process $\{U_t\}$ (i.e., $C_{\tau,OV} = 0$ and $C_{\tau,UV} = 0$ for all $\tau$). Note that, if $l < 0$, we can say that the process $\{Y_t\}$ *lags* the process $\{X_t\}$ by $-l\Delta t$ time units because both $X_{t+l}$ and $Y_t$ depend on $O_{t+l}$, and the index $t + l$ occurs before the index $t$. Conversely, if $l > 0$, we say that $\{Y_t\}$ *leads* the process $\{X_t\}$ by $l\Delta t$ time units.

If we denote the acvss for $\{O_t\}$, $\{U_t\}$ and $\{V_t\}$ by, respectively, $\{C_{\tau,O}\}$, $\{C_{\tau,U}\}$, and $\{C_{\tau,V}\}$, then the ccvs for $\{X_t, Y_t\}$ is given by

$$C_{\tau,XY} = E(X_t Y_{t+\tau}) = E[(O_t + U_t)(O_{t+l+\tau} + V_{t+\tau})] = C_{l+\tau,O}$$

i.e., the ccvs is just a shifted version of the acvs for $\{O_t\}$, and the amount of the shift is just given by the lead or lag factor $l$. The cross spectrum for $\{X_t, Y_t\}$ is just $S_{XY}(f) = e^{i2\pi fl \Delta t} S_O(f)$. Note that this cross spectrum expresses the commonality between $\{X_t\}$ and $\{Y_t\}$; i.e., it does not depend on either $\{U_t\}$ (which is part of $\{X_t\}$ but not of $\{Y_t\}$) or $\{V_t\}$ (part of $\{Y_t\}$ but not $\{X_t\}$).

Because $S_{XY}(f) = e^{i2\pi fl \Delta t} S_O(f)$ in this simple example, we can deduce that, whereas the sdf is always real valued, the cross spectrum is in general complex valued. This fact makes it difficult to interpret $S_{XY}(\cdot)$ and leads us to search for quantities related to it, but easier to deal with. If we express the conjugate of $S_{XY}(f)$ in terms of its real and imaginary components, i.e., $S_{XY}(f) = R_{XY}(f) - iI_{XY}(f)$, we can define the *cospectrum* $R_{XY}(\cdot)$ and the *quadrature spectrum* $I_{XY}(\cdot)$; on the other hand, if we express $S_{XY}(f)$ as its absolute value times a complex exponential, i.e., $S_{XY}(f) = A_{XY}(f)e^{i\phi_{XY}(f)}$ with $A_{XY}(f) = |S_{XY}(f)|$, we can define the *cross amplitude spectrum* $A_{XY}(\cdot)$ and the *phase spectrum* $\phi_{XY}(\cdot)$. Note that, if in fact $|A_{XY}(f)| = 0$, then $\phi_{XY}(f)$ is ill-defined; if $|A_{XY}(f)| > 0$, then $\phi_{XY}(f)$ is defined only up to an integer multiple of $2\pi$. In contrast to $S_{XY}(\cdot)$, the derived functions $R_{XY}(\cdot)$, $I_{XY}(\cdot)$, and $A_{XY}(\cdot)$ are all real valued and hence can at least be plotted easily. The function $\phi_{XY}(\cdot)$ is also real valued, but the "$2\pi$" ambiguity in its definition makes plotting it somewhat problematic (see the discussion in the next section).

Let us now see what these derived functions are for our simple ocean wave example. Since then $S_{XY}(f) = e^{i2\pi fl \Delta t} S_O(f)$, we have $R_{XY}(f) = \cos(2\pi fl \Delta t)S_O(f)$, $I_{XY}(f) = -\sin(2\pi fl \Delta t)S_O(f)$, $A_{XY}(f) = S_O(f)$, and assuming that $A_{XY}(f) > 0$, $\phi_{XY}(f) = 2\pi fl \Delta t$. For this example, the cross amplitude spectrum and the phase spectrum are most easily interpreted: the former is just the sdf for the ocean wave process $\{O_t\}$, while the latter is linear with a slope proportional to the lead or lag factor $l$. In general, if it exists, the function that by $-(1/2\pi) \cdot (d\phi_{XY}(f)/df)$ is called the *group delay* (or the *envelope delay;* see [3], p. 664). In our example, the group delay is equal to $-l\Delta t$ at all frequencies, but in general it depends upon the frequency $f$.

Whereas the cospectrum, quadrature, cross amplitude, and phase spectra and the group delay can all be derived solely from the cross spectrum, the *complex coherency,* defined as $w_{XY}(f) = S_{XY}(f)/\sqrt{S_X(f)S_Y(f)}$, depends on both the cross spectrum and the sdf's for $\{X_t\}$ and $\{Y_t\}$. If we write the spectral representations for $\{X_t\}$ and $\{Y_t\}$ as

$$X_t = \int_{-f_{(N)}}^{f_{(N)}} e^{i2\pi ft\,\Delta t}\,dZ_X(f) \quad \text{and} \quad Y_t = \int_{-f_{(N)}}^{f_{(N)}} e^{i2\pi ft\,\Delta t}\,dZ_Y(f)$$

(see Eq. (11.1), it can be shown that

$$w_{XY}(f) = \frac{\text{Cov}[dZ_X(f), dZ_Y(f)]}{\sqrt{V[dZ_X(f)]\,V[dZ_Y(f)]}}$$

(see [3], p. 661). The complex coherency is thus a complex-valued frequency domain correlation coefficient that measures the correlation in the random amplitudes assigned to the complex exponentials with frequency $f$ in the spectral representations for $\{X_t\}$ and $\{Y_t\}$. Since it is a complex-valued correlation coefficient, we must have $0 \leq |w_{XY}(f)|^2 \leq 1$. The quantity $|w_{XY}(f)|^2$ is called the *magnitude squared coherence* (msc) at the frequency $f$. Note that

$$|w_{XY}(f)|^2 = \frac{|S_{XY}(f)|^2}{S_X(f)S_Y(f)} = \frac{A_{XY}^2(f)}{S_X(f)S_Y(f)}$$

i.e., the msc is a normalized version of the square of the cross-amplitude spectrum. The msc essentially captures the "amplitude" part of the cross spectrum, but completely ignores its phase, so the msc and the phase spectrum together are useful real-valued summaries of the "information" in the complex-valued cross spectrum.

To derive the msc for our ocean wave example, we first note that, because $\{O_t\}$, $\{U_t\}$, and $\{V_t\}$ are pairwise uncorrelated processes, the sdfs for $\{X_t\}$ and $\{Y_t\}$ can be expressed as, respectively, $S_X(f) = S_O(f) + S_U(f)$ and $S_Y(f) = S_O(f) + S_V(f)$. The msc for $\{X_t, Y_t\}$ is thus given by

$$|w_{XY}(f)|^2 = 1/[1 + S_U(f)/S_O(f)]\,[1 + S_V(f)/S_O(f)]$$

which tells us that the msc is

1. equal to 0 if $S_O(f) = 0$ while either $S_U(f) > 0$ or $S_V(f) > 0$ (because then the variability in $\{X_t\}$ and $\{Y_t\}$ at frequency $f$ is due to instrumentation noise, assumed to be uncorrelated between gauges);
2. close to 0 if either $S_U(f)$ of $S_V(f)$ is large compared to $S_O(f)$ (because then the variability at $f$ in at least one of the component processes of $\{X_t, Y_t\}$ is mainly due to instrumentation noise);
3. close to unity if both $S_U(f)$ and $S_V(f)$ are small compared to $S_O(f)$ (because then the variability in both $\{X_t\}$ and $\{Y_t\}$ at $f$ is mainly due to the variability in $\{O_t\}$;
4. unity if, for example, both $U_t = 0$ and $V_t = 0$ for all $t$ (because then $\{Y_t\}$ is just a time-shifted version of $\{X_t\}$).

All of this lends credence to the interpretation of the msc as a measure of the correlation between $\{X_t\}$ and $\{Y_t\}$ at particular frequencies.

### 11.3.2 Bivariate Spectral Estimation

We now turn to the problem of estimating the cross spectrum and various functions derived from it. Our estimates are based upon a bivariate time series that can be regarded as a realization of a portion $X_t, Y_t, t = 1, \ldots, n$, of the bivariate stationary process $\{X_t, Y_t\}$ with cross spectrum $S_{XY}(\cdot)$ and sdfs $S_X(\cdot)$ and $S_Y(\cdot)$. In view of Eq. (11.20), an "obvious" estimator for $S_{XY}(f)$ is the *cross periodogram* given by

$$\hat{S}_{XY}^{(p)}(f) = \Delta t \sum_{\tau=-(n-1)}^{n-1} \hat{C}_{\tau,XY}^{(p)} e^{-i2\pi f \tau \Delta t}$$

(see the periodogram of Eq. (11.5), where $\hat{C}_{\tau,XY}^{(p)} = \frac{1}{n} \sum_t X_t Y_{t+\tau}$ (here the summation ranges from $t = 1$ to $t = n - \tau$ for nonnegative values of $\tau$ and from $t = 1 - \tau$ to $t = n$ for negative values of $\tau$). Note that this expression for $\hat{C}_{\tau,XY}^{(p)}$ implicitly assumes that both $\{X_t\}$ and $\{Y_t\}$ are known a priori to be zero mean processes—if this is not the case, the common practice is to use $X_t - \bar{X}$ and $Y_t - \bar{Y}$ in place of $X_t$ and $Y_t$ in all computational formulae, where $\bar{X}$ and $\bar{Y}$ are the sample means. In analogy to Eq. (11.6), the cross periodogram can also be written as

$$\hat{S}_{XY}^{(p)}(f) = \frac{\Delta t}{n} \left( \sum_{t=1}^{n} X_t e^{-i2\pi f t \Delta t} \right)^* \left( \sum_{t=1}^{n} Y_t e^{-i2\pi f t \Delta t} \right) \quad (11.21)$$

where the asterisk denotes complex conjugation.

Unfortunately, the bias and variance properties of the cross periodogram are as poor as those of the periodogram, so $\hat{S}_{XY}^{(p)}(f)$ must be used with great caution. For example, suppose we use the cross periodogram along with the individual periodograms to form the "obvious" estimator for the msc, namely, $|\hat{S}_{XY}^{(p)}(f)|^2 / \hat{S}_X^{(p)}(f) \hat{S}_Y^{(p)}(f)$. Because of Eq. (11.21) and (11.6), we have

$$|\hat{S}_{XY}^{(p)}(f)|^2 = \frac{(\Delta t)^2}{n^2} \left| \sum_{t=1}^{n} X_t e^{-i2\pi f t \Delta t} \right|^2 \left| \sum_{t=1}^{n} Y_t e^{-i2\pi f t \Delta t} \right|^2 = \hat{S}_X^{(p)}(f) \hat{S}_Y^{(p)}(f)$$

so that the "obvious" estimator for the msc is in fact always *unity!* Priestley [3], p. 708, gives a reasonable explanation for this seemingly unreasonable result; namely, that the frequency domain correlation coefficient $|w_{XY}(f)|^2$ is essentially being estimated using just a single observation of a bivariate frequency domain process at frequency $f$. It is thus vital to reduce the inherent variability in the cross periodogram if the corresponding msc estimator is to make any sense. In principle, variance reduction can be achieved using straightforward extensions of the approaches discussed in Section 11.2.3 for univariate sdf estimation; namely, smoothing across frequencies, WOSA, and multitapering. In practice, however, WOSA and multitapering are the methods of choice because of subtle (but surmountable) problems that can arise when smoothing across frequencies (in addition, Walden [20] demonstrates an improvement in msc estimation using multitapering instead of smoothing across frequencies). For brevity, we consider just the multitaper estimator (see Carter [13] for details on WOSA; Thomson and Chave [17] discuss WOSA, multitapering, and a combination of the two methods).

The simplest multitaper estimator of the cross spectrum is given by

$$\hat{S}_{XY}^{(mt)}(f) = \frac{\Delta t}{K} \sum_{k=0}^{K-1} \left( \sum_{t=1}^{n} h_{t,k} X_t e^{-i2\pi ft\Delta t} \right)^* \left( \sum_{t=1}^{n} h_{t,k} Y_t e^{-i2\pi ft\Delta t} \right)$$

where, as before, $\{h_{t,k}\}$ is the $k$th-order dpss data taper for a sequence of length $n$ and a fixed resolution bandwidth $2W$. As in the univariate case, each data taper is designed to prevent leakage, whereas the use of multiple tapers yields an estimator of the cross spectrum having less variability than the cross periodogram. The number of equivalent degrees of freedom for $\hat{S}_{XY}^{(mt)}(f)$ is $\nu = 2K$. The corresponding multitaper estimators for the cospectrum and related quantities follow directly from their definitions. For example, the estimators for the phase spectrum and the msc are given by $\hat{\phi}_{XY}^{(mt)}(f) = \arg(\hat{S}_{XY}^{(mt)}(f))$ and $|\hat{w}_{XY}^{(mt)}(f)|^2 = |\hat{S}_{XY}^{(mt)}(f)|^2/\hat{S}_{X}^{(mt)}(f)\hat{S}_{Y}^{(mt)}(f)$. In the expression for $\hat{\phi}_{XY}^{(mt)}(f)$, we can assume that the "arg" function returns a value between $-\pi$ and $\pi$ so that $\hat{\phi}_{XY}^{(mt)}(f)$ assumes values modulo $2\pi$ and hence can be discontinuous as phases pass over the $\pm \pi$ boundaries. These discontinuities can hinder ascertaining whether or not the phase spectrum varies linearly with frequency (an indication of a lead or lag relationship between $\{X_t\}$ and $\{Y_T\}$). To avoid these difficulties, Priestley [3], p. 709, advocates plotting $\hat{\phi}_{XY}^{(mt)}(f)$, $\hat{\phi}_{XY}^{(mt)}(f) + 2\pi$ and $\hat{\phi}_{XY}^{(mt)}(f) - 2\pi$ versus $f$ all together.

Figure 10 summarizes a bivariate spectral analysis of the wire wave gauge time series $\{X_t\}$ of Figure 1(a) and the infrared wave gauge time series $\{Y_t\}$ of Figure 1(b). The thick solid curve in Figure 10(c) is a reproduction of the multitaper spectral estimate $\hat{S}_{X}^{(mt)}(f)$ shown previously in Figure 9(b). The thin solid curve is the corresponding estimate $\hat{S}_{Y}^{(mt)}(f)$ for the infrared wave gauge time series. The two spectral estimates agree almost perfectly between the frequencies $f = 0.04$ Hz and $f = 0.34$ Hz (these frequencies are marked by two thin vertical lines) and

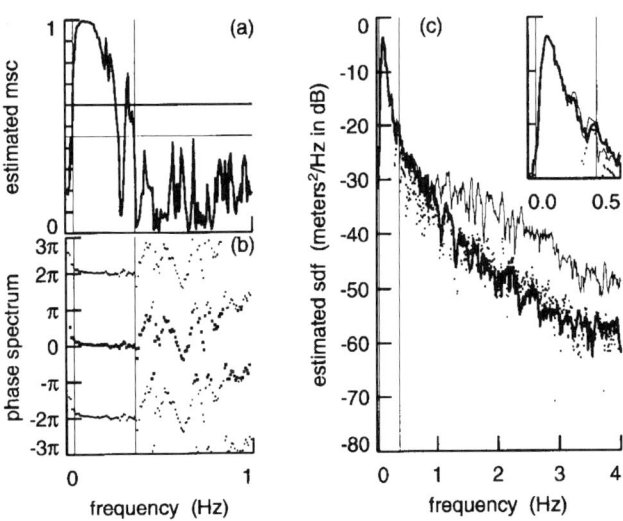

FIG. 10. Bivariate spectral analysis (see text for details).

reasonably well between $f = 0.34$ and $f = 0.97$, beyond which the level of $\hat{S}_Y^{(mt)}(f)$ is consistently higher than that of $\hat{S}_X^{(mt)}(f)$. This finding is in agreement with the fact that the infrared wave gauge has substantially more measurement noise at high frequencies than does the wire wave gauge. The dots in Figure 10(c) show the multitaper estimate $\hat{A}_{XY}^{(mt)}(\cdot)$ of the cross amplitude spectrum. Recall that, in terms of the simple lead or lag model for this data, the cross-amplitude spectrum is just the sdf for the common ocean wave process. Again there is very good agreement between $\hat{A}_{XY}^{(mt)}(\cdot)$ and either $\hat{S}_X^{(mt)}(\cdot)$ or $\hat{S}_Y^{(mt)}(\cdot)$ at the low frequencies delineated by the two thin vertical lines, after which $\hat{A}_{XY}^{(mt)}(\cdot)$ is in general agreement with $\hat{S}_X^{(mt)}(\cdot)$ but not $\hat{S}_Y^{(mt)}(\cdot)$. This finding is consistent with the measurement noise at high frequencies for the wire wave gauge being smaller than that of the other gauge. There are some small (less than 10 dB) differences between $\hat{S}_X^{(mt)}(\cdot)$ and $\hat{A}_{XY}^{(mt)}(\cdot)$, particularly between the frequencies $f = 0.34$ and $f = 1.0$, where $\hat{A}_{XY}^{(mt)}(f)$ is consistently lower than $\hat{S}_X^{(mt)}(\cdot)$. Our simple lead or lag model would suggest that the common ocean wave process $\{O_t\}$ is weak at these frequencies compared to local effects, all of which are lumped together with instrumentation noise in our simple model.

The solid curve in Figure 10(a) shows the corresponding estimated msc $|\hat{w}_{XY}^{(mt)}(f)|^2$ for $0 \leq f \leq 1$ Hz (the two thin vertical lines mark the same frequencies as the corresponding lines in Figure 10(c)). We can test, at the level of significance $\alpha$, the null hypothesis that the true msc is 0 at frequency

$f$ by comparing $|\hat{w}_{XY}^{(mt)}(f)|^2$ to the value $1 - \alpha^{2/(\nu-2)}$ and rejecting the null hypothesis if $|\hat{w}_{XY}^{(mt)}(f)|^2$ exceeds this value (for details, see [2], p. 284). As before, $\nu$ represents the number of equivalent degrees of freedom associated with the spectral estimates. Because we have used $K = 6$ tapers in the multitaper estimates $\hat{S}_X^{(mt)}(\cdot)$, $\hat{S}_Y^{(mt)}(\cdot)$, and $\hat{S}_{XY}^{(mt)}(\cdot)$, we have $\nu = 2K = 12$ degrees of freedom. The thick and thin horizontal lines on Figure 10(a) define the rejection regions for levels of significance of, respectively, $\alpha = 0.01$ and $\alpha = 0.05$. We see that the region of nonzero msc is bounded by the thin vertical lines marking $f = 0.04$ Hz and $f = 0.34$ Hz, between which $\hat{S}_X^{(mt)}(\cdot)$, $\hat{S}_Y^{(mt)}(\cdot)$, and $\hat{A}_{XY}^{(mt)}(\cdot)$ are all in good agreement (for $1.0 < f < 4.0$, the estimated msc is such that, except at a few isolated frequencies, we cannot reject the null hypothesis of zero msc). In terms of our simple lead or lag model, the small msc at frequencies higher than $f = 0.34$ Hz can be attributed to the large instrumentation noise associated with the infrared wave gauge.

Finally, the dots in Figure 10(b) show the estimated phase spectrum $\hat{\phi}_{XY}^{(mt)}(f)$ versus $f$ for $0 \leq f \leq 1$ Hz, plotted per Priestley's recommendation (the dots for $\hat{\phi}_{XY}^{(mt)}(f) \pm 2\pi$ are smaller than the dots for $\hat{\phi}_{XY}^{(mt)}(f)$). Again, vertical lines delineate the frequencies of apparent nonzero msc, between which the estimated phase spectrum increases approximately linearly. A linear least squares fit to $\hat{\phi}_{XY}^{(mt)}(f)$ versus $f$ at these frequencies yields an estimated slope of $-1.02$ rad per Hz. In terms of the lead or lag model, this slope translates into a lead or lag factor of $l = -4.9 \approx -5$ units or $-1/6$ sec since one unit is equal to $1/30$ sec. Because $l$ is negative, the prevailing ocean waves are arriving at the infrared wave gauge approximately $1/6$ sec behind of the wire wave gauge; moreover, because the two gauges were 6 m apart, the prevalent direction of the ocean waves is approximately perpendicular to a line drawn between the two gauges. Note that, in regions where the estimated msc is small, the estimated phase spectrum is erratic. This pattern is consistent with well-known approximations to the statistical properties of $\hat{\phi}_{XY}^{(mt)}(f)$, which indicate that it has large variability when the true msc is 0 ([3], p. 703).

Detailed discussions on bivariate spectral analysis can be found in Koopmans [2] and Priestley [3]. Thomson and Chave [17] describe resampling schemes for assessing variability in bivariate spectral estimates.

## Acknowledgments

R. Spindel and J. Harlett of the Applied Physics Laboratory, University of Washington, graciously provided discretionary funding to support writing this chapter. The author would also like to thank Andrew Walden, Christopher Chatfield, and the editors for helpful critiques.

## References

1. Cramér, H. (1942). "On Harmonic Analysis in Certain Functional Spaces." *Arkiv för Matematik, Astronomi och Fysik,* **28B,** 1–7.
2. Koopmans, L. H. (1974). *The Spectral Analysis of Time Series.* Academic Press, New York.
3. Priestley, M. B. (1981). *Spectral Analysis and Time Series.* Academic Press, London.
4. Percival, D. B., and Walden, A. T. (1993). *Spectral Analysis for Physical Applications: Multitaper and Conventional Univariate Techniques.* Cambridge University Press, Cambridge.
5. Jessup, A. T., Melville, W. K., and Keller, W. C. (1991). "Breaking Waves Affecting Microwave Backscatter: 1. Detection and Verification." *Journal of Geophysical Research,* **96** (20), 547–559.
6. Thomson, D. J. (1982). "Spectrum Estimation and Harmonic Analysis." *Proceedings of the IEEE* **70,** 1055–1096.
7. Walden, A. T. (1989). "Accurate Approximation of a Zeroth Order Discrete Prolate Spheroidal Sequence for Filtering and Data Tapering." *Signal Processing* **18,** 341–348.
8. Press, W. H., Flannery, B. P., Teukolsky, S. A., and Vetterling, W. T. (1992). *Numerical Recipes: The Art of Scientific Computing,* 2d ed. Cambridge University Press, Cambridge.
9. Percival, D. B., and Kerr, R. K. (1986). "$P_tTSS_A$—A Time Series Analysis System Embedded in LISP." *Proceedings of the 20th Symposium on the Interface of Computer Science and Statistics,* pp. 321–330.
10. Best, D. J., and Roberts, D. E. (1975). "The Percentage Points of the $\chi^2$ Distribution: Algorithm AS 91." *Applied Statistics* **24,** 385–388.
11. Parzen, E. (1961). "Mathematical Considerations in the Estimation of Spectra." *Technometrics* **3,** 167–190.
12. Welch, P. D. (1967). "The Use of Fast Fourier Transform for the Estimation of Power Spectra: A Method Based on Time Averaging over Short, Modified Periodograms." *IEEE Transactions on Audio and Electroacoustics* **15,** 70–73.
13. Carter, G. C. (1987). "Coherence and Time Delay Estimation." *Proceedings of the IEEE* **75,** 236–255.
14. Thomson, D. J. (1977). "Spectrum Estimation Techniques for Characterization and Development of WT4 Waveguide—I." *Bell System Technical Journal* **56,** 1769–1815.
15. Slepian, D. (1983). "Some Comments on Fourier Analysis, Uncertainty and Modeling." *SIAM Review* **25,** 379–393.
16. Bronez, T. P. (1992). "On the Performance Advantage of Multitaper Spectral Analysis." *IEEE Transactions on Signal Processing* **40,** 2941–2946.
17. Thomson, D. J., and Chave, A. D. (1991). "Jackknifed Error Estimates for Spectra, Coherences, and Transfer Functions," in *Advances in Spectrum Analysis and Array Processing,* S. Haykin, (ed.), **1,** pp. 58–113. Prentice-Hall, Englewood Cliffs, NJ.

18. Fisher, R. A. (1929). "Tests of Significance in Harmonic Analysis." *Proceedings of the Royal Society of London, Series A* **125,** 54–59.
19. Siegel, A. F. (1980). "Testing for Periodicity in a Time Series." *Journal of the American Statistical Association* **75,** 345–348.
20. Walden, A. T. (1991). "Wavelet Estimation Using the Multitaper Method." *Geophysical Prospecting* **39,** 625–642.

# 12. WEAK PERIODIC SIGNALS IN POINT PROCESS DATA

David A. Lewis

Department of Physics and Astronomy
Iowa State University, Ames

## 12.1 Introduction

Data in physics and astronomy often consist of lists of times at which events took place. This arises both because of the nature of the detection process, often using the photoelectric effect, and from the nature of the experiments themselves. In some instances only the average rate of events is important and not the detailed time structure. For instance, in both high-energy physics and nuclear physics, the prototype measurement is a scattering experiment in which an accelerated beam of particles strikes a target producing new particles that are detected and counted (see Chapter 15). The physically interesting quantity is the cross section for the process, and it can be calculated from the average particle production rate.

In a second type of measurement, the rate at which events take place is modulated at one (or perhaps more) frequencies. The strength, frequency, phase and perhaps waveform of the modulation must be inferred from the list of event times. An even more fundamental question is whether or not a signal actually *exists* in sparse or noisy data. The measurement sensitivity is usually limited by either a small number of events or by background events (i.e., noise). These limitations, and methods of digging into the noise as far as is reasonable are the topics considered here.

There are many examples of the second type of measurement in atomic and laser physics that fall under the broad category of "resonance" experiments. In particular, "double resonance experiments" in which optical photons are detected via photoelectric or "resonance ionization" experiments in which single ions are detected generate data consisting of lists of event times. In the language of Chapter 3, such data are "point process" data. These often have periodic signals, e.g., quantum beats, embedded in them. There are examples of these measurements described in [1]. The abundance of rare isotopic or atomic species can be measured in this way and atomic lifetimes, hyperfine structure and magnetic field energy-level splittings can all be determined. A major advantage of these methods is that both optical photons and ions can be detected with high efficiency. In resonance experiments utilizing the Hanle effect, an external magnetic field is used to cause excited atoms to precess, which modulates the rate of detected photons [2]; analogous measurements in nuclear physics are perturbed angular correlations as found in gamma-ray spectroscopy [3].

However, astrophysics is the field in which the techniques discussed here have largely been developed and are applied. In astronomy one almost always deals with detected photons, and there is a zoo of cosmic objects that exhibit periodic behavior on different time scales. For instance isolated neutron stars may have rotational periods as short as a few milliseconds, X-ray emitting neutron stars in binary systems typically have neutron star spin periods of a few seconds and orbital periods of a few days and white dwarf stars may have vibration periods comparable to a day. To a large extent, the availability of new telescopes both on the ground and on satellites has driven work on statistical methods for detecting weak periodic signals. Prompted by the advent of the two high-energy "Great Satellite Observatories," the Compton Gamma-Ray Observatory and the Advanced X-Ray Astrophysics Facility,[1] there is presently much work in this area.

In this chapter, we concentrate on techniques that are useful and used by working scientists. The two basic questions are (1) "Is there a signal in the data?" and (2) "If so, what are its strength, phase, etc.?" We will spend most of this chapter concentrating on the first question. It is usually formulated statistically in terms of two hypotheses. The first hypothesis, $H_0$, is that the data consist simply of noise; and the second, $H_1$, is that there is a signal present along with the noise. The standard procedure is to estimate the probability that the apparent signal is really just a fluctuation arising from background noise. Let us say that this probability is 0.001. In the literature this would be phrased as something like "the chance probability for this effect is 0.001." Unfortunately these normally refer only to the probability for *rejecting* the null hypothesis ($H_0$) when it is true. If the experimental apparatus and the background were perfectly understood, this would be adequate; this is never the case. Statistical tests to show that an apparent signal has the right characteristics are rare, though one is discussed here.

The analysis techniques described here grow in complexity as the chapter proceeds. It begins in the next section with a discussion of white noise that,

---

[1]The Compton Gamma-Ray Obsetrvatory, which was launched in April 1991, has already yielded several remarkable astrophysical surpises; we hope the Advanced X-Ray Astrophysics Facility will be launched within this decade.

by assumption, constitutes the background. The concept of a light curve is introduced in the same section. In the subsequent section various statistical tests for sniffing out a signal in a large background are described and compared. The "power" of a test for finding a signal is strongly dependent on the signal light curve, and some comparisons are made on this basis. The section ends with a discussion of a self-consistency test useful in establishing the reality of a possible signal. In the first part of the chapter it is assumed that the frequency of the possible signal is known a priori; frequency searches and their effect on test statistics are described in Sec. 12.5. Finally, the issue of repeated experiments or observations is considered in Sec. 12.6.

## 12.2 White Noise and Light Curves

Consider a sequence of events at random times such as detected photons from a telescope pointed at blank sky. We assume that whatever stars happen to be within the field of view of the telescope have approximately constant brightness. If one observes for a time, $T$, and the average rate of detected photons is $r$ then the probability for finding $N$ events in a particular measurement is given by the Poisson distribution

$$\text{Prob}(N \text{ events}) = \frac{(rT)^N}{N!} e^{-rT} \qquad (12.1)$$

Suppose that these $N$ photons were emitted at times at $t_1 \ldots t_N$. At any period, $p$, or corresponding frequency, $\nu = 1/p$, the times are mapped into phases, $\theta_1 \ldots \theta_N$, through

$$\theta_i = [2\pi\nu t_i] \bmod 2\pi \qquad (12.2)$$

The phases are uniformly distributed over the interval $(0, 2\pi)$ provided that $T \gg p$. A rigorous statement of this is Poincare's theorem, as described in Mardia's book [4]. The probability density of phases for "white noise" is simply $f(\theta) = 1/2\pi$ with normalization $\int_0^{2\pi} f(\theta) d\theta = 1$. A convenient pictorial representation of one measurement is shown in Fig. 1, in which each $\theta_i$ is shown as a line segment pointing in the appropriate direction.

Now let us assume that a small fraction of the light that the telescope sees is modulated at period $p$. An concrete example is light from the Crab pulsar, which is modulated at about 30 msec, the neutron star rotational period. Assume that this object is viewed with a telescope for time $T$, which might be a few minutes. The probability density $f_s(\theta)$ for the phases of photons from the pulsar might look something like that shown in Fig. 2(a). This is called the *light curve* of the signal, i.e., it is the basic pulse shape

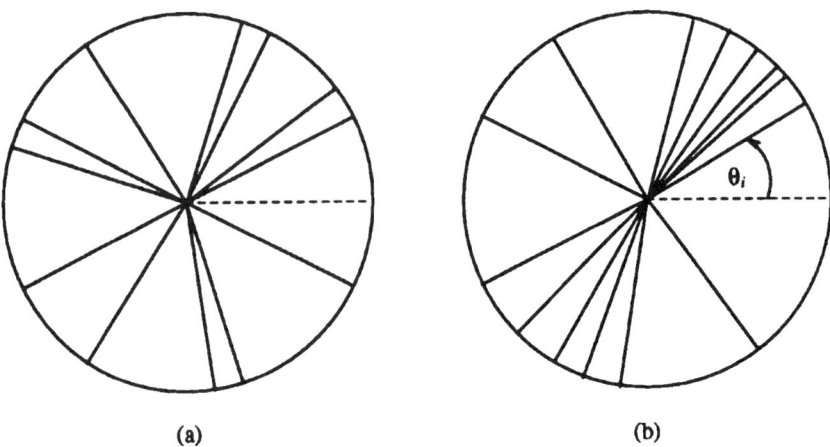

FIG. 1. Part (a) shows a distribution of phases that appear to be consistent with a uniform probability density, and part (b) shows a distribution for which the phases seem to fall preferentially near $\theta = \pi/4$ and $\theta = 5\pi/4$.

of the modulating signal. We will always use the term *light curve* to refer to the *signal*. In this example it would appear that the light might come from two hot spots of unequal temperature with one on each side of the neutron star. In a realistic measurement in X-ray or gamma-ray astronomy, the recorded data is dominated by background, which is ideally uniform at the period of interest. In this case the $f(\theta)$ for measured data would be much flatter. If $p_s$ is the long-run fraction of "signal" events, i.e., photons from pulsar, then for the measured data, $f(\theta)$ is given by

$$f(\theta) = p_s f_s(\theta) + \frac{(1 - p_s)}{2\pi} \qquad (12.3)$$

This is illustrated in Fig. 2b. Note that $f(\theta)$ is proportional to the average rate of events occurring at phase $\theta$.

The next question to be addressed is how one can recognize a weak signal in data and how to quantify the probability that it is not just an unfortunate statistical fluctuation.

## 12.3 Tests for Uniformity of Phase

How does one determine whether or not a set of phases, $\theta_1 \ldots \theta_N$, is uniformly distributed on the interval $(0, 2\pi)$? The book by Mardia [4]

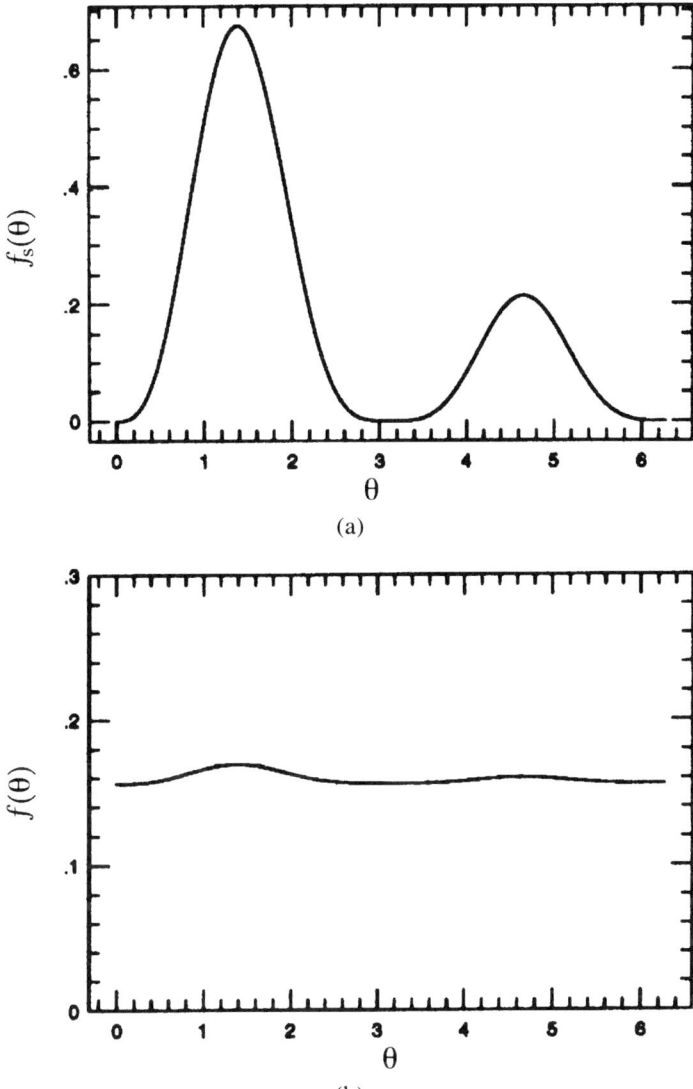

FIG. 2. Part (a) shows a possible light curve (signal probability density function) for X-rays from a neutron star with two hot spots with different intensity separated by $\pi$ rad about the axis of rotation of the star. Part (b) shows the probability density that one might measure if the signal were a small fraction of the total number of events recorded.

describes the classical tests, but newer ones have been developed that are often better suited to situations encountered in experimental physics and observational astronomy. Some of these are described here.[2] It should be emphasized that a nonuniform distribution of phases does not *necessarily* mean a signal, e.g., instrumental effects might also distort a flat (background) light curve. A consistency test is described in Sec. 12.4.

### 12.3.1 Epoch folding

The most direct approach is simply to bin the phases[3] and determine if the resulting plot is consistent with a flat probability density. If not, the null hypothesis can be rejected.[4] A sample phase histogram is shown in Fig. 3. Assuming only background, the number of events in a the $j$th bin, $N_j$, should be Poisson distributed with a mean that is the same for each bin. An appropriate test statistic for uniformity of phase is

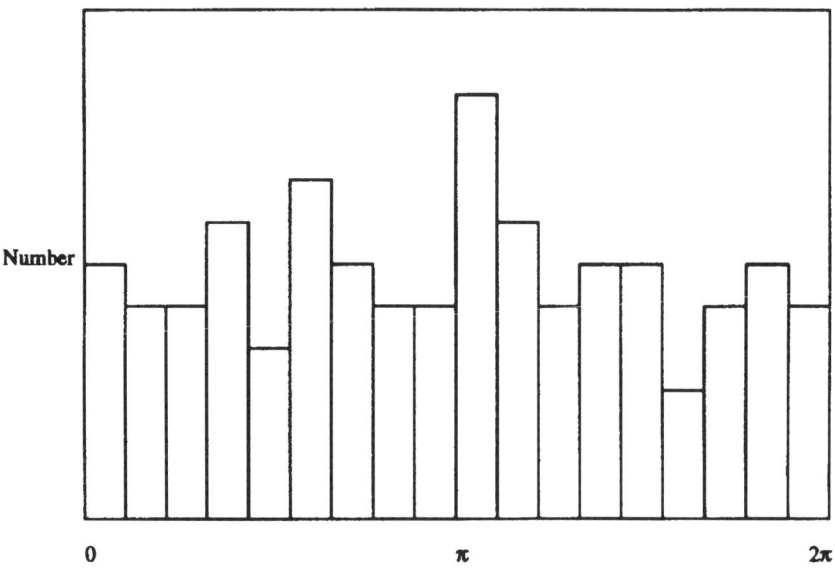

FIG. 3. A set of binned event phases.

[2]There are also newer goodness of fit tests described in Chapter 7 that might prove useful.
[3]This approach is analogous to "boxcar integration" in electrical engineering.
[4]In astronomy this is called *epoch folding* in that the times are folded modulo the suspected period and the resulting phases are binned.

$$S = \sum_{j=1}^{n} \frac{(N_j - N/n)^2}{N/n} \qquad (12.4)$$

where $n$ is the number of bins and $N = \Sigma\, N_j$. The statistic $S$ is a measure of the goodness of fit of the measured data $N_j$ by a flat distribution; anomalously large values of $S$ indicate a bad fit and hence a possible signal. In the limit where $N_j$ is large, the Poisson distribution for $N_j$ becomes approximately Gaussian.[5] In this case $S$ is approximately distributed as $\chi^2_{n-1}$ for a flat probability density. A thorough and well-written discussion of this test is an astrophysical context is given in the paper by Leahy et al. [5].

Epoch folding is relatively straightforward, the binned phase distribution has a clear interpretation providing an estimate of the light curve and the Pearson $\chi^2$ test is easy to apply. In the event that the shape of the light curve is known beforehand, the maximum likelihood method (Chapter 8) can be used to find the "best" fit for a measured set of binned phases by treating the long-run fraction of signal events $(p_s)$ as a parameter that maximizes the likelihood function.

Epoch folding has significant disadvantages in that both the bin width and the placement of the bin edges are up to the person doing the analysis. If there is some evidence of a peak in the phase distribution, there is a strong psychological tendency to choose the bin width and edge positions to maximize the number of counts in the bin near the peak. Thus the apparent significance of a possible signal in a data set depends on the details of the analysis. For instance, there were several detections of the X-ray emitting star, Cygnus X-3, at photon energies greater than $10^{12}$eV that were based on epoch folding. These have been vigorously criticized by Chardin and Gerbier [6] in that relatively insignificant effects in data can appear quite significant if the bin width and edge locations are chosen to optimize evidence for a putative effect.

### 12.3.2 Rayleigh Test

The Rayleigh test [7] is free of the binning uncertainties associated with epoch folding. Furthermore, it is simple to apply and computationally fast.[6] It is also a good conceptual starting point for introducing the $Z^2_m$-test and $H$-test described later. The most complete reference is Mardia's book, and the test was introduced in a astronomical context by Gibson et al. [9]. The

---

[5]See Chapter 2. In practice it is usually sufficient to have the mean $N_j$ greater than about 10.
[6]It is somewhat analogous to lock-in detection [8] in electrical engineering.

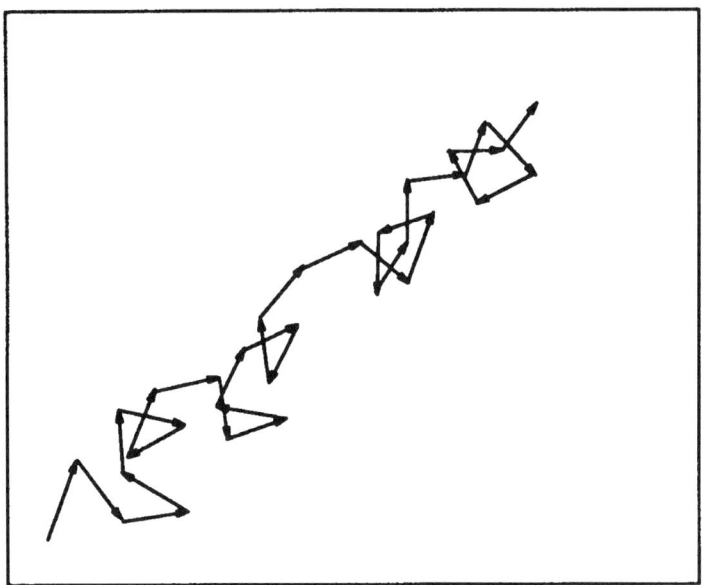

FIG. 4. The sum of 30 phasors.

concept is illustrated in Fig. 4, where phasors (unit two-dimensional vectors) representing the $\theta_i$ are summed vectorially.

If there is a preferred phase, the sum of the phasors will tend to point in the corresponding direction, and if the phase probability density is flat, the displacements are a uniform two-dimensional random walk. Thus for the no-signal (uniform phase) case this is essentially a two-dimensional diffusion problem with fixed step length. The Cartesian components of the sum are given by

$$c = \sum_{i=1}^{N} \cos(\theta_i) \qquad s = \sum_{i=1}^{N} \sin(\theta_i) \qquad (12.5)$$

and $R$, the net displacement from the origin, is given by

$$R = \sqrt{c^2 + s^2} \qquad (12.6)$$

The means and variances of $c$ and $s$ depend on the Fourier transform sine and cosine amplitudes of $f(\theta)$:

$$a_k = \int_0^{2\pi} \cos(k\theta)f(\theta)d\theta, \qquad b_k = \int_0^{2\pi} \sin(k\theta)f(\theta)d\theta \qquad (12.7)$$

These are derived in Mardia [4] and given by

$$E(c) = N a_1 \qquad (12.8)$$

$$E(s) = N b_1 \qquad (12.9)$$

$$\text{Var}(c) = (N/2)(1 + a_2 - 2a_1^2) \qquad (12.10)$$

$$\text{Var}(s) = (N/2)(1 - a_2 - 2a_1^2) \qquad (12.11)$$

$$\text{Cov}(c, s) = (N/2)(b_2 - 2a_1 b_1) \qquad (12.12)$$

For a uniform phase distribution, ($f(\theta) = 1/2\pi$), the Fourier coefficients with $k > 0$ vanish, so we have

$$E(c) = E(s) = 0 \qquad (12.13)$$

$$\text{Var}(c) = \text{Var}(s) = (N/2) \qquad (12.14)$$

$$\text{Cov}(c, s) = 0 \qquad (12.15)$$

Suppose that a particular set of measurements gives a point on the $(c, s)$ plane that is far from the origin, i.e., $R = \sqrt{c^2 + s^2}$ appears to be anomalously large for a particular measurement. This might indicate a signal; how can it be quantified? If there is no signal (i.e., a uniform phase distribution is assumed) then $c$ and $s$ are sums of independent random variables with means and variances given here. We can appeal to the central limit theorem to conclude that $c$ and $s$ are approximately normally distributed if $N$ is large. The Rayleigh power is $P = R^2/N$. Hence twice the Rayleigh power, $2P$,

$$2P = \frac{c^2}{N/2} + \frac{s^2}{N/2} \qquad (12.16)$$

is the sum of the squares of two independent standard normal random variables with unit standard deviation if a uniform phase distribution is assumed. The random variable $2P$ is therefore approximately distributed as $\chi_2^2$ since there are two terms in the sum. It turns out that this particular $\chi^2$ distribution is an exponential and that the probability that $P$ exceeds a threshold $P_0$ is then[7]

---

[7]This also follows from integrating the two-dimensional Gaussian distributions of $(c, s)$ points from $R_0$ out to infinity.

$$\text{Prob}(P > P_0) = e^{-P_0} \tag{12.17}$$

The Rayleigh power, $P$, is the test statistic used to assess the significance of any putative signal.

How does a signal change the distribution of Rayleigh powers? We can see this by writing $f(\theta)$ as the sum of signal and background portions as in Eq. (12.3). It follows that

$$a = a_1 = p_s \int_0^{2\pi} f_s(\theta) \cos(\theta) d\theta = p_s a_s \tag{12.18}$$

$$b = b_1 = p_s \int_0^{2\pi} f_s(\theta) \sin(\theta) d\theta = p_s b_s \tag{12.19}$$

where $a_s$ and $b_s$ are the Fourier amplitudes of the normalized light curve, $f_s(\theta)$. The overall magnitude is given by

$$g_s = \sqrt{a_s^2 + b_s^2} \tag{12.20}$$

In the large background limit where $p_s$ is small, it follows that

$$E(P) \cong 1 + \mu \tag{12.21}$$

$$\sigma(P) \cong \sqrt{2\mu + 1} = \sqrt{2E(P) - 1} \tag{12.22}$$

where

$$\mu = N(p_s g_s)^2 \tag{12.23}$$

Thus, both the mean Rayleigh power and its standard deviation are increased by the presence of a signal. The increase in average Rayleigh power ($\mu$) is proportional to the square of the long-run proportion of signal events, $p_s$, and to the square of the overall Fourier amplitude, $g_s$, of the light curve ($f_s(\theta)$). It follows from its definition that $g_s$ is 1, $2/\pi$ and $1/2$, for Dirac delta function, square-wave and sinusoidal $f_s(\theta)$ shapes, respectively. It also follows that this test is ineffective for finding signals with light curves having small first-harmonic Fourier amplitudes. This is not true of epoch folding, nor of the other tests described later.

### 12.3.3 Protheroe Test

If the light curve has a symmetric, double-peaked structure the first harmonic Fourier amplitude, $g_s$, may be quite small. For instance if the two peaks in Fig. 2(a) were of equal height, $g_s$ would be 0 and the Rayleigh test would indicate no signal. The Protheroe test [10, 11] does not have this weakness and it is also particularly appropriate for light curves with sharp structure (narrow peaks) and is free of the binning uncertainties inherent in epoch folding. Its disadvantage is that it is computationally very intensive and therefore appropriate only for data with very few events (usually less than 200). It is used most extensively in ultra-high-energy gamma-ray astronomy.

The basic idea goes as follows. For a uniform distribution of phases the phasors are spread roughly equally over a unit circle as illustrated in Fig. 1. The Rayleigh test picks out the tendency for phasors to point in the same direction. The Protheroe test tends to pick out phases that are correlated pairwise with other phases regardless of where they lie on the circle, i.e., clumpy phase distributions. The deviation, $\Delta_{i,j}$, between any two phases is defined as

$$\Delta_{i,j} = \frac{1}{2} - \left| \frac{1}{2\pi} |\theta_i - \theta_j| - \frac{1}{2} \right| \qquad (12.24)$$

and the Protheroe statistic is

$$Y_N = \frac{2}{N(N-1)} \sum_{i=1}^{N-1} \sum_{j=i+1}^{N} \frac{1}{\Delta_{i,j} + \frac{1}{N}} \qquad (12.25)$$

A large value of $Y_N$ indicates a signal.

The time required to compute $Y_N$ is thus proportional to the square of the number of data points. The distribution of $Y_N$ for a uniform phase distribution must in general be found from Monte Carlos simulations.

### 12.3.4 $Z_m^2$-Test and H-Test

Since the average Rayleigh power depends only on $a_1^2 + b_1^2$ the Rayleigh test is sensitive only to the first harmonic of the light curve. Note, however, that a sharp peak in the light curve would produce large Fourier amplitudes in many harmonics. A generalization of the Rayleigh test with greater sensitivity for light curves with narrow structures and multiple peaks is the $Z_m^2$-test based on the statistic

$$Z_m^2 = 2 \sum_{k=1}^{m} P(kv) \qquad (12.26)$$

which is twice the sum of Rayleigh powers for the first $m$ harmonics. (Remember that $v$ is the assumed frequency of the signal.) A Dirac delta-function light curve has equal Fourier power in all harmonics and a pure sinusoidal light curve has all of its power in a single $k$. Thus $m = 1$ is appropriate for a broad single-peak light curve, $m = 2$ is appropriate for a broad light curve with two peaks and large $m$ is appropriate for light curves with one or more very narrow peaks.

The distribution of the $Z_m^2$ statistic can be found from a simple extension of the argument for the Rayleigh statistics. Since, for a large number of events drawn from a uniform phase population, $2P$ is distributed approximately as $\chi_2^2$, it follows that $Z_m^2$ is distributed approximately as $\chi^2$ with $2m$ degrees of freedom. In this case $E(Z_m^2) = 2m$. Unfortunately, in practice the light curve is often not known a priori, and it is typically not clear which value of $m$ should be chosen. This is the same problem encountered with binning for epoch folding: choosing $m$ a posteriori invalidates the calculation of statistical significance. An experimenter who has worked hard to find a signal and who sees a marginal hint of one in the data would naturally choose the $m$ that makes the signal seem most real.

De Jager, Swanepoel and Raubenheimer [12] propose a way around this difficulty: they effectively include the effect of a zealous observer in the calculation of significance that must be assessed through Monte Carlo simulations. Specifically, they recommend using

$$H = \text{Max}(Z_m^2 - 4m + 4) \tag{12.27}$$

as the test statistic.[8] In effect, the value of $m$ is automatically adapted to give a large value of the test statistic for any light curve implicit in data. Through massive simulations they have shown under the hypothesis of uniformity of phase that the distribution of $H$ has the approximate form

$$\text{Prob}(H > h) = e^{-0.3982h} \tag{12.28}$$

for $N > 100$ events and ($0 < h < 23$). Approximate distributions for other cases are given in the De Jager et al. paper.

### 12.3.5 Comments on Power of the Statistical Tests

An experimenter would obviously like to choose a test that is likely to find a genuine signal in the data, if it is there. This is normally quantified as the "power" of the test. The Rayleigh test is quite powerful for sinusoidal single-peaked light curves and the $Z_m^2$ test with $m \sim 10$ is quite powerful for very narrow light curves.

Before the power of a test can be quantified, it is important to understand that it depends on the significance level selected for the test. For instance, one might require that the probability of wrongly deciding that there is a signal when there is none (i.e., deciding that the data set is inconsistent with a white noise null hypothesis) is 0.001. One can then ask, "What is the probability that a genuine signal of fixed strength and light curve will

---

[8]Only the first 20 harmonics ($1 \leq m \leq 20$) are normally considered in practice.

be identified as such using a statistical test?" This probability is the power of the test.

The general problem of the power of tests for uniformity of phase is addressed in the paper by Beran [13] and also described in Mardia [4]. A very useful recent Monte Carlo study of the power of the tests described in this section is given in De Jager et al. [12]. Simulated data sets containing a small number of "signal" events along with "background" events from a uniform distribution were subjected to a set of tests. The signal detection threshold was taken to be 5%, i.e., results with values of the test statistic more extreme than 95% of values produced under white noise were taken as indicating a signal present in the data set. Three types of light curves were considered: (*i*) single-peak Gaussian light curve having different peak widths with the signal strength fixed at 10% of the total number of events, (*ii*) double peaked light curves with variable relative intensities and (*iii*) evenly spaced multiple peaks with constant relative intensities. The principle conclusions from this work are that (1) the Protheroe test is very effective for narrow light curves and few events, (2) the Rayleigh test is well suited for wide or sinusoidal light curves and (3) if nothing is known about the light curve, the $H$-test is the best choice because it works well with essentially all light curve shapes. Quantitative values for test power are given in [12].

## 12.4 dc Excess vs. Periodic Strength

In many measurements it is possible to obtain an estimate of the number of signal events that does not depend on the periodicity. For example, in a resonance experiment it may be possible to repeat the measurement with no sample present in order to estimate the number of background events expected in actual measurements. In astronomical observations it may be possible to use the telescope to measure events from a comparison "background" region of the sky. In these cases the number of signal events can be estimated from the difference in number of events from the signal region and from the background region in a fixed time period. We call this difference the *dc excess* in that it is independent of the periodic "ac" character of the signal. The dc excess can be used to (*i*) improve the test for signal existence, (*ii*) check for consistency between the size of the dc excess and the periodic strength of the suspected signal and (*iii*) improve the estimate of the size of the signal.

The second point, consistency between the dc excess and periodic strength, is particularly important with signals that cannot be easily reproduced. Emission from many astronomical objects is sporadic, i.e., there are episodes of emission that do not repeat on a regular time scale or perhaps not at all. Examples include X-ray emission from massive accreting binary star systems that consist of a neutron star with a more normal companion and γ-ray bursters which emit intense γ-ray radiation for a short time and then disappear. In addition, many observatories and observatory instruments are available only for strictly limited times, thus effectively making the observations unique. Examples are satellites that have a limited lifetime and crowded observing schedules. If an episode of apparently sporadic emission has been seen that may or may not be confirmed in subsequent observations, it is essential that the reality of the signal be checked in every possible way; the comparison of dc excess with periodic strength can therefore be an important test.

The three issues, signal existence, consistency and signal size estimate, can be addressed in terms of three statistics, $\alpha$, $\beta$ and $\gamma$ [14]. As before, let $N$ be the total number of events in a measurement and $n_{dc}$ the dc excess. The $n_{dc}$ is usually normally distributed if $N$ is sufficiently large.[9] Since twice the Rayleigh power, $2P$, is distributed as $\chi_2^2$ for a uniform phase distribution and, assuming that $n_{dc}$ is normally distributed about 0 and independent of $P$ in the no-signal case, the statistic

$$\alpha = 2P + n_{dc}^2/\text{Var}(n_{dc}^2) \qquad (12.29)$$

is then distributed as $\chi_3^2$. Large values of $\alpha$ indicate a possible signal; the significance can be assessed simply through the $\chi_3^2$ distribution.

If a data set has a large value of $\alpha$ indicating a possible signal, the second and third points (consistency and signal estimate) can be addressed as follows. Using the form for $f(\theta)$ given in Eq. (12.3) and the assumptions of Section 12.3.2 (in particular that the signal proportion is small: $p_s \ll 1$), it follows that [14]

$$E(c) = Np_s a_s = Np_s g_s \cos(\phi_s) \qquad (12.30)$$

$$E(s) = Np_s b_s = Np_s g_s \sin(\phi_s) \qquad (12.31)$$

$$\text{Var}(c) = \text{Var}(s) = (N/2) \qquad (12.32)$$

$$\text{Cov}(c, s) = 0 \qquad (12.33)$$

where $\phi_s$ is the phase of the signal relative to the recording clock. Thus $c$ and $s$ are asymptotically normally and independently distributed with variance equal to the number of events divided by 2. If one thinks of a $(c, s)$

---

[9]Methods for obtaining $n_{dc}$ and its variance in cases where observation time for the signal region differs from that for the background region are given in the paper by Li and Ma [15].

plane, the density of $c$ and $s$ values in particular measurements is a two-dimensional Gaussian with a width determined by the number of events, $N$, and a centroid at $(Np_sg_s \cos(\phi_s), (Np_sg_s \sin(\phi_s))$ determined by the signal.

The values of $c$, $s$ and $n_{dc}$ are known for a particular measurement whereas $p_s$ and $\phi_s$ are desired. These can be estimated using the maximum likelihood method (Chapter 8) as follows. If it is assumed that the long-run proportion of signal events is $p$ and that the correct relative phase is $\phi$, then the probability of obtaining the measured values of $c$, $s$ and $n_{dc}$ is proportional to

$$\exp\left[-\frac{(c - pNg_s \cos(\phi))^2}{N}\right]$$
$$\times \exp\left[-\frac{(s - pNg_s \sin(\phi))^2}{N}\right] \exp\left[-\frac{(n_{dc} - pN)^2}{2\mathrm{Var}(n_{dc})}\right] \quad (12.34)$$

The sum of the exponents in this expression would be distributed as $\chi_3^2$ if $p$ and $\phi$ were known a priori. If they are chosen to maximize the likelihood function (i.e., minimize the exponent) then the exponent is distributed approximately as $\chi_1^2$. If the value of $Np$ that maximizes the likelihood function is called $\beta$ then

$$\beta = (Np)_{opt} = \frac{\mathrm{Var}(n_{dc})\sqrt{NP}/g_s + (N/(2g_s^2))n_{dc}}{\mathrm{Var}(n_{dc}) + N/(2g_s^2)} \quad (12.35)$$

$$\gamma = \frac{(n_{dc} - \sqrt{NP}/g_s)^2}{\mathrm{Var}(n_{dc}) + N/(2g_s^2)} \quad (12.36)$$

where $\gamma$ is the minimum exponent which is distributed as $\chi_1^2$. Thus $\beta$ and $\gamma$ are convenient measures of the signal strength and consistency. A large value of $\gamma$ indicates that the measurements are inconsistent with *any* combination of signal and uniform phase noise. This might occur, for instance, if $n_{dc}$ is less than 0 and the Rayleigh power is large. The obvious conclusion in this case is that something is wrong with the data!

Values of $N$, $n_{dc}$ and $P$ come directly from the data; a value for $g_s$ does not. Hence $\beta$ and $\gamma$ cannot be calculated directly unless the shape of the light curve (or at least its Fourier amplitude) is known. If the light curve is not known beforehand it is still possible to estimate $g_s$ by folding and binning the events as described in the section on epoch folding (Sec. 12.3.1) and performing an numerical Fourier transform. In any case, the maximum value $g_s$ can have is 1, which corresponds to a Dirac delta function light curve.

## 12.5 Frequency Searches

Often the frequency of a possible signal is not known. For instance if an an X-ray satellite discovers a dc excess from a point source, the object is likely to be binary star system consisting of a rapidly rotating neutron star and a more normal companion. In this case the X-ray arrival rate would be modulated by the neutron star rotational frequency that could be found from timing analysis of the observations. Other examples can be found in resonance experiments alluded to in the introduction to this chapter. For instance the characteristic frequencies of the radiating atom or molecule might not be known exactly due to nuclear shifts of atomic levels [2]. Indeed, finding the shifted frequencies is often the point of the measurements.

The most direct route in searching for a periodic signal with unknown frequency is to simply compute one of the test statistics described in the last section at assumed frequencies, spanning a range determined by the physics of the system. An alternative approach is to use the fast Fourier transform technique to find the Fourier power at essentially all relevant frequencies and calculate test statistics (e.g., De Jager's et al. H) from the Fourier powers. General algorithms and actual code for computing fast Fourier transforms are given and described in *Numerical Recipes in C* [16]. A good discussion of the use of FFTs in an astrophysical context is given in the paper by Leahy et al. [5], and we will not discuss FFTs further here. If the physically reasonable frequency search range is small, it is more efficient to directly compute the test statistic over selected frequencies within that range.

There are two questions that immediately arise in applying a signal test over a range of frequencies: (1) how fine a grid of frequencies should be used in searching for possible signals and (2) if a possible signal is found, how does one assess its significance? The second question can be restated as follows. As one searches on frequency, there are additional trials or opportunities for the test statistic to assume improbable values by chance. How does one accommodate these additional trials?

### 12.5.1 Independent Fourier Frequencies

If the test statistic is to be computed over a range of frequencies, how closely spaced should they be? An example Rayleigh power spectrum (plot of $P$ vs. $v$) for 0.99 Hz to 1.0 Hz computed for 5000 background events distributed over a $T = 1800$ sec interval is shown in Fig 5. At a *particular* frequency, the Rayleigh power will be exponentially distributed with a mean of 1. The question addressed here is how it changes as the frequency changes. It is obvious from Fig. 5 that the power at two very closely spaced frequencies

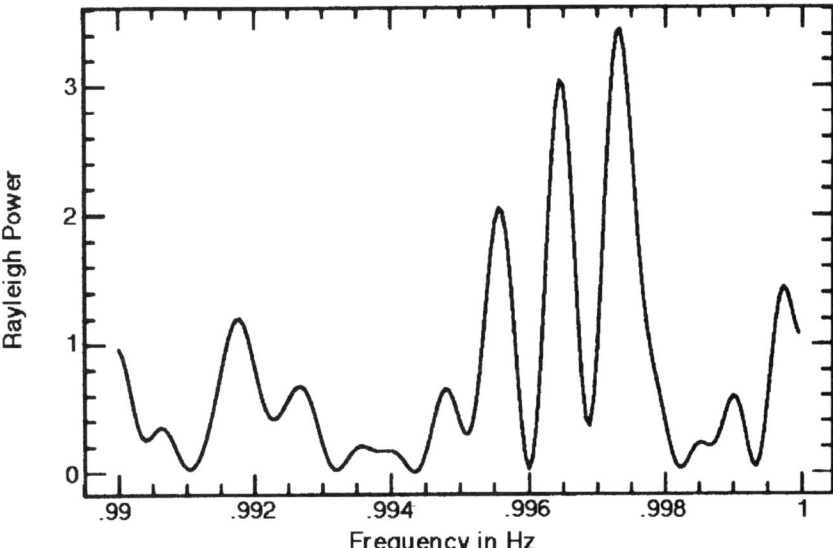

FIG. 5. The Rayleigh power spectrum over the frequency interval 0.99 to 1.0 Hz for 5000 background events spread over $T = 30$ min.

are highly correlated. Closer inspection shows that the characteristic peak width in the figure is about 1/1800 Hz, i.e., the inverse of the length of the data set. One would expect that frequencies far apart would be uncorrelated.

Now consider the following Fourier representation of a series of random noise pulses used by Einstein [17] to represent light from independent oscillators emerging from a black body cavity. It is assumed that the pulses come at random times uniformly distributed over a time interval of length $T$.[10]

$$I(t) = \frac{c_0}{2} + \sum_{k>0} [c_k \cos(2\pi k \Delta v\, t) + s_k \sin(2\pi k \Delta v\, t]  \quad (12.37)$$

If the frequency spacing is given by[11]

$$\Delta v = 1/T \quad (12.38)$$

then, for a sufficiently large number of events, the Fourier coefficients $c_k$ and $s_k$ are uncorrelated, i.e.,

---

[10] One can think of $I(t)$ as the electric current in a detector due to the emitted photons.
[11] The FFT algorithm automatically gives Fourier amplitudes at exactly these frequencies.

$$E(c_k c_l) = E(s_k s_l) = E(c_k s_k) = 0 \qquad (12.39)$$

where $k$ and $l$ are different indices. A proof is outlined in the excellent article by Rice [18].

The reason this works is that, at the independent frequencies, the times are exactly mapped on onto $(0, 2\pi k)$ via $\theta_i = 2\pi k \Delta \nu t_i$. Thus, for times spanning the interval $(0, T)$, the corresponding set of *relative* phases between the $k$'th and $k + 1$'th frequencies exactly span the interval $(0, 2\pi)$. This makes terms like $c_k c_l$ phase average to zero for a large number of times. If $\Delta \nu$ were chosen to be slightly larger or smaller than $1/T$, this would not be true. The Fourier amplitudes $c_k$ and $s_k$ are also normally distributed about a mean of zero; the proof of this is somewhat involved and outlined in Rice's extensive article [18].

There is a direct connection with the Rayleigh test in that, at frequency $\nu = k \Delta \nu$, the Fourier amplitude $c_k$ is proportional to $c(\nu) = \sum_{i=1}^{N} \cos(2\pi \nu t_i)$, where $c(\nu)$ is the Rayleigh cosine amplitude encountered before. The constant of proportionality depends on how the noise current $I(t)$ is defined.[12]

How can the independent Fourier frequencies (*IFF*) be used in assessing the statistical significance of a possible signal? We will do this by example. Suppose that an experimenter has a list of arrival times of X-rays from a suspected X-ray binary pulsar. She chooses a frequency range containing 100 *IFF* and computes the Rayleigh power at these 100 frequencies finding a maximum Rayleigh power of $P_0 = 10$. Does this power indicate a signal? Since the 100 Rayleigh powers are spaced by $\Delta \nu$, they constitute $N_t = 100$ independent trials. The probability of finding *no* powers greater than $P_0$ in all $N_t$ trials computed supposing that there is no real signal is $[1 - \exp(-P_0)]^{N_t} = 0.995$. Thus the probability of finding one or more powers greater than $P_0$ by chance from a uniform distribution is thus

$$1 - [1 - \exp(-P_0)]^{N_t} \sim N_t \exp(-P_0) = 0.005 \qquad (12.40)$$

indicating a possible signal.

### 12.5.2 Oversampling

There is a danger in examining the spectrum of a test statistic (e.g., Rayleigh power) only at the *IFF:* the signal peak may fall approximately between two of these and consequently be missed. Consequently one "oversamples" in frequency space, i.e., computes the test statistic at finer

---

[12] If $I(t)$ is written formally in the dimensionless fashion $I(t) = T \sum_{i=1}^{N} \delta(t - t_i)$ then the proportionality constant is unity.

intervals than the *IFF* spacing. Since the peak width is about one *IFF* spacing, it is sufficient to oversample by a factor of 5 or so to find all strong peaks.

How does one accommodate oversampling when accessing the significance of a possible signal? The first studies of this were made using Monte Carlo simulations [19, 12], and there is also an analytical approach that has been used for the Rayleigh test [20]. The basic result from these studies is that oversampling can be included by multiplying the pretrial probability by a factor that is approximately 2–4 for the Rayleigh test and about an order of magnitude for the Protheroe test. More details are given in De Jager et al. [12].

Let us reconsider the last example. When the X-ray astronomer oversamples her data she finds the maximum Rayleigh power is 13 instead of 10 because the nearest *IFF* was in fact on the shoulder of a large peak. Does this indicate a signal? Assuming an "oversampling factor" of 3, the pretrial probability of finding this power is $3e^{-13}$. There are again 100 *IFF*s within her a priori search range and the overall chance probability that a uniform distribution will produce a maximum Rayleigh power of 13 or greater is

$$1 - [1 - 3\exp(-P_0)]^{N_t} \sim N_t \, 3 \exp(-13) = 0.0007 \quad (12.41)$$

In this case, there is clear advantage in oversampling.

## 12.6 Multiple Data Sets

So far, we have considered the analysis of one data set consisting of a single list of events, i.e., one observation. In many instances, it is possible to repeat measurements; this is standard procedure in some long-term projects. In fact, there are possible gamma-ray sources (X-ray binary systems) that have been routinely monitored for years! In ground-based astronomy the measurements are offset in time by one or more days due to the Earth's rotation; in physics the offsets often have to do with the number of hours required for an experimenter's sleep. These "gaps" in the data stream make the analysis more complex. In this section we consider the analysis of many "segments" of data separated by "gaps" during which no data were taken.

### 12.6.1 Coherence

A fundamental question is whether or not the signal is coherent. By *coherent* we mean that the signal phase is either fixed or changes smoothly in a well-defined way.[13] This leads to the idea of a coherence time. The light from a sodium lamp is coherent for a few nanoseconds (~ the lifetime of the atomic transition) whereas light from a narrow-band laser may have coherence times that are orders of magnitude longer. Radio pulsars are isolated rotating neutron stars that may emit coherently for years and then abruptly undergo a "starquake" changing frequency and phase. If the coherence time is long compared with the total time span of all measurements, the signal should clearly be considered coherent.

We have tacitly assumed coherence in the analysis of single segments of data discussed so far. It may or may not be reasonable to assume coherence for longer times, e.g., for all data segments. Gamma rays or X-rays from the Crab pulsar are quite coherent; in fact the Crab pulsar gamma-ray light curve is actually used to correct errors in the clock aboard the Compton Gamma-Ray Observatory!

If the signal is coherent with known frequency and, even better, if the shape of the light curve is known as well, then epoch folding has many advantages. The entire set of times for all of the measurements can be converted into phases and binned. The maximum-likelihood method (see Chapter 8) can then be used to fit the folded data with a linear combination of background and signal events. In this case the long-run proportion, $p_s$, is a free parameter usually adjusted to fit the data (see Chapter 8). This generally results in a $\chi_n^2$ distributed statistic and the significance of the result is easily assessed. The disadvantages of epoch folding, e.g., uncertainties in binning the data were discussed earlier.

However other signals, e.g., X-rays from binary star systems that are quasi-periodic oscillators (QPOs) have coherence times less than days. In addition, even when it is suspected that the source may emit coherent radiation, if the frequency or its derivatives are unknown then it is sometimes better not to assume that one segment is coherent with another. The observing segments are usually much shorter than the ~24 hr gaps in the data and searching on frequency or its derivatives can make a weak incoherent signal (or noise) masquerade as a strong coherent signal. The basic problem is that frequency searches automatically tend to align the phases of widely separated segments whether or not they are actually coherent. The difficulties in treating segmented data as coherent is outlined in [21], where a test for true coherence is proposed. We will not pursue this further here.

---

[13]The phase–time relationship may be more complex yet change smoothly in a well-defined way, e.g., $\theta_i = 2\pi(\nu t_i + \frac{1}{2}\dot{\nu}t_i^2 + \ldots)$. This can introduce additional unknown parameters into the analysis.

### 12.6.2 Using All the Data

With large data sets consisting of many segments, there are powerful methods for finding weak signals buried in the noise. Following the discussion of the last subsection, we assume that possible signals are coherent within a single segment but not necessarily from one segment to the next.

To begin, let us assume that there are many measurements of a steady but weak signal; the list of times for each measurement forms a segment. In order to make the discussion concrete, we assume that Rayleigh power (see Sec 12.3.2) is used as the test statistic, and the signature of the weak signal will be a distortion of the distribution of Rayleigh powers from that corresponding to normal "background."

Let us be more specific about background. Some of the segments may have been taken with no sample in the apparatus in a "resonance" measurement; these are then "background" segments. Similarly, in astrophysics the telescope may have been trained on a comparison region of the sky with similar brightness characteristics to the region containing the suspected source to produce background or "off-source" observations. The basic tests are comparisons of the *full* distribution of Rayleigh powers for segments in the signal region and the background region.

The distribution of "background" Rayleigh powers is simply $e^{-P}$. How does this change with a weak signal? Note that $c$ and $s$ are normally distributed about 0 with variances given by Eqs. (12.13)–(12.15). If we view the distribution in the $(c,s)$ plane, then the probability that a threshold $P_0$ is exceeded is the same as the probability that $R = \sqrt{c^2 + s^2}$ exceeds $R_0 = \sqrt{NP_0}$. By integrating from $R_0$ to infinity in the $(c, s)$ plane we find

$$\text{Prob}(P > P_0) = \int_{P_0}^{\infty} I_0(2\sqrt{P\mu}) \, e^{-(P+\mu)} dP \quad (12.42)$$

where $\mu = N(p_s g_s)^2$ as before. By expanding the modified Bessel function the following rapidly converging series is found.

$$\text{Prob}(P > P_0) = e^{-(P_0+\mu)} \sum_{k=0}^{\infty} A_k \quad (12.43)$$

The coefficients can be found from recursion.

$$A_0 = 1, \quad A_k = \frac{\mu}{k} A_{k-1} + \left(\frac{1}{k!}\right)^2 (\mu P_0)^k \quad (12.44)$$

The resulting single parameter family of curves is shown in Fig. 6. The signal causes the distribution of Rayleigh powers to rise above the white

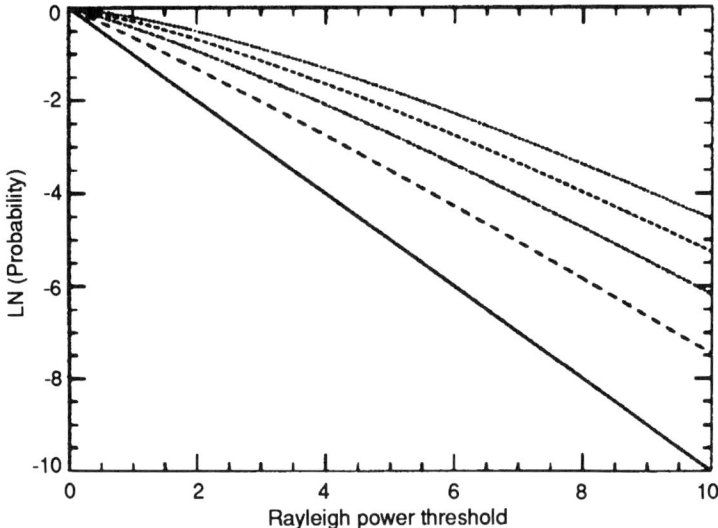

FIG. 6. The Rayleigh power distributions for several values of the signal strength $\mu$. As $\mu$ increases, the distribution of Rayleigh powers is distorted and rises above the white noise limit ($\mu = 0$).

noise distribution of $e^{-P}$ at all powers. A sensitive test would utilize the full distribution of powers because the entire distribution is distorted by the presence of a signal.

There are several standard methods for testing whether two distributions differ. Among the most widely used by physicists and astronomers are the Kolmogorov test and Smirnov–Cramers–Von Mises test [22] (see also Chapter 7). These are based on the maximum difference between the two distributions and the integrated mean square of the difference at each point in the distributions, respectively. A different and physically reasonable test is to compare the sum of the Rayleigh powers for segments with a possible signal present and background segments:

$$P_{\text{tot}} = \sum_{k=1}^{n_{\text{seg}}} P_k \qquad (12.45)$$

where $n_{\text{seg}}$ is the number of segments. For background, $2P_k$ is distributed as $\chi_2^2$; hence $2P_{\text{tot}}$ is distributed as distributed as $\chi_{2n_{\text{seg}}}^2$. The mean value and variance of $P_{\text{tot}}$ for weak signals can be found from by summing Eqs. (12.21) and (12.22).

$$E(P_{\text{tot}}) = n_{\text{seg}}(1 + \mu) \qquad \text{Var}(P_{\text{tot}}) = n_{\text{seg}}(1 + 2\mu) \qquad (12.46)$$

Note also that since $P_{tot}$ is the sum of many powers drawn from the same distribution, the central limit theorem implies that its distribution will move toward a normal distribution as $n_{seg}$ becomes large.

The test for signal existence based on the Rayleigh power sum has been compared with a variety of standard tests such as Kolmogorov and found to have superior power in this application. This holds not only for signals that are weak but steady, but also for signals that are highly sporadic and on only a fraction of the time. See [23] for further details.

There are other possibilities which should be considered. For instance, the De Jager et al. $H$ statistic is generally a more powerful test statistic for unknown light curves. Perhaps it should be used instead of Rayleigh power. A second possibility is that it may be necessary to search for the frequency of the signal. In this case, an appropriate test statistic for each segment might be the *maximum H*-value found within the search range for that segment. Another possibility is that the signal source (e.g., of gamma-rays from a sporadic X-ray binary star system) may turn on only occasionally but intensely. In this case significant signal would be present in only a small fraction of the data segments. In this case *only* the high-power end of the Rayleigh power distribution would be significantly modified, and only large Rayleigh powers should be included search strategy. It is not appropriate to go into these complications here; most have been at least briefly discussed in the literature [23–26]. However, there is a generalization of the sum of Rayleigh powers that provides a universal approach with any test statistic.

The approach goes as follows. We assume that the test statistic for each segment is $q$ and that its value is $q_i$ for segment $i$. Large values of $q$ indicate a signal. Assuming no signal, the chance probability that $q_i$ is found or exceeded for a particular segment is

$$f_i = \text{Prob}(q > q_i) \qquad (12.47)$$

It has been shown by Fisher [27] that the sum

$$F = -2 \sum_{i=1}^{n_{\text{seg}}} \log_e(f_i) \qquad (12.48)$$

is then distributed approximately as $\chi^2_{2n_{\text{seg}}}$. If $q$ is the Rayleigh power, this equation becomes Eq. (12.45) given previously. All that is required is that the distribution of $q$ is known for the no-signal, uniform phase case. It is, of course, possible to get the distribution of $q$ directly from the background segments. If these do not exist it may still be possible to obtain the distribution of $q$ from considering frequencies well outside the signal search range [26].

# References

1. Hall, J. L., and Carlsten, J. L. (eds.) (1977). "Laser Spectroscopy III." *Proc. of Third Internat. Conf. on Laser Spectroscopy.* Springer-Verlag, Berlin.
2. Corney, A. (1977). *Atomic and Laser Spectroscopy,* Chapters 15 and 16. Oxford University Press, Oxford.
3. Siegbahn, K. (1965). *Alpha-, Beta-, and Gamma-Ray Spectroscopy,* Chapter 19. North-Holland, Amsterdam.
4. Mardia, K. V. (1972). *Statistics of Directional Data.* Academic Press, London.
5. Leahy, D. A., Darbro, W., Elsner, R. F., Weisskopf, M. C., Sutherland, P. G., Kahn, S., and Grindlay, J. E. (1983). "On Searches for Pulsed Emission with Application to Four Globular Cluster X-Ray Sources: NGC 1851, 6441, 6624, and 6712." *Astrophysical Journal* **266,** 160.
6. Chardin, G., and Gerbier, G. (1987). "Cygnus X-3 at High Energies, A Critical Review." *Proc. 20th Internat. Cosmic Ray Conf.* (Moscow) **1,** 236.
7. Lord Rayleigh. (1894). *Theory of Sound,* **1,** p. 35. Macmillan London.
8. Malmstadt, H. V., Enke, C. G., and Crouch, (1974). *Electronic Measurements for Scientists.* W. A. Benjamin, Menlo Park, CA.
9. Gibson, A. I., Harrison, A. B., Kirkman, I. W., Lotts, A. P., Macrae, J. H., Orford, K. J., Turver, K. E., and Walmsley, M. (1982). "The University of Durham Gamma-Ray Facility at Dugway." *Proc. of the Internat. Workshop on Very High Gamma-Ray Astronomy,* (Ootacamund, India), P. V. Ramana Murthy and T. C. Weekes (eds.), p. 97.
10. Protheroe, R. J. (1985). "A New Statistic for the Analysis of Circular Data with Application in Ultra-High Energy Gamma-ray Astronomy." *Astronomy Express* **1,** 137.
11. Protheroe, R. J. (1985). "A New Statistic for the Analysis of Circular Data in Gamma-Ray Astronomy." *Proc. 20th Internat. Cosmic Ray Conf.* **3,** 485.
12. De Jager, O. C., Swanepoel, J. W. H., and Raubenheimer, B. C. (1989). "A Powerful Test for Weak Periodic Signals with Unknown Light Curve Shape in Sparse Data." *Astronomy and Astrophysics* **221,** 180.
13. Beran, R. J. (1969). "Asymptotic Theory of a Class of Tests for Uniformity of a Circular Distribution." *Ann. Math. Stat.,* **40,** 1196.
14. Lewis, D. A. (1989). "Detection of Weak Signals in TeV Gamma-Ray Astronomy: dc Excess vs. Periodic Amplitude." *Astronomy and Astrophysics* **219.**
15. Li, T. P., and Ma, Y. Q. (1983). "Analysis Methods for Results in Gamma-Ray Astronomy." *Astrophysical Journal* **272,** 317.
16. Press, W. H., Flannery, B. P., Teukolsky, S. A., and Vetterling, W. T. (1989). *Numerical Recipes in C,* Chapter 12. Cambridge University Press, Cambridge.
17. Einstein, A. (1915). "Ein Satz der Wahrscheinlich Keitsrechnung und Seine Anwendung auf die Strahlungstherie." *Ann. D. Physik* **47,** 879.
18. Rice, S. O. (1944, 1945). "Mathematical Analysis of Random Noise." *Bell System Technical Journal* **23,** 282 (1944), and **24,** 46 (1945). Reprinted in *Noise and Stochastic Processes,* Nelson Wax (ed.). Dover Publications, New York (1954).
19. De Jager, O. C. (1987). "The Analysis and Interpretation of VHE Gamma-Ray Measurements." Ph.D. thesis, University of Potchefstroom, South Africa.
20. Orford, K. J. (1991). "Analytical Treatment of Oversampling." *Experimental Astronomy* **1,** 305.

21. Lewis, D. A., Lamb, R. C., and Biller, S. D. (1991). "Phase Coherence for TeV/PeV Binary Sources." *Astrophysical Journal* **369,** 479.
22. Eadie, W. T., Drijard, D., James, R. E., Roos, J., and Sadoulet, B. (1971). *Statistical Methods in Experimental Physics.* North-Holland, Amsterdam.
23. Lewis, D. A. (1990). "Sensitivity and Consistency in Statistical Analysis of TeV Gamma-Ray Observations." *Nucl. Physics B,* **14A,** 299.
24. Marshak, M. L. (1990). "Periodicity Analysis of Radiation from VHE and UHE Sources." *Proc. 20th Internat. Cosmic Ray Conf.* **3.**
25. Macomb, D. J., et al. (1991). "Search for TeV Emission from 4U0115+63." *Astrophysical Journal,* **376,** 738.
26. Reynolds, P. T., Cawley, M. F., Fegan, D. J., Hillas, A. M., Kwok, P. W., Lamb, R. C., Lang, M. J., Lewis, D. A., Macomb, D. J., O'Flaherty, K. S., Vacanti, C., and Weekes, T. C. (1991). "Search for TeV Gamma-Ray Emission from Hercules X-1." *Astrophysical Journal* **382,** 640.
27. Fisher, R. A. (1958). *Statistical Methods for Research Workers.* Oliver and Boyd, London.
28. Malmstadt, H. V., Enke, C. G., and Crouch, (1974). *Electronic Measurements for Scientists.* W. A. Benjamin, Menlo Park, CA.

# 13. STATISTICAL ANALYSIS OF SPATIAL DATA

Dale L. Zimmerman

Department of Statistics and Actuarial Science
University of Iowa, Iowa City

## 13.1 Introduction

Most data obtained from scientific investigations in the physical sciences have spatial or temporal labels associated with them. Moreover, such data often have a temporally or spatially coherent structure, the characterization of which may be scientifically important. One prevalent type of spatial structure is that in which measurements taken at sites in close spatial proximity to one another tend to be more alike than measurements taken at sites further apart. The previous three chapters have dealt with temporally referenced data; in the analysis of such data, spatial labels, if they are available, are ignored in the belief that the data's spatial structure is unimportant. The present chapter is concerned with the analysis of spatially referenced data; the use of temporal labels in the analysis is not considered here. Of course, if the data possess both temporal and spatial structure, then the methods of analysis discussed in these four chapters are likely to be inadequate. Some remarks on possible methods of analysis for this case are given in the concluding section of this chapter.

There are many kinds of spatial data, each requiring its own method of analysis. This great variety arises because measurements may be univariate or multivariate, categorical or continuous, real valued or not real valued (e.g. set valued); the spatial labels (henceforth called *locations* or *sites*) may be best regarded as points or regions (or something else), they may belong to Euclidean space or non-Euclidean space, and their union may be a region or a countable set of points (or something else); and the mechanism that generates the site locations may be known or unknown, random or nonrandom, related or unrelated to the processes that determine the values of the response variate. Some important types of spatial data—namely, geostatistical data, lattice data, and spatial point patterns—were described in Chapter 4. The present chapter focuses on the analysis of geostatistical data, which are measurements on a real-valued response variate taken at known sites in some region $D$ in two- or three-dimensional Euclidean space and which occur perhaps more often than any other kind of spatial data in the physical sciences. Geostatistical data constitute an incomplete sample of a "spatial surface" $\{Z(\mathbf{s}): \mathbf{s} \in D\}$, generally assumed to be continuous, representing the value of the response variate over $D$. Typically, the physical dimensions of the sites (which may be weather stations, drillholes into an ore body, wells tapped into an aquifer, etc.) are very small relative to both the size of $D$ and the intersite distances, in which case it is entirely reasonable to regard the sites as points. In this case, which is the one considered here, the data can be written generically as $Z(\mathbf{s}_1), Z(\mathbf{s}_2), \ldots, Z(\mathbf{s}_n)$, where $\mathbf{s}_1, \mathbf{s}_2, \ldots, \mathbf{s}_n$ are points in $D$ representing the $n$ sites where measurements are taken.

The prefix *geo* in geostatistics reflects the subject's origins in the earth sciences and may lead the reader to presume that methods of analyzing geostatistical data are relevant only to physical scientists in these fields. This is not so, for data that are reasonably regarded as samples from spatial surfaces arise in a broad spectrum of physical sciences. An example from atmospheric science is considered in detail in this chapter.

Two main objectives are common to most analyses of geostatistical data: (1) identification of the data's spatial structure, i.e., the nature of the response variate's dependence on spatial location and the nature of the dependence among values of the response variate at neighboring sites; (2) optimal interpolation, or prediction, of the response variate at unsampled locations (such as gridding irregularly sampled raw satellite footprint data onto a latitude–longitude mesh). Statistical procedures for achieving both of these objectives are presented in this chapter.

In the next section I introduce a specific set of geostatistical data that will be analyzed in subsequent sections. In Section 13.3 I review a general model that serves as the basis for the analysis of geostatistical data. Sections 13.4 and 13.5, which constitute the bulk of the chapter, present statistical methods for characterizing the data's spatial structure; and Section 13.6 describes how to use these characterizations to predict values of the response variate at unsampled locations. A number of extensions and related issues are discussed in Section 13.7.

## 13.2 Sulfate Deposition Data

One of the important environmental problems of our time is acid deposition, which, if it continues unabated, may have disastrous effects on aquatic and terrestrial ecosystems in various regions of the world. Acid deposition results mainly from the emission into the atmosphere, and subsequent atmospheric alteration, of sulfur and nitrogen pollutants produced by industrial, power, and transportation sources. Acid deposition occurs in two forms: dry deposition, in which pollutants are removed from the air by direct contact with the earth, and wet deposition—so-called acid rain—in which pollutants are deposited via rain, snow, or fog.

In this section, I briefly describe a set of wet acid deposition data that is used for illustrative purposes in subsequent sections; a more complete description of the data can be found in [1]. The data are total wet deposition amounts, in 1987, of $SO_4$ at sites of the National Atmospheric Deposition Program/National Trends Network (NADP/NTN). This network, whose operation is a cooperative effort among several U.S. government agencies, is the only precipitation chemistry monitoring network in the United States that has national coverage. The network has been in operation since 1979, though many sites have been added, and a few sites have been dropped, since then. Deposition amounts of several ion species, including $SO_4$, are recorded weekly at these sites. I chose to aggregate the data into annual totals to eliminate weekly and seasonal trends; the year 1987 was chosen arbitrarily. Obviously, if annual totals for two or more years were to be analyzed, methods that account for temporal structure as well as spatial structure would likely be required.

Figure 1 depicts the annual $SO_4$ wet deposition amounts (in grams per square meter) for 1987, at locations of the 198 NADP/NTN sites in the conterminous United States that were operating and met NADP/NTN quality assurance criteria that year. It is clear that wet sulfate deposition is highest in the mideast and northeast United States and lowest in the Rocky Mountain states and southwest. Figure 2 displays a stem-and-leaf diagram of the same data (refer to Chapter 6 for an explanation of stem-and-leaf diagrams). Though it ignores the locational information, this diagram is useful for summarizing nonspatial aspects of the data, clearly revealing the smallest and largest values (to the nearest 0.1 $g/m^2$) and a lack of symmetry about a central value, for instance. However, since the data are not properly regarded as a random sample from a single distribution, this stem-and-leaf diagram should not be given the classical statistical interpretation as an estimate of the distribution from which the data are derived.

These data can be used to address many important questions. In our analysis we shall specifically address the following:

- Is the wet sulfate deposition surface over the United States "smooth" or do some sites have unusually high levels relative to nearby sites?
- At what scale does spatial dependence exist?
- Does the answer to the previous question depend on direction, i.e., on the relative orientation of sites?

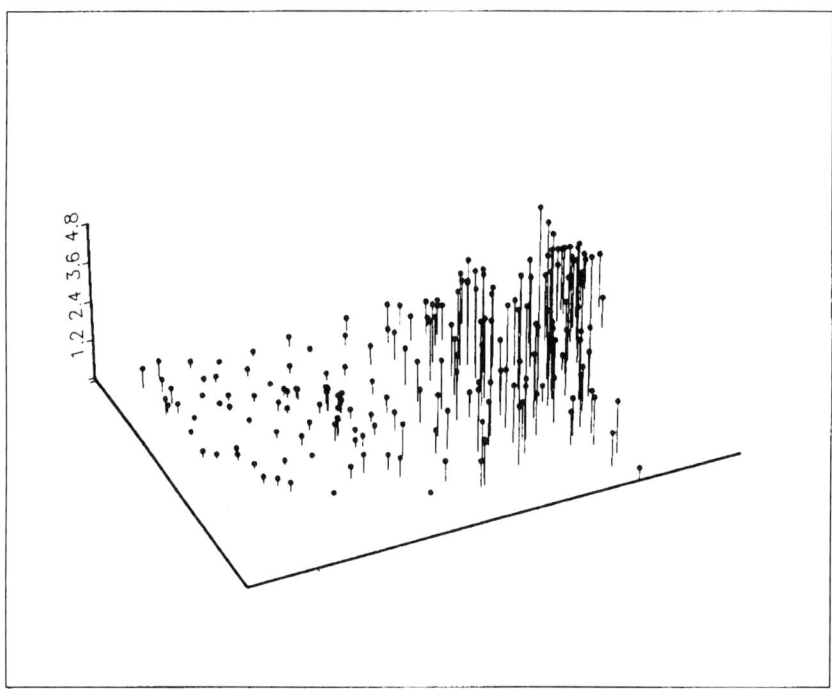

FIG. 1. Scatter plot of wet sulfate deposition data. Vertical lines are sited at the spatial locations of 198 NADP/NTN sites in the conterminous United States that were operating during 1987 and met quality assurance criteria; their heights are equal to the annual wet sulfate deposition (in $g/m^2$). Viewpoint of observer is from the southwest.

- What is the "best" estimate of the annual $SO_4$ deposition in 1987 at a selected unsampled location, or over a selected subregion?

## 13.3 The Geostatistical Model

Statistical inference for geostatistical data rests on the assumption that the response variate over the region of interest $D$ can be regarded as a realization of a random process (an infinite collection of random variables) $\{Z(\mathbf{s}): \mathbf{s} \in D\}$. Because only a single realization is available, and the data are merely an incomplete sample from it at that, the random process must satisfy some form of stationarity assumption for inference to be possible.

```
0 : 0001111111111111111
0 : 2222222222222222222223333333333333
0 : 444444445555555555555
0 : 666666666777
0 : 888999999
1 : 000000111111
1 : 22222333333
1 : 444445555
1 : 666666777
1 : 888888888999999999
2 : 0001111111
2 : 222222233
2 : 4555555
2 : 66777
2 : 8889999
3 : 01
3 : 2333
3 :
3 : 7
3 :
4 :
4 :
4 : 4
```

FIG. 2. Stem-and-leaf diagram of $SO_4$ deposition data. 1 : 7 corresponds to 1.7 g/m$^2$. The data were originally recorded to the nearest .001 g/m$^2$, but were rounded to the nearest 0.1 g/m$^2$ for this diagram.

Various stationarity assumptions are introduced in Chapter 4 but are reviewed again briefly here.

An assumption of second-order stationarity specifies that

$$E[Z(\mathbf{s})] = \mu, \quad \text{for all } \mathbf{s} \in D \qquad (13.1)$$

$$\text{cov}[Z(\mathbf{s}), Z(\mathbf{t})] = C(\mathbf{s} - \mathbf{t}), \quad \text{for all } \mathbf{s}, \mathbf{t} \in D \qquad (13.2)$$

In words, this asserts that the process's mean is constant and that the covariance between any two members of the process exists and depends on only the *relative* spatial locations, or the *displacement,* of the members. The function $C(\cdot)$ defined in (13.2) is called the covariance function. Observe that nothing is assumed about higher order moments of $\{Z(\mathbf{s}): \mathbf{s} \in D\}$ or about its joint distribution. Analysts of geostatistical data often prefer to make a slightly weaker assumption than second-order stationarity, called *intrinsic stationarity,* which specifies that (13.1) holds and that

$$\frac{1}{2}\text{var}[Z(\mathbf{s}) - Z(\mathbf{t})] = \gamma(\mathbf{s} - \mathbf{t}), \qquad \text{for all } \mathbf{s}, \mathbf{t} \in D \qquad (13.3)$$

The function $\gamma(\cdot)$ defined by (13.3) is called the *semivariogram*. A second-order stationary random process with covariance function $C(\cdot)$ is intrinsically stationary, with semivariogram

$$\gamma(\mathbf{h}) = C(\mathbf{0}) - C(\mathbf{h}) \qquad (13.4)$$

but the converse is not true in general.

Both the semivariogram and the covariance function characterize the (second-order) spatial dependence of the process $\{Z(\mathbf{s}): \mathbf{s} \in D\}$. If, as is often the case, it is reasonable to suppose that measurements at two sites displaced from one another in a given direction tend to be more dissimilar as the distance between those sites (in the given direction) increases, then we see from (13.2) and (13.3) that this requires the covariance function to be a monotone decreasing function, or the semivariogram to be a monotone increasing function, of the intersite distance in that direction.

Intrinsic stationarity specifies that the semivariogram is a function of only the lag vector $\mathbf{h} = \mathbf{s} - \mathbf{t}$ between sites $\mathbf{s}$ and $\mathbf{t}$. A stronger property, not needed for making inference from a single realization but often of scientific interest, is that of isotropy, which specifies that the semivariogram is a function of only the distance between sites, not of their relative orientation. More precisely, an intrinsically stationary random process with semivariogram $\gamma(\cdot)$ is said to be isotropic if $\gamma(\mathbf{h}) = \gamma^0(\|\mathbf{h}\|)$, where $\|\mathbf{h}\| = (\mathbf{h}'\mathbf{h})^{1/2}$; otherwise the process is anisotropic. One relatively simple kind of anisotropy is geometric anisotropy, for which $\gamma(\mathbf{h}) = \gamma^0((\mathbf{h}'\mathbf{B}\mathbf{h})^{1/2})$, where $\mathbf{B}$ is a positive definite matrix. Isotropy can be regarded as a special case of geometric anisotropy in which $\mathbf{B}$ is an identity matrix. Contours along which the semivariogram is constant (called *isocorrelation contours* in the second-order stationary case) are $d$-dimensional spheres in the case of isotropy but are $d$-dimensional ellipsoids in the more general case of geometric anisotropy.

In many practical situations it may be unreasonable to make the constant-mean assumption (13.1), and a more general model is needed. The geostatistical model adopted here assumes that

$$Z(\mathbf{s}) = \mu(\mathbf{s}) + \delta(\mathbf{s}) \qquad (13.5)$$

where $\mu(\mathbf{s}) \equiv E[Z(\mathbf{s})]$ is the (deterministic) mean function, assumed to be continuous, and $\delta(\cdot)$ is a zero-mean, intrinsically stationary random "error" process. The assumption of an intrinsically stationary $\delta(\cdot)$ means that

ESTIMATION OF FIRST-ORDER STRUCTURE

$$\frac{1}{2} \text{var}[\delta(\mathbf{s}) - \delta(\mathbf{t})] = \gamma(\mathbf{s} - \mathbf{t}) \qquad (13.6)$$

where $\gamma(\cdot)$ is the semivariogram of the error process (and of $Z(\cdot)$). Model (13.5) purports to account for large-scale spatial variation (trend) through the mean function $\mu(\cdot)$ and for small-scale spatial variation (spatial dependence) through the process $\delta(\cdot)$. This decomposition into large-scale and small-scale effects is not unique and is conceptually no different than the decomposition implicit in Chapter 10 for the analysis of time series (where large-scale time effects, or trends, were removed by detrending and the structure of the remaining temporal variation was then studied). A plausible, but admittedly nonunique, decomposition of spatial variation into large-scale and small-scale components can usually be determined with the aid of exploratory data analysis and model diagnostics.

In addition to capturing the small-scale spatial dependence, the error process $\delta(\cdot)$ in model (13.5) accounts for measurement error that may occur in the data collection process. This measurement error component typically has no spatial structure, and hence for some purposes it may be desirable to explicitly separate this component from the spatial dependence component. Such a decomposition is discussed in more detail in Section 13.7.

The objectives of an analysis of geostatistical data, which were spelled out in the previous two sections, can be reexpressed in terms of model (13.5). The first objective, that of characterizing the spatial structure, is tantamount to the estimation of $\mu(\cdot)$ (first-order structure) and the estimation of $C(\cdot)$ or $\gamma(\cdot)$ (second-order structure). The second objective, that of optimal interpolation, can be reexpressed as predicting the value of $Z(\mathbf{s}_0) = \mu(\mathbf{s}_0) + \delta(\mathbf{s}_0)$ at a specified, but unsampled, site $\mathbf{s}_0$.

## 13.4. Estimation of First-Order Structure

The first stage of a spatial analysis of geostatistical data is the estimation of the unspecified mean function $\mu(\cdot)$ in model (13.5). Several methods for accomplishing this have been proposed; surveys of available techniques can be found in [2] and [3]. Here I present two of the simplest yet most useful methods: trend surface analysis and median polish.

### 13.4.1 Trend Surface Analysis

*Trend surface analysis* is the name given to the fitting of a smooth parametric family of models for $\mu(\cdot)$ to the data. Ideally, the family of models should be flexible enough to approximate closely surfaces of various shapes yet not be dependent on a large number of unknown parameters. Perhaps the most suitable, and certainly the most commonly used, family of models is the polynomial family of order $q$ (where $q$ is user specified), for which

$$\mu(\mathbf{s}) = \mu(\mathbf{s}; \boldsymbol{\beta}) = \beta_0 f_0(\mathbf{s}) + \beta_1 f_1(\mathbf{s}) + \ldots + \beta_p f_p(\mathbf{s}) \quad (13.7)$$

here $\boldsymbol{\beta} = (\beta_0, \beta_1, \ldots, \beta_p)'$ are unknown parameters, $f_0(\mathbf{s}) \equiv 1$, and the set $\{f_1(\mathbf{s}), \ldots, f_p(\mathbf{s})\}$ consists of pure and mixed monomials of degree $\leq q$ in the coordinates of $\mathbf{s}$. (Throughout this chapter, $\mathbf{A}'$ denotes the transpose of a matrix $\mathbf{A}$.) For example, a (full) quadratic surface in $R^2$ is given by

$$\mu(\mathbf{s}) = \beta_0 + \beta_1 x + \beta_2 y + \beta_3 x^2 + \beta_4 xy + \beta_5 y^2$$

where $\mathbf{s} = (x, y)'$. Other parametric families of methods for $\mu(\cdot)$ are possible. For instance, periodic functions could be used if the surface appears to exhibit periodic behavior. However, families other than the polynomials are not often used, probably because they generally have larger numbers of unknown parameters or because they may depend nonlinearly on some of their parameters.

In fitting a polynomial surface, a Euclidean coordinate system in $R^d$ must be specified. The origin and orientation of this system can be chosen for convenience, but to ensure that the fitted surface is invariant to these choices only "full" polynomial surfaces, i.e., models of the form (13.7) in which *all* pure and mixed monomials of degree $\leq q$ are present, should be fit.

Having chosen a polynomial family of order $q$ for the mean function, how should the surface actually be fit to the data? One obvious possibility is to use ordinary least squares (OLS), i.e., choose $\boldsymbol{\beta}$ to minimize $\sum_{i=1}^{n} [Z(\mathbf{s}_i) - \mu(\mathbf{s}_i; \boldsymbol{\beta})]^2$. The OLS estimator of $\boldsymbol{\beta}$ is $\hat{\boldsymbol{\beta}}_{OLS} = (\mathbf{X}'\mathbf{X})^{-1}\mathbf{X}'\mathbf{Z}$, where $\mathbf{X}$ is the $n \times (p + 1)$ matrix (assumed to have full column rank) whose $i$th row is given by $[f_0(\mathbf{s}_i), f_1(\mathbf{s}_i), \ldots, f_p(\mathbf{s}_i)]$ and $\mathbf{Z} = [Z(\mathbf{s}_1), \ldots, Z(\mathbf{s}_n)]'$ is the data vector. Assuming that model (13.7) holds, the OLS estimator is unbiased (that is, $E(\hat{\boldsymbol{\beta}}_{OLS}) = \boldsymbol{\beta}$) regardless of the model's correlation structure but it is not the most efficient estimator if the data are spatially correlated. If the covariance structure were known, then the best (minimum-variance) linear unbiased estimator of $\boldsymbol{\beta}$ would be the generalized least-squares (GLS) estimator

$$\hat{\boldsymbol{\beta}}_{GLS} = (\mathbf{X}'\boldsymbol{\Sigma}^{-1}\mathbf{X})^{-1}\mathbf{X}'\boldsymbol{\Sigma}^{-1}\mathbf{Z} \quad (13.8)$$

where $\boldsymbol{\Sigma}$ is the covariance matrix of $\mathbf{Z}$, i.e., the $n \times n$ matrix whose $ij$th element is $\text{cov}[Z(\mathbf{s}_i), Z(\mathbf{s}_j)]$, which is assumed to be nonsingular. In reality, of course, the data's covariance structure is unknown, so in practice $\boldsymbol{\Sigma}$ is estimated, possibly by one of the methods described in Section 13.5, and the resulting estimate $\hat{\Sigma}$ substituted for $\Sigma$ in (13.8). This yields the "estimated" GLS estimator

$$\hat{\boldsymbol{\beta}}_{EGLS} = (\mathbf{X}'\hat{\boldsymbol{\Sigma}}^{-1}\mathbf{X})^{-1}\mathbf{X}'\hat{\boldsymbol{\Sigma}}^{-1}\mathbf{Z}$$

Properties of $\hat{\boldsymbol{\beta}}_{EGLS}$, which is a nonlinear estimator, are not as well understood as those of $\hat{\boldsymbol{\beta}}_{OLS}$ or $\hat{\boldsymbol{\beta}}_{GLS}$; it is known, however, that $\hat{\boldsymbol{\beta}}_{EGLS}$ is unbiased under very mild conditions [4]. Intuitively, one would expect the variability of $\hat{\boldsymbol{\beta}}_{EGLS}$ to be greater than that of $\hat{\boldsymbol{\beta}}_{GLS}$; this is known to be true if the process is Gaussian and $\hat{\Sigma}$ is a "reasonable" estimator [5].

Whichever estimator of $\boldsymbol{\beta}$ is used, the fitted surface is

$$\hat{\mu}(\mathbf{s}) = \mu(\mathbf{s}; \hat{\boldsymbol{\beta}}) = \hat{\beta}_0 f_0(\mathbf{s}) + \hat{\beta}_1 f_1(\mathbf{s}) + \ldots + \hat{\beta}_p f_p(\mathbf{s})$$

Residuals from the fitted surface at the data locations are defined by

$$\hat{\delta}(\mathbf{s}_i) = Z(\mathbf{s}_i) - \hat{\mu}(\mathbf{s}_i) \quad (i = 1, \ldots, n)$$

Estimation of the data's second-order variation can be based on these residuals; see Section 13.5.

This parametric modeling and fitting approach is a multidimensional generalization of polynomial regression on $R$, and as such it has the same shortcomings, some of which are even more severe in this higher dimensional context; details can be found in [2]. Because of these shortcomings, trend surface analysis should be performed with care. Nevertheless, its relative simplicity and the ease with which it can be carried out using widely available computing software make the method usually worth considering.

### 13.4.2 Median Polish

A second approach to estimating the first-order structure of the data is median polish, which was proposed for general use with two-way data layouts by Tukey [6] and adapted for use with geostatistical data by Cressie [7, 8]. Though the method can be applied to data in either $R^2$ or $R^3$, the discussion here will be limited to the two-dimensional context. Median polish is based on a model in which $\mu(\cdot)$ decomposes additively into orthogonal directional components. The model (in $R^2$) is

$$\mu(\mathbf{s}) = a + r(x) + c(y)$$

where $\mathbf{s} = (x, y)'$. For general spatial configurations of sites, this model will often be overparameterized; for instance, in the case of irregularly spaced data there are potentially as many as $2n + 1$ unknown parameters (and only $n$ observations). However, if the sites form a rectangular grid $\{(x_l, y_k): k = 1, \ldots, p; l = 1, \ldots, q\}$, then the model for the mean at these sites is, using obvious notation,

$$\mu(x_l, y_k) = a + r_k + c_l$$
$$(k = 1, \ldots, p;\ l = 1, \ldots, q) \tag{13.9}$$

and there are only $p + q + 1$ parameters to estimate: an "all" effect, $p$ "row" effects, and $q$ "column" effects. The data–model situation in this case is such that row effects can be estimated by exploiting replication across columns, with a similar result for column effects. Of course, we are interested in modeling the entire surface, not just its values at data locations: a way to construct the entire surface from the fit at data locations will be given shortly.

First, however, let us describe the median-polish algorithm by which model (13.9) is fit to the data. I give only an informal description here; a formal algebraic presentation can be found in [9]. Suppose again that the data locations form a $p \times q$ rectangular (but not necessarily regular) grid, and regard these sites as cells in a two-way table. The number of observations in each cell need not be equal nor, in fact, must every cell have an observation. To the table we append an extra row at the bottom, an extra column at the right, and an extra cell in the lower right corner (for a total of $p + q + 1$ extra cells). Single zeros are placed in the extra cells. We then operate iteratively on the data table, alternately subtracting row medians and column medians and accumulating these medians in the extra cells. More precisely, on the first iteration we subtract row medians from values in the corresponding rows and add these medians to the extra column of cells. On the second iteration, we subtract the column medians from values in the corresponding columns, including the extra column of row medians. Thus, the value in the $(p + 1, q + 1)$th cell after two iterations is the median of the original data's row medians. The entire process can be repeated, if desired, until some stopping criterion is satisfied, e.g., until another iteration changes each entry in the table by no more than a small positive number $\epsilon$.

The final entries in the $(p + 1, q + 1)$th cell, the $(p + 1)$th row, and the $(q + 1)$th column of the table are the median polish estimates of, respectively, $a$, $c_1, \ldots, c_q$, and $r_1, \ldots, r_p$. Using obvious notation, the fitted values at data locations are

$$\hat{\mu}(x_l, y_k) = \hat{a} + \hat{r}_k + \hat{c}_l \tag{13.10}$$

and residuals from the fit, which are defined by

$$\hat{\delta}(x_l, y_k) = Z(x_l, y_k) - \hat{\mu}(x_l, y_k)$$

are the final entries in the first $p$ rows and first $q$ columns of the table. The fitted surface at sites other than data locations is obtained from (13.10) by interpolating between, and extrapolating beyond, the data locations. For $\mathbf{s} = (x, y)'$ in the rectangular region between the four gridpoints $(x_l, y_k)'$,

$(x_{l+1}, y_k)'$, $(x_l, y_{k+1})'$, $(x_{l+1}, y_{k+1})'$, where $x_l < x_{l+1}$ and $y_k < y_{k+1}$, the fit is given by the planar interpolant

$$\hat{\mu}(\mathbf{s}) = \hat{a} + \hat{r}_k + \left(\frac{y - y_k}{y_{k+1} - y_k}\right)(\hat{r}_{k+1} - \hat{r}_k) + \hat{c}_l \\ + \left(\frac{x - x_l}{x_{l+1} - x_l}\right)(\hat{c}_{l+1} - \hat{c}_l) \tag{13.11}$$

Similar formulas are applicable for extrapolation when $x < x_1$, $x > x_q$, $y < y_1$, or $y > y_p$. The overall result is a continuous surface defined over all of $D$.

The median-polish algorithm just described begins with the removal of row medians, but it could also have begun with the removal of column medians. The two starting points can result in different fitted values and residuals, but experience has shown that the differences are usually small.

With a few modifications, median polish may be applied to irregularly spaced data as well as gridded data. For irregularly spaced data, rows and columns of the two-way table can be formed by overlaying a rectangular grid on $D$ and assigning observations to the nearest gridpoint. Of course, the grid spacings should be chosen small enough to preserve broad spatial relationships, yet not so small that a large number of cells are empty. Median polish and interpolation or extrapolation of $\mu(\cdot)$ over all of $D$ can then proceed entirely as with genuinely gridded data. However, for more accurate estimation of second-order structure and for purposes of spatial prediction, the median-polish residuals should be defined as

$$\hat{\delta}(\mathbf{s}_i) \equiv Z(\mathbf{s}_i) - \hat{\mu}(\mathbf{s}_i)$$

where $\hat{\mu}(\mathbf{s}_i)$ is obtained by (13.11); that is, for the purpose of obtaining residuals, the data should be reassigned from gridpoints to their original locations.

Estimation of the data's first-order variation by median polish avoids many of the pitfalls of trend surface analysis. First of all, median polish is less parametric and hence more flexible than trend surface analysis, in the sense that polynomial trend (linear, quadratic, etc.) or periodic behavior is not imposed. Second, median polish is much more resistant to outliers [9]. Third, there is evidence that the residuals from a median polish are less biased than those from a trend surface analysis [10]; this is important because the data's second-order variation is estimated from these residuals. One limitation of median polish, however, is the row–column additivity implicit in model (13.9); a procedure for verifying the (approximate) additivity of rows and columns is described in [8].

### 13.4.3 Application to Sulfate Deposition Data

In this subsection, trend surface analysis and median polish are illustrated using the sulfate deposition data. Admittedly these data, having been taken at the earth's surface over a significant geographic region, do not strictly lie in a Euclidean space. This causes no difficulties, however, and throughout the analysis of the $SO_4$ data we shall let $(x, y)$ denote the $(-1 \times$ longitude, latitude) coordinates of a site.

To illustrate a trend surface analysis, full polynomial surfaces of successively higher order were fit to the data by ordinary least squares until a standard size-.05 $F$-test of the hypothesis that all terms of highest order are nonsignificant could be accepted. This resulted in a fifth-order surface, which is shown in Figure 3. The fitted surface appears to reproduce the scatterplot in Figure 1 quite well; however, a plot (not shown) of the residuals versus the fitted mean revealed some problems: there was strong evidence that the variance of the residuals increased with the mean, and there were several outliers among the residuals. Rather than trying to remedy the

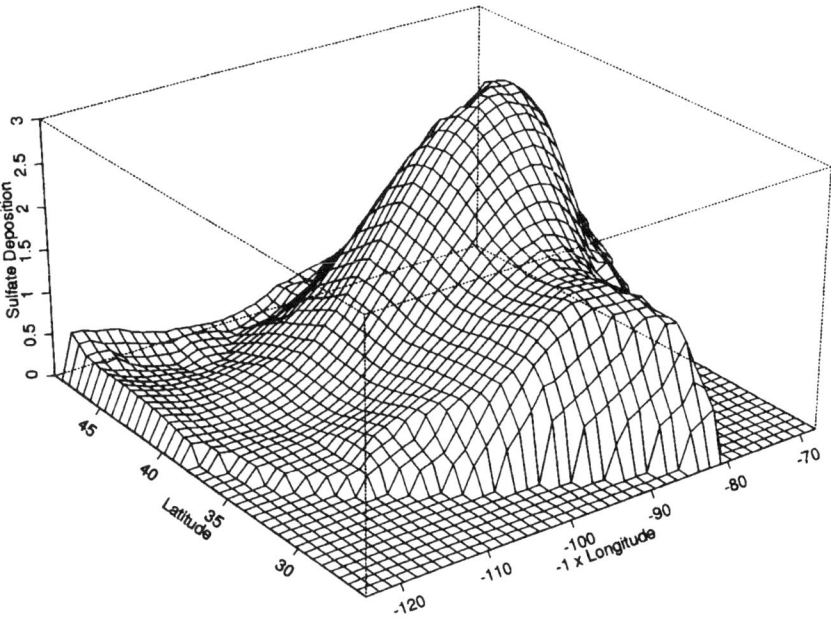

FIG. 3. Three-dimensional view of fitted fifth-order polynomial surface from the southwest. Units on the vertical axis are in $g/m^2$.

situation (e.g., by making a transformation or using a robust regression technique), we shall sidestep these problems and perform a median polish analysis.

To carry out median polish, a two-way table was obtained by assigning the irregularly spaced data to points forming a 6 × 15 rectangular grid with grid spacing of 4° in both latitude and longitude. With this choice of spacing, 27 (30%) of the 90 cells are empty, which is regarded as acceptable. Median polish (beginning with row median removal) was performed, with iteration continuing until each entry in the table changed by no more than .0001 g/m² on successive iterations. The fitted all, row (in order of increasing latitude, i.e., from south to north), and column (moving from east to west) effects were as follows:

$\hat{a} = 1.0578$

$\hat{r}_1 = -0.1488$  $\hat{c}_1 = -0.7478$  $\hat{c}_9 = 0.3432$
$\hat{r}_2 = -0.0570$  $\hat{c}_2 = -1.0068$  $\hat{c}_{10} = 0.7862$
$\hat{r}_3 = 0.1350$   $\hat{c}_3 = -0.9548$  $\hat{c}_{11} = 0.8172$
$\hat{r}_4 = 0.2855$   $\hat{c}_4 = -0.9568$  $\hat{c}_{12} = 0.7041$
$\hat{r}_5 = 0.0570$   $\hat{c}_5 = -0.8293$  $\hat{c}_{13} = 0.9862$
$\hat{r}_6 = -0.1440$  $\hat{c}_6 = -0.8729$  $\hat{c}_{14} = 1.0467$
                      $\hat{c}_7 = -0.5819$  $\hat{c}_{15} = 0.0102$
                      $\hat{c}_8 = 0.0000$

(When the algorithm was repeated beginning with column median removal, the results were not appreciably changed, as all effects and residuals were within .02 g/m² of their corresponding values for the original analysis.) As an example of (13.10), the fitted value in the first column and fourth row is given by $\hat{\mu}(x_1, y_4) = 1.0578 + 0.2855 - 0.7478 = 0.5955$. Fitted values at all 90 grid points were computed in this fashion, and the entire surface was then obtained by interpolation and extrapolation. Figure 4 shows this surface.

To lay the groundwork for the estimation of second-order structure, median-polish residuals were obtained for all observations according to the prescription given for irregularly spaced data in Section 13.4.2. Figure 5 shows a stem-and-leaf plot of these residuals. By comparing Figure 5 with Figure 2, the effect that removal of trend by median polish has on the shape of the stem-and-leaf plots is evident. In particular, the residuals in the latter plot appear to be distributed symmetrically about 0. A normal probability plot (not shown) indicated that the residuals come from a distribution more heavy tailed than the normal distribution; indeed, Figure 5 reveals that there is at least one site (the one with a residual of 2.3 g/m²) with an unusually high sulfate deposition. Consequently, a robust procedure for estimating the semivariogram is advisable.

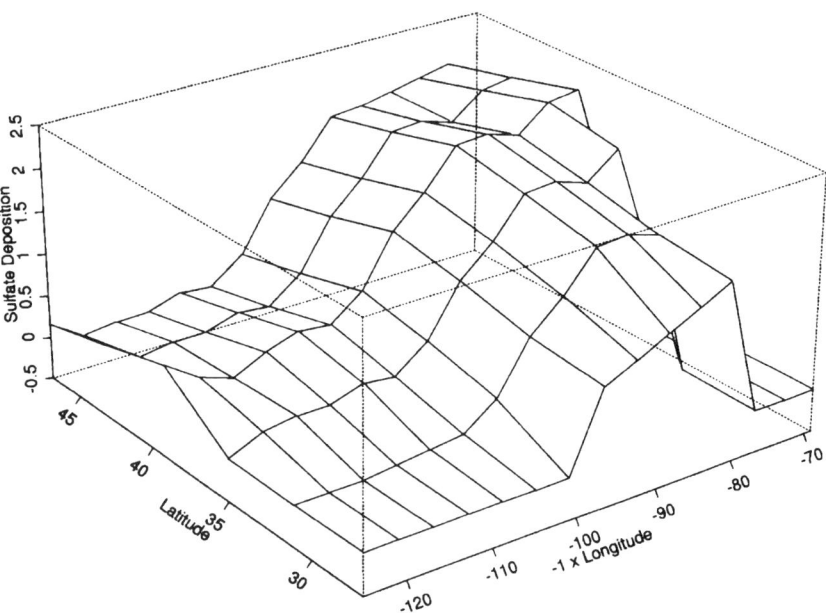

FIG. 4. Three-dimensional view of median-polish surface from the southwest. Units on the vertical axis are in grams per square meter.

## 13.5 Estimation of Second-Order Structure

In this section it is assumed that the data's large-scale (first-order) structure or trend has been estimated, and we turn our attention to estimating the data's small-scale (second-order) structure, i.e., the spatial dependence. Since the basis for our analysis is model (13.5), in which the error process $\delta(\cdot)$ is intrinsically stationary, the semivariogram (13.6) of $\delta(\cdot)$ is an appropriate mode of description of the second-order variation. If $\delta(\cdot)$ is not merely intrinsically stationary but also second-order stationary, then the second-order variation can be described by either the semivariogram or the covariance function; even in this case, however, there are good reasons for modeling and estimating the former rather than the latter [3, 11].

The classical paradigm for estimating the semivariogram $\gamma(\cdot)$ consists of two main steps: first $\gamma(\mathbf{h})$ is estimated nonparametrically for each of several values of $\mathbf{h}$, and then a valid (i.e., conditionally nonpositive definite; see Chapter 4 for a definition) parametric family of models is fit to the nonparametric estimate. Two important decisions that must be made in the course

```
-1 : 7
-1 : 55
-1 : 2
-1 :
-0 : 88888
-0 : 7777666666666
-0 : 5544444
-0 : 333333333333333222222222222222
-0 : 111111111111111111110000000000000
 0 : 00000000000000111111111111111111111111
 0 : 22222222222222233333333
 0 : 444444444444444555555
 0 : 667777777
 0 : 888999
 1 : 001
 1 : 33
 1 :
 1 : 6
 1 :
 2 :
 2 : 3
```

FIG. 5. Stem-and-leaf diagram of the sulfate median-polish residuals.

of this analysis are (1) whether the second-order variation is isotropic, i.e., direction invariant, and (2) what family of models to use for $\gamma(\cdot)$. These steps are described in the next two subsections and then illustrated with the sulfate deposition data. In this section it is assumed for simplicity that the index set $D$ of sites is two-dimensional and that only one measurement is taken at each site. Modifications for three-dimensional data are straightforward, and modifications to accommodate replicate measurements at individual sites are discussed in Section 13.7.

### 13.5.1 Nonparametric Estimation of the Semivariogram

Under the intrinsic stationarity assumption, the semivariogram of $\delta(\cdot)$ can be expressed as a function of the lag or displacement vector $\mathbf{h} = \mathbf{s} - \mathbf{t}$ between two sites $\mathbf{s}$ and $\mathbf{t}$ in $D$. Let $H$ represent the set of all possible lags generated by points in $D$ whose corresponding displacement angle belongs to the interval $[-\pi/2, \pi/2)$. (Attention can be restricted to lags with these displacement angles because $\gamma(-\mathbf{h}) = \gamma(\mathbf{h})$ for all lags $\mathbf{h}$.) Next, let $\{H_1, \ldots, H_k\}$ be a partition of $H$ into sets, or "windows," within which the values of **h** are approximately equal (in a well-defined sense). For each $u = 1, \ldots, k$, let $\mathbf{h}_u$ represent a "typical" value in $H_u$, e.g., its centroid, and let $N(\mathbf{h}_u)$ be the set of site pairs that are displaced approximately by $\mathbf{h}_u$. Thus, for any fixed $u = 1, \ldots, k$, the elements of $N(\mathbf{h}_u)$ are site pairs approximately the same distance apart with approximately the same relative orientation. Finally, let $|N(\mathbf{h}_u)|$ denote the number of elements of $N(\mathbf{h}_u)$.

Now suppose that $\hat{\mu}(\cdot)$ is any estimate of the data's first-order variation, and define the residuals at data locations by $\hat{\delta}(\mathbf{s}_i) = Z(\mathbf{s}_i) - \hat{\mu}(\mathbf{s}_i)$. Then the classical nonparametric estimator of $\gamma(\mathbf{h})$ at lag $\mathbf{h}_u$ is one-half the average squared difference of residuals

$$\hat{\gamma}(\mathbf{h}_u) = \frac{1}{2|N(\mathbf{h}_u)|} \sum_{N(\mathbf{h}_u)} [\hat{\delta}(\mathbf{s}_i) - \hat{\delta}(\mathbf{s}_j)]^2 \qquad (13.12)$$

This estimator is approximately unbiased (it would be exactly unbiased if the unobservable process $\delta(\cdot)$ could be used in place of $\hat{\delta}(\cdot)$ and if $\mathbf{s}_i - \mathbf{s}_j$ were equal for every element of $N(\mathbf{h}_u)$), but because it is a function of a sum of squares it is not resistant to outliers. A more robust estimator, proposed by Cressie and Hawkins [12], is

$$\overline{\gamma}(\mathbf{h}_u) = \frac{\left( \frac{1}{|N(\mathbf{h}_u)|} \sum_{N(\mathbf{h}_u)} |\hat{\delta}(\mathbf{s}_i) - \hat{\delta}(\mathbf{s}_j)|^{1/2} \right)^4}{.914 + (.988/|N(\mathbf{h}_u)|)} \qquad (13.13)$$

Typically, the partition of $H$ used in the nonparametric estimation of $\gamma(\cdot)$ is a "polar" partition, i.e., the lags are classified on the basis of their lengths and displacement angles. This partition effectively yields estimated "directional semivariograms" in each of several directions. If the second-order variation is judged to be isotropic (by one of the graphical methods described later), then the estimation of the semivariogram is simplified somewhat, for in this case an omnidirectional semivariogram can be estimated, based on a partition of $H$ into distance classes only.

An important practical question in this estimation scheme is how finely to partition $H$, or equivalently, how large to make the lag classes. Clearly, there is a trade-off involved: the smaller are the class widths, the better the lags in $N(\mathbf{h}_u)$ are approximated by $\mathbf{h}_u$, but the fewer the number of site pairs belonging to $N(\mathbf{h}_u)$ (with the consequence that the nonparametric estimator's sampling variation at that lag is larger). A rule of thumb for choosing the class widths is given in the next section.

### 13.5.2 Fitting a Semivariogram Model

The second step of semivariogram estimation is the fitting of a parametric model to the nonparametric estimator of $\gamma(\cdot)$. Several examples of valid parametric models for $\gamma(\cdot)$, e.g, the spherical and exponential models, were given in Chapter 4. In practice, the family of models to be fit has often been chosen "by eye" after looking at a plot of the nonparametric estimator versus $\mathbf{h}_u$. Some less subjective model selection procedures are mentioned at the end of this subsection.

Together with the selection of a model for $\gamma(\cdot)$, the way in which the semivariogram depends on the relative orientation of data locations should be determined prior to fitting a model. Some graphical diagnostics are available for checking for isotropy or geometric anisotropy. One diagnostic consists of simply a plot of several (at least three, preferably more) estimated directional semivariograms overlaid on the same graph. If these semivariograms lie roughly on top of each other, then isotropy can reasonably be assumed. A second diagnostic is obtained by connecting, with a smooth curve, lag vectors $\mathbf{h}_u$ for which the nonparametric semivariogram estimates are roughly equal (in effect plotting estimated isocorrelation contours, in the case of a second-order stationary process). If the contours are roughly elliptical (circular), then geometric anisotropy (isotropy) is a reasonable assumption. One might suspect that the second diagnostic would be better than the first at identifying the strength and orientation of geometric anisotropy when such an anisotropy exists, but in practice it may not be so because there is often too much sampling variation in the estimated semivariogram for different contours to have a consistent shape and orientation.

Once a semivariogram model has been chosen and the issue of direction–dependence has been addressed, the chosen model can be fit to the nonparametric estimator of $\gamma(\cdot)$. There are several model fitting procedures available for this, some more practical than others. Seven procedures are summarized and compared in [13]; here I will briefly describe one widely used procedure—the nonlinear weighted least squares (WLS) procedure of Cressie [14]—and mention two others. Let $\{\tilde{\gamma}(\mathbf{h}_u): u = 1, \ldots, k\}$ denote a nonparametric semivariogram estimator, either $\hat{\gamma}$ or the more robust $\bar{\gamma}$; let $\gamma(\mathbf{h}; \boldsymbol{\theta})$ denote a parametric model to be fit to this estimator; and let $\Theta$ denote the parameter space for the parameter vector $\boldsymbol{\theta}$. Cressie's WLS estimator of $\boldsymbol{\theta}$ is defined as a value $\hat{\boldsymbol{\theta}} \in \Theta$ that minimizes the weighted residual sum of squares function

$$w(\boldsymbol{\theta}) = \sum_{u \in \mathcal{U}} \frac{|N(\mathbf{h}_u)|}{[\gamma(\mathbf{h}_u; \boldsymbol{\theta})]^2} [\hat{\gamma}(\mathbf{h}_u) - \gamma(\mathbf{h}_u; \boldsymbol{\theta})]^2 \quad (13.14)$$

where $\mathcal{U}$ is a specified subset of lag classes believed to yield reliable estimates of $\gamma(\mathbf{h})$. Observe that the weights, $|N(\mathbf{h}_u)|/[\gamma(\mathbf{h}_u; \boldsymbol{\theta})]^2$, are small if either $|N(\mathbf{h}_u)|$ is small or $\gamma(\mathbf{h}_u; \boldsymbol{\theta})$ is large. Since, in practice, $|N(\mathbf{h}_u)|$ tends to be small for large lags and since the model $\gamma(\mathbf{h}_u; \boldsymbol{\theta})$ is typically nondecreasing in $\mathbf{h}/\|\mathbf{h}\|$, nonparametric estimates at large lags tend to receive relatively less weight. In spite of this, the sampling variation of estimates at large lags and at other lags for which $|N(\mathbf{h})|$ is too small is so great that these estimates are not sufficiently downweighted in (13.14); this is usually accounted for by taking $\mathcal{U}$ in (13.14) to be of the form $\mathcal{U} = \{u: |N(\mathbf{h}_u)| \geq G_1, \|\mathbf{h}_u\| \leq G_2\}$. Journel and Huijbregts [15] suggest the choices $G_1 = 30$ and $G_2 = r_m/2$ (half the largest possible lag).

Two other widely used methods for fitting semivariogram models are maximum likelihood (ML) and restricted maximum likelihood (REML); details on these can be found in [16] and [17]. These methods are generally applied to the estimation of the process's first- and second-order variation simultaneously. Consequently, the estimation of the semivariogram by ML or REML does not strictly require one to first obtain a nonparametric estimator such as $\bar{\gamma}(\mathbf{h}_u)$. Nevertheless, the calculation of a nonparametric estimator is recommended for its diagnostic value even if the semivariogram is to be estimated by a likelihood-based method; it can be helpful in selecting an appropriate semivariogram model and it provides a benchmark to which the ML or REML estimate can be compared.

Of the available semivariogram model fitting procedures, which should be used? Although likelihood-based estimation is more efficient under the assumption that the random process is Gaussian, it may not be robust to departures from this assumption. Moreover, estimation by ML or REML is a large computational problem for even moderately sized data sets. In contrast, WLS estimation, if based on the use of $\bar{\gamma}(\mathbf{h})$ in (13.14), is robust to departures from the Gaussian assumption, and presents a much simpler computational problem. Furthermore, WLS estimation has been shown to be nearly as efficient as likelihood-based estimation even when the process is Gaussian [13]. Consequently, WLS estimation appears to represent a reasonable compromise between statistical efficiency and practicality.

Though the family of models fit to a nonparametric semivariogram estimator $\tilde{\gamma}(\mathbf{h}_u)$ is often chosen on the basis of a visual inspection of a plot of $\tilde{\gamma}(\mathbf{h}_u)$ versus $\mathbf{h}_u$, more formal model selection procedures are possible. A natural model selection criterion for use in conjunction with WLS estimation is the weighted residual sum of squares evaluated at the estimate of $\boldsymbol{\theta}$, viz. $w(\hat{\boldsymbol{\theta}})$. That is, to select a model from among two or more candidates, fit each by WLS and choose the one with the smallest weighted residual sum of squares. Provided that each candidate model has the same number of unknown parameters, this criterion is sensible. If some model(s) under consideration has (have) more parameters than others, however, a criterion that has some kind of penalty for increased model complexity, such as Akaike's information criterion, is more appropriate; see [18]. Cross-validation of the estimated semivariogram [19] is another possibility.

### 13.5.3 Application to Sulfate Decomposition Data

Now we estimate the second-order structure of the sulfate deposition data, making use of some of the methodology just discussed. Since the error process $\delta(\cdot)$ is unobservable, the estimation of second-order structure is based on residuals from the fit of the first-order structure. The analysis here utilizes the median polish residuals $\{\hat{\delta}(s_i): i = 1, \ldots, 198\}$ obtained at the conclusion of Section 13.4.3.

The first step of the analysis, that of nonparametrically estimating the semivariogram of $\delta(\cdot)$, requires that the set $H$ of all possible lags with displacement angle in $[-\pi/2, \pi/2)$ be partitioned. Here, as is common practice, a polar partition into distance and angle classes was used. Distances between sites were measured as great-arc lengths, in miles, along the earth's surface, i.e.,

$$\|s_i - s_j\| = 3969.665 \cos^{-1}\left\{\cos\left(\frac{(90-y_i)\pi}{180}\right)\cos\left(\frac{(90-y_j)\pi}{180}\right)\right.$$

$$\left. + \sin\left(\frac{(90-y_i)\pi}{180}\right)\sin\left(\frac{(90-y_j)\pi}{180}\right)\cos\left(\frac{|x_i - x_j|\pi}{180}\right)\right\}$$

where $s_i = (x_i, y_i)'$ ($i = 1, \ldots, 198$), and the distance classes were taken to be intervals of 100 mi, with each interval identified by its midpoint. Displacement angles, measured as $\phi_{ij} = \arctan[(y_i - y_j)/(x_i - x_j)]$, were assigned to the four angle classes $\{\phi: -\pi/2 \leq \phi < -3\pi/8 \text{ or } 3\pi/8 \leq \phi < \pi/2\}$, $\{\phi: -3\pi/8 \leq \phi < -\pi/8\}$, $\{\phi: -\pi/8 \leq \phi < \pi/8\}$, and $\{\phi: \pi/8 \leq \phi < 3\pi/8\}$, which will henceforth be called the north–south (N–S), northwest–southeast (NW–SE), east–west (E–W), and northeast–southwest (NE–SW) directions, respectively.

Based on this partition, the robust nonparametric semivariogram estimator $\{\overline{\gamma}(\mathbf{h}_u): u = 1, \ldots, k\}$, given by (13.13), was obtained. The four estimated directional semivariograms are shown in Figure 6. Estimates at lags up to 1250 mi are shown for the NW–SE and NE–SW directions, while for the N–S and E–W directions estimates are shown for lags up to 1050 mi and 1450 mi, respectively. Distance classes beyond these cutoffs contained too few lags for the corresponding estimates to be regarded as reliable. Although the estimated semivariograms in Figure 6 exhibit a certain amount of variability, for small distances they generally increase with distance, indicating that small-scale spatial dependence exists. With the exception of the NE–SW semivariogram, they tend to level out at about the same height (sill value) as distance increases, and as distance decreases they seem to be tending to roughly the same value (nugget effect). There is a perceptible difference in their ranges, however, with the spatial dependence in the N–S

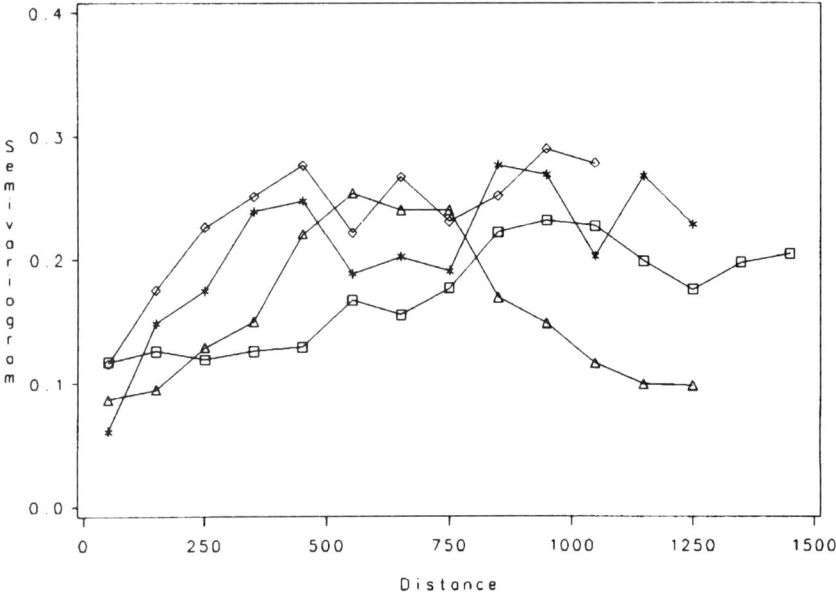

FIG. 6. Plot of nonparametric semivariogram estimates in the four directions N–S (◊), NE–SW (△), E–W (□), and NW–SE (*).

direction appearing to vanish at about half the distance it vanishes in the E–W direction (the ranges in the other two directions being intermediate). A possible explanation for this may be directional differences in wind currents. This suggests that instead of using an isotropic semivariogram model it is more reasonable to adopt a geometrically anisotropic model, with elliptical isocorrelation contours that are aligned with the compass points and twice as long in the E–W direction as in the N–S direction. Equivalently, we can halve the scale in the E–W direction (i.e., halve the difference in longitude between sites) and use an isotropic model. In what follows we take the latter course of action.

After halving the scale in the E–W direction, the semivariogram was re-estimated (again by the robust nonparametric estimator), utilizing the same distance classes but without regard to displacement angle. This estimate is shown in Figure 7. The next step is to fit a valid parametric model to $\bar{\gamma}$. Based on the shape of $\bar{\gamma}$, a spherical or exponential model with a nugget effect (a discontinuity at the origin) appears to be reasonable. Both of these models were fit by WLS, with the spherical resulting in a weighted residual sum of squares that was 29% smaller. The fitted spherical semivariogram is

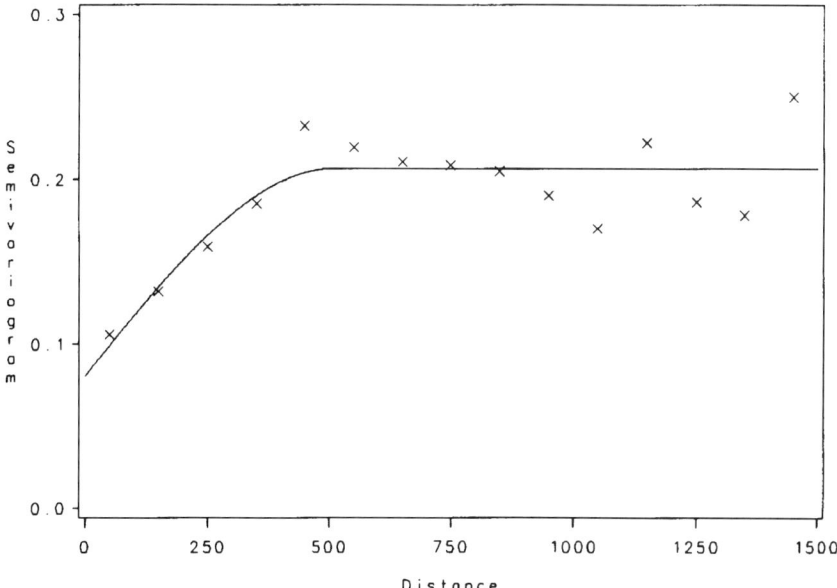

FIG. 7. Plot of nonparametric semivariogram estimator, assuming isotropy after halving the scale in the E–W direction. A plot of the fitted spherical semivariogram model (13.15) is superimposed.

$$\gamma(\|\mathbf{h}\|) = \begin{cases} 0, & \|\mathbf{h}\| = 0 \\ .0802 + .1263 \left\{ \frac{3}{2}\left(\frac{\|\mathbf{h}\|}{507.97}\right) - \frac{1}{2}\left(\frac{\|\mathbf{h}\|}{507.97}\right)^3 \right\}, & 0 < \|\mathbf{h}\| \leq 507.97 \\ .2065, & \|\mathbf{h}\| > 507.97. \end{cases}$$

(13.15)

The fitted semivariogram, which is shown in Figure 7, broadly reproduces the nonparametric estimate, especially at small lags, where it is most important to do so (see [20]). The estimated range of 507.97 indicates that the small-scale spatial dependence vanishes at a distance of about 500 miles in the N–S direction and about 1000 miles in the E–W direction.

In this section we have presented techniques for characterizing the spatial dependence remaining in the data after the first-order variation (trend) has been removed. The end-product of these techniques, for the sulfate deposition data, was the estimated semivariogram (13.15). It remains to utilize (13.15) in the prediction of $SO_4$ deposition at unsampled sites.

## 13.6 Spatial Prediction (Kriging)

Prediction of the response variate at one or more sites is often the ultimate objective of a geostatistical analysis. Many spatial prediction procedures have been devised; see [21] for a review of 14 of them. Those methods known as *kriging* yield minimum variance linear unbiased predictors under an assumed model of the form (13.5). There are several kriging methods, which differ (among other ways) according to what is assumed about the mean function $\mu(\cdot)$ and how it is estimated. We shall discuss two kriging methods, universal kriging and median-polish kriging, which correspond to the two methods of estimating first-order variation that were described in Section 13.4.

### 13.6.1 Universal Kriging

The basic concepts of universal kriging were given in Chapter 4, and we merely review them here. Suppose that model (13.5) holds, that the semivariogram $\gamma(\cdot)$ is known, and that $\mu(\cdot)$ is of the form (13.7). The universal kriging predictor of $Z(\mathbf{s}_0)$ at an unsampled location $\mathbf{s}_0$ is given by

$$\hat{Z}_U(\mathbf{s}_0) = [\boldsymbol{\gamma} + \mathbf{X}(\mathbf{X}'\boldsymbol{\Gamma}^{-1}\mathbf{X})^{-1}(\mathbf{x}_0 - \mathbf{X}'\boldsymbol{\Gamma}^{-1}\boldsymbol{\gamma})]'\boldsymbol{\Gamma}^{-1}\mathbf{Z} \quad (13.16)$$

where $\boldsymbol{\gamma} = [\gamma(\mathbf{s}_1 - \mathbf{s}_0), \ldots, \gamma(\mathbf{s}_n - \mathbf{s}_0)]'$, $\boldsymbol{\Gamma}$ is the $n \times n$ symmetric matrix with $ij$th element $\gamma(\mathbf{s}_i - \mathbf{s}_j)$, $\mathbf{X}$ is defined as in Section 13.4.1, and $\mathbf{x}_0 = [f_0(\mathbf{s}_0), \ldots, f_p(\mathbf{s}_0)]'$. This predictor minimizes the mean squared prediction error (MSPE) $E[(Z(\mathbf{s}_0) - \boldsymbol{\lambda}'\mathbf{Z})^2]$ among all linear predictors $\boldsymbol{\lambda}'\mathbf{Z}$ that satisfy the unbiasedness condition $\boldsymbol{\lambda}'\mathbf{X} = \mathbf{x}_0'$. The minimized value of the MSPE is called the kriging variance and is given by

$$\sigma_U^2(\mathbf{s}_0) = \boldsymbol{\gamma}'\boldsymbol{\Gamma}^{-1}\boldsymbol{\gamma} - (\mathbf{X}'\boldsymbol{\Gamma}^{-1}\boldsymbol{\gamma} - \mathbf{x}_0)'(\mathbf{X}'\boldsymbol{\Gamma}^{-1}\mathbf{X})^{-1} \\ (\mathbf{X}'\boldsymbol{\Gamma}^{-1}\boldsymbol{\gamma} - \mathbf{x}_0) \quad (13.17)$$

The kriging variance can be used to construct a nominal $100(1 - \alpha)\%$ prediction interval for $Z(\mathbf{s}_0)$, which is given by

$$I_U = \hat{Z}_U(\mathbf{s}_0) \pm z_{\alpha/2}\sigma_U(\mathbf{s}_0)$$

where $0 < \alpha < 1$ and $z_{\alpha/2}$ is the upper $\alpha/2$ percentage point of a standard normal distribution. Under the assumption that $Z(\cdot)$ is Gaussian and $\gamma(\cdot)$ is known, $Pr[Z(\mathbf{s}_0) \in I_U] = 1 - \alpha$.

In practice, two modifications are usually made to the universal kriging procedure just described. First, the prediction of $Z(\mathbf{s}_0)$ may be based not on the entire data vector $\mathbf{Z}$ but on only those observations that lie in a specified neighborhood around $\mathbf{s}_0$; guidelines for choosing this neighborhood are given in [3]. When this modification is made, the formulas for the universal kriging predictor and its associated kriging variance are of the same form as (13.16) and (13.17) but with $\boldsymbol{\gamma}$ and $\mathbf{Z}$ replaced by the subvectors, and $\boldsymbol{\Gamma}$ and $\mathbf{X}$ replaced by the submatrices, corresponding to the neighborhood. Second, since in reality the semivariogram is unknown, it is common practice to substitute $\hat{\boldsymbol{\gamma}} = \boldsymbol{\gamma}(\hat{\boldsymbol{\theta}})$ and $\hat{\boldsymbol{\Gamma}} = \boldsymbol{\Gamma}(\hat{\boldsymbol{\theta}})$ for $\boldsymbol{\gamma}$ and $\boldsymbol{\Gamma}$ in (13.16) and (13.17), where $\hat{\boldsymbol{\theta}}$ is an estimate of $\boldsymbol{\theta}$ obtained by, say, one of the methods presented in Section 13.5.2. The effects of this on kriging are considered in [22].

In some cases a predictor of the average value $Z(B) \equiv \int_B Z(\mathbf{s}) d\mathbf{s}/|B|$ over a region (block) $B \in D$ of positive $d$-dimensional volume $|B|$ is desired, rather than a predictor at a single point; this is called *block kriging*. Formulas for the universal block kriging predictor of $Z(B)$ and its associated kriging variance can be found in [3].

### 13.6.2 Median-Polish Kriging

Again suppose that model (13.5) holds and $\gamma(\cdot)$ is known, but now take $\mu(\cdot)$ to be of the form (13.9) (where, of course, if the data locations are irregularly spaced they are assigned to the nodes of an overlaid rectangular grid). Let $\tilde{\mu}(\cdot)$ represent the fitted median-polish surface (13.11). The median-polish kriging predictor of $Z(\mathbf{s}_0)$ is defined as

$$\hat{Z}_{MP}(\mathbf{s}_0) = \tilde{\mu}(\mathbf{s}_0) + \tilde{\delta}(\mathbf{s}_0)$$

where $\tilde{\delta}(\mathbf{s}_0) = [\boldsymbol{\gamma}' + (1 - \boldsymbol{\gamma}'\boldsymbol{\Gamma}^{-1}\mathbf{1})(\mathbf{1}'\boldsymbol{\Gamma}^{-1}\mathbf{1})^{-1}\mathbf{1}']\boldsymbol{\Gamma}^{-1}\hat{\boldsymbol{\delta}}$ is the *ordinary* kriging predictor of $\delta(\mathbf{s}_0)$, $\boldsymbol{\gamma}$ and $\boldsymbol{\Gamma}$ are defined as in Section 13.6.1, $\hat{\boldsymbol{\delta}}$ is the vector of median-polish residuals, and $\mathbf{1}$ is an $n \times 1$ vector of ones [8]. (Ordinary kriging is the special case of universal kriging, corresponding to a constant mean function.) The median-polish kriging variance is

$$\sigma^2_{MP}(\mathbf{s}_0) = \boldsymbol{\gamma}'\boldsymbol{\Gamma}^{-1}\boldsymbol{\gamma} - (\mathbf{1}'\boldsymbol{\Gamma}^{-1}\boldsymbol{\gamma} - 1)^2/(\mathbf{1}'\boldsymbol{\Gamma}^{-1}\mathbf{1}) \qquad (13.18)$$

A nominal $100(1 - \alpha)\%$ prediction interval for $Z(\mathbf{s}_0)$ is $I_{MP} = \hat{Z}_{MP}(\mathbf{s}_0) \pm z_{\alpha/2}\sigma_{MP}(\mathbf{s}_0)$. The same practical modifications usually made to universal kriging are also usually made to median-polish kriging.

### 13.6.3 Application to Sulfate Deposition Data

Finally, we apply the two kriging methods just described to the prediction of wet sulfate deposition at two unsampled sites, taken for illustration to be Des Moines, Iowa (latitude 41.58°, longitude 93.62°), and Portland, Maine (latitude 43.65°, longitude 70.26°). Note that Des Moines lies near the center of the spatial domain and Portland lies on its boundary. For both methods, prediction was based on all observations (i.e., a kriging neighborhood was not employed) and on the fitted spherical semivariogram (13.15). With $\mathbf{X}$ and $\mathbf{x}_0$ defined to correspond to a full fifth-order polynomial surface in the coordinates of $\mathbf{s}$ (the surface fitted in Section 13.4.3), the universal kriging predictors of wet $SO_4$ deposition in 1987 are 1.693 g/m$^2$ at Des Moines and 1.559 g/m$^2$ at Portland. The associated nominal 95% prediction intervals are (1.023, 2.363) and (0.887, 2.232). With $\hat{\boldsymbol{\delta}}$ defined as the vector of median-polish residuals obtained in Section 13.4.3, the corresponding median-polish kriging predictors are 1.638 g/m$^2$ and 1.528 g/m$^2$, and the associated nominal 95% prediction intervals are (0.969, 2.307) and (0.864, 2.192). Observe that, at each of these two sites, the two kriging methods give nearly the same predictors and kriging variances.

## 13.7 Extensions and Related Issues

In attempting to lay out the basic elements of the geostatistical method in this chapter, several extensions and related issues have been glossed over. In this concluding section I mention some of these that I consider important, giving references for the reader who wishes more information.

As noted in Section 13.1, data in the physical sciences are often collected over time as well as space. A natural approach to the analysis of such data is to extend existing methodology for spatial data into the space–time domain. That is, regard the data as a realization of a random process $\{Z(\mathbf{s},t): \mathbf{s} \in D, t \in T\}$, where $T$ is an index set for time, adopt a model $Z(\mathbf{s},t) = \mu(\mathbf{s}, t) + \delta(\mathbf{s}, t)$ analogous to (13.5) and treat time as simply another spatial dimension in estimation and prediction procedures. Despite the straightforward appearance of this extension, some important qualitative differences between temporal and spatial phenomena must be reckoned with; see [23]. For instance, a spatio-temporal phenomenon's temporal and spatial second-order variation generally are fundamentally different and not comparable on the same scale. To account for this the space–time semivariogram could be split into either a product [24] or a sum [25] of space and time component semivariograms.

Throughout this chapter it was assumed that a single response variate was of interest. In some situations, however, there may be two or more variables of interest, and one may wish to study how these variables covary across the spatial domain or to predict their values at unsampled locations. These problems can be handled by a multivariate generalization of the univariate geostatistical approach I have described. In this multivariate approach, $\{\mathbf{Z}(\mathbf{s}) \equiv [\mathbf{Z}_1(\mathbf{s}), \ldots , Z_m(\mathbf{s})]': \mathbf{s} \in D\}$ represents the $m$-variate "spatial surface" of interest and a model $\mathbf{Z}(\mathbf{s}) = \boldsymbol{\mu}(\mathbf{s}) + \boldsymbol{\delta}(\mathbf{s})$ analogous to (13.5) is adopted in which the second-order covariation is characterized by a function called the cross-semivariogram; this function is defined most satisfactorily as

$$\gamma_{jj'}(\mathbf{h}) = \frac{1}{2} \text{var}[\delta_j^*(\mathbf{s} + \mathbf{h}) - \delta_{j'}^*(\mathbf{s})] \tag{13.19}$$

where $\delta_j^*(\cdot)$ and $\delta_{j'}^*(\cdot)$ are versions of $\delta_j(\cdot)$ and $\delta_{j'}(\cdot)$ [the $j$th and $j'$th elements of $\delta(\cdot)$] normalized to have the same units [26]. The procedure that yields a best (in a certain sense) linear unbiased predictor of $\mathbf{Z}(\mathbf{s}_0)$ at an unsampled location $\mathbf{s}_0$ is called *cokriging;* see [27].

In Section 13.5 it was assumed that measurements were taken at *distinct* sites and made without error. This was in keeping with the classical geostatistical approach, which was not originally set up to handle replicate measurements or to identify measurement error (even though it has sometimes been recognized as a potentially important source of variation). Replication can be accommodated and measurement error can be identified, however, by extending model (13.5) and modifying the definition of the semivariogram. Let $Z(\mathbf{s}, k)$ represent the $k$th replicate measurement taken at site $\mathbf{s} \in D$, and suppose that

$$Z(\mathbf{s}, k) = \mu(\mathbf{s}) + \delta(\mathbf{s}) + \epsilon(\mathbf{s}, k)$$

where $\mu(\mathbf{s})$ and $\delta(\mathbf{s})$ are defined as in model (13.5) and $\epsilon(\mathbf{s}, k)$ is a $(d + 1)$-dimensional zero-mean white noise process with constant variance $v$, independent of $\delta(\mathbf{s})$. Now the semivariogram for the total error process, $\delta(\mathbf{s}) + \epsilon(\mathbf{s}, k)$, is defined as

$$\gamma'(\mathbf{h}, \zeta_{kl}) = \frac{1}{2} \text{var}[Z(\mathbf{s}, k) - Z(\mathbf{s} + \mathbf{h}, l)]$$

$$= \begin{cases} 0 & \text{if } \mathbf{h} = 0 \text{ and } \zeta_{kl} = 1 \\ v & \text{if } \mathbf{h} = 0 \text{ and } \zeta_{kl} = 0 \\ v + \gamma(\mathbf{h}) & \text{if } \|\mathbf{h}\| > 0 \end{cases} \tag{13.20}$$

where $\gamma(\cdot)$ is the semivariogram of $\delta(\cdot)$ and $\zeta_{kl} = 1$ if $k = l$, 0 otherwise. Note that (13.20) is a particular kind of cross-semivariogram. This cross-semivariogram can be estimated from data $Z(\mathbf{s}_1), \ldots, Z(\mathbf{s}_n)$ (where $\mathbf{s}_1, \ldots, \mathbf{s}_n$ are not necessarily distinct) by nonparametric estimators similar to (13.12) and (13.13); one simply replaces $\hat{\delta}(\mathbf{s}_i)$ and $\hat{\delta}(\mathbf{s}_j)$ in those formulas with $Z(\mathbf{s}_i, k) - \hat{\mu}(\mathbf{s}_i)$ and $Z(\mathbf{s}_j, l) - \hat{\mu}(\mathbf{s}_j)$, respectively. Moreover, the quantity

$$\frac{1}{2|N(0)|} \sum_{N(0)} [Z(\mathbf{s}_i, k) - Z(\mathbf{s}_j, l)]^2$$

is a model-free estimate [i.e., an estimate that does not depend on the validity of the assumed model for the mean function $\mu(\cdot)$] of the measurement–error variance $v$.

One very important practical issue is that of "design," i.e., the selection of sites where measurements are to be taken. Because the kriging variance (either (13.17) or (13.18)) depends on the semivariogram and the data locations, but not on the data themselves, one can, in principle, prospectively determine the data locations that will minimize the kriging variance corresponding to any given potential site. From a more global perspective, one may wish instead to minimize the average, or perhaps the maximum, kriging variance over all of the study area $D$. Some practical solutions to these and similar design problems can be found in [28–31]. Design criteria for precise estimation of the semivariogram have also been considered; see e.g., [32] and [33].

A second practical issue pertains to the intrinsic stationarity assumption for the error process in model (13.5), upon which the entire analysis is based. Though experience suggests that this assumption is well-satisfied locally, i.e., within small regions, it may be of questionable validity globally. A natural way to check for nonstationarity in the second-order structure, and to account for it if it is found, is to estimate a "local semivariogram" at each site, using only those data contained in a moving window centered on the site; see [34]. Spatial prediction at a site could then be based on the corresponding local semivariogram, the predictor being a function of only those data in the corresponding window.

## References

1. National Atmospheric Deposition Program. (1988). *NADP/NTN Annual Data Summary: Precipitation Chemistry in the United States. 1987.* National Resource Ecology Laboratory, Colorado State University, Fort Collins.
2. Ripley, B. D. (1981). *Spatial Statistics.* John Wiley and Sons, New York.
3. Cressie, N. (1991). *Statistics for Spatial Data.* John Wiley and Sons, New York.
4. Zimmerman, D. L., and Harville, D. A. (1989). "On the Unbiasedness of the Papadakis Estimator and Other Nonlinear Estimators of Treatment Contrasts in Field-Plot Experiments."*Biometrika* **76,** 253.
5. Harville, D. A. (1985). "Decomposition of Prediction Error." *J. Amer. Statist. Assoc.* **80,** 132.
6. Tukey, J. W. (1977). *Exploratory Data Analysis.* Addison-Wesley, Reading, Massachusetts.
7. Cressie, N. (1984). "Towards Resistant Geostatistics," in *Geostatistics for Natural Resources Characterization, Part 1,* G. Verly, M. David, A. G. Journel, and A. Marechal (eds.), p. 21. Reidel, Dordrecht.
8. Cressie, N. (1986). "Kriging Nonstationary Data." *J. Amer. Statist. Assoc.* **81,** 625.

9. Emerson, J. D., and Hoaglin, D. C. (1983). "Analysis of Two-Way Tables by Medians," in *Understanding Robust and Exploratory Data Analysis,* D. C. Hoaglin, F. Mosteller, and J. W. Tukey (eds.), p. 166. John Wiley and Sons, New York.
10. Cressie, N., and Glonek, G. (1984). "Median Based Covariogram Estimators Reduce Bias." *Statist. Probab. Letters* **2,** 299.
11. Robinson, G. K. (1990). "A Role for Variograms." *Austral. J. Statist.* **32,** 327.
12. Cressie, N., and Hawkins, D. M. (1980). "Robust Estimation of the Variogram, I." *J. Inter. Assoc. Math. Geol.* **2,** 115.
13. Zimmerman, D. L., and Zimmerman, M. B. (1991). "A Comparison of Spatial Semivariogram Estimators and Corresponding Ordinary Kriging Predictors." *Technometrics* **33,** 77.
14. Cressie, N. (1985). "Fitting Variogram Models by Weighted Least Squares." *J. Inter. Assoc. Math. Geol.* **17,** 563.
15. Journel, A. G., and Huijbregts, C. J. (1978). *Mining Geostatistics.* Academic Press, London.
16. Mardia, K. V., and Marshall, R. J. (1984). "Maximum Likelihood Estimation of Models for Residual Covariance in Spatial Regression." *Biometrika* **71,** 135.
17. Kitanidis, P. K. (1983). "Statistical Estimation of Polynomial Generalized Covariance Functions and Hydrologic Applications." *Water Res. Research* **19,** 909.
18. Webster, R., and McBratney, A. B. (1989). "On the Akaike Information Criterion for Choosing Models for Variograms of Soil Properties." *J. Soil Sci.* **40,** 493.
19. Davis, B. M. (1987). "Uses and Abuses of Cross-Validation in Geostatistics." *Math. Geol.* **19,** 241.
20. Stein, M. L. (1988). "Asymptotically Efficient Prediction of a Random Field with a Misspecified Covariance Function." *Ann. Statist.* **16,** 55.
21. Cressie, N. (1989). "The Many Faces of Spatial Prediction," in *Geostatistics,* M. Armstrong (ed.), **1,** p. 163. Kluwer, Dordrecht.
22. Zimmerman, D. L., and Cressie, N. (1992). "Mean Squared Prediction Error in the Spatial Linear Model with Estimated Covariance Parameters." *Ann. Inst. Statist. Math.* **44,** 27.
23. Rouhani, S., and Myers, D. (1990). "Problems in Space–Time Kriging of Geohydrological Data." *Math. Geol.* **22,** 611.
24. Rodriguez-Iturbe, I., and Mejia, J. M. (1974). "The Design of Rainfall Networks in Time and Space." *Water Res. Research* **10,** 713.
25. Rouhani, S., and Hall, T. J. (1989). "Space–Time Kriging of Groundwater Data," in *Geostatistics,* M. Armstrong (ed.), **2,** p. 639. Kluwer, Dordrecht.
26. Clark, I., Basinger, K. L., and Harper, W. V. (1989). "MUCK: A Novel Approach to Co-Kriging," in *Proceedings of the Conference on Geostatistical, Sensitivity, and Uncertainty Methods for Ground-Water Flow and Radionuclide Transport Modeling,* B. E. Buxton (ed.), p. 473. Battelle Press, Columbus, OH.
27. Ver Hoef, J. M., and Cressie, N. (1993). "Multivariable Spatial Prediction." *Math. Geol.* **25,** 219.
28. Bras, R. L., and Rodriguez-Iturbe, I. (1976). "Network Design for the Estimation of Areal Mean of Rainfall Events." *Water Res. Research* **12,** 1185.
29. McBratney, A. B., Webster, R., and Burgess, T. M. (1981). "The Design of Optimal Sampling Schemes for Local Estimation and Mapping of Regionalized Variables—I." *Computers and Geosciences* **7,** 331.

30. Barnes, R. J. (1989). "Sample Design for Geological Site Characterization," in *Geostatistics,* M. Armstrong (ed.), **1,** p. 809. Kluwer, Dordrecht.
31. Cressie, N., Gotway, C. A., and Grondona, M. O. (1990). "Spatial Prediction from Networks." *Chemometrics and Intelligent Laboratory Systems* **7,** 251.
32. Warrick, A. W., and Myers, D. E. (1987). "Optimization of Sampling Locations for Variogram Calculations." *Water Res. Research* **23,** 496.
33. Zimmerman, D. L., and Homer, K. E. (1991). "A Network Design Criterion for Estimating Selected Attributes of the Semivariogram." *Environmetrics* **2,** 425.
34. Haas, T. C. (1990). "Lognormal and Moving Window Methods of Estimating Acid Deposition." *J. Amer. Statist. Assoc.* **85,** 950.

# 14. BAYESIAN METHODS

Harry F. Martz and Ray A. Waller

Los Alamos National Laboratory

Methods of statistical inference can be loosely partitioned into two broad classes: classical and Bayesian statistical methods. Although these two classes are fundamentally quite different (as we shall soon see), both have found wide acceptance and utility in the physical sciences. Classical statistics are epitomized by such methods as maximum likelihood (Chapter 8), least squares (Chapter 9), confidence intervals (Chapter 6), and significance testing (Chapter 6). Bayesian alternatives to such methods are considered in this chapter.

Papers in a variety of international journals illustrate a wide breadth of recent Bayesian research activity and application. For example, Bayesian research applications and areas include nuclear magnetic resonance (NMR) spectroscopy, macromolecular crystallography, isotopic dating of geological age differences, shot noise, ground-water contaminant transport, phase diagrams, ion cyclotron resonance time-domain signals, climate change, space–time distribution of sunspot groups, image recovery–reconstruction, neutron reflectivity and scattering, small signal analysis, observations aboard a spacecraft, fatigue in composite laminates, Bell's theorem, and peak fitting for gamma-ray spectra, econometrics, astronomy, condensed matter physics, search for extraterrestrial intelligence, foundations of quantum mechanics, expert systems, pharmacology, medicine, DNA sequencing, machine translation of languages, neural networks, data classification, and theoretical nuclear physics. We consider four such applications in Sec. 14.4.

## 14.1 Bayesian Statistical Inference

Bayesian methods are named for the philosophical approach embodied in the 18th-century work of Thomas Bayes (1702–1761), a Presbyterian minister and mathematician. Bayes's original manuscript on inverse probability (including Bayes's theorem) for the binomial distribution was posthumously published in 1763 [1]. Laplace [2] stated the theorem on inverse probability in general form.

Bayesian and classical methods of inference share the same goals. Both methods attempt to infer (learn) something about an assumed distribution or model (or its parameters) based on a sample of data that provides only partial and inconclusive information about the model or its parameters. We use the term *parameter* in the broadest sense, and it may, in fact, denote an entire unknown distribution or spectrum, such as an unknown neutron scattering law in a neutron scattering experiment (see Sec. 14.4.1). The unknown parameters in the assumed *data (or sampling) model* of interest are considered to be random variables having a so-called *prior distribution* (Sec. 14.2) that expresses the analyst's knowledge about the parameters prior to observing the experimental (or sample) data.

The widespread use of Bayesian methods indicates that physical scientists use an assortment of relevant prior information (a Bayesian approach) to supplement current experimental data (the classical approach) in data analysis. The information may be in the form of data or results from previous experiments; known conservation laws or default models; known characteristics of the assumed model (such as symmetry, positivity or additivity); known data smoothing functions or filters, known phase information, scientific conjecture, engineering knowledge, and experience; or other objective or subjective data sources.

Figure 1 depicts the Bayesian method of statistical inference. A data (sampling) distribution is postulated that relates the unknown parameters to the experimental data, and a prior distribution is postulated for the

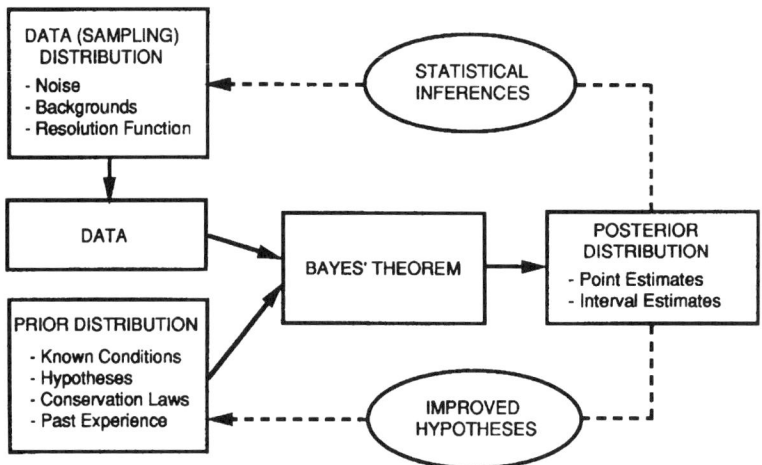

FIG. 1. Bayesian statistical inference

unknown parameters. The experimental data and the prior distribution are then combined using Bayes's theorem (see later) to produce a so-called *posterior distribution*. The posterior distribution is then used to make all the desired inferences, usually in the form of point and interval estimates of the unknown parameters (see Sec. 14.3). These posterior-produced estimates of the parameters may also lead to improved hypotheses regarding the parameters. Bayesian statistical inference thus represents a logically correct and deductive paradigm that has been demonstrated to be useful in a wide variety of scientific experiments.

Bayesian and classical methods differ in both practical and philosophical aspects. The major practical distinction is that Bayesian methods permit the formal incorporation of supplementary knowledge, belief, and information beyond that contained in the observed data in the inference process. This additional information is embodied in the prior distribution.

Now consider the philosophical distinctions. Classical methods are rooted in the well-known relative frequency notion of probability, where the probability of an event is defined as the limiting relative frequency of occurrence of the event in a series of repeated trials. In contrast to this notion of probability, the cornerstone of the foundation of Bayesian methods is the notion of *subjective probability*. Bayesian methods consider probability to be a subjective assessment of the knowledge (sometimes called degree of belief) about propositions of interest, given any available evidence. When the available evidence is mostly empirical, both views often arrive at the same numerical value for the probability in question. For example, a Bayesian would say his subjective probability is one-third that a fair die comes up 1 or 2 when tossed because of what is known about the physics of tossing a die, the uniform mass distribution of the die, and past experience in rolling a die. Subjective probabilities are philosophically appropriate for events that cannot be repeated under similar and controlled conditions, such as a severe core melt in a nuclear power reactor. It has been shown that subjective probability has a sound theoretical foundation and obeys the axioms of probability.

Bayesian and classical methods also differ in their logical approach. Classical statistics uses inductive methods of reasoning, e.g., inductive reasoning is used to construct confidence intervals (see Chapter 6). On the other hand, Bayesian methods use a formal deductive method of reasoning employing Bayes's theorem, which we now introduce and illustrate with a simple example.

To illustrate Bayes theorem, consider a process that produces electronic sensors for radar detectors. In manufacturing the sensors, $n$ automated production lines, $A_1, A_2, \ldots, A_n$ are operated. Further, we suppose that the supply of sensors for shipment is a composite of sensors produced by all lines, that the portion of the total sensors produced by line $A_i$ is given by $P(A_i)$, and that each production line has a defective rate given by $P(D \mid A_i)$. A question of interest follows: A shipment is inspected and a defective sensor is observed, what is the probability that the defective sensor was produced by line $A_i$? Or what is the $P(A_i \mid D)$?

**Bayes's Theorem:** *Consider a set of disjoint events,* $A_1, A_2, \ldots, A_n$, *for which* $P(A_i)$, $i = 1, 2, \ldots, n$, *are known and a proposition* D *such that* $P(D) > 0$. *Then*

$$P(A_i \mid D) = \frac{P(D \mid A_i)P(A_i)}{P(D)}, \quad i = 1, 2, \ldots, n \quad (14.1)$$

*where* $P(D) = \sum_i P(D \mid A_i)P(A_i)$.

The results of Bayes's theorem are sometimes referred to as *inverse probabilities*, which follows from using the *prior probabilities* $P(A_i)$ and the *conditional (or sampling) probabilities* $P(D \mid A_i)$ to obtain the *posterior (inverse) probabilities* $P(A_i \mid D)$.

**Example:** XYZ Sensors, Inc., has four production lines, $A_1, A_2, A_3$, and $A_4$, producing radar detector sensors for bulk shipments to customers. It is known that the lines produce 25%, 20%, 35%, and 15% of the sensors, respectively. Historical operations data indicate that the defective rates for the lines are as follows: $P(D \mid A_1) = 0.010$, $P(D \mid A_2) = 0.005$, $P(D \mid A_3) = 0.012$, and $P(D \mid A_4) = 0.007$. What percent of the defective sensors are produced by line $A_3$? By Bayes's theorem,

$$P(D) = 0.010(0.25) + 0.005(0.20) + 0.012(0.35) + 0.007(0.15)$$
$$= 0.00875$$

$$P(A_3 \mid D) = 0.012(0.35)/0.00875 = 0.480.$$

Thus, 48.0% of all defectives are produced by line $A_3$. Similarly, lines $A_1$, $A_2$, and $A_4$ produce 28.6%, 11.4%, and 12.0%, respectively.

The foregoing discussion and illustration of Bayes's theorem for the discrete parameter–discrete data case can be extended to the continuous parameter–continuous data case in the following manner. Suppose that the continuous random variable $T$ is distributed according to the probability density function $f(t; \theta)$, where $\theta$ is an unknown parameter that is of interest to be estimated, and that the prior probability density function of $\theta$ is $\pi(\theta)$. Here $f(t; \theta)$ is the so-called sampling distribution of $T$ for any specified value of $\theta$. We note that the sampling distribution is also known as the *likelihood function (or likelihood)* when considered as a function of $\theta$ for any observed (sample) value of $t$. The likelihood, which we usually denote by $l(\theta \mid t)$, summarizes the information contained in the sample regarding $\theta$. These notions also apply to the discrete parameter–discrete data case as well.

The continuous version of Bayes's theorem is as follows:

$$\pi(\theta \mid t) = \frac{f(t; \theta)\pi(\theta)}{f(t)}, \qquad (14.2)$$

where $f(t) = \int f(t; \theta)\pi(\theta)d\theta$ is the *marginal (unconditional) distribution* of $T$. There also exist other "mixed" cases of the preceding discrete and continuous cases. We see that the posterior is proportional to the product of the likelihood and the prior, the proportionality constant being the reciprocal of the marginal distribution evaluated at $t$. The proportionality constant ensures that the posterior distribution integrates to 1.

## 14.2 The Prior Distribution

The necessity to identify an appropriate prior distribution that describes prior knowledge (or ignorance) is the heart of Bayesian statistics. Thus, in applying Bayesian methods, we must explicitly identify and summarize what prior knowledge we have regarding the quantity to be estimated and somehow reflect this knowledge in the form of an appropriately chosen prior distribution. While this task can be difficult, it is nearly always well worth the effort, as illustrated by the examples in Sec. 14.4. The prior distribution can be used to express *any* desired prior state of knowledge, ranging from virtual ignorance (such as noninformative priors in Sec. 14.2.2), through limited information (such as maximum entropy priors subject to partial prior information in Sec. 14.2.5), to quite informative knowledge (such as conjugate priors in Sec. 14.2.4).

### 14.2.1 Consequences of Prior Choice

As given by Bayes's theorem, the posterior distribution of the parameter is proportional to the product of the prior and the likelihood. In general, if the prior is relatively flat where the likelihood is appreciable, then the likelihood dominates the prior in the sense that the posterior largely reflects only the information regarding the model (or model parameter(s)) of interest contained in the data (as expressed by the likelihood). This is known as a *dominant likelihood* situation. The resulting posterior distribution is then proportional to the likelihood and, in particular, the value of $\theta$ that maximizes the likelihood (the so-called maximum likelihood estimator in Chapter 6) will be the same as the value that maximizes the posterior, the so-called Bayesian *maximum a posteriori* (or MAP) estimator. Otherwise, if the prior is sharply peaked relative to the likelihood (i.e., a strongly informative prior), then the prior dominates the likelihood and the posterior largely reflects only the information contained in the prior.

In most cases, the likelihood will increasingly tend to dominate the prior as the number of observations increases. For large sample sizes, the posterior will thus be approximately equal to the normalized likelihood (having unit area or mass), and the difference between Bayesian and classical inferences is insignificant. In many practical problems, a moderate sample size will yield a dominant likelihood if the sample results are consistent with the prior distribution. If the sample data are inconsistent with the prior assumptions, then the likelihood will not tend to dominate the prior, and a weighted combination of both will be reflected in the posterior.

The notion of dominant likelihood is illustrated by a simple coin-tossing experiment. Suppose a coin is flipped $n$ times and heads appear $x$ times, but we do not know whether the coin was fair. Our problem is to *infer* the probability $p$ of a head (the coin's bias weighting for heads). Let $p = 0$ denote a double-tailed coin; $p = 0.5$, a fair coin; and $p = 1$, a double-headed coin. Figure 2(a) shows three different possible prior states of knowledge about the coin: a uniform (or ignorant) prior (Prior A), a prior that assumes the coin is most likely either double-headed or double-tailed (Prior B), and a prior that assumes the coin is probably fair (Prior C).

The likelihood for our coin-tossing experiment is given by the binomial distribution. Figures 2(b)–2(d) show how the posterior for each of the three priors in Fig. 2(a) changes as we collect more and more data. Note as we increase the size of $n$, we become more confident in our inferred value for $p$ (i.e., the variance of each posterior decreases) and the influence of the prior distribution decreases (i.e., the likelihood increasingly dominates the prior and the posteriors converge to the same answer when sufficient data are available). In other words, no matter what our prior state of knowledge, the data force us to the same conclusion. If the data are inaccurate, few in number, or inappropriate for the parameter of interest, then the posterior information will depend crucially on our prior knowledge.

### 14.2.2 Noninformative Priors

There are times when the scientist wants to complete a Bayesian analysis but does not wish to show a preference for any particular values of the parameter a priori. In those cases, the desired prior distribution must give equal probability to each possible value of the parameter(s). For a discrete variable with mass assigned to n points, say $A_1, A_2, \ldots, A_n$, a noninformative prior is

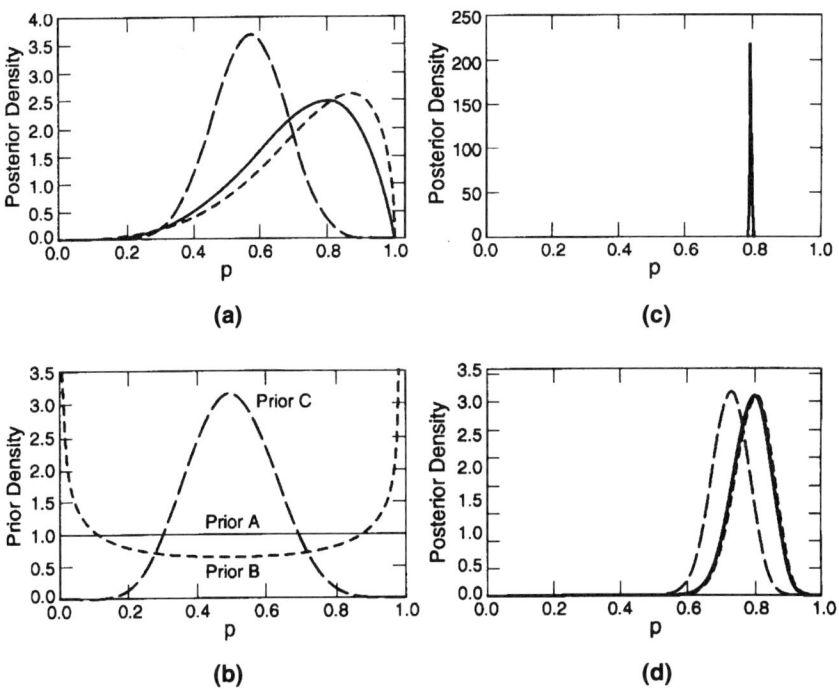

FIG. 2. Prior and posterior distributions for the coin-tossing experiment. (a) Three different prior distributions. (b) Posterior distributions given 4 heads in 5 tosses. (c) Posterior distributions given 40 heads in 50 tosses. (d) Posterior distributions given 40,000 heads in 50,000 tosses.

$P(A_i) = 1/n, i = 1, 2, \ldots, n$. For a normal variable with mean $\mu$, a noninformative prior for $\mu$ is $\pi(\mu)$ = constant, $-\infty < \mu < \infty$. In this case, $\pi(\mu)$ is an improper prior as discussed in Sec. 14.2.3. A non-informative (also improper) prior for a scale parameter, $\theta$, is found to be $\pi(\theta) = 1/\theta$, $0 < \theta < \infty$. The derivation for both of these is presented in [3], p. 83. For further examples and discussion, also see [4], p. 223.

### 14.2.3 Improper Priors

As implied throughout the preceding discussions of prior distributions, the role of the prior in any Bayesian analysis is to assign probabilities, densities, or "weights" to the parameter space. Many priors are either probability mass functions or density functions (probability measures) and, thus, are proper priors that distribute unit mass to the parameter space. That is, the prior distributions, $P(A_i)$ or $\pi(\theta)$, sum or integrate to 1 over the parameter space. There exist certain situations in which weighting functions for the parameter space, other than proper probability measures, yield useful analytic results through Bayes's theorem. This is true even when the measure is infinite, as in the case of the noninformative priors for the location and scale parameters in the preceding section.

### 14.2.4 Conjugate Priors

It is often difficult to perform the required analysis necessary to obtain a closed form solution for the posterior distribution $P(A_i \mid D)$ or $\pi(\theta \mid t)$. Numerical approximations and computer simulations are sometimes used to provide approximate answers. However, so-called conjugate families of distributions are an exception to the general case. Loosely stated, conjugate families of distributions are those distributions for which the prior and posterior distributions are members of the same family. For a more precise definition and discussion, see [3]. A discussion and further references are available in [4]. When conjugate priors exist for a Bayesian analysis, it is usually trivial to obtain moments (or other summary measures of interest) from the posterior distribution, as they are often a simple function of the prior moments and the data. The following example illustrates a common conjugate family of distributions.

Consider a gamma prior distribution on $\lambda$, the unknown rate of occurrence of events in a Poisson process, where the Poisson data consist of observing $x$ events in a given period of time $t$. Bayes's theorem gives a gamma posterior distribution on $\lambda$ as follows:

Gamma prior: $\pi(\lambda; \alpha, \beta) = \dfrac{\beta^\alpha}{\Gamma(\alpha)} \lambda^{\alpha-1} e^{-\beta\lambda}, \quad \lambda, \alpha, \beta > 0$

Poisson data: $p(x; \lambda, t) = e^{-\lambda t}(\lambda t)^x / x!, \quad x = 0, 1, \ldots; \lambda, t > 0$

Gamma posterior: $\pi(\lambda \mid x; \alpha, \beta, t) = \dfrac{(t + \beta)^{x+\alpha}}{\Gamma(x + \alpha)} \lambda^{x+\alpha-1} e^{-(\beta+t)\lambda},$
$\lambda > 0$

For convenience, we denote a gamma distribution on $\lambda$ with parameters $\alpha$ and $\beta$ as Gamma($\lambda; \alpha, \beta$). The moments of the gamma prior are functions of the parameters $\alpha$ and $\beta$; for example, the mean is $\alpha/\beta$. Similarly, the parameters of the gamma posterior distribution are $(x + \alpha)$ and $(\beta + t)$ and the posterior mean is $(x + \alpha)/(\beta + t)$. Note that the posterior parameters are a function of the prior parameters $\alpha$ and $\beta$ and the data $x$ and $t$.

It may be shown that the beta family of distributions is the conjugate family for binomial sample data, while the gamma family is conjugate for exponential data [4].

### 14.2.5 Maximum Entropy Priors

Often when choosing a prior our knowledge lies somewhere between complete ignorance (i.e., the use of a noninformative prior) and strong prior knowledge (i.e., the use of a conjugate prior). Frequently, only partial prior information is available, in which case we may want to use a prior that is as noninformative as possible given the partial prior information constraint.

A useful method for dealing with this situation is through the concept of entropy [5, 6]. Let $\pi$ denote a prior on the discrete parameter space $\Theta$ in which $\pi_j = \pi(\theta_j) = \Pr(\Theta = \theta_j), j = 1, 2, \ldots$ . The Shannon–Jaynes *entropy* of this distribution is defined as

$$S(\pi) = - \sum_j \pi_j \log \pi_j \qquad (14.3)$$

in which, if $\pi_j = 0$, the quantity $\pi_j \log \pi_j$ is defined to be 0. Entropy inversely measures the information inherent in a distribution [7]; as the inherent information decreases, the entropy increases. Thus, it seems reasonable to choose as our prior distribution that particular distribution, say, $\bar{\pi}$, which maximizes the entropy (and thus is conservative), subject to any specified prior constraints regarding $\pi$. If there are no constraints and $\Theta$ contains $n$ values, then the prior that maximizes the entropy is the "flat" (or noninformative) prior $\pi_j = 1/n, j = 1, 2, \ldots, n$.

Now suppose that we have $m$ constraints on $\pi$ of the form

$$\sum_j \pi(\theta_j) g_k(\theta_j) = \mu_k, \qquad k = 1, \ldots, m \qquad (14.4)$$

in which $g_k$ and $\mu_k$ are known. Provided $\sum \pi_j = 1$, the corresponding maximum entropy prior subject to the restrictions in Eq. (14.4) is known to be

$$\bar{\pi}(\theta_j) \propto \exp\left[\sum_k \lambda_k g_k(\theta_j)\right], \qquad j = 1, 2, \ldots \qquad (14.5)$$

where the Lagrange multipliers $\lambda_k$ are constants that are determined from the constraints in Eq. (14.4).

If $\Theta$ is a continuous random variable with prior density function $\pi(\theta)$, then the concept and use of maximum entropy becomes more complicated [5]. However, it is still possible to obtain and use maximum entropy priors.

Suppose that we wish to maximize the generalized Shannon–Jaynes entropy (also known as *cross entropy* or *relative entropy*) functional defined as

$$S(\pi, \pi_0) = -\int \pi(\theta) \log\left\{\frac{\pi(\theta)}{\pi_0(\theta)}\right\} d\theta \qquad (14.6)$$

subject to m prior constraints of the form

$$\int g_k(\theta)\pi(\theta)d\theta = \mu_k, \qquad k = 1, \ldots, m \qquad (14.7)$$

In Eq. (14.6), $\pi_0(\theta)$ is an appropriately chosen noninformative reference prior (or default model) to which the maximum entropy solution will default in the absence of prior constraints. In the absence of constraints on the prior information, the maximum of Eq. (14.6) occurs when $\pi(\theta)$ equals $\pi_0(\theta)$. The negative of the entropy in Eq. (14.6) is also known as the *Kullback–Leibler distance* between $\pi(\theta)$ and $\pi_0(\theta)$.

When it exists, the maximum entropy prior, say $\bar{\pi}$, which maximizes Eq. (14.6), subject to the constraints in Eq. (14.7), is given by

$$\bar{\pi}(\theta) \propto \pi_0(\theta) \exp\left[\sum_k \lambda_k g_k(\theta)\right] \qquad (14.8)$$

where the Lagrange multipliers $\lambda_k$ are determined from the constraints in Eq. (14.7).

Consider the following example found in [3]. Suppose we are interested in the maximum entropy prior $\bar{\pi}(\theta)$ of a location parameter $\theta$, $-\infty < \theta < \infty$, having known mean $\mu$ and variance $\sigma^2$. The noninformative (default) prior is thus $\pi_0(\theta) = 1$. Also, $g_1(\theta) = \theta$, $\mu_1 = \mu$, $g_2(\theta) = (\theta - \mu)^2$, and $\mu_2 = \sigma^2$ in Eq. (14.7). From Eq. (14.8), the corresponding maximum entropy prior density is thus

$$\begin{aligned}\bar{\pi}(\theta) &\propto \exp\left[\lambda_1\theta + \lambda_2(\theta - \lambda)^2\right] \\ &= \exp\left\{\lambda_2[\theta - (\mu - \lambda_1/2\lambda_2)]^2\right\}\end{aligned} \qquad (14.9)$$

where $\lambda_1$ and $\lambda_2$ are chosen so that the two constraints are satisfied. Choosing $\lambda_1 = 0$ and $\lambda_2 = -1/(2\sigma^2)$ satisfies the constraints; thus, the least informative prior on $\theta$ having known mean $\mu$ and variance $\sigma^2$ is a Gaussian distribution.

Bayesian methods in the physical sciences often use maximum entropy priors when estimating an unknown distribution *f*. In this case, *f* is vector valued when considered at a finite number of points on its domain. Areas of application include the neutron scattering law, the electron density in a crystal, incoherent light intensity as a function of position in an optical image, ion cyclotron resonance spectral analysis, radio-frequency brightness of an astronomical source, and many others. The definition of entropy in Eq. (14.6) can be generalized to include such cases. This will be illustrated in the neutron scattering example in Sec. 14.4.1.

## 14.3 Bayesian Estimation

The most common Bayesian point estimators are the mean, median, and mode of the posterior distribution. When it exists, the posterior mode is the MAP estimator discussed in Sec. 14.2.1, and it is widely used in physical science.

The Bayesian analog of a classical confidence interval is called a Bayesian *credibility interval* (or sometimes a *probability interval*). However, unlike classical confidence intervals, credibility intervals have a natural probability interpretation and are obtained directly from the posterior distribution. A $100(1 - \alpha)\%$ credibility interval $(\theta_1, \theta_2)$ is a probabilistic statement about $\theta$ given the sample data $x$ and the prior distribution; that is, $P(\theta_1 \leq \theta \leq \theta_2 \mid x) = 1 - \alpha$. This direct probability statement avoids the inherent problems in interpreting classical confidence intervals ([8], p. 30).

The notion of *highest posterior density* (HPD) is often used to determine a particular credibility interval. Using this notion, we seek a $100(1 - \alpha)\%$ credibility interval $(\theta_1, \theta_2)$ so that two conditions are satisfied: (1) $P(\theta_1 \leq \theta \leq \theta_2 \mid x) = 1 - \alpha$; and (2) the posterior density for $(\theta_1, \theta_2)$ is greater than that for any other interval satisfying (1). Thus, condition (2) ensures that a HPD credibility interval is the shortest interval among the class of all intervals having the desired coverage probability. Such intervals are usually easy to obtain in practice, particularly when $\theta$ is a univariate parameter (see [3, 4, and 8] for examples).

Predictive distributions are statistical distributions of some unknown future quantity of interest and play an important role in Bayesian estimation. While such distributions pose a problem for classical methods, they pose no fundamental difficulties when using Bayesian methods. The predictive distribution is defined as follows. Suppose we have a sample $x$ from some specified distribution $f(x; \theta)$ that depends on an unknown parameter $\theta$.

Now suppose that we wish to predict a new (future) observation $Y$. The *predictive distribution* for $Y$ is

$$f(y \mid x) = \int f(y; \theta)\pi(\theta \mid x)d\theta \qquad (14.10)$$

and predictive estimates of $y$ (both point and interval) are obtained from this distribution. Note that the predictive distribution is free of $\theta$. A good example is given in [8], p. 58.

## 14.4 Examples

We present four examples in neutron scattering, nuclear magnetic resonance, climate change, and rocket photometry to illustrate the broad applicability of Bayesian estimation methods in the physical sciences.

### 14.4.1 Neutron Scattering

The modern maximum entropy method [9–12] is an optimal Bayesian method that is appropriate for making inferences about positive and additive distributions. Recently, it has been used to analyze neutron scattering data [13–15], which we now illustrate.

Consider the one-dimensional neutron scattering law $f(x)$, where $x$ may represent energy transfer, time of flight, or any other quantity appropriate to one of the many types of neutron scattering spectrometers. In a given experiment $f$ is a positive distribution because it is proportional to the number of neutrons scattered with, say, energy transfer between $x$ and $x + dx$. It is additive because the number of neutrons scattered in a large interval is equal to the sum of the neutrons scattered in smaller intervals that comprise the large interval.

In the traditional classical approach, the scattering law is represented by a model having parameters that are subsequently estimated by least squares. Using the maximum entropy method, there is no need to model the scattering law, and the reconstruction usually shows finer detail (increased resolution) with minimal artifacts of the noise, as we shall see.

The data in a neutron scattering experiment can be represented mathematically as

$$D(x) = \int_{-\infty}^{\infty} R(x - y) f(y) \, dy + B(x) + N(x) \tag{14.11}$$

where $D(x)$ are the observed sample data, $R(x)$ is the instrument response (or resolution) function, $f(x)$ is the unknown scattering law, $B(x)$ is the background, and $N(x)$ is random noise. Note that the integral in Eq. (14.11) represents the *convolution* of $f$ and $R$. Given the data $D$ and the instrument response function $R$, the problem is to infer the scattering law $f$; to perform a *deconvolution* of the data to recover $f$. This represents a so-called *inverse problem*. In general, however, the problem is ill-posed (incomplete) because an infinity of values of $f$ will fit the data. By using Bayesian methods, we can accomplish the required deconvolution and infer the unknown $f$ in the form of the MAP estimate of $f$.

Given only the prior information $I$ that $f$ is positive and additive, we must assign a prior probability distribution $\pi(f; I)$ to $f$. While the appropriate prior for $f$ is not immediately obvious, many different arguments, including logical consistency, combinatorics, information theory, and coding theory, lead us to believe that the prior has the rather special form

$$\pi(f; I, \alpha, m) \propto \exp[\alpha S(f, m)] \qquad (14.12)$$

where $S$ is the generalized Shannon–Jaynes entropy

$$S(f, m) = \sum_i \{ f(x_i) - m(x_i) - f(x_i) \log [f(x_i)/m(x_i)] \} \qquad (14.13)$$

Here $\alpha$ is a dimensional constant (a statistical "regularization constant" required to make the exponential argument unitless) that must be estimated from the data, and $x_i$ is the $i$th discretized value of $x$ on the support of $f$.

In the absence of sample data the maximum of Eq. (14.13) occurs at $f(x_i) = m(x_i)$. The function $m(x_i)$ is, therefore a *default model* analogous to $f_0$ in Eq. (14.6), and it is usually taken to be uniform; that is, $m(x_i) =$ constant.

The other quantity we need to infer $f$ is the likelihood $l(f \mid D)$ that expresses the likelihood of $f$ given by the sample data $D$. In the usual case in which the noise $N(x)$ is Gaussian (really a Gaussian approximation for a Poisson) and the observed neutron counts are independent from bin to bin, then

$$l(f \mid D) \propto \exp(-\chi^2/2) \qquad (14.14)$$

where $\chi^2$ is the usual misfit statistic (which measures how well a trial distribution $f$ fits the observed data):

$$\chi^2 = \sum_{i=1}^{N} \frac{(D_i - F_i)^2}{D_i} \qquad (14.15)$$

Here $D_i$ is the number of neutron counts in the $i$th bin, $F_i$ is the value for the $i$th bin that a trial distribution $f$ would have produced in the absence of noise, and $N$ is the number of bins (or channels) into which the neutron counts have been accumulated (i.e., the number of data points).

Combining Eqs. (14.12) and (14.14) according to Bayes's theorem, the posterior probability distribution for $f$ becomes

$$\pi(f \mid D; I, \alpha, m) \propto \exp(\alpha S - \chi^2/2) \qquad (14.16)$$

The MAP estimate of $f$ maximizes the exponent $\alpha S - \chi^2/2$. This procedure can be interpreted as maximizing the entropy $S$ subject to a constraint on the value of the misfit statistic $\chi^2$, where $\alpha$ is interpreted as a Lagrange multiplier—hence, the name *maximum entropy* (ME) *method* (also known as *MaxEnt*).

Although it is common to choose $\alpha$ such that $\chi^2 = N$, Bayesian arguments can also be used to select $\alpha$. Excellent additional information and insights regarding the ME method are given in [16], while an interesting discussion of the statistical issues surrounding the ME approach is given in [17].

We now illustrate the results in applying the ME method by means of an example considered in [13]. Figure 3 illustrates the results of using the

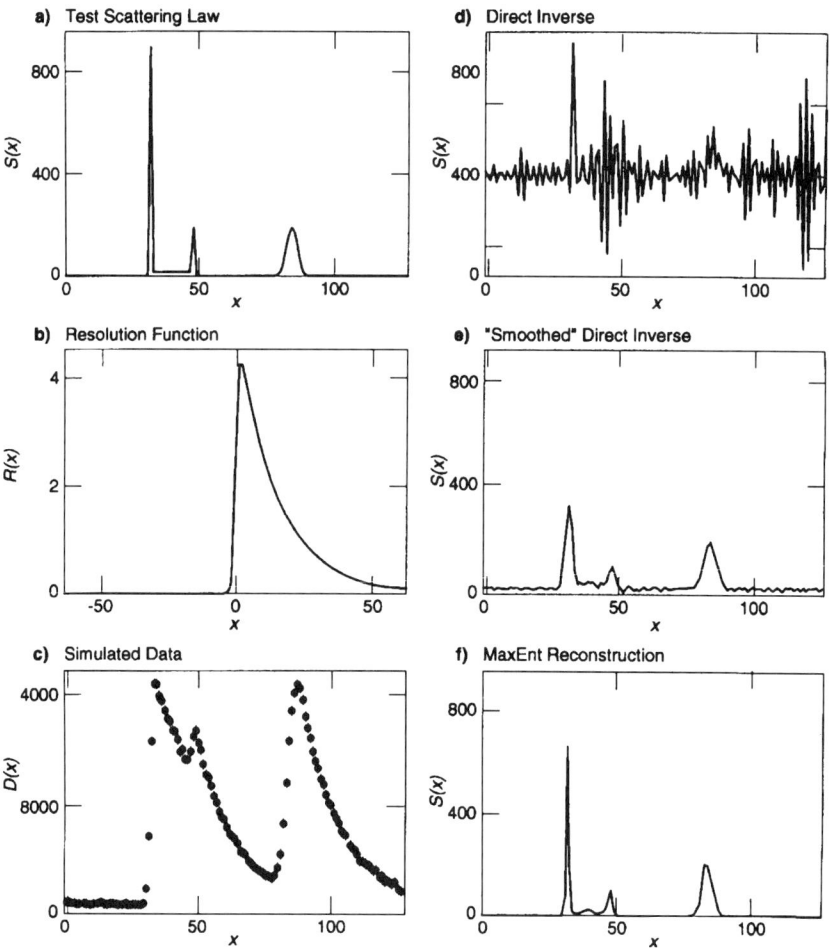

FIG. 3. Deconvolution of simulated neutron scattering data. (Reprinted with permission from D. S. Sivia, "Bayesian Inductive Inference Maximum Entropy and Neutron Scattering," *Los Alamos Science* [Summer 1990], Fig. 7, p. 193.)

ME method for simulated data (computed on a 128-point grid) similar to the transmission spectrum of the filters used in the filter-difference spectrometer (FDS) at the Manuel Lujan, Jr., Neutron Scattering Center (LANSCE) at the Los Alamos National Laboratory. The "true" neutron scattering law, consisting of two spikes on the left and a broader peak on the right, is shown in Fig. 3(a). The noisy data (Fig. 3(c)) are generated by convolving the true law with the FDS resolution function in Fig. 3(b) and then adding a small background count and random noise. Note that a large single spike can produce data similar to that of a smaller, broad peak.

Figure 3(d) gives the deconvolution produced by direct use of the Fourier transform. This method is equivalent to the MLE using a uniform prior. Note that this method produces a reconstruction having a great deal of high-frequency ringing. To overcome this difficulty, a "smoothed" version of the direct inverse, a procedure known as *Fourier filtering*, can be applied. Figure 3(e) illustrates the results in applying this procedure. This classical result may be directly compared with the ME solution shown in Fig. 3(f). We observed that ME has suppressed the level of artifacts without sacrificing as much detail in the deconvolution as does Fourier filtering. Typical ME results for real scattering spectra are illustrated in [13–15].

Finally, the ME procedure may be implemented in practice by using any of the commercial computer programs, such as MEMSYS3 [18].

### 14.4.2 Nuclear Magnetic Resonance

This example is from [19] and illustrates the use of Bayesian methods in analyzing time-domain nuclear magnetic resonance (NMR) signals. This approach is an alternative to the widely used discrete Fourier transform and is particularly useful for analyzing a sum of decaying sinusoids such as time-domain signals associated with pulsed magnetic resonance spectroscopy. The Bayesian approach removes nuisance parameters and simplifies the search process. Also, Bayesian analysis takes advantage of any available prior information (frequently substantial) about the "true" NMR signal, usually expressed in terms of a model. Iterative model-fitting techniques allow interactive residual analyses to continue until all coherent characteristics are represented by the model and removed from the residuals.

The Bayesian spectrum analysis postulates a data model to be a sum of exponentially decaying sinusoids as follows:

$$f(t) = C + \sum_{j=1}^{r} [B_j \cos(\omega_j t) + B_{j+r} \sin(\omega_j t)] e^{-\alpha_j t} \quad (14.17)$$

where $B_j$ and $B_{j+r}$ are effectively the amplitude and phase of the $j$th sinusoid, $\omega_j$ is the frequency, $\alpha_j$ is the decay rate, and $r$ is the total number of sinusoids or resonance frequencies. For this example, an adequate fit to the data is obtained for $r = 4$.

Thus, for the Bayesian procedure, only four frequencies $\omega_j$ and four decay rates $\alpha_j$ need to be determined in the analysis while a least squares analysis has a noise variance and eight additional amplitudes (nuisance parameters) to determine. Our goal is to determine the posterior probability given by

$$\pi(\omega, \alpha \mid D; I) \propto \left[1 - \frac{m\overline{h^2}}{N\overline{d^2}}\right]^{(m-N)/2}, \quad \overline{d^2} = \frac{1}{N}\sum_{i=1}^{N} d_i^2 \quad (14.18)$$

where $m = 2r + 1$ is the total number of amplitudes (including $C$) appearing in Eq. (14.17), $\overline{d^2}$ is the mean square of all $N$ digitized data values $d_i$, and $\overline{h^2}$ is a "sufficient statistic" for making inferences about the frequencies and decay rates.

The following illustrates the output of the Bayesian spectrum analysis for $^{13}$C-{$^1$H} free-induction decay (FID, 75.14 MHz) of a sample of 40% 1,4-dioxane and 60% benzene-d$_6$ and compares the results with Fourier analysis for two data sets. Figure 4 presents the data, the Bayesian spectrum analysis, and the residuals for 256 data points. Note the close agreement between the data and the model.

Figure 5 presents the frequency-domain results for Fourier transformation, Bayesian analysis, and Bayesian line spectrum for 4096 points. The agreement between the Fourier and Bayesian analyses is good for these data, which is true for data with high signal-to-noise ratios.

The capability of the Bayesian analysis to produce good model estimates for poor signal-to-noise ratio data is a strong reason to use the Bayesian procedure in analyzing NMR signals. To illustrate this property, white noise is added to the segment of data shown in Fig. 4(a). Figure 6 presents the modified data, the Bayesian reconstruction, and the residuals. The reconstructed data (model) in Fig. 6(b) matches the original model in Fig. 4(b).

Figure 7 provides the Fourier transformation, Bayesian spectrum analysis, and Bayesian line spectrum for the decreased signal-to-noise ratio data. The impact of the poor signal-to-noise data on the Fourier analysis is considerably greater than on the Bayesian analysis. This property and the reduced complexity provided by decreasing the number of parameters to be determined for adequate fit are two advantages in favor of the Bayesian procedure over the widely used Fourier analysis.

### 14.4.3 Climate Change

The example presented in this section is based on [20], which should be consulted for additional details and discussion. The problem is to use

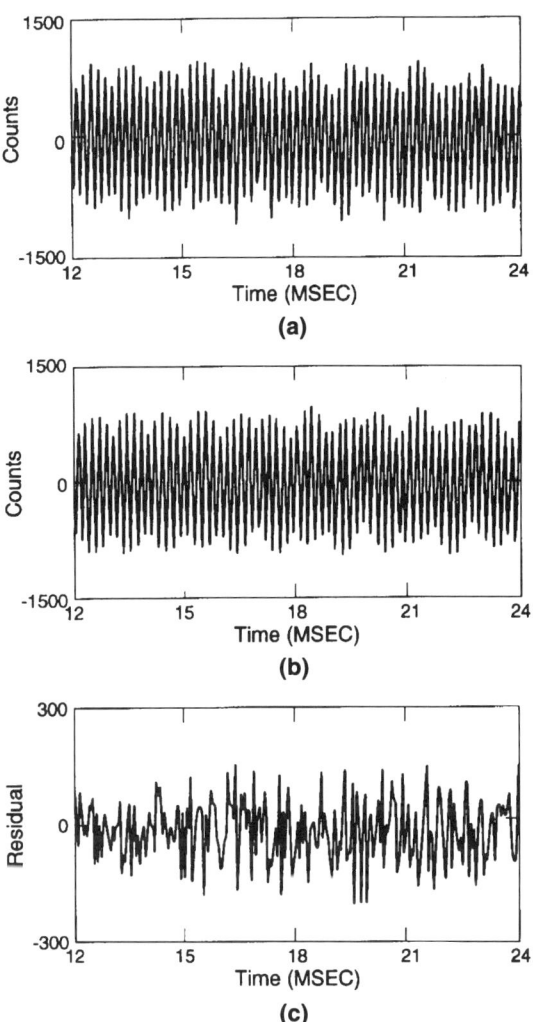

FIG. 4. Bayesian spectrum analysis for high signal-to-noise ratio. (a) Data. (b) Bayesian optimized model. (c) Residuals. (Reprinted with permission from G. L. Bretthorst, C.-C. Hung, D. A. D'Avignon, and J. J. H. Ackerman, "Bayesian Analysis of Time-Domain Magnetic Resonance Signals," *Journal of Magnetic Resonance,* **79** (1988), Fig. 1, pp. 372–375.)

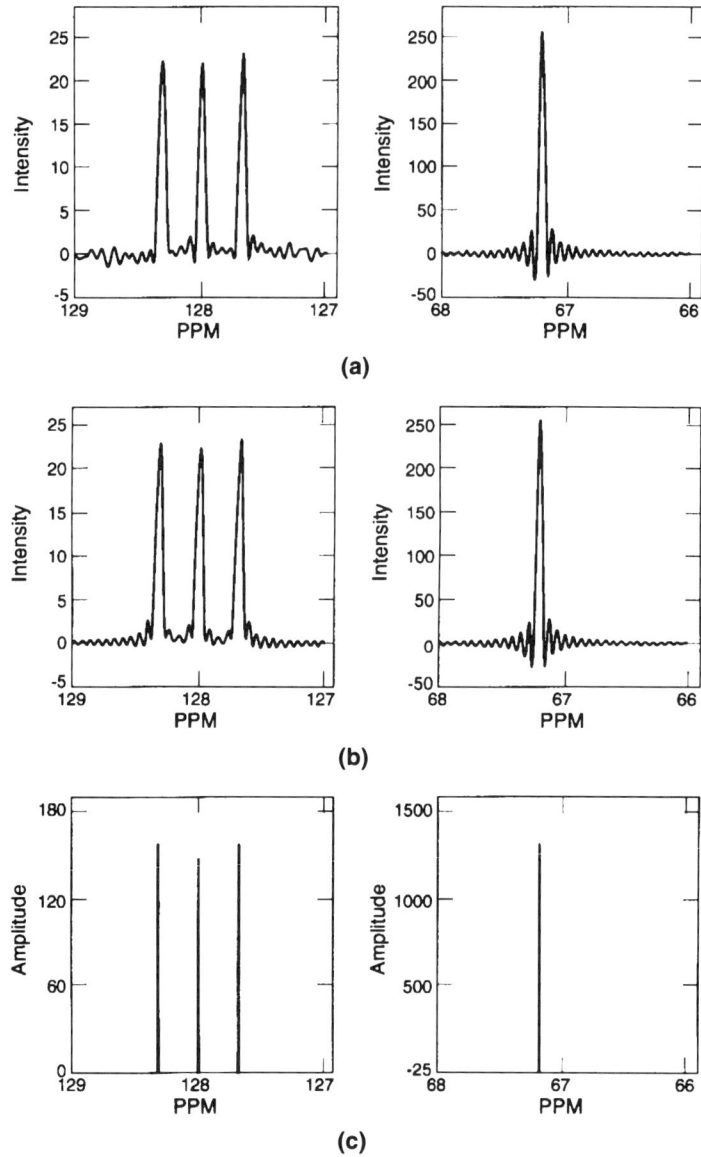

FIG. 5. Frequency-domain results corresponding to Fig. 4. (a) Fourier transform of the data. (b) Fourier transform of the optimized Bayesian model. (c) Line spectrum of the Bayesian model. (Reprinted with permission from G. L. Bretthorst, C.-C. Hung, D. A. D'Avignon, and J. J. H. Ackerman, "Bayesian Analysis of Time-Domain Magnetic Resonance Signals," *Journal of Magnetic Resonance*, **79** (1988), Fig. 2, pp. 372–375.)

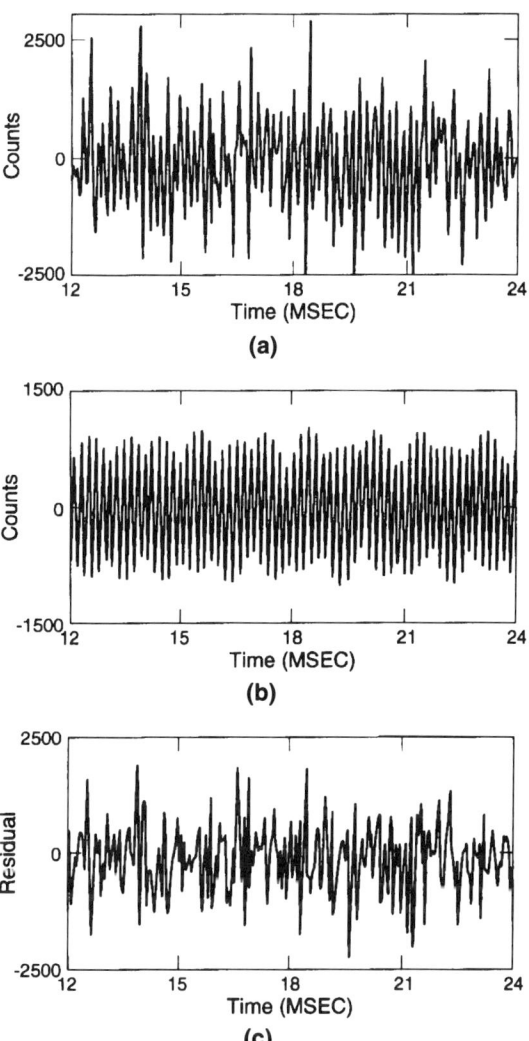

FIG. 6. Bayesian spectrum analysis for low signal-to-noise ratio. (a) Data with white noise added. (b) Bayesian optimized model. (c) Residuals. (Reprinted with permission from G. L. Bretthorst, C.-C. Hung, D. A. D'Avignon, and J. J. H. Ackerman, "Bayesian Analysis of Time-Domain Magnetic Resonance Signals," *Journal of Magnetic Resonance,* **79** (1988), Fig. 3, pp. 372–375.)

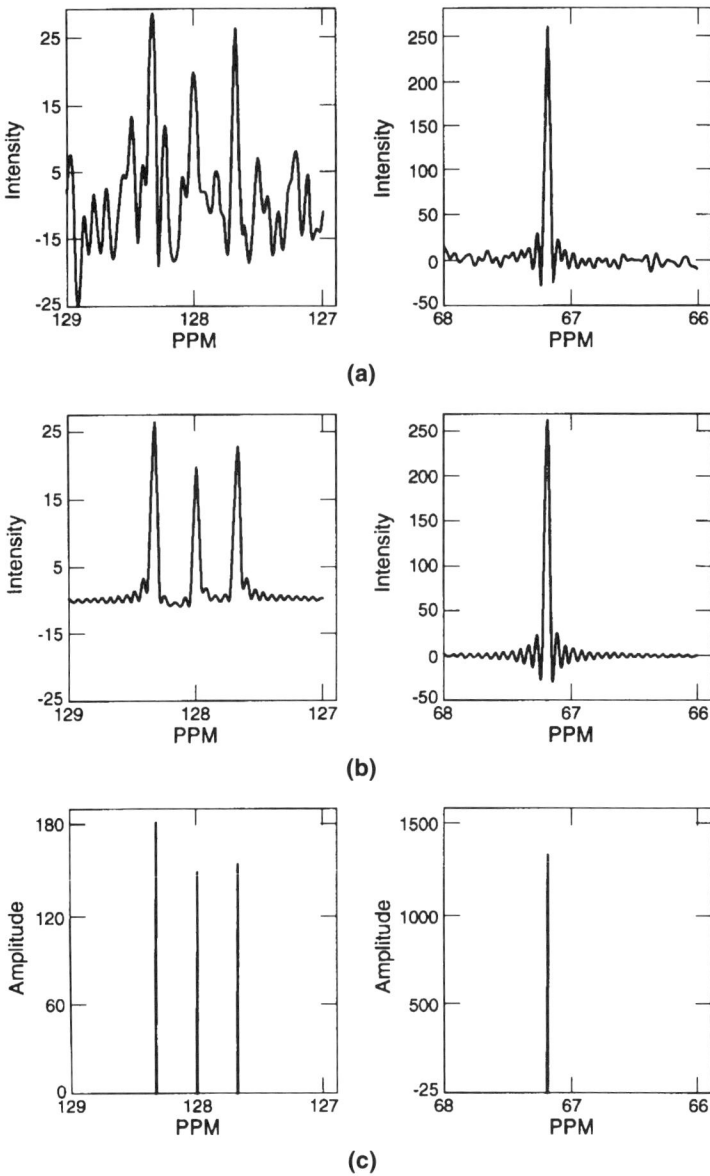

FIG. 7. Frequency-domain results corresponding to Fig. 6. (a) Fourier transform of the data with white noise added. (b) Fourier transform of the optimized Bayesian model. (c) Line spectrum of the Bayesian model. (Reprinted with permission from G. L. Bretthorst, C.-C. Hung, D. A. D'Avignon, and J. J. H. Ackerman, "Bayesian Analysis of Time-Domain Magnetic Resonance Signals," *Journal of Magnetic Resonance,* **79** (1988), Fig. 4, pp. 372–375.)

annual temperature records to investigate a hypothesis of no climate change (no annual temperature change). In particular, if there is evidence that a change has occurred, what is the best estimate of the time (year) of the change?

Consider a series of observed annual temperatures, $T_i$, $i = 1, \ldots, n$, and define a two-phase linear regression model as follows:

$$T_i = \begin{cases} a_0 + b_0 i + e_i, & i = 1, \ldots, r \\ a_1 + b_1 i + e_i, & i = r+1, \ldots, n \end{cases} \quad (14.19)$$

where $e_i$, $i = 1, \ldots, n$, are normally and independently distributed error terms with mean 0 and variance $\sigma^2$.

Our focus is on the intersection of the two regression lines, say, $c$, that represents the change point. For analytic reasons, $c$ is required to be in the interval $(r, r+1)$ and is given by $c = (a_0 - a_1)/(b_1 - b_0)$. Using $c$, it is convenient to write the two-part equation as a regression equation with two variables $i$ and $(i - c)\text{IND}_c(i)$ as follows:

$$T_i = a_0 + b_0 i + b(i - c)\,\text{IND}_c(i) + e_i, \quad i = 1, 2, \ldots, n \quad (14.20)$$

where

$$\text{IND}_c(i) = \begin{cases} 0, & \text{if } i < c \\ 1, & \text{if } i > c \end{cases} \quad (14.21)$$

and $b = b_1 - b_0$. Then we can express the observations as a linear model

$$T = X_c a + e \quad (14.22)$$

where

$$X_c^T = \begin{Bmatrix} 1 & 1 & \ldots & 1 & 1 & \ldots & 1 \\ 1 & 2 & \ldots & r & r+1 & \ldots & n \\ 0 & 0 & \ldots & 0 & 1 & \ldots & n-r \end{Bmatrix}, \quad T = (T_1, T_2, \ldots, T_n)^T$$

$$a = (a_0, b_0, b)^T, \quad e = (e_1, e_2, \ldots, e_n)^T \quad (14.23)$$

$$a_c^* = (X_c^T X_c)^{-1} X_c^T T, \quad \text{RSS}_c = (T - X_c a_c^*)^T (T - X_c a_c^*)$$

in which $a_c^*$ is the least squares estimate of $a$, and $\text{RSS}_c$ is the residual sum of squares.

The data for the application of Bayes's theorem are annual temperature deviations from the Southern Hemisphere from 1858 to 1985 shown in

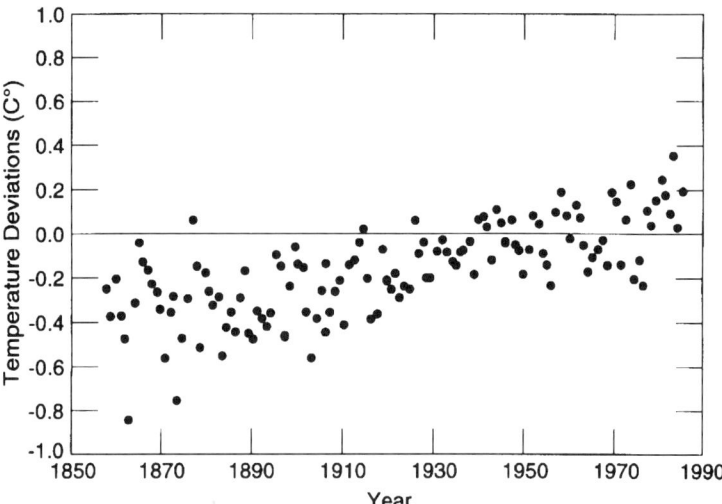

FIG. 8. Southern Hemisphere surface air temperature deviations, 1858–1985. (Reprinted with permission from A. R. Solow, "A Bayesian Approach to Statistical Inference About Climate Change," *Journal of Climate* (published by the American Meteorological Society), **1** (1988), Fig. 1, p. 512.)

Fig. 8. We assume that $c$ does not occur in either the first four or the last five years of the record making it possible a priori for $c$ to assume any one of 119 values (years). Reference [20] discusses distributional issues associated with the likelihood $l(T; c) \propto (\text{RSS}_c)^{-n/2}$, with proportionality constant independent of $c$. This likelihood is combined with the following three prior distributions for $c$:

$$\pi_1(c) = 1/119, \qquad 1862 \leq c \leq 1980$$

$$\pi_2(c) = \begin{cases} (0.25)1/60, & 1862 \leq c \leq 1921 \\ (0.75)1/59, & 1921 < c \leq 1980 \end{cases} \qquad (14.24)$$

$$\pi_3(c) = \begin{cases} (0.25)1/89, & 1862 \leq c \leq 1950 \\ (0.75)1/30, & 1950 < c \leq 1980 \end{cases}$$

The three priors and corresponding posteriors $\pi(c|T)$ are plotted in Fig. 9.

Note that each posterior distribution has two local maxima, 1887 and 1976. Further, the global maximum for $\pi_1$ and $\pi_2$ occurs at 1887 and for $\pi_3$ at 1976, illustrating the role played by the different priors. Also, [20] develops the following 50% highest posterior density (HPD) credibility intervals for $c$:

# EXAMPLES

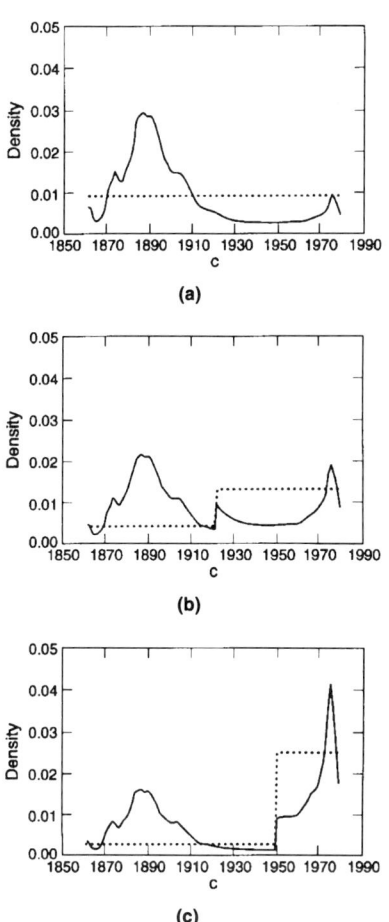

FIG. 9. Posterior distribution of c for three prior distributions (shown by dashed lines). (a) $\pi_1(c)$ and $\pi_1(c/T)$. (b) $\pi_2(c)$ and $\pi_2(c/T)$. (c) $\pi_3(c)$ and $\pi_3(c/T)$. (Reprinted with permission from A. R. Solow, "A Bayesian Approach to Statistical Inference About Climate Change," *Journal of Climate* (published by the American Meteorological Society), **1** (1988), Fig. 2, p. 512.)

$$(1879, 1899) \text{ for } \pi_1; \quad (1879, 1904) \text{ and } (1973, 1978) \text{ for } \pi_2;$$
$$\text{and } (1884, 1893) \text{ and } (1967, 1980) \text{ for } \pi_3 \tag{14.25}$$

The posterior medians are approximately 1892, 1902, and 1951, respectively. We note that the median is of limited value in bimodal distributions; for example, the median for $\pi_3$ does not occur in the 50% HPD interval.

### 14.4.4 Rocket Photometry

The object of the analysis in [21] is to estimate the altitude distribution of airglow as measured by rocket observations. Rocket photometry observations measure the column emission rate by integrating volume emission rates along the line of sight. That is,

$$I(z) = \int_z^\infty J(y)dy \qquad (14.26)$$

where $I$ denotes vertical column rate, $J$ is the volume emission rate, and $z$ is the altitude. Random fluctuations and spurious modulations make the inversion problem difficult and subject to large error. The inference of interest is to reconstruct an altitude profile of $J$ given $I$. The Bayesian procedure presented removes both the random noise and periodic modulations.

We suppose that the observed data $I^T = (I_1, I_2, \ldots)$ to be composed of three components: trend $T$, period $S$, and error $e$, so that

$$I_j = T_j + S_j + e_j \qquad (14.27)$$

Given the observations $I$ we want to obtain the most likely model of $T_j$ and $S_j$. As prior information, we assume a trend $T$ with locally smooth variation and a periodic function $S$ with sinusoidal variation. An earlier reference [22] provides details for representing this prior information as constraints given by minimizing the second-order difference expressions $\Sigma (T_{j+1} - 2T_j + T_{j-1})^2$ and $\Sigma (S_{j+1} - 2cS_j + S_{j-1})^2$, where the constant $c$ relates to the period of the sinusoidal variation. Two *hyperparameters*, $s$ and $d$, are introduced. First, $s$ controls the balance between the $T$ and $S$ (large $s$ being more sinusoidal). Second, $d$ controls the balance between the systematic function $T + S$ and the error component $e$ (large $d$ implies strong constraint on the systematic part and leads to large residuals). For given values of $s$ and $d$, constrained least squares is used to choose the best model of $T$ and $S$. The data distribution is completed by assuming Gaussian models for the second-order expressions of the components $T$, $S$, and $e$. The final solution is determined by selecting values of $s$ and $d$ optimally for given data $I$. Let $L(d, s, I)$ denote the expected value of the data $I$ (or, similarly, $T$ and $S$) for fixed values of the hyperparameters $d$ and $s$. The optimal values of $d$ and $s$ are those values that minimize a Bayesian information criterion $-2 \log L(d, s, I)$.

Figure 10 presents 156 data values taken at intervals of 0.2 km from rocket measurements of vertical column emission rates of atomic oxygen airglow. Figure 11(a) presents the MAP solution for $T$, $S$, and $e$; and Fig. 11(b) presents the MAP estimate of the volume emission rate from the Bayesian analysis.

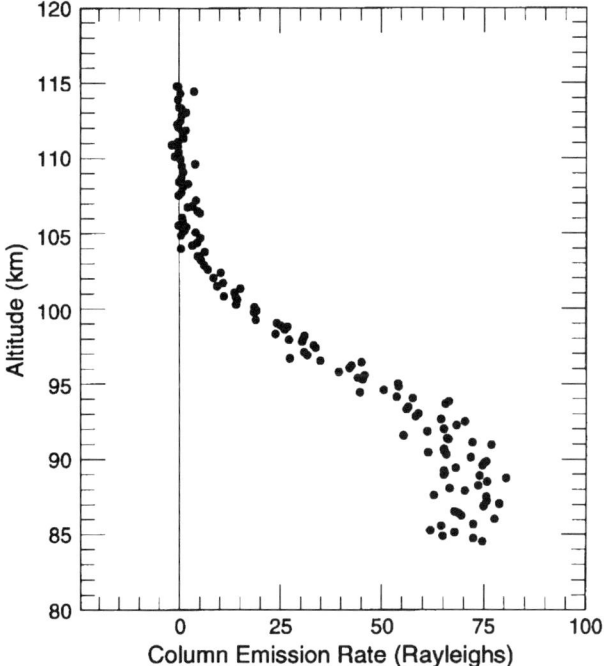

FIG. 10. Vertical column atomic oxygen night airglow emission rate data. (Reprinted with permission from *Planetary and Space Sciences,* **37**, K. Kita, T. Higuchi, and T. Ogawa, "Bayesian Statistical Inference of Airglow Profiles from Rocket Observational Data: Comparison with Conventional Methods," Fig. 1. Copyright 1989, Pergamon Press, Inc.)

## 14.5 The Gibbs Sampler

A major task in applying Bayesian methods is the necessity to calculate the joint posterior distribution (and usually the marginal posterior distributions) of a set of parameters of interest. In many cases, however, the required integrations are difficult to perform, either analytically or numerically. The Gibbs sampler is a technique that can be used to indirectly generate a random sample from the joint posterior distribution (and hence the marginal posterior distributions). The desired Bayesian point and interval estimates can thus be directly computed from the corresponding sample observations obtained via the Gibbs sampler without the need for tedious analytical or numerical calculations.

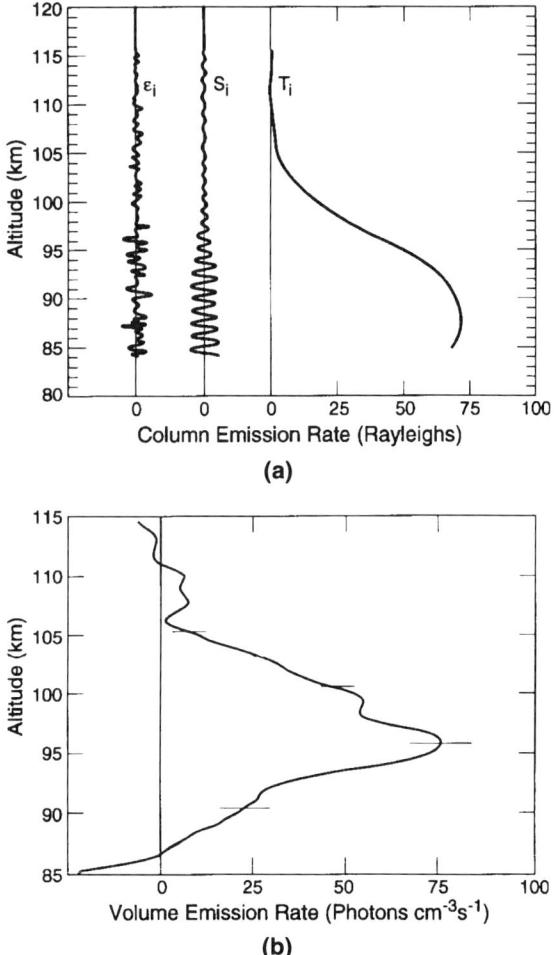

FIG. 11. Bayesian estimates of the model components and volume emission rates for data in Fig. 10. (a) Best estimate of the model components. (b) Estimated volume emission rates. (Reprinted with permission from [*Planetary and Space Sciences,* **37**, K. Kita, T. Higuchi, and T. Ogawa, "Bayesian Statistical Inference of Airglow Profiles from Rocket Observational Data: Comparison with Conventional Methods," Fig. 2. Copyright 1989, Pergamon Press, Inc.].)

Based on a Monte Carlo algorithm [23], Geman and Geman [24] use an adaptive version of this algorithm for image restoration, which they subsequently refer to as the *Gibbs sampler*. It is basically a Monte Carlo integration technique that is implemented using a Markovian updating scheme. Successful statistical (and often Bayesian) applications of the Gibbs sampler may be found in [25–28], while a good introductory paper is Casella and George [29].

We briefly summarize the method. Suppose that we have a collection of $p$ random variables (which may be vector valued) $\theta_1, \ldots, \theta_p$ whose joint distribution $\pi(\theta_1, \ldots, \theta_p)$ is unknown. However, suppose that their full conditional distributions $\pi(\theta_i \mid \theta_j, j \neq i)$ $i = 1, \ldots, p$, are known in the sense that sample values of $\theta_i$, conditioned on values of $\theta_j, j \neq i$, may be generated by some method. Under mild conditions, these marginal conditional distributions uniquely determine the required joint distribution $\pi(\theta_1, \ldots, \theta_p)$; hence, all unconditional marginal distributions, $\pi(\theta_i), i = 1, \ldots, p$, as well.

The Gibbs sampler generates samples from the required joint distribution as follows:

1. Select an arbitrary starting set of values $\theta_1^0, \ldots, \theta_p^0$.
2. Draw $\theta_1^1$ from $\pi(\theta_1 \mid \theta_2^0, \ldots, \theta_p^0)$, then $\theta_2^1$ from $\pi(\theta_2 \mid \theta_1^1, \theta_3^0, \ldots, \theta_p^0)$, and so on up to $\theta_p^1$ from $\pi(\theta_p \mid \theta_1^1, \ldots, \theta_{p-1}^1)$ to complete one iteration of the sampler.
3. After $n$ such iterations, we have obtained the sample $(\theta_1^n, \ldots, \theta_p^n)$.

Reference [24] shows that, under mild conditions, this $p$-tuple converges in distribution to the unknown joint distribution $\pi(\theta_1, \ldots, \theta_p)$ as $n \to \infty$. The set of simulated observations can then be used to estimate $\pi(\theta_1, \ldots, \theta_p)$, its marginal densities, marginal moments, and so forth.

**Example:** In a recent year, $x = 3$ failures of a certain continuously operating pump were observed at a certain U.S. commercial nuclear power plant. We assume that $x$ follows a Poisson distribution with parameter $\lambda t$, where $\lambda$ is the unknown pump failure rate (in failures per operating hour) that is to be estimated and $t = 8760$ hrs (one year). Using a Bayesian estimation approach, we place a Gamma$(\lambda; 2, \beta)$ prior distribution on $\lambda$; however, we are uncertain about the value of $\beta$. Compatible with the Bayesian approach, we further express our uncertainty about $\beta$ by placing a Gamma$(\beta; 0.01, 0.5 \times 10^{-6})$ *hyperprior distribution* on $\beta$. The use of such hyperprior distributions characterizes statistical methods known as *hierarchical Bayes* [3]. We desire the marginal posterior distribution $\pi(\lambda \mid 3; 8760)$, from which we can then estimate $\lambda$.

Because we cannot obtain $\pi(\lambda \mid 3; 8760)$ in closed form, we use the Gibbs sampler to approximate it as follows. From the Poisson process results in Sec. 14.2.4, we find that $(\lambda \mid 3, \beta; 8760)$ has a Gamma$(\lambda; 5, \beta + 8760)$

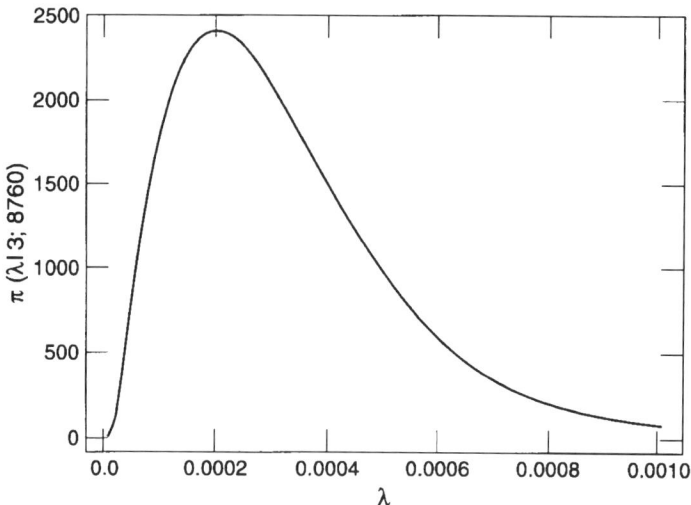

FIG. 12. Gibbs sampler-produced marginal posterior distribution of $\lambda$ given $s = 3$ failures in 8760 operating hours.

distribution. By direct use of Bayes's theorem, we also find that $(\beta \mid \lambda)$ has a Gamma$(\beta; 2.01, \lambda + 0.5 \times 10^{-6})$ distribution. These are the two required conditional distributions that we need to iterate on $\lambda$ and $\beta$ in Steps 2 and 3 in the Gibbs sampler.

We ran the Gibbs sampler to $n = 2600$ observations, discarded the first 100 pairs of values to ensure convergence, and saved every fifth pair of values thereafter as a sample of size 500 from the joint posterior distribution $\pi(\lambda, \beta \mid 3; 8760)$. A smoothed histogram of the 500 marginal values of $\lambda$ is given in Fig. 12, and this is our desired approximation to $\pi(\lambda \mid 3; 8760)$.

The posterior mean of $\lambda$ is estimated (by simply averaging the 500 values of $\lambda$) to be $\hat{\lambda} = 3.3 \times 10^{-4}$ failures per hour. We also obtain the 90% Bayesian posterior credibility interval on $\lambda$ given by $[1.0 \times 10^{-4}, 7.3 \times 10^{-4}]$. These are our desired point and interval estimates of $\lambda$.

## References

1. Bayes, T. (1958). "An Essay Towards Solving a Problem in the Doctrine of Chances." *Biometrika* **45**, 296 (reprint of 1763 publication).
2. Laplace, P. S. (1774). *Mem. Acad. R. Sa. Presentes par Divers Savans*, **6**, 621.
3. Berger, J. O. (1985). *Statistical Decision Theory and Bayesian Analysis*, 2d ed. Springer-Verlag.

4. Martz, H. F., and Waller, R. A. (1991). *Bayesian Reliability Analysis.* Krieger, Melbourne, FL.
5. Jaynes, E. T. (1968). "Prior Probabilities." *IEEE Trans. on Sys. Sci. and Cyber.* **SSC-4,** 227.
6. Jaynes, E. T. (1983). *E. T. Jaynes: Papers on Probability, Statistics and Statistical Physics,* R. D. Rosenkrantz, (ed.). Reidel, Dordrecht.
7. Rosenkrantz, R. D. (1977). *Inference, Method, and Decision: Towards a Bayesian Philosophy of Science.* Reidel, Boston.
8. Press, S. J. (1989). *Bayesian Statistics: Principles, Models, and Applications.* John Wiley and Sons, New York.
9. Skilling, J. (1989). *Maximum Entropy and Bayesian Methods.* Kluwer Academic Press.
10. Gull, S. F., and Skilling, J. (1984). "Maximum Entropy Method in Image Processing." *IEE Proc.* **F131,** 646.
11. Skilling, J., and Gull, S. F. (1989). In *Maximum Entropy and Bayesian Methods,* J. Skilling, (ed.). Kluwer Academic, Dordrecht.
12. Bretthorst, G. L. (1988). *Bayesian Spectrum Analysis and Parameter Estimation.* Springer-Verlag.
13. Sivia, D. S. (1990). "Bayesian Inductive Inference Maximum Entropy and Neutron Scattering." *Los Alamos Science* (Summer), 180.
14. Mukhopadhyay, R., Carlile, C. J., and Silver, R. N. (1991). "Application of Maximum Entropy to Neutron Tunneling Spectroscopy." *Physica B,* **174,** 546.
15. Sivia, D. S., Vorderwisch, P., and Silver, R. N. (1990). "Deconvolution of Data from the Filter Difference Spectrometer: From Hardware to Maximum Entropy." *Nuclear Instruments and Methods in Physics Research,* **A290,** 492.
16. Meier, J. E., and Marshall, A. G. (1991). "Pure Absorption-Mode Spectra from Bayesian Maximum Entropy Analysis of Ion Cyclotron Resonance Time-Domain Signals." *Anal. Chem.* **63,** 551.
17. Donoho, D. L., Johnstone, I. M., Hoch, J. C., and Stern, A. S. (1992). "Maximum Entropy and the Nearly Black Object." *J. R. Statist. Soc. B* **54,** 41.
18. MEMSYS3, produced by Maximum Entropy Data Consultants, Ltd., 1989; for details on MEMSYS1, see J. Skilling and R. K. Bryan, *Mon. Not. R. Astron. Soc.* **211,** 114 (1984).
19. Bretthorst, G. L., Hung, C.-C., D'Avignon, D. A., and Ackerman, J. J. H. (1988). "Bayesian Analysis of Time-Domain Magnetic Resonance Signals." *J. Magn. Reson.* **79,** 369.
20. Solow, A. R. (1988). "A Bayesian Approach to Statistical Inference About Climate Change." *J. Climate.* **1,** 512.
21. Kita, K., Higuchi, T., and Ogawa, T. (1989). "Bayesian Statistical Inference of Airglow Profiles from Rocket Observational Data: Comparison with Conventional Methods." *Plant. Space Sci.* **37,** 1327.
22. Higuchi, T., Kita, K., and Ogawa, T. (1988). Bayesian statistical inference to remove periodic noise in the optical observations aboard a spacecraft, *Applied Optics,* 27, 4514.
23. Metropolis, N., Rosenbluth, A. W., Rosenbluth, M. N., Teller, A. H., and Teller, E. (1953). "Equation of State Calculations by Fast Computing Machines." *J. Chem. Phys.* **21,** 1087.

24. Geman, S., and Geman, D. (1984). "Stochastic Relaxation, Gibbs Distributions, and the Bayesian Restoration of Images." *IEEE Trans. Pat. Anal. and Mach. Int.* **PAMI-6,** 721.
25. Gelfand, A. E., and Smith, A. F. M. (1990). "Sampling-Based Approaches to Calculating Marginal Densities," *J. of the Am. Stat. Assoc.* **85,** 398.
26. Gelfand, A. E., Hills, S. E., Racine-Poon, A., and Smith, A. F. M. (1990). "Illustration of Bayesian Inference in Normal Data Models Using Gibbs Sampling." *J. of the Am. Stat. Assoc.* **85,** 972.
27. Zeger, S. L., and Karim, M. R. (1991). "Generalized Linear Models with Random Effects: A Gibbs Sampling Approach." *J. of the Am. Stat. Assoc.* **86,** 79.
28. Carlin, B. P., and Polson, N. G. (1991). "An Expected Utility Approach to Influence Diagnostics," *J. of the Am. Stat. Assoc.* **86,** 1013.
29. Casella, G., and George, E. I. (1992). "Explaining the Gibbs Sampler." *American Statistician,* **46,** 167.

# 15. SIMULATION OF PHYSICAL SYSTEMS

John M. Hauptman

Department of Physics and Astronomy
Iowa State University, Ames

## 15.1 Introduction

Simulations are now used in both experimental and theoretical problems, both complex and simple. Complex systems such as large detectors in high-energy and nuclear physics, surface interactions in condensed matter physics, radiation dose calculations in medicine, and fundamental quantum mechanical interactions are readily simulated numerically. In the design of expensive experiments, it is much cheaper to simulate the experiment than to build and test protoytpes in the laboratory. In the calculation of properties of highly coupled or complex systems, simulation methods are often the only available method, since analytic methods are sometimes intractable. This chapter is concerned with statistical simulations of physical systems in which the inherent fluctuations are random. In the case of the Metropolis or simulated annealing algorithms, the randomness of a system is employed for the solution of a problem.

The statistical simulation of physical systems is conceptually quite simple. The fundamental interactions in physics are simple and well understood, but it is the complicated interplay of many interactions among many elements that leads to the sometimes complicated (but sometimes simple) behavior or response of a system. In high-energy physics, a detailed simulation of particle interactions from first principles is fairly simple and the computational burden is managable, and thus simulations are used extensively in detector design and physics analysis of detector data. On the other hand, in condensed matter or atmospheric physics, the computational burden of even a small fraction of Avagadro's number of particles is at present not managable, and such simulations are not usually attempted. A detailed simulation requires a knowledge of the elemental physical processes occurring in the system, the sequence of processes and their probabilities of occurrence, the fluctuations in each of these processes, and knowledge of the measurement procedures by which the response of the system is finally ascertained. A complete simulation would therefore simply mimic the microscopic interactions of all elements in this chain of processes. Whenever a process is exactly predictable or contains negligibly small fluctuations, this mimicking can be analytic, with a consequent reduction in simulation effort and computation time.

### 15.1.1 Historical Introduction

Early simulations of physical systems were employed both to facilitate difficult theoretical calculations and to aid in the design of instruments. One example of the former is the problem of thermalization, capture and loss of neutrons in a nuclear pile. Another is the well-known Metropolis algorithm [1] for problems in statistical mechanics, and algorithms discussed by Linden [2] for problems in the quantum mechanics of many body systems. Yet another example [3] is the calculation of the energy distribution of residual nucleons upon neutron bombardment as well as the angle and energy distribution of emerging particles, in which neutrons of 86.6 MeV kinetic energy were incident upon an atomic nucleus described by a potential depth, a radius, and a Fermi level. The essential steps in the simulation were (i) sample the interaction mean free path, (ii) sample the target momentum from the Fermi motion and (iii) sample the scattering angle according to recent measurements made at the Berkeley cyclotron, with energy–momentum conservation, the Pauli principle and the isotropy of the nucleon–nucleon interaction imposed. Goldberger writes

> In the computation carried out in this paper, 100 incident particles were followed. . . . The actual calculations were carried out primarily by graphical means; analytical calculations would have required an exhorbitant amount of time if done with ordinary desk calculating machines. The calculation described here required about two weeks full-time work by two people.

An example of the design of instruments [4] is a simulation study by Robert R. Wilson of the highly stochastic fluctuations in electromagnetic shower development in materials for the purpose of electromagnetic calorimeter design:

> The procedure used was a simple graphical and mechanical one. The distance into the lead was broken into intervals of one-fifth of a radiation length (about 1 mm). The electrons or photons were followed through successive intervals and their fate in passing through a given interval was decided by spinning a wheel of chance; the fate being read from one of a family of curves drawn on a cylinder. . .
> A word about the wheel of chance: The cylinder, 4 in. outside diameter by 12 in. long, is driven by a high speed motor geared down by a ratio of 20 to 1. The motor armature is heavier than the cylinder and determines where the cylinder stops. The motor was observed to stop at random and, in

so far as the cylinder is concerned, its randomness is multiplied by the gear ratio. . .

In the early bubble chamber days, when it seemed that a new resonance was being discovered every week, Luis Alvarez [5] tested the statistical credibility of a claim with a simulation:

> In my work as a nuclear physicist before World War II, I had often been skeptical of the significance of the "bumps" in histograms, to which importance was attached by their authors. I developed my own criteria for judging statistical significance, by plotting simulated histograms, assuming the curves to be smooth; I drew several samples of "Monte Carlo distributions," using a table of random numbers as the generators of the samples. . . . All would contain the same number of events as the real experiment, . . . The standard procedure is to ask a group of physicists to leaf through the 100 histograms—with the experimental histogram somewhere in the pile—and vote on the apparent significance of the statistical fluctuations that appear. The first time this was tried, the experimenter—who had felt confident that his bump was significant—didn't know that his own histogram was in the pile, and didn't pick it out as convincing; he picked out two of the computer-generated histograms as looking significant, and pronounced all others—including his own—as of no significance!

### 15.1.2 Motivation for Simulations

The preceding examples of simulations are from modern physics experiments in which the intrinsic quantum mechanical fluctuations in the underlying process are large, often much larger than the precision of the measurements. In addition to these intrinsic fluctuations characteristic of the physics of the system, there are statistical fluctuations in the event sample recorded by a given experiment. Typically, if $N$ events are obtained, then expected Poisson statistical fluctuations in the number that would be obtained in repeated experiments is $\sqrt{N}$. Any quantity derived from this sample will also be subject to fractional statistical fluctuation uncertainties of order $1/\sqrt{N}$. All of these statistical effects can be simulated.

In modern physics experiments, measurements of momentum, energy, mass or time are often made on individual particles sampled from a quantum mechanical ensemble, and these measurements are in turn subject to instrumental resolution. The measurement of the time of arrival of a photon from an excited atomic state, for example, depends first on the quantum mechanical probability of the transition that yields the photon and second on the time resolution of the instrument used to measure the arrival time, such as a fast photomultiplier tube. The measured invariant mass of the decay products of a resonant state in nuclear or particle physics depends upon both the quantum mechanical width or lifetime of the state and fluctuations in the measurements of particle momenta.

Purely classical simulations of systems that do not involve quantum statistical fluctuations are employed in such practical matters as automobile, airline and telephonic traffic flow, queueing time in line at the bank, multidimensional integrations and in such recreational matters as puzzle solving and game playing. These processes involve random times of arrival and random sampling from an ensemble and are therefore simulations of stochastic, or randomly fluctuating, processes. Another class of problems involve, for example, the numerical integration of the trajectories of planets in the solar system, or stars in the collisions of galaxies or fluid flow in various contexts. These simulations do not involve random processes, and are not discussed here.

Finally, the advent of reduced-instruction-set-computing (RISC) processors several years ago, and the subsequent reduction in cost per million instructions per second (MIPS) to about $50/MIPS, has greatly reduced the cost of simulations. Nowadays, computer time is a small expense compared to human time, and detailed simulations from first principles in many areas of physics generally require relatively little human effort.

## 15.2 Basic Techniques in Simulation

In the simplest terms, the simulation of a system allows the probability of any particular outcome to be calculated by essentially counting the number of times this outcome occurs divided by the number of tries. More completely, a simulation allows one to calculate the probability distribution of any variable in the simulation [6].

### 15.2.1 Examples of Simple Simulations

The following are examples of simple "hit-or-miss" simulations in which mere counting is required.

**15.2.1.1 The Value of $\pi$.** The value of $\pi$ can be estimated by the Monte Carlo method after a light rainfall on a sidewalk square. Inscribe a circle inside the square and count the number of raindrops $n$ inside the square and the number $c$ inside the circle. Then an estimate of $\pi$ is

$$\pi \approx 4\frac{c}{n}$$

and the expected fluctuation in this estimate is binomially distributed with root mean square (rms) $\sigma_\pi \approx 1.64/\sqrt{n}$. This is just a numerical integration by the Monte Carlo method and can be easily simulated on a computer by choosing $n$ pairs of random numbers on the unit square and counting the number $c$ that fall inside the inscribed circle.

```
Function Pi(n,Seed)                           Simulation of the value of π
c=0.
do i=1,n
    x=ran(Seed)                               Uniformly sample the unit square
    y=ran(Seed)
    if(sqrt(x**2+y**2).lt.1.) c=c+1.          Inside the 1/4 circle
end do
pi=4.*c/float(n)                              Estimate of π
return
end
```

Simple simulations of counter arrays or the geometrical acceptance of a detector to event ensembles are no more complicated than this simulation, although the bookkeeping for many, possibly thousands, of detectors is more elaborate. The essential point here is that a Monte Carlo integration is merely a counting procedure, and computers are good at counting. The simulator must take care that the populations are properly generated, i.e., that the basic physics of the system is understood.

**15.2.1.2 Another Estimate of π.** In the 18th century Conte de Buffon implemented a simulation of π, now referred to as *Buffon's needle*. A needle of length $l$ is dropped onto a page with parallel lines ruled a distance $s$ apart, where $s > l$ so the needle may cross at most one line. For $N$ total drops, the number of times the needle crosses a line $N_l$ is counted, and this probability is $N_l/N \approx 2l/\pi s$, from which π is estimated.

**15.2.1.3 $d$-Dimensional Integration.** In a detector this procedure is equivalent to calculating the acceptance or the efficiency of the apparatus. It is essentially a numerical integration, so why not just integrate exactly? In one dimension with step size $h$, the simplest numerical integration has an error $\mathcal{O}(h^2)$, proportional to $1/n^2$ over an interval $nh$, whereas the simulation integration has a Poisson fluctuations of $1/\sqrt{n}$. However, in $d$ dimensions characterizing a multipurpose detector system, the error in the simplest numerical integration [7] is proportional to $n^{-2/d}$, whereas the simulation fluctuation is still Poisson and is $1/\sqrt{n}$. Furthermore, the specification of the $2d$ boundaries in a numerical integration can be difficult, whereas in a simulation it is usually simple, and hence Monte Carlo integrations are not only often easier but more accurate than numerical integrations. A very sophisticated algorithm for multidimensional integration *Vegas* is described in Press et al. [8].

### 15.2.2 Sampling Any Distribution

In the detailed simulation of physical systems, probability distributions inherent to the system must be sampled by repeated calls to a function or subroutine. Several common distributions that appear in the sciences, the Gaussian, binomial, Poisson and exponential distributions, are treated in Chapter 5. When the probability distribution is itself calculated from a simulation, derived from data or a complex analytic form, then the following algorithm can be used to sample such a probability distribution. Clifford gives a general discussion of this inversion method in Sec. 5.2.

In the case of an analytic function $p(x)$ from which values of $x$ are to be sampled over the $x$-interval $(a, b)$, one forms the ratio of the partial integral of $p(x)$ over $(a, x')$ to the full integral over $(a, b)$, and sets this ratio equal to a uniform random number $r$ on $(0, 1)$:

$$r = \frac{\int_a^x p(x')dx'}{\int_a^b p(x')dx'}$$

Solving this for $x$ in terms of $r$ yields a formula for $x$ that when repeatedly sampled reproduces the distribution $p(x)$. This is illustrated in Fig. 1. A uniform sampling in $r$ will preferentially populate the regions of $x$ where $r$ has a large slope, i.e., those regions of $x$ where $p(x)$ is large.

If the indefinite integral cannot be solved in closed form for $x$ as a function of $r$, one may construct a table of the ratio of integrals ($r$ above) as a function of $x$. This table can be randomly sampled in $r$, and the value of $x$ interpolated from the table.

The distribution function itself may not be analytic; for example, it may have been derived from experimental data or calculated by simulation. In this case a simple table of $r$ versus $x$ can be constructed by replacing the integrals above by sums over the table of $p(x)$ values. Table look-up can be computationally rapid, and for time-critical simulations may be used in preference to an analytic form.

The following routine can be called with an external function reference to sample any FORTRAN function.

```
Subroutine Sample_Function (PDF,r,a,b,n,PDF_cumul,x)
```

The external function PDF is the probability density function to be sampled, r is the uniform random number supplied to the routine, [a,b] is the closed interval over which PDF is to be sampled, n is the length of the working area PDF_cumul supplied to the routine, in which is returned the cumulative probability density function, and x is the sampled value of the abscissa corresponding to r.

```
External PDF
Real PDF_cumul(n), a,b, r, x
```

```
Integer n, i1, i2
Logical First/.true./
if(First) then
  PDF_cumul(1) = 0.
  do i = 2,n
    x=a+(b-a)*float(i-1)/float(n-1)
    PDF_cumul(i)=PDF_cumul(i-1)+PDF(x)
  end do
  do i = 1,n
    PDF_cumul(i) = PDF_cumul(i) / PDF_cumul(n)
  end do
  First = .false.
end if
i1 =1                                              Bisection algorithm
i2 = n
10 if(i2-i1.gt.1) then                                   Bisect again
  i = (i1 + i2)/2
  if(r.gt.PDF_cumul(i).) then
    i1 = i                                         Replace lower index
  else
    i2 = i                                         Replace upper index
  end if
  go to 10
end if
f = (r - PDF_cumul(i1)) / (PDF_cumul(i2) - PDF_cumul(i1))
x = a+(b-a)*(float(i1-1)+f)/float(n-1)              Linearly interpolate
return
end
```

### 15.2.3 Variance Reduction

The statistical precision of a simulation can be improved by removing from the sampling space a volume that is simple and known exactly and proceeding to simulate only the remaining part. In a hit-or-miss simulation, suppose one wanted to sample the area of an irregular but nearly circular shape. One could inscribe a circle inside the shape, calculate its area exactly, and then proceed to sample only the much smaller space remaining between the inscribed circle and the irregular shape. This simple idea can be applied in more complicated settings, and is described in several of the basic references in this chapter.

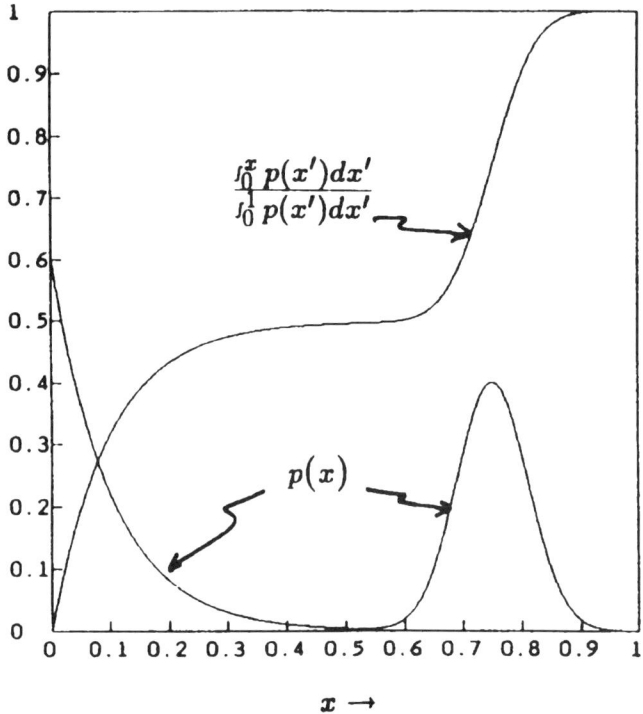

FIG. 1. The probability distribution function $p(x)$ and the cumulative probability distribution function, $r$.

## 15.3 Finding Nonalgebraic Solutions

In the simulations discussed previously there are alternative analytic methods of solution. For some classes of problems there is no possibility of an analytic solution, and in these cases simulations become necessary.

### 15.3.1 Simulated Annealing

Annealing is the process of cooling a thermodynamic ensemble, during which the atoms bounce around, most making transitions to lower energy states, but some making less probable transitions to higher energy states. Ultimately the ensemble cools to its lowest energy state. This minimization of the energy or variance of a system is central to many problems in physics and statistics: a least squares fit requires the minimization of $\chi^2$, the equilibrium configuration of any system requires the minimization of the potential energy function, where the variables in these problems are continuous. Simulated annealing is a procedure for finding the minimum of some function of the state of the system in a space of discrete variables, rather than continuous variables. One example is finding the best configuration of circuit elements on a chip requiring the minimization of the number of interconnections. Loosely put, simulated annealing allows the system to search for a minimum not just along an analytic valley, but also allows random excursions up out of the valley into neighboring valleys in a search for even better minima. As the system "cools" these excursions become less probable.

**15.3.1.1 The Traveling Salesman Problem.** This long-standing problem in combinatorial minimization (finding the shortest route around $N$ cities, each visited once) has been solved easily by simulated annealing. The function to be minimized is the total path, and the various states of the system are the sequences of cities. An algorithm to solve this problem, along with a good discussion, is given by Press et al. [8]

**15.3.1.2 Background Rejection.** In many problems in astrophysics, particle physics, and generally in any problem of statistical pattern recognition, there often happens to be a set of measurements of, for example, coordinate hits in a tracking system or stars in a field of view. Not all of the stars or hits belong to the hypothesized ensemble of stars in a galaxy or hits on a track. These background data constitute a problem in combinatorial minimization, since the inclusion or exclusion of a datum from the ensemble is discrete. These quantized backgrounds can be removed by a simulated annealing algorithm in which the function to be minimized is an overall goodness-of-fit criterion.

### 15.3.2 Simple Roulette.

An impoverished student once suggested that one could make steady money at roulette by playing even–odd, and doubling one's wager after each loss. If there are 18 locations each for even and odd, and both a 0 and a 00 slot, the probability to win by wagering a bet on even, say, is 18/38 ≈ 0.4737 . . . Doubling the wager after each loss ensures that each win after a string of $n$ losses nets a profit of 1. One loses everything if $n$ is so large that $2^n$ is larger that one's holdings. The program ROULETTE simulates this strategy.

```
Program Roulette
Integer Holdings, Wager, Spin, Trial, Seed/4176353/
Parameter Even = 18., Odd = 18., Zero = 1., Double_Zero = 1.
Parameter p_Win = Even / (Even+Odd+Zero+Double_Zero)
```

```
do Trial = 1, 100
  Holdings=100                          Initial stake
  Wager=1                     Wager one unit on first spin
  Spin=0                      Count spins of the wheel
  do while(Holdings.gt.0)
    Spin=Spin+1
    if(ran(Seed).lt.p_Win) then                  Win
      Holdings=Holdings+Wager
      Wager=1
    else                                         Lose
      Holdings=Holdings-Wager
      Wager=min0(2*Wager,Holdings)   Double wager, but do not
    end if                                exceed holdings
  end do
end do
stop
end
```

The result of the first trial (for a stake of 100) is displayed in Fig. 2 insert, in which the player attains a maximum holding of 184, but loses everything at spin 182 after a very unfortunate string of eight consecutive loses. Playing four further trials yields the holdings histories in Fig. 2, in which the longest playing history held out for just over 950 spins. What is the probability

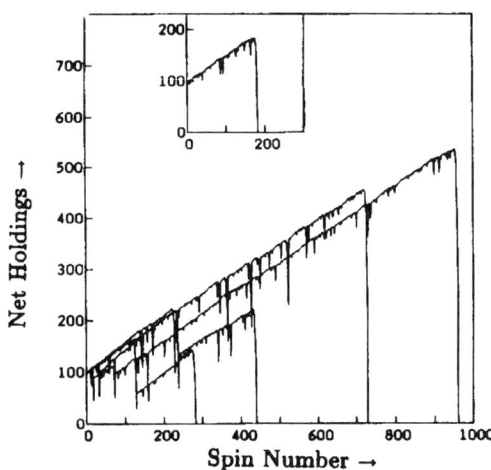

FIG. 2. Net holdings as a function of spin number. The inset is the first game; the main plot shows the next four games.

that a history will last for, say, 1000 spins or more? This is easily calculated from the simulation. The distribution of the spin number at which all is lost is shown in Fig. 3 for initial stakes of 100 and 200. This may be a lot faster and cheaper than actually doing the experiment. This has been a relatively simple simulation; but, it may be appreciated that the detailed histories and probabilities are sometimes more easily calculable in this simulation than they would be analytically.

### 15.3.3 Games

Some difficult mathematical puzzles requiring the finding of nonalgebraic solutions in many dimensions can be easily solved with simulations. Some extremely difficult probability problems can be solved relatively easily, and some problems with millions of possible combinations can be narrowed down by simulating the probability distribution of the possible answers, thereby reducing the number of configurations that must be calculated. Consider the following problem. On a 4 × 4 grid with 16 squares, block out any 7 squares, and fill the remaining 9 squares with the integers 1 through 9. Take the horizontal product in each row and add these products. Do the same vertically, and form the quotient of the vertical and horizontal products. Find the configuration that yields a quotient closest to 10, *but not equal* to 10. Posed in a popular games magazine, this problem can be easily solved by simulating the distribution of quotients for each configuration by

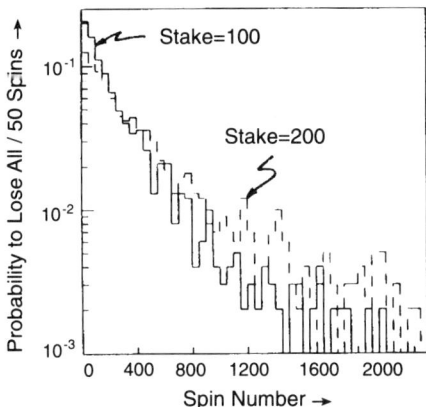

FIG. 3. Probability to lose everything during each 50-spin interval of play. The solid histogram is for an initial stake of 100, and the probability of losing everything within the first 50 spins is about 20%. The dashed histogram for a stake of 200, and the probability to lose everything within the first 50 spins is about 12%.

randomly filling the squares. Only those configurations with a probability distribution of quotients near 10 need be further investigated, and a directed search employed.

## 15.4 Simulation of Experiments

Many systems in condensed matter physics are now understood through simulation. The sputtering of atoms on a surface in an ultrahigh vacuum, the scattering and interactions of molecules on a surface, the energy loss and scattering of Auger electrons through atomic layers and the examination of magnetic lattices are studied by simulations on systems of several tens, hundreds or thousands of atoms. In this section we discuss one example of higher energy particles in a simple but conventional particle spectrometer.

### 15.4.1 Simulation of High-Energy Particle Interactions

Simulations are used extensively in the design and understanding of big detector systems in high-energy physics and in the interpretation of measurements made by these large facilities. The fundamental electroweak interactions and the approximately modeled strong interactions can be reliably simulated, and the subsequent response of complex detectors to ensembles of these interactions are also simulated. In this way, complex and costly experiments can be assessed and optimized.

For large detectors, it is the task of a simulation to make the overall response of a detector predictable for each event that the detector measures and to thereby have the capability to calculate the detector response distribution in any measurable variable for any ensemble of events. The stochastic fluctuations in the interactions of particles with materials are difficult to calculate analytically, whereas a simulation can mimic such fluctuations easily and with confidence. Such simulations have been in use for 40 years in high-energy physics, at least on small scales, and in the past decade sophisticated and extensive codes have been developed for applications to general multipurpose detectors. A typical general purpose detector facility may cost $50–500 million and consist of precision charged particle detectors near the beam collision point, surrounded by a larger tracking volume for the momentum measurement of charged particles and immersed in a strong magnetic field supported by a superconducting coil. These detectors are surrounded by an electromagnetic calorimeter for the energy measurement of electrons and photons and in turn surrounded by a hadronic calorimeter for the energy measurement of hadrons. The limitations on experimental identification of events derive from fluctuations in fundamental physics processes, fluctuations in event development and fluctuations in measurements; and they are usually complex in character and often involve the interplay of two or more detector subsystems. These processes can be reliably reproduced only by simulation. The design of such facilities requires simulations to assess the relationship of performance to cost.

**15.4.1.1 Energy Loss.** The physics of particle energy loss and scattering in materials is treated well by many authors [9]. A charged particle of mass $m$, velocity $v = \beta c$ and energy $E = \gamma m c^2$ traversing a medium imparts a Coulomb impulse to the atomic electrons, resulting in an energy loss rate per unit depth of material,

$$\left(\frac{dE}{dx}\right)_{\text{Mean}} \approx \frac{\alpha}{\beta^2}\left[\ln\frac{2m_e c^2 \gamma^2 \beta^2 \epsilon_{\max}}{I^2} - 2\beta^2\right] \quad (15.1)$$

where $\alpha = .1536\ z^2\ (Z/A)$ MeV/g-cm$^{-2}$, $\epsilon_{\max} \approx 2m_e c^2 \gamma^2 \beta^2$, and the mean ionization potential can be expressed to about 10% precision as $I = 16 Z^{0.9}$ eV for $Z > 1$ [10]. Both this mean value and fluctuations about this mean can be simulated by, for example, the algorithm of Ispirian et al. [11] from discrete atomic levels. The measurement requirements for some very accurate detectors result in designs containing a minimum of material, such as 300 μm-thick silicon wafers or a few millimeters of gas, in which the mean energy loss is so small that the fluctuations are large compared to the mean, yielding a long high-side tail extending to energy losses many times the mean, referred to as the *Landau tail*. For thicker materials, the fluctuations in energy loss are Gaussian, $\sigma \approx \sqrt{2\alpha m_e c^2 dx \gamma^2 (1 - \beta^2/2)}$ and so can be directly simulated. A simple simulation of a step $\Delta x$ would deposit an energy $\Delta E = (dE/dx)\Delta x + \sigma Z$, where $Z$ is a random number drawn from a unit normal Gaussian.

**15.4.1.2 Multiple Scattering.** A charged particle of momentum $p$ will suffer many small-angle Coulomb scatters from the nuclei of the medium. For a sufficiently thick absorber and very many scatters, the transverse distribution in angle of the emerging particle after a depth of s radiation length is nearly Gaussian with an rms width of

$$\theta_{\text{rms}}^{\text{plane}} \approx \frac{13.6\ \text{MeV}}{p\beta}\sqrt{s}[1 + 0.20\ \ln(s)]$$

and such a simple description in an experimental simulation is adequate for most purposes [10]. An exact analytic treatment as implemented in the EGS [12] code and as an option in the GEANT code [13] is the Molière theory, reviewed by Scott [14]. A simple simulation of multiple scattering would replace the $x$ and $y$ scattering angles by $\theta_x \rightarrow \theta_x + \theta_{\text{rms}}^{\text{plane}} Z$ and $\theta_y \rightarrow \theta_y + \theta_{\text{rms}}^{\text{plane}} Z$.

### 15.4.2 A Simple Particle Detector

The required degree of detail in a simulation will depend upon the good judgment of the experimenter, the level of understanding of the elemental processes and the amount of computer time available. In principle, any simulation can become arbitrarily detailed, and one usually stops when the level of detail in the simulation is comparable to the level of measurement detail and precision.

A detailed simulation would transport a particle in steps sufficiently small that the multiple scattering angle, the energy loss and the probability of an interaction with secondary particle production were all small. At each step the trajectory of the particle would be altered by the multiple scattering angle, the energy reduced by the energy loss, and if secondaries were produced they would also be subsequently followed. Such detail is often required, and is implemented in the large standard codes referenced in Section 15.4.4.

**15.4.2.1 Tracking Detectors.** The momentum precision of a charged particle tracking system has been studied in detail [15] and the essentials are readily described. The magnitude of the momentum $p$ of a charged particle moving perpendicular to a magnetic field of strength $B$ is proportional to the radius of curvature of the circle $R$ and is given by

$$p(\text{GeV}/c) = 0.3 \, B(T) \, R(m)$$

For high momentum, the radius of curvature is large compared to the dimension of the tracking system and the measurement of a track consists of points on just an arc of length $L$. The sagitta $s$ of the arc is calculated as the distance between the center of the arc and the midpoint of the two ends of the arc, and the radius of curvature is related to the sagitta by $R \approx L^2/8s$. The fluctuations in spatial measurements are Gaussian and $s \propto 1/R \propto 1/p$, and therefore the inverse momentum, *not* the momentum, is Gaussian distributed. If $N$ spatial measurements along the arc are distributed such that one-half of the measurements is made at the center of the arc and one-quarter at either end, and if the rms spatial tracking precision per point is $\sigma_0$, then the rms variation of the sagitta is $\sigma_s \approx 2\sigma_0/\sqrt{N}$, and the rms inverse momentum resolution is

$$\sigma_{1/p} = \frac{8}{0.3BL^2}\sigma_s$$

where typical values of $\sigma_{1/p}$ are $\approx 10^{-3}\,(\text{GeV}/c)^{-1}$. The momentum resolution is improved for large $B$ and large $L$, and consequently large detectors usually maintain $B \approx 2T$ and $L \geq 1\,m$. Some detectors emphasize B whereas others emphasize L. At very high momenta the momentum distribution itself develops a high-side tail since $p \to \infty$ as $s \to 0$. A spatial mismeasurement resulting in an inverse momentum deviation of $\Delta(1/p)$ also results in a bending angle deviation of $\Delta\phi = (0.3BL/2)\,\Delta(1/p)$, using the correlation derived by Gluckstern, where the sign holds for a positive particle in the $x$–$y$ plane in a $B_z$ field.

A simple simulation of a tracking detector would perform the replacements $1/p \to 1/p + [8\sigma_s/0.3BL^2]Z \equiv 1/p + \Delta(1/p)$ and for the bending angle, $\phi \to \phi + [0.3BL/2]\Delta(1/p)$.

15.4.2.2 Calorimeter Detectors. The energy resolution of an electromagnetic sampling calorimeter [16] depends primarily upon the critical energy, radiation length and the thickness of the absorbing plates interleaved between the charge-sensing media. For an electromagnetic particle of energy $E$, the number of shower particles created is $N_{\text{total}} \approx E/E_c$, where each particle traverses roughly one radiation length. If the thickness of the absorbing plates is $t$ radiation lengths, then the sensitive medium will sample a total population of $N \approx E/E_c t$ particles, and Poisson fluctuations in this $N$ are $\sqrt{N}$ resulting in energy fluctuations of $\sigma_E \propto \sigma_N \approx \sqrt{E/E_c t}$, and an energy resolution of

$$\frac{\sigma_E}{E} \approx \frac{\sqrt{E_c t}}{\sqrt{E}} \equiv \frac{\kappa_{EM}}{\sqrt{E}} \approx \frac{8\%}{\sqrt{E}}$$

for Pb ($E_c = 6.9$ MeV) sampling of $t = 1.0$ radiation length (0.55 cm plates). Hadronic interactions have much larger intrinsic fluctuations, but a crude estimate of energy resolution may be obtained from the preceding example by substituting a hadronic "critical energy" of about $7m_\pi \approx 1$ GeV for $E_c$, and a sampling thickness of 0.25 nuclear absorption lengths for $t$. This results in a typical hadronic energy resolution of

$$\frac{\sigma_E}{E} \equiv \frac{\kappa_{Had}}{\sqrt{E}} \approx \frac{50\%}{\sqrt{E}}$$

A simple calorimeter simulation would replace the energies of particles by their energies plus Gaussian resolution fluctuations, $E \to E + \sigma Z$, where the appropriate $\sigma$ is chosen for electromagnetic or hadronic particles.

### 15.4.3 Example: $K_1^0 \to \pi^+\pi^-$ Decay

Simulation of the measured $\pi^+\pi^-$ invariant mass distribution in $K_1^0 \to \pi^+\pi^-$ decay may illustrate a simple simulation of both tracking and calorimeter detectors. The detector has a magnetic field volume filled with tracking chambers with sagitta resolution $\sigma_s$, $B$ perpendicular to the $\pi$ trajectories and the chamber materials constitute 1% of a radiation length ($X_0$). The $\pi$ energies are also measured by a hadronic calorimeter with energy resolution of $50\%/\sqrt{E}$. Repeated calls to the subroutine K1_Decay will sample the $\pi\pi$ mass spectrum in $K_1^0 \to \pi^+\pi^-$ decay at momentum $P_K$ using the formulae of Section 15.4.2.

subroutine k1_decay(pk,sigmas,kappaE,mass)

pk is the kaon momentum, sigmas is the rms sagitta error, kappaE is the rms calorimeter resolution at 1 GeV and mass is the invariant $\pi\pi$ mass. Grn is a Gaussian random number generator.

```
      real pl(2), e(2), pion(0:3,2), mass, kappaE
      parameter mk=.49767, mpi=.1395679, pcm=.20601, pi=3.14159
      parameter b=1., l=1., x0=.01
      ek=sqrt(pk**2+mk**2)
      gamma=ek/mk                             Lorentz boost parameters γ and β
      beta=pk/ek
      costh=2.*(ran(Seed)-.5)                 Uniform decay in $K^\cup$ cm
      phi=2.*pi*ran(Seed)
      plcm=pcm*costh                          Energy and momentum in cm
      ecm=sqrt(pcm**2+mpi**2)
      sc=8.*sigmas/(0.3*b*l**2)               Rms curvature error
      do i=1,2
        pl(i)=gamma*plcm+gamma*beta*ecm       Longitudinal momentum
        e(i)=sqrt(pl(i)**2+pt**2+mpi**2                    Energy
        dc=sc*grn(Seed)                       Curvature fluctuation
      pl(i)=1./(1./pl(i)+dc)                  p measurement
        dth=0.3*b*l*dc/2.
        pt=pcm*sin(acos(costh)+dth)
        if(kappaE.gt.0.) then                 Include calorimeter measurement
          e(i)=amax1(e(i)+kappaE*sqrt(e(i))*grn(Seed),mpi)
          wp=1./(sc*pl(i)**2)**2              Least squares fit to PL(i)
          we=(e(i)/pl(i)/kappaE)**2
          pl(i)=(pl(i)*wp+sqrt(e(i)**2-mpi**2)*we)/(wp+we)
        end if
        pion(0,i)=sqrt(pl(i)**2+pt**2+mpi**2)        π 4-vector:  E
        pion(1,i)=pt*cos(phi)                                     $P_x$
        pion(2,i)=pt*sin(phi)                                     $P_y$
        pion(3,i)=pl(i)                                           $P_z$
        phi=phi-pi                            Next π is back-to-back in cm
        plcm=-plcm
      end do
      e2=(pion(0,1)+pion(0,2))**2             Energy component of 4-vector
      do m = 1,3                              Subtract momentum components
        e2=e2-(pion(m,1)+pion(m,2))**2
      end do
      mass=sqrt(e2)                           Compute ππ invariant mass
      return
      end
```

Since the sagitta and energy measurements are Gaussian, one might expect that the $\pi\pi$ mass resolution would also be Gaussian, or nearly so. The distribution of the $\pi^+\pi^-$ invariant mass is shown in Fig. 4 for both high- and low-momentum $K^0$, and for two values of the sagitta resolution, $\sigma_s$.

Evidently, the effects of the non-Gaussian momentum resolution on each $\pi$, the random decay angular distribution in the $K^0$ center of mass, resulting in a spread in $\pi$ momenta and angles, and the relationship of the $\pi\pi$ invariant mass to the measured momenta has resulted in a distinctly non-Gaussian mass resolution at high $K^0$ momenta. Such a broad mass distribution with long tails may be confused with background combinations of random $\pi^+$

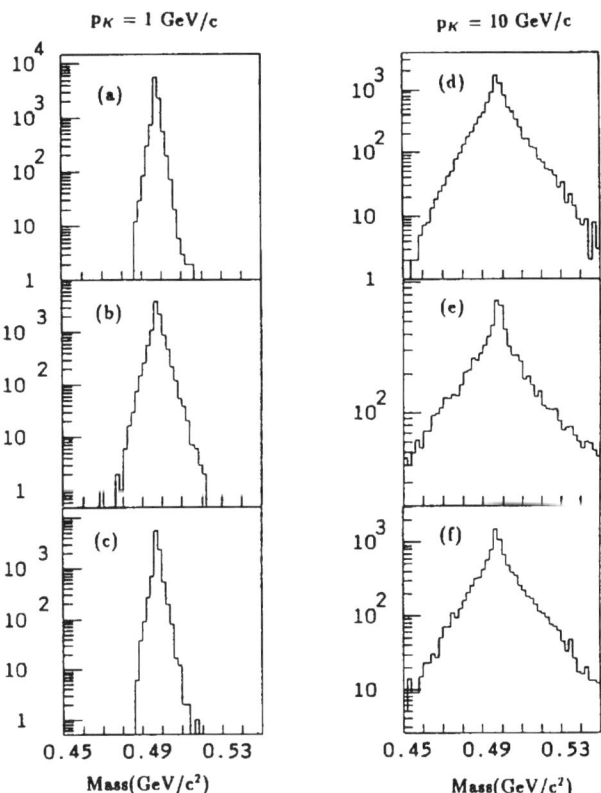

FIG. 4. The $\pi^+\pi^-$ mass resolution from $K_0$ decay for low $K_0$ momentum (1 GeV/c; a–c) and high momentum (10 GeV/c; d–f). The sagitta resolution, $\sigma_s$, is 400 μm in (a) and (d), and 1200 μm in (b) and (e). The addition of a calorimeter with energy resolution constant $\kappa_E = 50\%$ to a tracking system with sagitta resolution of 400 μm yields the resolutions in (c) and (f), for 1 and 10 GeV/c, respectively.

and $\pi^-$ in a real experiment, and the proper extraction of the $K^0 \to \pi^+\pi^-$ rate would depend upon the careful solution of this problem. Finally, the inclusion of calorimeter measurements does not appreciably improve the mass resolution.

Such a simulation may serve both in the design of the detector and in the analysis and understanding of the measurements. The choice of tracking precision, magnetic field and calorimetry may be compared both to the desired measurement precision and the associated costs of different detector designs.

### 15.4.4 Available Simulation Codes

There are two major codes widely used in the high-energy physics community for the simulation of detectors: the EGS (electron gamma shower) code [12] for the simulation of electromagnetic interactions, and the GEANT code [13] for the simulation of both electromagnetic and hadronic interactions. Both codes have been extensively employed in the simulation of experiments.

The ensemble of events for which a detector is to be optimized can be simulated by many physics codes for hadron–hadron, lepton–hadron and lepton–lepton interactions. Two codes for the simulation of proton–proton interactions at high energies are ISAJET [17] and PYTHIA [18]. Lepton–lepton and lepton–nucleon interactions can be simulated by the Lund codes JETSET [19] and LEPTO. These codes are now used in several areas of high-energy astrophysics [20], nuclear physics [21] and medical physics [22].

## 15.5 Validity Testing and Analysis

### 15.5.1 Simulation and Validity Testing

For large codes simulating complex systems, the probability of bugs and simulation deficiencies is nonzero. There are several strategies for lowering this probability.

1. The writing of modular code allows each subsystem of a multipurpose facility to be developed and tested independent of other subsystems. Each module may start out as very simple; for example, by using a $\delta$-function for a complicated distribution function and only later filling in each module with the full details and complications of the simulation.
2. The code should be tested by solving a problem to which the answer is known; for example, by turning off some interactions, replacing a

complicated stochastic procedure by a constant or sending a single particle through the detector on a known trajectory.
3. It is good procedure to check against data wherever possible, even to the extent of modeling someone else's experimental apparatus with your own code. Such checking may be laborious, but for complex problems in which intuition fails it may be essential.
4. The sensitivity of the simulation to variations in one parameter of the simulation at a time should make sense for the problem at hand.
5. Force the simulation to fail. You may set some values of the parameters to unphysical values and check to see that the simulation fails in an understandable way.
6. Exaggerate a particular process. In complex simulations there may be competing processes of comparable strength. A small deficiency in one process cannot be "seen" in the results, but may introduce biases in conclusions drawn from the simulation that are never found. By making one process at a time be, say, 1000 times more probable, and thereby exaggerating its importance in the overall system, deficiencies and errors may become more obvious.

It has been our experience that simulations from first principles are far more likely to be error free than simulations that employ parameterizations or approximations of intermediate processes. The principles we know in physics are often quite simple, and their simulation is trivial. Often the only reason for not doing a first principles simulation is lack of computer time, and approximate simulation of intermediate processes is necessary for economy, but it does introduce further risks.

### 15.5.2 Simulation in the Analysis of Experiments

The event ensemble collected by an imperfect experiment does not represent the true underlying physics distributions. Experimental resolutions and inefficiencies will change the shape of distributions and selectively deplete distributions in regions of lower efficiency. Consequently, directly measured distributions must be corrected for efficiency losses and resolution fluctuations, and ultimately the systematic uncertainties in these correction procedures must be estimated.

**15.5.2.1 Efficiency and Acceptance.** Let an experimental distribution in a measured variable $x$ for a purported event ensemble $\mathcal{E}$ be $(dN/dx)_{\text{Data}}$ in units of events per unit $x$. A fundamental physics event generator (e.g., PYTHIA) is used to generate an ensemble as similar as possible to $\mathcal{E}$, yielding a distribution $(dN/dx)_{\text{Generated}}$ and this generated ensemble is passed through the detector simulation, which records the simulated distribution

$(dN/dx)_\text{Simulated}$ as the detector would measure it. The true underlying physics distribution may be estimated as

$$\left(\frac{dN}{dx}\right)_\text{True} = \left(\frac{dN}{dx}\right)_\text{Data} \times \frac{(dN/dx)_\text{Generated}}{(dN/dx)_\text{Simulated}}$$

where it is understood that these distribution functions in $x$ can be subjected to any set of selections in other variables of the event.

**15.5.2.2 Physics Parameter Estimation.** Under the assumption that the generated ensemble is identical to the true ensemble, the physical assumptions of the physics generator expressed as parameters or algorithms can be varied, and these various ensembles can be passed through the detector simulation and directly compared to the data distribution. A $\chi^2$ comparison would evaluate $\chi^2$ between the data and each simulated physics ensemble, using estimated rms errors on the difference, and find the minimum in the parameter–algorithm space. For one parameter this $\chi^2$ is parabolic near the minimum, and the minimum can be found by interpolation between simulated parameter values.

**15.5.2.3 Estimation of Statistical Uncertainties (or Errors).** If an experiment is simulated and a result calculated as if the experiment had been performed once, then this simulation–experiment can be repeated many times with a different sequence of random numbers to find the distribution of the result. This is a direct means of estimating the fluctuation or error on an experimental result, and can be used as a check on more direct and conventional methods.

**15.5.2.4 Estimation of Systematic Uncertainties.** One of the uncomfortable aspects in the analysis procedures just described is the unknown ensemble $\mathcal{E}$, which represents the true physics distribution. We guess at this ensemble, and use codes (ISAJET, PYTHIA, etc.) to simulate the distribution functions of $\mathcal{E}$, but of course we are at best approximating $\mathcal{E}$, either well or poorly. Any physical measurement must not depend upon our assumptions about $\mathcal{E}$. One procedure is to repeat the entire analysis procedure for other simulators of $\mathcal{E}$ or to modify the parameters of one or more of the simulators to a degree that reflects our expected ignorance of these parameters. The variation in the results of these analyses is an estimate of the systematic error in the measurement. The interpretation of this estimated systematic error may be difficult, and it has become fashionable in recent years to quote both the statistical error and the systematic error on any measurement $\mu$, as $\mu \pm \sigma_\text{stat} \pm \sigma_\text{sys}$. Similarly, the analysis procedure can be repeated for variations in any of the assumptions: physics generator, detector efficiencies or operating conditions of any kind. The difficulty, however, still remains in making an honest estimate of the allowable variations in poorly known quantities.

## 15.6 Improbable Events and Small Effects

For processes such as Rutherford scattering or proton–proton scattering at high energy, the cross section or probability distribution function that must be sampled in a simulation is very small in the interesting regions, *viz.*, at large scattering angles. In these processes one wishes to avoid the simulation of a large number of uninteresting events for each interesting one. The sampling distribution may be weighted such that the interesting events are generated much more frequently, thereby gaining statistical precision in the region of interest.

### 15.6.1 Weighting Schemes

Some probability distribution functions in physics vary over orders of magnitude, and regions of low probability may be the regions of interest. Direct sampling as in Sec. 15.2.3 would be very inefficient since only very infrequently would an interesting event be sampled. Suppose we have such a rapidly varying function, $p(y)$,

$$p(y) = f(y)\, e^{-y^2}$$

where $f(y)$ may be complicated but slowly varying function, possibly found by simulation, and the rapid dependence is all in the exponential. Generating a weighted distribution and then weighting the simulated events ("importance sampling" [6], p. 165) allows a larger population of events to be generated in the interesting region. In this case, a reasonable and easily integrable weighting function would be the exponential part,

$$w(y) = e^{-y^2}$$

and one would sample $y$ from the distribution of the function

$$\frac{p(y)}{w(y)}$$

which in this simple case is just the function $f(y)$. No matter how complicated $f(y)$ may be, it can be sampled by the technique in Sec. 15.2.2. Each generated event whose $y$ value is sampled from this distribution is weighted by $w(y)$. This scheme is easy to implement and can be very useful.

### 15.6.2 Correlated Sampling

It is sometimes necessary to calculate and compare small effects. As an illustration, consider the effect of a small "dead" volume within a calorimeter medium that is insensitive to ionization; for example, a support bolt passing through the sensitive gap. Simulating this calorimeter with a bolt, and again without a bolt, and then comparing energy resolutions or shower shape functions would be imprecise even for large simulation samples, since shower fluctuations in these two statistically independent samples would mask the effect. Correlated sampling is a technique whereby, in this illustration, one set of showers is simulated in the calorimeter, but two sums of signal ionization are taken, one with the bolt, the other without the bolt. The shower-to-shower fluctuations are identical in these two sums and cancel when the difference in resolution or shape is taken. Even small simulation samples are sufficient to accurately calculate such small effects.

## 15.7 Simulations within Simulations

The examples in this chapter are small and single purpose. The simulation of a large or complex system, especially from first physics principles, will entail larger and more complex codes.

The simulation of tracks in a tracking system may consist of a single line of code (replacing the inverse momentum by a Gaussian smeared inverse momentum) or consist of the detailed following of the charged track through the detector medium, the simulation of the ionization loss and the drift and diffusion of electrons in the medium, the amplification and measurement of this signal, the electronic acquisition of these signals and the pattern recognition and reconstruction of these coordinates into a definable track.

Similarly, the simulation of particle energy measurements in a calorimeter may consist of one line (replacing the true energy by a value Gaussian distributed about the true value) or by detailed following of all tracks within the calorimeter medium, the energy loss of all particles, the generation of the measured response, the simulation of the output signals, and their pattern recognition and identification with the energy deposits of a single particle.

Each subprocess in these two examples is a separate modular simulation. The structure of such a program will typically have both nested simulations and simulations in series; that is, the output configuration from the simulation of one physical process is the input to a subsequent process. Large detectors and complex systems are simulated by such modular codes.

The advent of massively parallel machines and workstation supercluster networks [23] will result in vastly expanded usage due to lower computational costs. Since simulation of a physical system is merely a mimicking of nature, and the basic principles of physics are usually simple, it is expected that simulations in both theoretical and experimental work will become the methods of choice for the solutions of many problems in science and technology.

## References

1. Metropolis, N., Rosenbluth, A. W., Rosenbluth, M. N., Teller, A. H., and Teller, E. (1953). "Equation of State Calculations by Fast Computing Machines." *J. Chem. Phys.* **21,** 1087–1092.
2. Linden, W. v. d. (1992). "A Quantum Monte Carlo Approach to Many-Body Physics." *Physics Reports* **220,** 53–162.
3. Goldberger, M. L. (1948). "The Interaction of High Energy Neutrons and Heavy Nuclei." *Phys. Rev.* **74,** 1269–1277.
4. Wilson, R. R. (1951, 1952). "The Range and Straggling of High Energy Electrons." *Phys. Rev.* **84,** 100–103 (1951); and "Monte Carlo Study of Shower Production," *Phys. Rev.* **86,** 261–269 (1952).
5. Alvarez, L. W. (1968). Nobel lecture, December 11, 1968; reprinted in *Discovering Alvarez,* W. P. Trower (ed.), pp. 132–133. University of Chicago Press, Chicago, 1987.
6. J. M. Hammersley and D.C. Handscomb, *Monte Carlo Methods,* John Wiley & Sons, 1964; B.D. Ripley, *Stochastic Simulation,* John Wiley & Sons, 1987; R.Y. Rubinstein, *Simulation and the Monte Carlo Method,* John Wiley & Sons, 1981; and S. Jain, *Monte Carlo Simulations of Disordered Systems,* World Scientific, 1992. An early simulation of $\pi$ was by Hall, *Messenger Math.* **2,** 113–114 (1873).
7. Lyons, L. (1986). *Statistics for Nuclear and Particle Physicists.* Cambridge University Press, Cambridge.
8. Press, W. H., et al. (1992). *Numerical Recipes: The Art of Scientific Computing,* 2d ed. Cambridge University Press, Cambridge.
9. Leo, W. R. (1987). *Techniques for Nuclear and Particle Physics Experiments.* Springer-Verlag, Chu, W. K. (1981). "Energy Loss of High-Velocity Ions in Matter," in *Methods of Experimental Physics,* P. Richard (ed.) Academic Press, New York.
10. Particle Data Group, Lawrence Berkeley Laboratory. (1992). "Review of Particle Properties, Part II." *Phys. Rev.* **D45.** Updated and published every two years, this volume contains useful information on materials, detectors, and energy loss, in addition to short sections on statistics and Monte Carlo techniques.
11. Ispirian, K. A., *et al.* (1974). "A Monte Carlo Method for Calculation of the Distribution of Ionization Losses." *Nucl. Instr. Meths.* **117,** 125.
12. Nelson, W. R., Hirayama, H., and Rogers, D. W. O. (1985). "The EGS Code System." Report SLAC-265, Stanford Linear Accelerator Center.
13. Brun, R., Bruyant, F., Maire, M., McPherson, A., *et al.* (1991). *GEANT User's Guide, Version 3.15.* Code and documentation can be obtained from CERN Program Library Office, CERN-CN, CH-1211 Geneva 23, Switzerland, or BITNET to CERNLIB@CERNVM.

14. Scott, W. T. (1963). "The Theory of Small-Angle Multiple Scattering of Fast Charged Particles." *Rev. Mod. Phys.* **35**, 231–313. A useful practical discussion is given in W. R. Leo, [9], pp. 42–43.
15. Gluckstern, R. L. (1963). "Uncertainties in Track Momentum and Direction Due to Multiple Scattering and Measurement Errors." *Nucl. Instr. Meth.* **24**, 381.
16. Amaldi, U. (1987). "Fluctuations in Calorimetry Measurements," in *Experimental Techniques in High Energy Physics*, T. Ferbel (ed.). Addison-Wesley, Reading, MA.
17. Paige, F. E., and Protopopescu, S. D. (1986). "ISAJET 5.30: A Monte Carlo Event Generator for $pp$ and $pp$ Interactions." *Proc. of the Summer Study on Physics of the SSC* (Snowmass, Colorado), 1986, R. Donaldson and J. Marx (eds.).
18. Bengtsson, H. U. (1986). "PYTHIA: The Lund Monte Carlo for Hadronic Processes, CERN Program Library," in [17].
19. Lund JETSET, Andersson, B., Gustafson, G., Ingelman, G., and Sjöstrand, (1983). "Parton Fragmentation and String Dynamics." *Phys. Rep.* **97**, 33. Sjöstrand, T. (1986). *Computer Phys. Comm.* **39**, 347. Sjöstrand, T., and Bengtsson, M. (1987). *Computer Phys. Comm.* **43**, 367.
20. Casper, D. (1991). "Measurement of Atmospheric Neutrino Composition with the IMB-3 Detector, *et al.*" *Phys. Rev. Lett.* **66**, 2561. Hirata, K. S. (1988). "Experimental Study of the Atmospheric Neutrino flux, *et al.*" *Phys. Lett.* **B205**, 416. Hauptman, J. (1991). "Reconstruction of Simulated Long Baseline Events in DUMAND." *Workshop on Long Baseline Neutrino Oscillations*, Fermilab.
21. RHIC Letter of Intent. (1990). "An Experiment on Particle and Jet Production at Midrapidity." The STAR Collaboration, LBL-29651, Lawrence Berkeley Laboratory.
22. Nelson, R., and Bielajew, A. (1991). "EGS: A Technology Spinoff to Medicine." Stanford Linear Accelerator Center *Beam Line* **1**, 7.
23. Anderson, E. W., et al. (1992). "BATRUN (Batch After Twilight RUNing)." *Proc. of the Conference on Computing in High Energy Physics—1992*, Annecy, France.

# 16. FIELD (MAP) STATISTICS

John L. Stanford and Jerald R. Ziemke

Department of Physics and Astronomy
Iowa State University, Ames

## 16.1 Introduction

### 16.1.1 Fields (Maps) of Statistics

Many investigations in physical science involve spatial fields of measured or computed variables; an example is atmospheric temperature (or pressure, or concentration of some constituent such as ozone) measured at given altitudes and geographic positions on the earth. Atmospheric dynamical equations predict a variety of normal mode and instability wave motions, and a powerful way to investigate the existence of such perturbations is to use spatial fields of cross-correlation calculations. (The cross-correlation statistic is discussed in Sec. 16.1.5; see also Chapters 3 and 11.)

Suppose the field statistic is a two-dimensional map (array of points located at $i, j$) of correlations determined with respect to a chosen reference point, generated by calculating correlations of the data time series at some reference point with time series at all other grid points. An example involving time-lag correlation for TOMS (total ozone mapping spectrometer satellite instrument) total-column ozone is shown in Fig. 1.

Aspects of the patterns in Fig. 1 appear to be nonnoiselike, but the question is: To what extent are the patterns due to signals in the data and to what extent are they due to noise? The goal of this chapter is to provide an answer to this question. As will be shown later, the answer, for Fig. 1, is that the patterns *cannot be distinguished from noise*. It is perhaps appropriate to paraphrase Blackman and Tukey's famous statement (which they made regarding the study of spectra): *"All too often the practical assessment of field statistics requires care."*

After calculating an array such as Fig. 1, it should be *field* tested for statistical significance at some level. Unfortunately this test has sometimes been omitted, often to the detriment of conclusions drawn from the results.

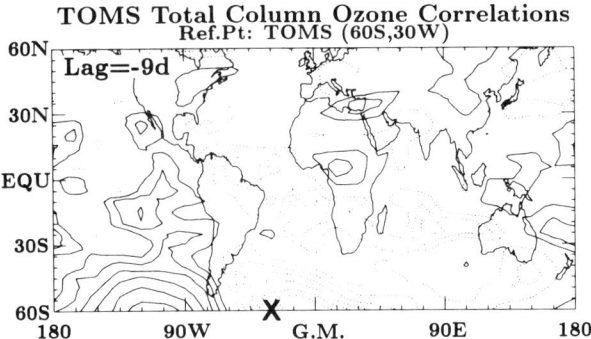

FIG. 1. Global cross-correlation map of total column ozone amounts in the atmosphere. Reference point (large $X$) is 60° S latitude, 30° W longitude. Solid and dashed contours begin at $r = 0.1$ and $-0.1$, respectively, and increment (decrement) by 0.1. The ordinate is latitude (60° N to 60° S); abscissa is longitude, centered on the Greenwich meridian. Lag = $-9d$ means the calculation is performed with grid point time series nine days behind the reference time series (see Sec. 16.1.5). Details of the ozone data set are given in Section 16.4

In their seminal treatment of the subject, Livezey and Chen [1] discuss several examples of papers with statistical conclusions that are either incorrect or in need of reevaluation.

To appreciate the problem it must be recognized that the testing of *individual* grid points for *local* statistical significance is not equivalent to testing for *field* significance. The reasons for this will be elaborated further later. By the terminology *field significance* we will refer to the statistical significance of the patterns of correlations that invariably are found in such maps. Field significance can also be called *global significance.*

The testing of statistical significance for fields (arrays) of variables is more involved than testing temporal data at a single grid location. A statistic field that appears to contain possible signals may in fact prove, under rigorous testing, to be statistically indistinguishable from random noise. The underlying difficulty in assessing the statistical significance of such a field is related to both sampling variation and possible interdependence of the data at different grid points on the array. Because of spatial interdependence, statistic values at nearby grid points may tend to be of similar magnitude and sign. A map of correlations may give the appearance of containing "signals" if it contains sizeable homogeneous regions. In reality, these features may be an artifact caused by spatial correlations in otherwise random data.

## 16.1.2 Finite Data Set Lengths and Temporal Interdependence

Consider a single map (field) of grid points, with each grid point having a corresponding single time series of data. Conclusions drawn globally from correlations must take into account uncertainties due to the inevitable finite length and possible temporal interdependence between values in time series. After proper preprocessing of each of the time series to remove time averages, trends, and perhaps other unwanted strong features (see Chapter 10), a significance level for correlation can be set, say, at 5%. Under a noise-only null hypothesis, only 5% of points would have series producing observed correlations larger than a chosen threshold. The null hypothesis that the two time series are each independently Gaussian (normally distributed) leads to a two-sided $t$ test and $n - 2$ (temporal) degrees of freedom (hereafter DOF). If the series contain temporal interdependence, the number of temporal DOF for this test is smaller than $n - 2$ but can still be estimated, for example, by a technique involving time lag autocorrelation calculations [2]. A different method is used here to compute approximate temporal DOF and will be discussed in Section 16.2.4.

## 16.1.3 Spatial Interdependence

In addition to possible temporal dependence, time series at different grid points may exhibit *spatial* interdependence. For example, atmospheric temperature fields always contain significant spatial correlation, physically due to the finite extent of air masses. The spatial scale of cold air masses is such that if the temperature on a cold winter day is $-20°C$ ($-4°F$) in Chicago, it is not likely to be $+40°C$ ($+104°F$) 500 km to the west in Iowa at that time. In fact, atmospheric fields usually contain both temporal and spatial interdependence, due to the finite size and propagation speed of more or less distinct air masses (*i.e.*, neither is it likely to be $+40°C$ at Chicago the *next* day).

Spatial interdependence has often been overlooked in statistical assessment. For that reason, special care and methods must be exercised.

Methods necessary to properly assess statistical field significance are addressed in the following sections.

## 16.1.4 If Spatial Degrees of Freedom Are Known

Envision all (25 latitudes by 72 longitudes) grid point time series in the time-lag correlation map of Fig. 1 as independent (spatially and temporally) and with identical probability density functions. If the probability of passing some chosen local test (*e.g.*, calculated correlation magnitude

> 0.4) between the reference point and any other arbitrary grid point is $p$, the corresponding probability distribution for the number of sites among the $m = 25 \times 72 - 1 = 1799$ independent grid points passing would be a binomial distribution $b(m,p)$. The expression *spatial degrees of freedom* used here refers to the effective number of independent grid points weighted with area (if array areas change with position, as for a sphere where longitude lines converge at the poles). Because of instrumental biases, natural large-scale features in the atmosphere, and the fact that the area represented by a grid point on a sphere is proportional to the cosine of latitude, the 1799 grid points in Fig. 1 will not be spatially independent and the number 1799 for the spatial DOF will not be valid; however, if an effective number for the spatial DOF is known or can be estimated with some knowledge of the physical system, the binomial distribution can still be used to test the field statistic significance [1]. The problem is that the spatial DOF are usually not known with certainty (although a rough estimate can sometimes by made and the field significance judged).

### 16.1.5 Calculation of Correlation Statistic Maps

Suitably preprocessed (see Chapter 10) data time series at map grid points $i, j$ are used to generate global correlation maps with respect to a common reference point $(i_{\text{ref}}, j_{\text{ref}})$ using (see Chapter 3, [7])

$$\hat{\rho}_{i,j}(\tau) = \frac{1}{n - \tau} \cdot \sum_{t=1}^{n-\tau} Y_{i_{\text{ref}}, j_{\text{ref}}}(t) \cdot Y_{i,j}(t + \tau) / [\hat{\sigma}_{i,j} \cdot \hat{\sigma}_{i_{\text{ref}}, j_{\text{ref}}}] \qquad (16.1)$$

$$\hat{\sigma}_{i,j}^2 = \frac{1}{n - \tau} \cdot \sum_{t=1}^{n-\tau} Y_{i,j}^2 (t + \tau) \qquad (16.2)$$

$$\hat{\sigma}_{i_{\text{ref}}, j_{\text{ref}}}^2 = \frac{1}{n - \tau} \cdot \sum_{t=1}^{n-\tau} Y_{i_{\text{ref}}, j_{\text{ref}}}^2 (t) \qquad (16.3)$$

Here $\hat{\rho}_{i,j}(\tau)$ is the calculated time-lag correlation between an arbitrary time series $Y_{i,j}$ and a "reference" time series $Y_{i_{\text{ref}}, j_{\text{ref}}}$. Data measurements exist for index times $t = 1, 2, 3, \ldots, n$.

A grid point is denoted by the ordered pair $(i, j)$, $\hat{\sigma}_{i,j}$ and $\hat{\sigma}_{i_{\text{ref}}, j_{\text{ref}}}$ are the calculated standard deviations of the two time series, and $\tau$ is a "lag" time. With this definition of lag $\tau$, the reference point time series is correlated with time series at other array locations stepped forward $\tau$ (or backward, for negative $\tau$) in time. (Such lagged-statistic maps are used to search for traveling wave features by examining maps at different lags.)

It is fundamental to note that the reference time series can be from either: (1) measured data of the same type, which would produce a map like Fig.

# INTRODUCTION

1 (or of a different type, as discussed in Sec. 16.5), or (2) a *randomly generated* data series. Use of the latter is a key element in the Monte Carlo simulations described in this chapter.

With bandpassed filtered data it is often computationally more efficient to do the correlation calculation in the frequency domain. The time-lag correlation between the reference point $(i_{ref}, j_{ref})$ and an arbitrary grid point $(i, j)$ is calculated in frequency space from the following:

$$\hat{\rho}_{i,j}(\tau) = \sum_{\omega=0}^{\pi/\Delta t} \{[a_{i,j}(\omega) \cdot a_{iref,jref}(\omega) + b_{i,j}(\omega) \cdot b_{iref,jref}(\omega)] \cos(\omega\tau)$$

$$+ [b_{i,j}(\omega) \cdot a_{iref,jref}(\omega) - a_{i,j}(\omega) \cdot b_{iref,jref}(\omega)] \sin(\omega\tau)\} \quad (16.4)$$

$$/ [\hat{\delta}_{i,j} \cdot \hat{\delta}_{iref,jref}]$$

$$\hat{\delta}_{i,j} = \sum_{\omega=0}^{\pi/\Delta t} [a_{i,j}^2(\omega) + b_{i,j}^2(\omega)]^{1/2} \quad (16.5)$$

$$\hat{\delta}_{iref,jref} = \sum_{\omega=0}^{\pi/\Delta t} [a_{iref,jref}^2(\omega) + b_{iref,jref}^2(\omega)]^{1/2} \quad (16.6)$$

Here $a_{i,j}(\omega)$ and $b_{i,j}(\omega)$ are the cosine and sine Fourier transform coefficients of the time series, at circular frequency $\omega$ and grid location $i, j$. Circular frequency $\omega$ is equal to $2\pi p/(n\Delta t)$ where the frequency index is $p = 0, 1, 2, \ldots, n/2$ and $\Delta t$ is the time interval between consecutive measurements, assumed constant. The Fourier cosine and sine coefficients are related to the time series $Y_{i,j}(t)$ (similar expression for $Y_{iref,jref}(t)$) by

$$Y_{i,j}(t, t-1, 2, \ldots, n) = \sum_{\omega=0}^{\pi/\Delta t} [a_{i,j}(\omega) \cdot \cos(\omega t \cdot \Delta t)$$

$$+ b_{i,j}(\omega) \cdot \sin(\omega t \cdot \Delta t)] \quad (16.7)$$

The forms of $Y_{i,j}$ and $Y_{iref,jref}$ from (16.7) were substituted directly into (16.1)–(16.3) to obtain (16.4)–(16.6). Note that (16.4)–(16.6) are approximate since they contain the implicit assumption that all time series are infinite sequences with period $n$. Additional details related to this technique may be found in the appendices of [3, 4]. Whether the calculations are performed in the time or frequency domain is a matter of preference and computational efficiency; for bandpass-filtered time series with time lags that are much smaller than the data set length $n$, the results are nearly equivalent because of the properties of the Fourier transform.

The remainder of this chapter demonstrates how field statistical assessment can be accomplished using Monte Carlo simulation techniques. Several examples will be given, based on actual data fields from atmospheric physics.

## 16.2 Field Statistic Assessment by Monte Carlo Simulation

With the rapid advances and availability of high-speed computing power it seems reasonable to go immediately to better statistical estimates based on intensive numerical simulations with the use of Monte Carlo methods. Evaluating field statistical significance of a statistic map is generally easy to accomplish with the Monte Carlo method, because the evaluation of spatial DOF is *not* required. We now turn our attention to such simulations, which are the heart of this chapter.

General Monte Carlo foundational aspects are presented in Chapter 5, and other applications are discussed in Chapter 15, of this book.

### 16.2.1 Overview

Our goal is to determine whether or not a statistic map can be said to be statistically significant. If it is, the map will be said to produce field (or global) statistical significance. The testing procedure that follows involves two null hypotheses and several steps. The first step involves tests of whether or not the correlation ($\hat{\rho}$) between a reference time series and that at each map grid point (called the *local* correlation value) exceeds some critical local correlation value $\rho_c$. The latter is based on the first of two null hypotheses, that correlations between two time series are due to random fluctuations (discussed in Sec. 16.2.4).

Next, a large number of random maps are computed, using random time series as reference. The percentage of map area ($A$) enclosing correlations $|\hat{\rho}| > \rho_c$ is determined for each of the random maps. For field testing at the 5% level of significance, we denote $A_c$ as the value of $A$ reached by only 5% of the random maps. $A_c$ is then compared with the area $A$ found for the map calculated with measured data.

At this point the second null hypothesis is made that the field statistic patterns are the result of only random fluctuations in the data. This hypothesis can be ruled out if $A \geq A_c$, and if so, the field (map) statistic is said to produce field (global) statistical significance.

The several steps in the testing are described in more detail in the following sections.

### 16.2.2 Use of Random Time Series

The Monte Carlo method of statistical analysis uses random time series generated with spectral characteristics similar to the actual data time series used in producing the statistic fields. It is important to adequately approximate the spectrum of the data series being investigated. In the atmosphere, data generally contain sufficient low-frequency (long-period temporal or large-scale spatial) variability to produce correlations that can be misinterpreted. For example, suppose that the actual global grid point data field contains large temporal "red noise" (low-frequency) spectrum components, and we wish to calculate a global map of lag correlation (for some arbitrarily chosen time lag) using the Monte Carlo method. If white noise (*i.e.,* variance spectrum independent of frequency) is used for the reference point time series, correlation values will often be found that are due, not to true signals (physical oscillations in the data), but to random, low-frequency background components in the spectrum of the data. However, as will be seen, appropriate statistical assessment is possible with the Monte Carlo method, providing the spectral behavior of the data is adequately modeled in the (random) test data sets used.

### 16.2.3 Simulation of the Field (Map) Statistic

A model of spectrum (discussed in Sec. 16.2.2) of the time series is used to generate the random time series used in the Monte Carlo simulations. A *red noise*-like spectrum of a time series can be fit, for example, using a first-order Markov model (also called a *first-order autoregressive process;* see Chapter 3). By this or other methods (several will be discussed) spectral amplitudes are obtained for the random time series. The random series is then used in place of the reference point time series of the real data, and a large number (typically hundreds or thousands) of correlation maps are generated, each utilizing a different random series for the correlation reference point. Each such map might be something like Fig. 1 in appearance. Specific examples will be discussed in Sections 16.3–16.5.

### 16.2.4 "Local" Critical Correlation Estimation

The first hypothesis-testing procedure concerns correlation between two time series, one at a given map grip point and the other a reference time series (as in Fig. 1). This is to reject the null hypothesis (that correlations are due to chance data variations) if $|\hat{\rho}| > \rho_c$, the local critical correlation value. For the examples in this chapter, we choose the "local test" of the temporal correlation test between two time series to be made at the 5% level of significance. (The local test level is the choice of the investigator.)

There are at least two methods for finding a suitable critical correlation value $\rho_c$ for the local test:

1. *Computational method.* One method is to use a Monte Carlo experiment, plotting percent area vs. correlation after combining a large number

(hundreds or thousands) of randomly generated correlation maps (each map derived from a simulated series at the reference point). This single plot of percent area vs. correlation will generally have a symmetric bell-shape about zero correlation, from which a value for $\rho_c$ can be easily found, *i.e.* that value of $|\hat{\rho}|$ exceeded in only 5% of test cases.

2. *Bivariate normal model method.* If one assumes that data time series taken from two fixed sites can be adequately modeled with a bivariate normal distribution, then it can be shown that the random variable $T = \hat{\rho}\sqrt{n - 2}/\sqrt{1 - \hat{\rho}^2}$ has a $t$ distribution with $n - 2$ (temporal) DOF under the null hypothesis that $\rho = 0$. The method used in this chapter for obtaining approximate temporal DOF assumes that the response of the filter used in data preprocessing (here a bandpass filter), plotted as response vs. frequency, is positive–definite and normalized such that the maximum filter value is 1. Because of the bandpass filtering, the temporal DOF will not be equal to $n - 2$. Temporal DOF are estimated by first multiplying $n - 2$ by the area of the filter and then dividing by $n/2$; $n/2$ is the area of a full spectrum response (no temporal filtering). This simple approach yields critical correlation values that are often quite close to those derived from method (1) [3]. Since the $t$ probability density function is symmetric about $t = 0$, a 5% local test becomes 2.5% on each tail, requiring the one-sided critical value $T_c$ such that the probability $T > T_c$ is 2.5% (that is, the $t$ distribution function has value 0.975 for $T = T_c$).

A main strength of method (2) is that $\rho_c$ can be calculated easily, particularly for small significance tests, for example, 0.01% local level. At such a small significance level, method (1) would require much more computation, viz, tens of thousands of randomly generated maps. Method (2) is used exclusively in this chapter for determining $\rho_c$ because it is straightforward and requires minimal effort.

In later sections we will use global fields (latitude vs. longitude) of temperature and ozone data in separate correlation studies. The temperature (ozone) data sets use a 40–50 day (27-day) period bandpass filter with normalized area calculated to be 22.5 (22), which, according to method (2), results in 45 (44) temporal DOF. For either 44 or 45 DOF, the 5% test value for $T_c$ is found from tables to be approximately 2.02, and from the definition of random variable $T$, the corresponding 5% critical correlation $\rho_c$ is $T_c/\sqrt{T_c^2 + \text{DOF}} = 0.29$ for both ozone and temperature analyses. The same critical correlation value for ozone and temperature studies is purely coincidental; two different filter responses will generally have two different computed values of $\rho_c$ for the same chosen local significance level.

## 16.2.5 Estimation of Critical Area $A_c$

With $\rho_c$ as the critical correlation for local testing, the reference time series and another grid point series are said to be correlated if the magnitude of the calculated correlation $\hat{\rho}$ exceeds $\rho_c$. Conducting this local test between a *randomly generated reference point time series* and each of the grid point series on a single map, weighting each test success ($|\hat{\rho}| > \rho_c$) by area, one then obtains a single value of area-weighted percentage of success, denoted here as percentage "area" $A$. Note that all of the examples in this chapter use latitude–longitude gridded data where longitude lines converge at the poles; it should be remembered that regardless of the map type (another example is altitude vs. latitude [3, 4]), proper spatial account should always be taken when grid points do not represent regions of equal geometric size. Doing this same procedure a large number of times, say, 1000, results in that many randomized values of area $A$. These values of $A$ are then used to test for field significance by first partitioning values of $A$ into small bins, counting the number of maps lying within each of these partitions, and then plotting these numbers vs. percentage area $A$. A 5% global significance test follows from the 50 largest areas plotted (of the 1000 generated). The smallest of these 50 areas defines a 5% critical area (denoted here $A_c$) that must be exceeded by area $A$ of a map (calculated with *observed data as the reference time series*) in order to pass the 5% field significance test.

## 16.2.6 Testing for Field Significance

The second null hypothesis is now made, that apparent patterns in the field (map) statistic obtained with the actual, observed data series are due only to random data fluctuations. This hypothesis can be ruled out at the 5% significance level if the area $A$ for the actual field statistic equals or exceeds the critical area $A_c$ (described in Sec. 16.2.5).

Such a result can be interpreted as meaning that if the statistic map were calculated from similarly measured data time series, obtained from 20 intervals of identical lengths, only once in the 20 cases would similarly strong correlation patterns be expected to occur if the data were random (contained no true signals). The choice of significance level is up to the investigator: a 10% significance level test is less difficult to achieve, but would yield a false positive about 1 time in 10, whereas a more stringent 1% level test would be expected to yield a false positive conclusion only once in 100 cases where no real signal was present.

It should be noted that this procedure cannot be said to test the statistical significance of individual pattern features on a map, but only of the significance of the map as a whole.

The method just outlined, utilizing a large number of randomly generated data sets to assess statistical significance, is commonly called a *Monte Carlo simulation*. We now give concrete examples of such tests using actual atmospheric physics data fields.

## 16.3 Example One: Atmospheric Temperature Fields

### 16.3.1 Data Description

Balloon and satellite instruments are used to measure the state of the atmosphere on a routine basis. After initial checks, and sometimes statistical inversion techniques, the satellite measurements yield atmospheric variables (such as temperature) as functions of altitude, geographical location, and time. By interpolating in space and time, 3D global grids are produced, usually twice daily (at 0000 and 1200 UTC). Such fields are important because they can be used to diagnose the atmospheric structure as a function of location and time and because they are used as initial conditions for large computer models based on complex equations of motion of the atmosphere–ocean system. Such "prognostic" models step forward in time to calculate (forecast) the expected state of the atmosphere or ocean at future times.

The calculations that follow involve time series of temperature at 2376 grid points (33 latitudes × 72 longitudes) on a geographical world map. An example of the daily time series at a specific grid point, covering 2920 consecutive days, is shown in Fig. 2.

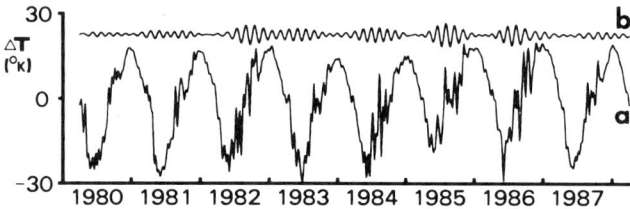

FIG. 2. Temperature deviations (Kelvins) from time series average of 255K, for 2 hPa pressure level (~43 km altitude) at map grid point 50° S latitude, 0° longitude. Eight years of daily data covering 1 April 1980 to 31 March 1988. (a) Filtered with low-pass nonrecursive filter with half-amplitude at 15-day period; (b) data in (a) have been further bandpass filtered (40–50 day periods) and displaced upwards from zero for clarity.

### 16.3.2 Data Preprocessing

It cannot be overemphasized that meticulous preprocessing of time series data is crucial. Unless care is exercised to remove the time means, trends, and hidden contaminants (such as harmonics from strong signals), a statistic such as correlation will often be misleading. Furthermore, discovery of signals in time series often requires judicious filtering to remove or reduce variance at unwanted frequencies that may complicate (or even completely obscure) the analysis through aliasing or leakage effects. Preprocessing is discussed in more detail in Chapters 10 and 11.

The time series shown in Fig. 2(b) was used as the reference time series for lag correlations in an investigation of low-frequency oscillations of the atmosphere [3]. Prior to the calculation, each of the 2376 grid point time series were preprocessed to remove time mean and long-term trends. Figure 2(a) (filtered to greatly reduce data variations having periods shorter than about 15 days) shows an annual cycle, along with apparent variability on a time scale of from 1 to 2 month periods. To investigate further the 1–2 month phenomenon, the time series in Fig. 2(a) was filtered with a digital 40–50 day bandpass filter. Trace $b$ shows the result.

As a preprocessing step the data were also spatially smoothed by filtering. This is because little information content exists in the satellite data at short spatial scales and also because the gridding procedure is known to introduce short-scale artifacts due to satellite orbit geometry [3].

### 16.3.3 Estimation of Map Statistical Significance

The correlation map to be tested statistically is made from time-lag correlations between all time series $Y_{i,j}(t)$ and a single reference point series $Y_{i_{ref}, j_{ref}}(t)$; time index $t$ may typically have thousands of values. Figure 3 shows an example of such a correlation map using the preprocessed time series discussed earlier. Eight years of daily values of $t$ were used. Examination of this plot reveals global patterns with maxima and minima that resemble waves. The question is this: Are these apparent wavelike patterns real or could they have been generated by a reference point time series of random noise in conjunction with a temperature field with small spatial DOF? To answer this question, we use Monte Carlo testing.

The Monte Carlo technique for estimating the map's statistical significance involves performing calculations in every regard identical to that described previously, with the exception that a *random* time series is substituted for the location $i_{ref}, j_{ref}$. Examining the distribution of the results from a large number of such calculations, each with a different random time series at $i_{ref}, j_{ref}$, allows a test of the statistical significance of the map in Fig. 3.

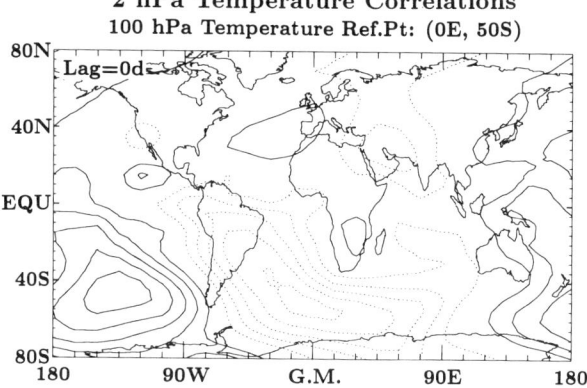

FIG. 3. Upper stratosphere global temperature correlation field for 40–50 day bandpassed data as in Fig. 2(a). Reference point is at 50° S latitude, 0° longitude at 100 hPa pressure level (~16 km altitude). Solid (dashed) contours of correlation begin at 0.1 (−0.1) and increment (decrement) by 0.1. A correlation of 1.0 is not found because the reference point is at a different pressure (altitude) than the other data. Map is calculated for no time lag between the reference point and the higher altitude time series. Maps with different lag values (not shown here) reveal large-scale features moving eastward and circling the globe every 40–50 days in the Southern Hemisphere [3].

**Random Time Series Generation Methods.** Three distinct methods were used to generate a *random* time series for the reference point series $Y_{i_{ref}, j_{ref}}(t)$ used in global correlation simulations.

Scheme 1 used a *Gaussian (white) noise time series.* The reference point time series is produced by Gaussian noise. Both $a_{random}$ and $b_{random}$ Fourier transform coefficients are taken from a random number generator yielding standard normal ordinates. Spectral features (for example, spectral "redness") of the reference point time series $Y_{i_{ref}, j_{ref}}(t)$ are not modeled in this white noise scheme. This is a viable approximation if the actual reference point series $Y_{i_{ref}, j_{ref}}(t)$ has a relatively flat spectrum or if, as in some of the examples that follow, a sufficiently narrow bandpass filter is employed.

Scheme 2 uses a *red noise spectrum.* In this scheme the reference point series $Y_{i_{ref}, j_{ref}}(t)$ is simulated with the aid of a first-order Markov process with coefficient α (Chapter 3). An estimator ($\hat{\alpha}$) for α can be obtained from the reference point time series $Y_{i_{ref}, j_{ref}}$ (after removing the time series average) as follows:

$$\hat{\alpha} = \frac{\frac{1}{n-1} \cdot \sum_{t=2}^{n} [Y_{i_{\text{ref}}, j_{\text{ref}}}(t) \cdot Y_{i_{\text{ref}}, j_{\text{ref}}}(t-1)]}{\frac{1}{n-1} \cdot \sum_{t=2}^{n} [Y_{i_{\text{ref}}, j_{\text{ref}}}^2(t-1)]} \quad (16.8)$$

The numerator and denominator in this equation represent estimators of lag one covariance and variance, respectively, for reference series $Y_{i_{\text{ref}}, j_{\text{ref}}}$. The random time series $Y_{\text{random}}(t)$ is then obtained from

$$Y_{\text{random}}(t) = \hat{\alpha} \cdot Y_{\text{random}}(t-1) + Z(t) \quad (16.9)$$

where $Z(t)$ is taken to be zero mean variance $\sigma_Z^2$ random Gaussian noise. To generate the random time series we need to know the variance of $Z(t)$. It follows from (16.9) that $\sigma_Z^2 = [1 - \alpha^2] \cdot \sigma_Y^2$, where $\sigma_Y^2$ is the variance of the time series $Y_{i_{\text{ref}}, j_{\text{ref}}}(t)$. An estimator $\hat{\sigma}_Z^2$ for $\sigma_Z^2$ is

$$\hat{\sigma}_Z^2 = (1 - \hat{\alpha}^2) \cdot \hat{\sigma}_Y^2 \quad (16.10)$$

where $\hat{\sigma}_Y^2$ is computed from

$$\hat{\sigma}_Y^2 = \frac{1}{n-1} \cdot \sum_{t=2}^{n} Y_{i_{\text{ref}}, j_{\text{ref}}}^2(t-1) \quad (16.11)$$

Scheme 2 has been used in a number of studies of low-frequency phenomena because of the tendency of the atmosphere to exhibit red-noise spectral behavior. A clear discussion of accounting for red noise is given in [5, 6]. "Blue" (larger amplitudes at higher frequencies) spectrum behavior can also be modeled by the method outlined here; the calculated first-order Markov coefficient $\hat{\alpha}$ will be negative for "blue" spectra. More complex spectra can also be modeled with higher order autoregressive processes. Chatfield [7] gives a concise, readable account. An alternate method is given next.

Scheme 3 uses a *measured amplitude with random phase (MARP) method*. To be most reliable, the Monte Carlo calculation should simulate the reference point spectrum as closely as possible. Modeling the reference point spectrum by a white- or red-noise spectrum may prove inadequate in cases where the reference point time series exhibits a sufficiently complicated spectral structure. A better approximation to the actual reference spectrum can be obtained by using the calculated Fourier transform amplitudes of the measured time series $Y_{i_{\text{ref}}, j_{\text{ref}}}(t)$.

First, random coefficients $a_{\text{random}}$ and $b_{\text{random}}$ are obtained as in Scheme

1. Next, they are adjusted to be exactly equal in amplitude with the corresponding Fourier coefficients of $Y_{i_{ref},j_{ref}}(t)$ at each circular frequency $\omega$:

$$a_{i_{ref},j_{ref}}(\omega) \to a_{i_{ref},j_{ref}}(\omega) \cdot \{A(\omega)/A_{random}(\omega)\} \quad (16.12)$$

$$b_{i_{ref},j_{ref}}(\omega) \to b_{i_{ref},j_{ref}}(\omega) \cdot \{A(\omega)/A_{random}(\omega)\} \quad (16.13)$$

where $A(\omega)$ is the actual measured amplitude at frequency $\omega$ of the measured reference-point time series $Y_{i_{ref},j_{ref}}(t)$,

$$A(\omega) = \{a_{i_{ref},j_{ref}}^2(\omega) + b_{i_{ref},j_{ref}}^2(\omega)\}^{\frac{1}{2}} \quad (16.14)$$

and $A_{random}(\omega)$ is the amplitude of the generated noise time series

$$A_{random}(\omega) = \{a_{random}^2(\omega) + b_{random}^2(\omega)\}^{\frac{1}{2}} \quad (16.15)$$

This method (hereafter denoted as the *measured amplitude with random phase method*) is an approximation to the true amplitude spectrum in that it uses the calculated amplitude from only a single realization of the time series (i.e., the measured data series) at the reference point. However, it seems reasonable that in many cases this method may be expected to provide a more accurate spectral simulation than Schemes 1 or 2. In several of the examples discussed later, the better spectral representation of the MARP method leads to a different (and presumably more rigorous) statistical assessment than either the white- or red-noise method.

**Field Significance Testing of Temperature Correlation Maps.** The temperature correlation map for zero lag shown in Fig. 3 reveals apparent large scale wavelike features. The statistical significance of this and other maps calculated for other lag values (not shown) was tested with the Monte Carlo simulation techniques described in Sec. 16.2.3 and 16.2.6. Figure 4 shows a histogram plot of the number of times a given map percentage area passed a 5% local test in 1000 simulation runs. Data at the 100 hPa reference point (50°S, 0°E) were generated randomly by Scheme 1. Nearly identical results were obtained by Schemes 2 and 3 (not shown), presumably due to the relatively narrow bandpass filter employed. The local temporal test is taken at the 5% significance level, with critical correlation magnitude equal to 0.29 (see Section 16.2.4).

With the 5% local test $|\hat{\rho}| \geq 0.29$, 95% of the 1000 simulation runs were found to have area $A \leq 6.7\%$. By definition, 6.7% represents the critical area $A_c$ described in Sec. 16.2.5. To be judged statistically significant at the 0.05 level, a map must exhibit $A \geq A_c$.

A key concept is that a random time series shifted by any lag $\tau$ is still random. Thus, because it was produced with a random time series at its reference point, the distribution in Fig. 4 can be used to test observed-data maps produced for *any lag*.

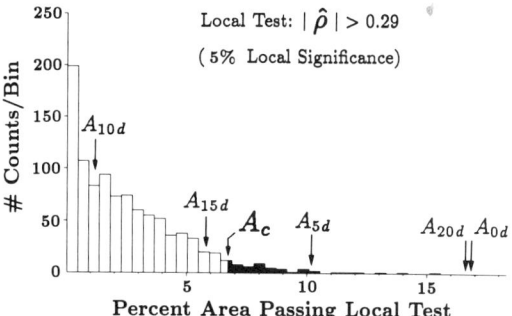

FIG. 4. Results of 1000 Monte Carlo simulations of eight years (1 April 1980–31 March 1988) of global 100 hPa pressure level (~16 km altitude) temperature correlations using 40–50 day bandpass-filtered daily data. Histogram of counts vs. % area (counts/bin) passing local 5% significance test ($|\hat{\rho}| \geq 0.29$). The random time series at the reference point was modeled by Scheme 1 Gaussian noise (see text). Arrows labeled $A_{10d}$, $A_{15d}$, . . . denote areas from maps (calculated with observed data using reference point time series) with 10, 15, . . . days behind the time series at other map locations (see text). Black regions denote areas exceeding 5% field significance value, indicated by critical area $A_c$.

In Fig. 4, the arrows labeling $A_{10d}$, $A_{15d}$, . . . give the map areas determined from maps calculated (with observed data as reference time series) with lags of 10 days, 15 days, . . ., respectively. It will be seen that lags of 0, 5, and 20 d have areas exceeding $A_c$, and pass the field statistical significance test at the 5% level. On the other hand, map areas of 10 and 15 d lags have areas less than $A_c$ and fail to produce field statistical significance. Physically this is because sinusoidal oscillations separated by a quarter period are independent, and at the predominant period of the bandpassed data (~40–50 day periods) lags of 10 and 15 days represent phase lags near a quarter period between the reference point and other map points. The correlations are thus sufficiently reduced that these maps are not statistically significant.

In summary, Monte Carlo simulations suggest statistically significant oscillations in the earth's stratosphere at periods near 40–50 days. Note that specific features cannot be said to be correlated, but only whether the entire map is statistically significant or not. Further details are given in [3].

## 16.4. Example Two: Global Ozone Data Fields

The total ozone mapping spectrometer (TOMS) satellite instrument has provided measurements of total column ozone over most of the earth since late 1978. TOMS data are important in the investigation of ozone depletion and have been used to produce the widely seen maps of the Antarctic "ozone hole" phenomenon. The data are in the form of total column ozone $\Omega$, the amount of ozone in the total atmospheric column above a given geographical location. Dobson units (D.U.) are used for $\Omega$, where 100 D.U. are equivalent to a 1 mm thick layer of pure ozone gas at sealevel standard pressure and temperature.

Figure 5 shows the global distribution of $\Omega$ for 10 March 1980. Similar daily plots reveal that the features move and change dynamically with time. Figure 6 shows an $\Omega$ time series at location 60°S, 30°W longitude, which lies within the Southern Hemisphere belt of relatively high ozone amounts seen in Fig. 5. Trace $a$ of Fig. 6 reveals annual as well as shorter period fluctuations. The spectrum [7] of this time series (Fig. 7) reveals a peak with periods near 25–30 days. To examine this phenomenon, the series were digitally filtered with a bandpass filter centered at 27 days. The resulting time series is shown in trace $b$ of Fig. 6.

The TOMS data were used to construct correlation maps using the procedure outlined in Sec. 16.1.5. Correlation was calculated using (16.4)–(16.6) by first multiplying all Fourier coefficients by the 27-day bandpass frequency response and then truncating the summation over frequency $\omega$ to only those values lying within the main lobe of the band-pass filter. To reduce noise,

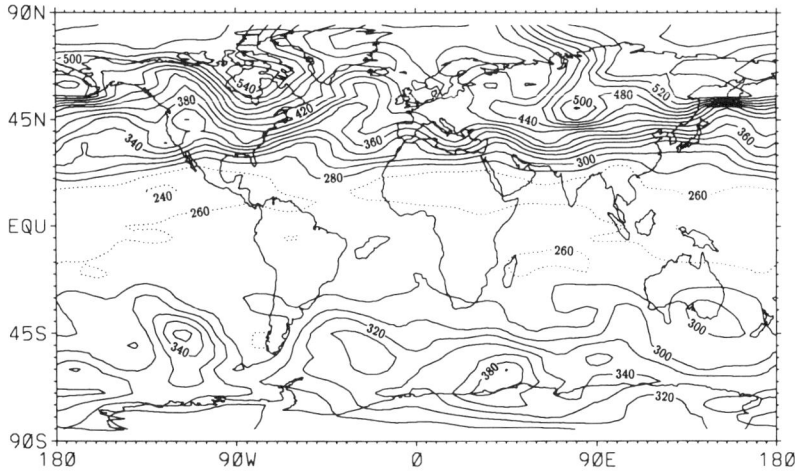

FIG. 5. Global plot of total column ozone ($\Omega$) for 10 March 1980. Ordinate is latitude (90° N to 90° S); abscissa is longitude. Units: Dobson units (D.U.). Solid (dashed) contours begin at 280 (260) and increment (decrement) by 20 D.U.

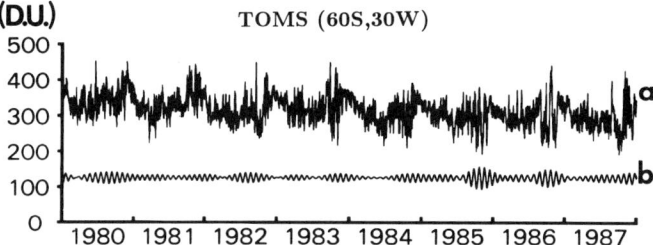

FIG. 6. (a) TOMS ozone time series at 60° S latitude, 30° W longitude. (b) Filtered with bandpass filter centered at 27 days and displaced upward from zero for clarity. Same units (Dobson units) as (a).

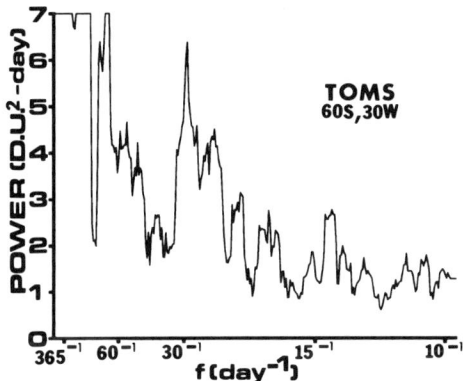

FIG. 7. Power spectral density plotted vs. frequency, for the TOMS data in Fig. 6(a). A 9-point running mean spectral estimator has been applied.

only zonal wavenumbers 1–11 were used in all correlation analyses with TOMS data. Figure 8 shows such a correlation map, obtained using the data of Fig. 6(b) as the reference time series. A large wavelike feature with positive and negative correlation regions is seen in the high Southern Hemisphere latitudes. This feature was found to propagate toward the east in time, as determined by maps at other lags (not shown).

To test for field statistical significance of the ozone correlation maps, 1000 Monte Carlo simulations were run in the manner described earlier. The bandpass filter used for the data exhibited 44 temporal DOF, as determined from method (2) described in Section 16.2.4. From this, a 5% local significance test is a success if $|\hat{\rho}| \geq 0.29$.

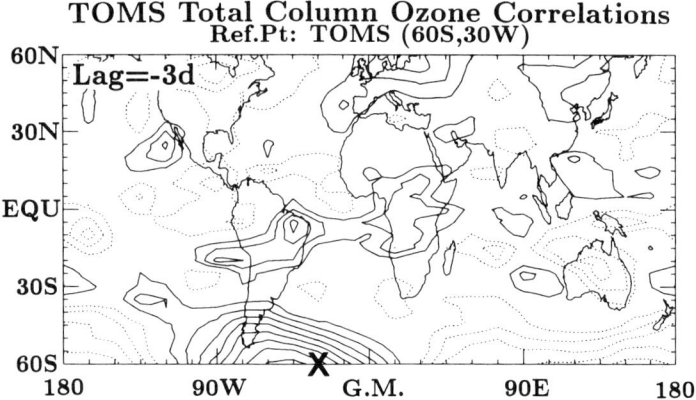

FIG. 8. Correlation map of TOMS total ozone $\Omega$. Reference point is 60° S latitude, 30° W longitude, indicated by X. Solid (dashed) correlation contours begin at 0.1 ($-0.1$) and increment (decrement) by 0.1. The ordinate is latitude (60° N to 60° S); abscissa is longitude, centered on the Greenwich meridian. In the calculation the reference point time series is 3 days ahead of other map grid point time series.

Similar to Fig. 4, the results of the Monte Carlo simulations are given in the histogram Fig. 9. The correlation map of Fig. 8, having lag of -3 days, is indicated by the -3 d arrow in Fig. 9. With the Gaussian white noise (Scheme 1) random time series (histogram $a$) the map just fails to meet field statistical significance at the 5% level. (A red-noise (Scheme 2) random time series yielded nearly equivalent results and is not shown here.) On the other hand, the map in Fig. 8 *would* meet the requirement for field statistical significance with histogram $b$, which employed the MARP method (Scheme 3) for generating the random reference point time series. The different results from these methods is presumably caused by a reference point ozone spectrum that deviates substantially from white noise (and Markov red noise), even with the rather narrow bandpass filter used here. The MARP method is thought to give the more reliable statistical test and the 27-day feature in the ozone data fields is judged to pass the field statistical significance test.

Reexamination and Explanation of Figure 1. We are now in a position to reexamine and understand Fig. 1, which, while exhibiting sizeable regions with similar correlation sign, was stated *not* to satisfy field statistical significance. The arrow labeled -9d in Fig. 9 represents Fig. 1, which was calculated with map point time series 9 days behind the ozone reference point series. The map in Fig. 1 is seen to fail to meet field statistical significance at the 5% level, with both white noise and MARP methods.

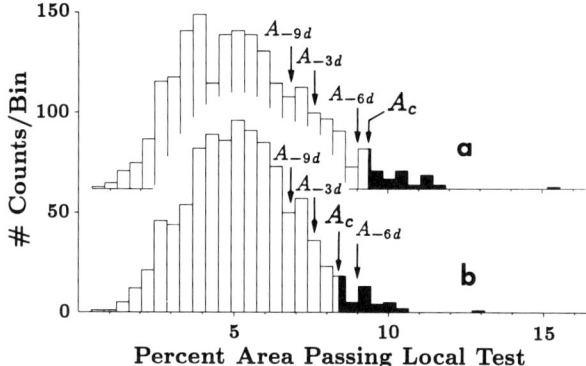

FIG. 9. Results of 1000 Monte Carlo simulations of eight-years of 27-day bandpassed global TOMS total ozone ($\Omega$) data: Correlation with reference point at 60° S latitude, 30° W longitude. Histogram: counts vs. % area passing local 5% significance test ($|\hat{\rho}| \geq 0.29$). Random time series at the reference point were modeled by (a) Gaussian white noise, and (b) MARP methods. Arrows labeled with negative lag ($-9d$, etc.) denote areas for maps calculated with the time series that many days behind the reference point time series (see Eq. (16.1)). Black regions denote simulations having areas passing 5% field significance.

The 9-day lag represents a 1/3 period lag of the relatively narrow filter used, which was centered on a 27-day period. That the correlation is weak is not unreasonable when it is recalled that a lag of 1/4 period will produce zero correlation in an infinitesimally narrow bandpass. Evidently the 9-day lag is sufficiently close to a quarter period that the correlations are weak.

## 16.5 Example Three: Cross Correlation between Ozone and Solar Flux Time Series

As a final example of the use of Monte Carlo simulations we examine possible correlation between the 27-day bandpassed ozone data series (discussed previously) and a 27-day spectral feature found in solar flux time series. Solar flux measurements at 10.7 cm (microwave) wavelength have been widely used as a proxy for sunspot activity. Figure 10(a) shows the unfiltered solar flux time series as a function of time for the years covered by the ozone data used in Fig. 6. Spectral amplitude is in Fig. 11. A prominent peak near a 27-day period is known to be associated with the rotation of the sun about its axis. The bandpass filter shown in the enlargement in Fig. 11 was applied to the data of Fig. 10(a), yielding trace $c$ of Fig. 10. [Trace $b$ shows an intermediate step of filtering to reduce effects

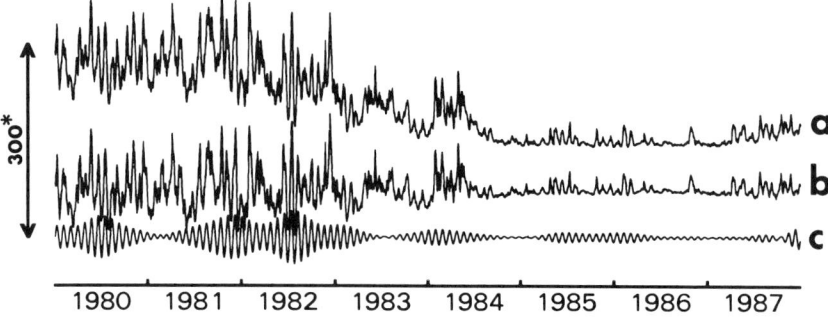

FIG. 10. (a) Time series of 10.7 cm solar flux (relative units). (b) As in (a) but with low frequencies removed using a low-pass (half-amplitude at 500-day period) digital recursive filter in time. (c) 27-day temporal bandpass filter applied to time series (b).

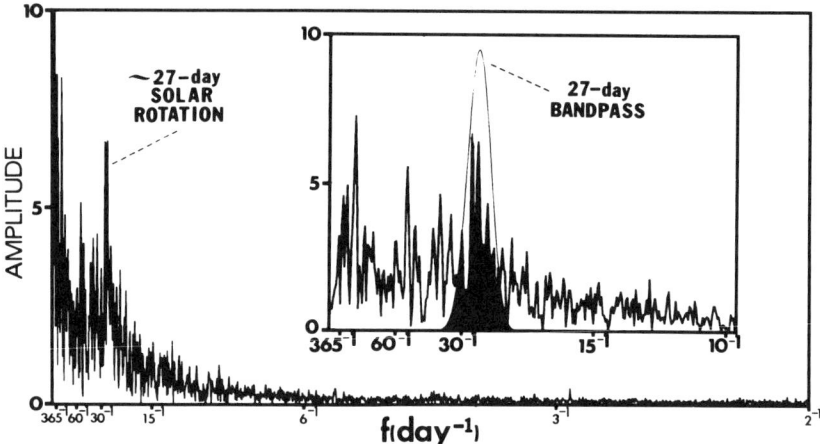

FIG. 11. Large frame: Raw (no smoothing in frequency) amplitude of 10.7 cm solar flux time series of Fig. 10(a) plotted against frequency. Insert: expanded view of (a), showing 27-day bandpass filter (darkened region) used to calculate correlations; this insert figure is identically the raw amplitude plot calculated from time series Fig. 10(b).

of the 11 year solar cycle evident in trace *a*, performed to reduce leakage (sidelobe) problems associated with time series end effects (see, for example, [8]).]

In Sec. 16.1.5 $Y_{ref}$ referred to a time series representing a point with the map data set. $Y_{ref}$ can also be taken from some other type of data that is to be compared with the original set. In the present example, $Y_{ref}$ is taken to be a time series of solar flux data, used to examine possible correlations between stratospheric ozone and solar effects. Global correlation maps (not shown) between total column ozone ($\Omega$) and modeled (random) reference time series for 10.7 cm solar flux were simulated for 1000 Monte Carlo data runs. Figure 12 shows the results, analogous to Figs. 4 and 9. In this example, sizeable differences exist between red noise (trace *a*) and MARP noise (trace *b*) methods of simulation. This difference is thought to be due largely to the spectral structure found within the bandpass of the solar flux data (Fig. 11, insert).

The correlation map derived from measured data 10.7 cm solar flux reference series and global TOMS ozone data yielded $A \simeq 9.2\%$. Based on the red-noise simulation (histogram *a* in Fig. 12), the map barely passes field statistical significance at the 5% level. But based on the MARP noise simulations (histogram *b*), it fails to pass by a sizable amount. Since in our judgment the MARP method is the more reliable statistical test, we judge the total ozone $\Omega$ correlation field with 10.7 cm solar flux to fail to produce 5% field statistical significance for the 8-year data set used here.

Note that a nonstationary association could have existed at times between ozone and solar flux, say in the sunspot maximum that occurred in the early years of the 8-yr record used here. However, the preceding results suggest that any such correlation is not sufficiently strong to be distinguishable from noise when the 8-yr record is considered as a whole.

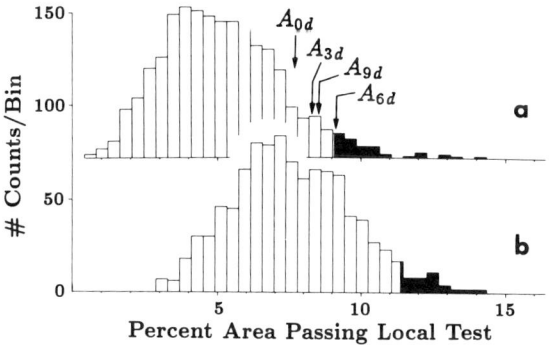

FIG. 12. Histogram for maps of correlations between total ozone and solar flux: counts vs. % area passing local 5% significance test ($|\hat{\rho}| \geq 0.29$). Results of 1000 Monte Carlo simulations with (a) red noise ($\hat{\alpha} = 0.9983$; see Sec. 16.3.3, Scheme 2) and (b) MARP method for generating random time series used. The black regions denote simulations with areas passing 5% field statistical significance.

## 16.6 Higher Dimensions

The field significance tests have described how to test two-dimensional fields of statistics. However, the method can be generalized to any number of dimensions using Monte Carlo techniques, given enough computing capacity and time [3,4].

## 16.7 Summary

Monte Carlo methods have been described for testing the statistical significance of *fields* of a statistic. Examples were given using several actual data time series from current atmospheric physics research. The tests can be important when, as is often the case, the measured data fields contain interdependencies that may give the illusion of nonrandomness. Three different schemes are discussed for generating the noise series used in the Monte Carlo tests, and the data examples utilized illustrate how the use of different schemes affect the outcome of the statistical tests.

This work has been supported in part by National Aeronautics and Space Administration Grant NAG 5-1519.

## References

1. Livezey, R. E., and Chen, W. Y. (1983). "Statistical Field Significance and Its Determination by Monte Carlo Techniques." *Mon. Wea. Rev.* **111**, 46–59.
2. Davis, R. E. (1976). "Predictability of Sea Surface Temperature and Sea Level Pressure Anomalies over the North Pacific Ocean." *J. Phys. Oceanogr.* **6**, 249–266.
3. Ziemke, J. R., and Stanford, J. L. (1991). "One-to-Two Month Oscillations: Observed High-Latitude Tropospheric and Stratospheric Response to Tropical Forcing." *J. Atmos. Sci.* **48**, 1336–1347.
4. Ziemke, J. R., and Stanford, J. L. (1990). "One-to-Two Month Oscillations in the Stratosphere During Southern Winter." *J. Atmos. Sci.* **47**, 1778–1793.
5. Madden, R. A., and Julian, P. R. (1972). "Description of Global Scale Circulation Cells in the Tropics with a 40–50 Day Period." *J. Atmos. Sci.* **29**, 1109–1123.
6. Graves, C. E., and Stanford, J. L. (1989). "Middle and High Latitude Southern Hemisphere Oscillations on the 35–60 Day Time Scale." *J. Atmos. Sci.* **46**, 1826–1837.
7. Chatfield, C. (1989). *The Analysis of Time Series*, 4th ed., Chapter 6. Chapman and Hall, London.
8. Jenkins, G. M., and Watts, D. G. (1986). *Spectral Analysis and Its Application*, Chapter 7. Holden-Day, San Francisco.

9. Madden, R. A., and Julian, P. R. (1971). "Detection of a 40–50 Day Oscillation in the Zonal Wind in the Tropical Pacific." *J. Atmos. Sci.* **28,** 702–708.
10. Hogg, R. V., and Craig, A. T. (1978). *Introduction to Mathematical Statistics,* Macmillan, New York.

# 17. MODERN STATISTICAL COMPUTING AND GRAPHICS

Frederick L. Hulting and Andrzej P. Jaworski

Alcoa Technical Center
Alcoa Center, Pennsylvania

## 17.1 Introduction

Advances in computing technology over the last 30 years have had an enormous impact on scientific practice. The modern computer has opened up new avenues of experimentation and has increased the amount, the variety, and the quality of experimental data. The analysis of that data has been similarly affected by modern computing technology. Advanced computer hardware and software have provided new environments for data analysis, enabling the development of new computer intensive statistical and graphical methods. These new developments, as well as a wide range of other topics at the interface of statistics and computer science, fall into an area known as *statistical computing*.

This chapter will survey some of the aspects of statistical computing that are of value to the experimental scientist. We begin in Section 17.2 with a discussion of some popular statistical computing environments. This includes an overview of state-of-the-art hardware and software, as well as an illustration of how data is analyzed within one of these environments, the S language on a Unix workstation. Section 17.3 briefly reviews some important numerical methods, while Section 17.4 looks at some modern topics in statistical computing by examining two useful computer intensive methods for model fitting and inference. Then, in Section 17.5, we use the example of fitting differential equation models to illustrate the collective application of these classical and modern techniques to solve a nonstandard statistical problem.

Graphical methods are both independent of and closely linked to the field of statistical computing. While many of the important graphical methods were developed independent of the computer, they have flourished since the advent of advanced computing and display technology. Additionally, modern statistical computing environments have inspired new methods, including interactive, dynamic techniques. Section 17.6 describes and illustrates a variety of computer-based graphical methods.

## 17.2 Statistical Computing Environments

An *environment* for statistical computing [1], refers to a combination of computing hardware, general-purpose software, and statistical software that empowers a user with effective tools for data analysis. Selection of an environment will depend on the needs and sophistication of the user. In this section we discuss the current state of statistical computing environments and describe one powerful environment that will be used as an example throughout this chapter.

### 17.2.1 Hardware Platforms and Operating Systems

While the distinctions between hardware platforms for scientific computing are fuzzy, one could currently group the platforms into three categories: supercomputers, workstations, and personal computers. At the high end of scientific computing are the supercomputers that offer the fastest CPUs, the largest amounts of random access memory (RAM), and the most sophisticated vectorizing compilers. At the low end of scientific computing are the personal computers, typically equipped as stand-alone machines, with limited amounts of RAM, small disk drives, and medium resolution graphics.

In the wide middle ground between these extremes fall the workstations, of which there are two major types: desktop machines and servers. Typically the desktop models are designed for personal use, with a small local disk (possibly diskless) and limited RAM, but with a high-resolution graphical subsystem. These machines are often connected via a network to a server. The servers are designed to provide computing power to a large number of users, and thus have more memory, several large disks, and upgraded network communications facilities.

The ubiquitous personal computer is probably the most common platform for statistical computing. However, the platform of choice for statistical computing is the workstation, which currently provides the best balance of power, convenience, access, graphics, and software availability for the price. Our own work, including that discussed in this chapter, uses a typical example of such a system: a DECStation 5000/240 equipped with 48 MByte of RAM, 2 GByte of disk space, a cartridge tape drive, and a standard 600 MByte CD-ROM drive. It has 8-bit color graphics with a 17 in. color monitor, keyboard, and mechanical mouse as a pointing device. The machine is connected to the local area network (ethernet), and through a gateway to the Internet.

The operating system is an integral part of any computing environment, as it can enhance or diminish our ability to make effective use of a hardware platform. This is particularly true in statistical computing, where the ability to directly interact with the computer, and the ease of obtaining and manipulating of graphical information, is of primary importance. For workstations the clear choice in operating systems is Unix, which on most workstations is bundled with a "point-and-click" graphical user interface based on X-windows. Thus, workstation statistical software typically takes advantage of the existence of the X-windows interface to offer flexible displays of numerical results, high-resolution graphical output, and interactive graphics.

### 17.2.2 Statistical Software

Statistical software—software that implements methods and tools for data analysis—can be categorized in many ways. One important distinction is between statistical *packages* and specialized statistical software. Statistical packages provide a comprehensive collection of data handling, numerical, and graphical techniques. Specialized software typically focuses on only one or two of these capabilities.

The best statistical packages have nice user interfaces, good graphics capabilities and numerical algorithms, and a wide array of "built-in" methods for data analysis. Thus, selecting a package is largely a matter of personal preference. However, these packages can be distinguished by the power and extensibility of their programming languages, that is, by their capabilities for implementing data analysis techniques that are not already built-in. These are useful criteria for readers of this volume, as the various chapters have introduced new and nonstandard, as well as standard, methods for analyzing data. For example, the likelihood-based methods of Chapter 8 are not available in standard packages, but can be easily implemented in an extensible statistical package. Thus, we will distinguish between three types of statistical software: standard statistical packages, extensible statistical packages, and specialized statistical software.

Many of the widely available and widely used statistical software packages fall into the "standard" category. These packages generally support a large suite of statistical methods and have interfaces that are designed to be "user friendly." Most of these packages are menu driven, making the software easy to learn. However, these packages typically do not have a built-in programming language that may be used to extend the software's capabilities, although they may have limited macro capabilities. Thus, new methods are difficult, or impossible, to implement. Still, these packages will generally meet the needs of a most users, and many of the methods covered in this book can be found in them.

The clear trend in software for the more sophisticated scientist or data analyst is toward the category of extensible statistical packages. These packages generally have a core set of standard methods, but also possess a powerful programming language that enables nonstandard methods to be implemented. Some packages even include an interface to programming languages such as C or FORTRAN, further increasing their flexibility. A consequence of this is that, in contrast to the standard packages, these packages take time to learn.

Two very popular packages in this category are S (see [2, 3]), and SAS (see [4]). In comparing the two, the SAS system possesses more strength in data handling, including the ability to handle large data sets. SAS also has a much larger set of core statistical and numerical routines and is available on a wider variety of platforms. On the other hand, the S system has a more flexible, object-oriented programming language, with a variety of data structures that make it an excellent choice for prototyping, testing, and implementing new methodology. S offers better control over graphical displays and does a better job of integrating graphics with analytical computations. Similar to S, but with a greater emphasis on graphical methods, is Lisp-Stat [5]. Because of its extensibility, including its easy interface to FORTRAN and C, we will use the S system in subsequent sections to illustrate the use of the computer in modern statistical and graphical analysis.

Specialized statistical software includes subroutine libraries and other programs that fill a specific niche by providing a special statistical tool to a narrow target audience. Developers of such software usually assume that the user will rely on other software for for data manipulation and standard analyses. A nice example is the publicly available software Xgobi, which implements leading-edge dynamic graphical methods. Information on such programs can be found in a variety of locations, including books, magazines, journals, computer bulletin boards, and network ftp sites. For example, the Statlib archive, at lib.stat.cmu.edu, maintains an extensive collection of special purpose statistical software (it can be reached using anonymous ftp, or by sending a "send index" electronic mail message to the archive).

### 17.2.3 Using S on a Unix Workstation

Our work is done primarily on a workstation, using a combination of SAS and S in order to capitalize on the strengths of both packages. We have chosen to use S to provide tangible illustrations of the concepts in this chapter because its high-level programming language is well-suited to the task of describing a variety of statistical and numerical computations. (Our use of S should not be viewed as an endorsement of this software over any other software for general use.) In this section we provide an introduction to S by using it to perform a fairly standard statistical analysis.

Antoine's equation, an empirical extension of the Clausius–Clapeyron equation, provides a simple but useful representation of vapor pressure vs. temperature relationship for pure compounds. The equation is usually presented as

$$P = \exp\left(A - \frac{B}{C + t}\right) \quad (17.1)$$

where $P$ denotes the pressure, $t$ is temperature, and $A$, $B$, $C$ are constants (parameters) specific to a given substance.

Equation (17.1) is used in several chemical engineering and physical calculations because vapor pressure constitutes a crucial component determining vapor–liquid equilibrium. It can be also used to approximate the heat of vaporization rather than measuring it directly by a calorimetric method. Thus, having obtained some experimental data on $P$ and $t$ for a compound, one might want to obtain estimates of the parameters $A$, $B$, and $C$ in Model 17.1. An example of such data is listed in Table 1, for benzene. To obtain the parameter estimates, we would use nonlinear least squares (Chapter 9) to fit Model (17.1) to the data. The S statements required to do the fitting, and the associated output is given in Figure 1. (In this figure, and other figures displaying S code, the S statements are preceded by the S prompt character > on the left margin, and output provided by the S system in response to these statements is preceded by a * on the left margin; comments are preceded by a #).

The process of fitting this nonlinear model nicely illustrates some important features of the S language. The first is the richness of the data structures in the S language. While in this simple analysis we do not need anything more than a matrix of numbers to represent the data set, we can still make use of the S *data frame*. A data frame is essentially a two-dimensional matrix with labeled rows and columns. However, each of the

TABLE I. Vapor Pressure of Benzene

| Row Label | $t$ (°C) | $P$ (mmHg) | Label | $t$ (°C) | $P$ (mmHg) |
|---|---|---|---|---|---|
| 1 | 14.548 | 57.41 | 11 | 54.832 | 324.93 |
| 2 | 17.720 | 67.22 | 12 | 60.784 | 402.42 |
| 3 | 20.594 | 77.28 | 13 | 67.135 | 500.69 |
| 4 | 23.270 | 87.75 | 14 | 74.028 | 627.93 |
| 5 | 26.886 | 103.64 | 15 | 78.891 | 732.07 |
| 6 | 31.004 | 124.67 | 16 | 79.413 | 744.04 |
| 7 | 35.191 | 149.43 | 17 | 79.898 | 755.23 |
| 8 | 39.078 | 175.89 | 18 | 80.442 | 767.94 |
| 9 | 44.284 | 217.16 | 19 | 80.922 | 779.34 |
| 10 | 49.066 | 261.75 | | | |

```
> benzene.fit <- nls(formula = P ~ exp(A - B/(t + C)),
    data = benzene.dat, start = list( A = 15.87, B = 2770,
    C = 220)).

> summary(benzene.fit)        # summarize nonlinear regression
*   Formula: P   exp(A - B/(t + C))
*
*   Parameters:
*         Value Std. Error t value
*   A   15.8675 0.00448526 3537.70
*   B 2769.2500 2.50131000 1107.12
*   C  219.7900 0.12547100 1751.72
*
*   Residual standard error: 0.0224287 on 16 degrees of freedom
*
*   Correlation of Parameter Estimates:
*        A      B
*   B 1.000
*   C 0.998 0.999

> coef(benzene.fit)           # extract coefficient estimates from
                                benzene.fit
*           A       B        C
*    15.86747 2769.25 219.7897

> residuals(benzene.fit)      # extract residuals from benzene.fit
*    [1]  0.005612169  0.001781895  0.007027014  0.027651762
        -0.027647414
*    [6]  0.018215291 -0.002122055 -0.010558357 -0.038656155
         0.014815704
*   [11]  0.004312370 -0.006521336  0.002916390  0.013978707
         0.010683526
*   [16]  0.042904726  0.005035055 -0.036215065 -0.026781509

> fn.antcoefs <- function(compound.data)   # define new function
  {
    start <- list(A = 15.87, B = 2770, C = 220)
    compound.fit <- nls(P ~ exp(A - B/(t + C)), compound.data,
      start)
    return(coef(compound.fit))
  }
> fn.antcoefs(benzene.dat)   # use function to get coefficient ests
*           A       B        C
*    15.86747 2769.25 219.7897
```

FIG. 1. S code for fitting Antoine's model to the benzene data.

columns may hold data of a different type, e.g., character strings, real numbers, or logical values. For our example, we will store the data of Table I in a data frame called benzene.dat. We refer to the two columns within our data frame by the column labels *P* and *t* (see Table I).

A second important feature of the S system is its usage of formulas to specify models. One can specify a model almost in the same way one would write it on a piece of paper. In the case of Antoine's equation we simply write

```
P ~ exp(A - B/(C + t)),
```

with the tilde in place of equals being the only difference from standard notation.

A third feature is the range of built-in functions for numerical and statistical computation. To fit Antoine's model to the data in Table I, we only require one S statement, the first one in Figure 1. This statement is a call to the built-in S function nls, which performs nonlinear least squares. We have passed three arguments to the function: (1) a formula describing the model, (2) a data frame holding the data to be fit, and (3) a list of starting values (a *list* is another S data structure). The output of our function is assigned to an *object* called benzene.fit (the symbol <- is the assignment operator).

Such objects represent probably the most significant feature of the S language. Essentially all S functions create objects as their output, and these objects can be assigned names for future use. These objects can be of any type; the object created by the function nls is a list whose components include the final values of the parameters, the formula used in the fit, the vector of residuals, and the vector of fitted values. After calling nls in the example, all of this information is contained in the object that we have named benzene.fit. S objects are usually passed as arguments to other functions to be manipulated in some useful way. For example, to see a summary of our nonlinear least squares analysis, we would pass benzene.fit to the summary function. Doing so produces the output shown in Figure 1. All "interesting" S objects have a summary function associated with them, although the form of the summaries will differ depending on the type of object. A variety of functions also exist to extract specific components from objects; for example, the function coef extracts the vector of coefficient estimates and the function residuals extracts the vector of residuals, from the object produced by nls—see Figure 1.

One way of extending the S system is to build onto the language with user-defined functions written in the S language. For example, we could define a function to take a set of data for a compound, fit Antoine's model to it, and return the three coefficients in a vector. Figure 1 shows how to define such a function. The definition begins with the keyword `function` and is followed by the arguments in parentheses and the body of the function within the curly brackets. The body of the function consists of a series of S statements, with the `return` statement defining the output of the function. The function we have defined is also an S object (of type function), and using the assignment operator, we have given our function a name—`fn.antcoefs`—for future use (see Section 17.4.1.4). Now, the statement `fn.antcoefs(benzene.dat)` produces the coefficient estimates for the benzene data (as shown in Figure 1).

## 17.3 Computational Methods in Statistics

According to Thisted ([6]) there are four different kinds of statistical computations: (1) *numerical*—classical number crunching applications where majority of calculations are floating point; (2) *seminumerical*—where majority of calculations are integer or string manipulations (e.g., the Monte Carlo methods of Chapter 5); (3) *graphical*—where the goal of the computation is to construct a symbolic or pictorial representation of some aspect of the problem (see Section 17.6); and (4) *symbolic*—where symbols are directly manipulated by the computer. "Classical" statistical computing has focused primarily on numerical computations, and it is those methods that will be reviewed in this section. The review will rely heavily on references, and since the literature on numerical methods is vast, only selected references will be mentioned.

In most statistical packages the numerical methods are hidden from the user. Sometimes a choice of a method is given as an option, but usually the methods used are embedded in the code and not directly accessible to the user. This creates the tendency to take numerical methods for granted, that is, to assume that the packages produce the expected results with negligible error. Although this is often true, there exist situations where the results of numerical computations can be completely wrong. A classic example of such a situation is multiple linear regression calculation when the model matrix has "almost" linearly dependent columns. It turns out that the numerical error in calculated parameter estimates is proportional to the condition number of the design matrix, and this condition number can be very large (see, for example, [7] for discussion of condition numbers and numerical stability of the least squares problem). In cases like this, an appreciation for, and rudimentary knowledge of, numerical mathematics can be very helpful.

Although numerical methods have been used for centuries, the field of numerical mathematics developed when modern computers became available. This field deals with the peculiar situation created by finite precision of computer arithmetic. All numerical problems are therefore discrete and constitute only approximations to real-life continuous problems. The hope is that, due to very fine resolution of computer arithmetic, the solution of such discretized problems is close to the solution of their continuous counterparts. In other words, we hope that the inevitable round-off errors occurring when the input data as well as all the intermediate results are translated into their machine representations, will not accumulate to such a degree that the validity of the final result might be in question.

Note that we distinguish between numerical mathematics and numerical methods. The field of numerical mathematics is concerned with ways of controlling and estimating round-off errors. Numerical methods deal primarily with the algorithms used to solve different classes of problems on a computer. Both aspects of numerical computing are important; for example, a good algorithm may fail because of poor numerical mathematics. These two aspects of numerical computing are usually presented together, but are really separate; good algorithms do not imply good numerical mathematics. That is, it is possible to develop a nice algorithm that can fail badly because of poor numerical mathematics. For the user of numerical methods, the knowledge of numerical mathematics is at least as important as the knowledge of particular algorithms. For example, in the statistical computing example given previously, it is more important to know why, when solving the linear regression problem, one should not even form the $X^TX$ matrix than to know details of the specific numerical algorithms. In this case the condition number of the $X^TX$ matrix is the square of the condition number of the $X$ matrix itself, so solving the so-called normal equations becomes numerically unstable (see, for example, reference [7] for in-depth discussion of these issues). Moreover, if the $X$ matrix is ill-conditioned (i.e., has nearly colinear columns), the $X^TX$ matrix may become numerically singular. Methods such as the QR decomposition (see [8]) enable us to avoid forming the $X^TX$ matrix during the least squares computations, thus circumventing the problem.

The literature on numerical computing is vast. Most textbooks and monographs present a mixture of numerical mathematics and specific algorithms with emphasis on one or the other. An example of a book presenting mostly algorithms is [9], although it contains some sound advice on numerical mathematics. A good example of a numerical mathematics book is [10]. There are several books on statistical computing, where the discussion of numerical methods is tailored towards statistical applications. The two books worth looking into are [6] and [11]. Finally there are several books discussing individual fields of numerical mathematics. For example there are several books dealing with numerical linear algebra. A classic example is [8]. Similar situation exists with optimization. Books by Luenberger [12] and Gill, Murray, and Wright [13] are excellent examples of textbooks presenting both theory and practice of optimization methods.

Most of areas of numerical mathematics have some relevance to statistical computing, but five areas seem to be of particular importance: linear algebra, optimization, integration, approximation, and special functions. We finish this short review by describing these five areas and their importance in statistical computing.

1. *Linear algebra.* This is arguably the most important and widely used area in statistics. Since a matrix is a natural structure for representing data, linear algebra is used in most formulations of statistical methods. The matrix formalism is so prevalent in statistical theory, that sometimes definitions in matrix notation are taken literally and converted into algorithms (see the preceding comment on $X^TX$ matrix). Linear regression and ANOVA use linear algebra algorithms for finding the estimates, calculating diagnostics, and detecting multicollinearity. In multivariate analysis several methods (principal components for example) rely on the linear eigenvalue–eigenvector problem.

2. *Optimization.* Ideas of optimization permeate the whole field of statistics, because most statistical procedures may be derived as solutions to optimization problems. For example, most estimators are defined as "best" with respect to a certain criterion. Some methods use numerical optimization explicitly. For example, maximum likelihood estimates (Chapter 8) are often calculated by directly maximizing the likelihood function. Another example is parameter estimation in nonlinear regression models. Here specialized methods are used that take into account specific form of the objective function. Direct numerical optimization is also used in the field of designed experiments to compute, for example, the so-called $D$-optimal designs. In Bayesian methodology the estimators are defined as modes (maxima) of suitable posterior distributions (see Chapter 14) and often explicit optimization is employed. Closely related to optimization is the area of solution of nonlinear equations (optimality conditions are often presented as sets of nonlinear equations). Certain robust estimators ($M$-estimators) are defined as solutions of sets of nonlinear equations.

3. *Integration.* Both single and multiple integration play a special role in statistics because, for continuous probability distributions, probabilities are defined as definite integrals. All Bayesian methodology is based on (often multiple) integration to calculate marginal probability densities. Integration can be considered a special case of differential equation

solving, which is required for fitting complex mechanistic models that are postulated in terms of differential equations (see Section 17.5).
4. *Approximation.* There are three kinds of approximation: parametric fitting (linear and nonlinear regression), interpolation, and smoothing. Smoothing methods are employed in many nonparametric estimation situations, including nonparametric density estimation and nonparametric modeling.
5. *Special functions.* These are typically functions representing cumulative distribution functions (CDF), inverse CDFs, and probability density functions of different distributions. The CDFs are often used in calculating tail probabilities of different distributions used in hypothesis testing and judging significance of estimators. Although tail probabilities can be calculated by integration, it is often more efficient to use specialized formulas (see, for example, Chapter 26 in [14]). Inverse CDFs are often used to generate the random deviates used in Monte Carlo work.

## 17.4 Computer-Intensive Statistical Methods

The term *computer-intensive methods* refers to a variety of procedures that replace some of the standard assumptions of classical statistics with computing power. The primary motivation for these techniques is the desire to be free of parametric or distributional assumptions that are not appropriate for a given problem—see Thisted [6] for a review. In this section we will discuss two types of computer intensive procedures: bootstrap methods for assessing the accuracy of estimators, and nonparametric methods for fitting "smooth" models to data.

### 17.4.1 Bootstrap Methods

**17.4.1.1 Motivation.** Throughout this book, the estimation of unknown parameters has been considered in a variety of contexts. Those discussions have typically focused on the choice of a "good" estimator, and on the evaluation of the standard error associated with the selected estimator (i.e., the standard deviation of the estimator). In many standard estimation problems, once an estimator is selected there exists a good estimator of its standard error, often derived from theory. For example, for a sample $x_1, \ldots, x_n$ of size $n$ from a population, we would estimate the mean of the population using the sample average $\bar{x} = \sum_{i=1}^{n} x_i/n$. The well-known estimate of its standard error is $[s^2/n]^{1/2}$, where $s^2 = \sum_{i=1}^{n} (x_i - \bar{x})^2/(n - 1)$ is the sample variance. However, in nonstandard estimation problems, it is often not obvious how to obtain the standard error for a given estimator. For example, suppose we are interested in estimating the median of our population using the sample median $\tilde{x}$. Even under the assumption of a normal distribution, the formula for the standard error of this estimator is not simple, nor is it widely known.

This type of situation is common for the practitioner. In some cases the theory gets quite complicated, and no formula exists. In other cases, a standard error estimate may exist in the literature, yet the scientist does not have access to it (as in the case of median). For this reason, it is desirable to have a general purpose procedure for obtaining a reasonable standard error estimate.

Consider one such procedure that is made possible by the computer. Recall that a statistic can be viewed as a random variable, and its behavior can be described by a probability distribution, often referred to as the *sampling distribution* of the statistic. Given a set of data, and some assumptions about the underlying population, one could use simulation to generate the sampling distribution and use it to construct an estimate of the standard error. This general procedure is known as the *bootstrap,* after Efron [15 and 16]. Good general discussions of the bootstrap may be found in [17–19]. Because the bootstrap generates an estimate of the sampling distribution, it could also be used to create confidence intervals or hypothesis tests.

Note that the bootstrap is the most recently developed general method for computing standard errors. Older methods, such as the jacknife, the "delta method," and cross-validation, may also be used. See the paper by Efron and Gong [18] for a comparative review. As with any of these general methods, the potential for misuse of the bootstrap is very real. The reader is encouraged to use care in its application and refer to references that give some practical guidelines for using it (see [20, 21]).

**17.4.1.2 Methodology.** For a given situation, the specifics of the bootstrap procedure will depend on the structure of the data and the distributional assumptions that are made about the underlying population. The basic idea can be illustrated for the problem of estimating the median of a population based on a sample of size $n$ from that population. Consider three sets of assumptions that we might make about the population from which we are sampling:

Case 1: The population is normal, with unknown mean $\mu$ and known variance $\sigma^2$.
Case 2: The population is normal, with unknown mean and variance $\mu$ and $\sigma^2$.
Case 3: Nothing is known about the distribution.

In the first two cases, we assume we know the general form of the distribution, but we have little or no knowledge about the parameters of that distribution. In the last case, we make no assumption about the form of the distribution.

How do we estimate the sampling distribution of the median under these three cases? For Case 1 we proceed as follows:

(A1) Use the sample average $\bar{x}$ to estimate $\mu$, and let $F$ represent the cumulative normal $(\bar{x}, \sigma^2)$ distribution function.

For $j = 1, \ldots, B$ repeat the following two steps:

(B) Generate a sample of $n$ random variables $X_{1j}^*, \ldots X_{nj}^*$ from the distribution $F$ (see Chapter 5). This is called a *bootstrap sample*.
(C) Let $\tilde{x}_j^*$ be the sample median of the $X_{1j}^*, \ldots X_{nj}^*$.

This alogrithm yields a sample of "bootstrap" medians $\tilde{x}_1^*, \ldots, \tilde{x}_B^*$, where $B$ is some large number (see [16] for a discussion of how to select $B$). The empirical cumulative distribution function (ECDF—see Chapter 7) of the $B$ bootstrap medians can be used to characterize the sampling distribution of the sample median. Using this distribution, the estimate of the standard error of the median would be

$$\hat{SE}(\tilde{x}) = \left[ \left( \sum_{j=1}^{B} (\tilde{x}_j^* - \tilde{x}_{\cdot}^*) \right) / (B - 1) \right]^{1/2} \quad (17.2)$$

where $\tilde{x}_{\cdot}^* = (1/B) \sum_{j=1}^{B} \tilde{x}_j^*$ is the mean of the bootstrap medians.

Case 2 is handled similarly, but Step (A1) is replaced by step

(A2) Use the sample average $\bar{x}$ to estimate $\mu$ and the sample variance $s^2$ to estimate $\sigma^2$, and let $F$ represent the cumulative normal$(\bar{x}, s^2)$ distribution function.

Otherwise, we proceed as previously, with Eq. (17.2) again yielding the standard error estimate. Note that we are relying on the sample to provide more information about the population than in Case 1.

In Case 3 we are making no assumptions about the distribution of the underlying population. In this case we must use the ECDF of the original sample to define the distribution $F$ on which our simulations are based. That is, we replace Step (A1) in the algorithm with the step

(A3) Let $F$ represent the ECDF of the sample $x_1, \ldots, x_n$.

In practice we sample from the ECDF by sampling with replacement, and with equal probability, from the original data.

We refer to the general method used all three cases as the *bootstrap*. The first two applications use a *parametric* bootstrap, because of the distributional assumptions. The third application uses a *nonparametric* bootstrap, which is the method discussed by Efron in his original article [15]. The bootstrap method can be applied to many problems. As in this simple example, the specifics of the procedure will depend on (1) the type of data, (2) the assumptions about the underlying population (which determine how $F$ is defined), and (3) the estimator itself. To illustrate the wide applicability of the method, we will use the bootstrap in Section 17.4.1.4 to estimate the standard errors of the coefficient estimates for Antoine's model.

**17.4.1.3 Implementation.** The bootstrap is not typically available in standard statistical packages, but it is easy to implement in S, and Figure 2 gives one such implementation. The function fn.bootstrap provides the framework for the bootstrap method, and it adapts to different situations through two user-defined functions that are supplied as arguments. These two functions handle all operations that must recognize the structure of the data. The argument SAMPLE.FUN refers to a function that defines the method of drawing the bootstrap sample and returns that sample (Steps A and B

```
> fn.bootstrap <- function(dataset, nobs, SAMPLE.FUN, EST.FUN,
      nboot)
  {
    bootstat <- rep(0, nboot)
    for(i in 1:nboot) {
            bootdata <- SAMPLE.FUN(dataset, nobs)  # Steps A and B
            bootstat[i] <- EST.FUN(bootdata)        # Step C
    }
    return(bootstat)
  }
> fn.int.sample <- function(nobs)
      { return(sample(1:nobs, replace = T)) }

> fn.npara <- function(dataset, nobs)
      { return(dataset[fn.int.sample(nobs)]) }

> fn.gauss <- function(dataset, nobs)
  {
    mn <- mean(dataset)                 # estimate mean
    sd <- sqrt(var(dataset))            # estimate std dev
    bootdata <- rnorm(nobs, mn, sd)     # generate nobs Normal RV's
  }
```

FIG. 2. S code for bootstrapping the median.

in our algorithm). Examples of such functions for the last two cases of the median problem are `fn.gauss` (Steps A2 and B) and `fn.npara` (Steps A3 and B). The argument `EST.FUN` refers to a function that returns the value of the estimator for the bootstrap sample (Step C). For the median example, we would use `fn.median`.

To illustrate these functions, we generated a pseudo-random sample of size 100 from a normal(1, 1) distribution to be our sample data. The sample median was $\tilde{x} = 1.104$. Standard errors and histograms of the bootstrap medians for the two cases with $B = 200$ were generated by the S statements

```
> boot.gauss <- fn.bootstrap(dataset, 100, fn.gauss, fn.median,
    200)
> sqrt(var(boot.gauss))       # Case 2
> hist(boot.gauss)            # Make histogram of Case 2 bootstrap
                                sample
> boot.npara <- fn.bootstrap(dataset, 100, fn.npara, fn.median,
    200)
> sqrt(var(boot.npara))       # Case 3
> hist(boot.npara)            # Make histogram of Case 3 bootstrap
                                sample
```

These statements yielded $\hat{SE}(\tilde{x}) = 0.129$ for Case 2, and $\hat{SE}(\tilde{x}) = 0.117$ for Case 3. The histograms are pictured in Figure 3.

**17.4.1.4 Applications.** When fitting nonlinear regression models, such as Eq. (17.1), asymptotic theory can be used to obtain estimated standard errors for the model coefficients (see Chapter 9, and output in Figure 1). Alternatively, we might use the bootstrap. In this section we consider two bootstrap procedures for obtaining the estimated standard error of the C coefficient.

In Section 17.2.3.2 we obtained the fitted model

$$\hat{P} = \exp\left(15.87 - \frac{2769.25}{219.79 + t}\right) \quad (17.3)$$

Let $t_i$ and $P_i$, where $i = 1, \ldots, 19$, denote the raw data values used to obtain this fit (Table I). Then let $\hat{P}_i$ be the predicted pressure at temperature $t_i$, computed from Eq. (17.3), and let $e_i = P_i - \hat{P}_i$ be the associated residual.

Any bootstrap approach to this problem will have the general form:

(A) Define a method by which to sample the *(t, P)* pairs.

For $j = 1, \ldots, B$, repeat the following two steps:

(B) Generate a sample of 19 pairs $(t^*_{1,j}, P^*_{1,j}), \ldots, (t^*_{19,j}, P^*_{19,j})$ from the method in Step (A). This is the *j*th *bootstrap sample*.

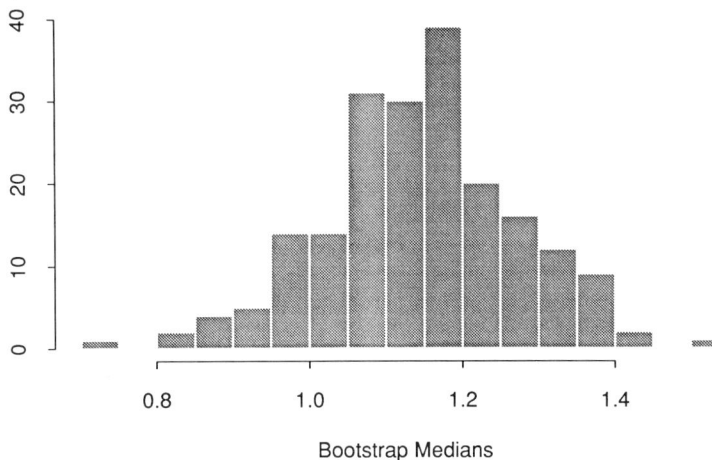

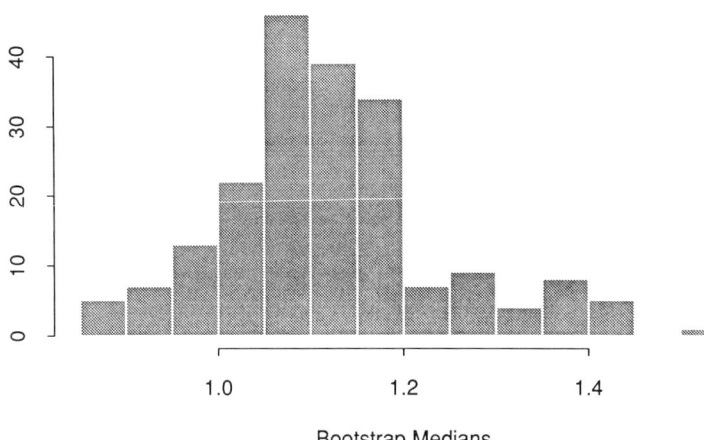

FIG. 3. Histograms of bootstrap medians.

(C) Let $\tilde{C}_j^*$ be the estimate of the coefficient $C$ obtained by fitting Eq. (17.1) to the $j$th bootstrap sample.

This yields a sample of bootstrap coefficients $\tilde{C}_1^*, \ldots, \tilde{C}_B^*$. We use the $\tilde{C}_1^*, \ldots, \tilde{C}_B^*$, to compute a standard error, in the manner of Eq. (17.2).

One approach defines Step (A) as follows. Draw a sample of size 19, with replacement, from the residuals $e_i$. Denote this sample by $e_1^*, \ldots, e_{19}^*$. Let $t_{k,j}^* = t_k$ and $P_{k,j}^* = \hat{P}_k + e_k^*$, for $k = 1, \ldots, 19$. While our sampling of the residuals is similar to the nonparametric bootstrap described earlier, we are assuming that the functional form of Antoine's model is correct and using that fact to generate the bootstrap samples. That is, we are making a specific assumption about the underlying population of the $(t, P)$ pairs. Thus, this approach has a parametric flavor to it.

We can also define Step (A) in a truly nonparametric, and simpler, manner. Draw a sample of size 19, with replacement, from the integers $1, \ldots, 19$; call these numbers $I_k^*$, $k = 1, \ldots, 19$. Then, let $t_{k,j}^* = t_{I_k^*}$, and $P_{k,j}^* = P_{I_k^*}$. That is, we are just treating each pair $(t_i, P_i)$ as a data point, and applying the nonparametric approach described in Section 17.4.1.2. Note that we do not make assumptions about the underlying population of $(t, P)$ pairs.

To implement either approach, we can use the general bootstrap function fn.bootstrap of Figure 2. Specifically, we replace the argument SAMPLE.FUN with fn.special in the "parametric" approach, and fn.simple in the simpler bootstrap procedure. Both approaches use the function fn.antC as the argument EST.FUN. These three new functions are listed in Figure 4

```
> fn.simple <- function(dataset, nobs)
      { return(dataset[fn.int.sample(nobs), ]) }
> fn.special <- function(dataset, nobs)
  {
    t <- dataset$t                              # get t from data frame
    resid <- residuals (benzene. fit)           # get residuals from fit
    P <- dataset$P + resid[fn.int.sample(nobs)] # compute new P's
    return(data.frame(t, P))                    # return new (t,P) pairs
  }
> fn.antC <- function(bootdata)
  {
    coeffs <- f.antcoefs(bootdata)     # Get vector of coefficients
    return(coeffs[3])                  # return the third element, which is C
  }
```

FIG. 4. S code for bootstrapping nonlinear regression coefficients.

As discussed by Efron and Gong [18], these two approaches are asymptotically similar. However, in this example the two approaches yield slightly different answers because of the very small sample size. Using the parametric bootstrap, we obtain a standard error for $C$ of 0.12; using the nonparametric approach, that estimate is 0.10. The asymptotic standard error is 0.13.

### 17.4.2 Smoothing

#### 17.4.2.1 Overview.
Given a plot of $y$ vs. $x$, as in Figure 5(a), it is natural to ask: what to do the data suggest is the relationship of $y$ to $x$? Chapter 9 looked at one method of answering that question. In that chapter, a parametric model $y = g(x; \boldsymbol{\beta}) + \varepsilon$ (possibly nonlinear) was postulated and fit to the data using least squares. These models are called *parametric* because the function $g(x; \boldsymbol{\beta})$ is assumed known except for a fixed number of parameters (the $\boldsymbol{\beta}$ vector).

An alternative to these parametric models is the wide class of nonparametric models in which the form of the function $g(x)$ is completely unknown, except for some requirement on its degree of "smoothness." Such models can be attractive because they require fewer assumptions on the part of the experimenter or analyst. For example, nonparametric models may be desirable when there is little or no information about the form of $g(x)$.

In the nonparametric context our estimate of $g(x)$ is an estimate of the function itself. That is, for a set of values $x_1, \ldots, x_I$, we would compute the estimates $\hat{g}(x_1), \ldots, \hat{g}(x_I)$. How do we compute the estimates $\hat{g}(x_i)$? As discussed in Chapter 9, $g(x)$ can be viewed as the middle of the distribution of $y$, conditional on $x$. In the absence of any information about the form of $g(x)$, a logical estimate at some point $x_i$ would be a function of the $(x, y)$ pairs of $x$ values in a neighborhood of $x_i$. The neighborhood may be defined as a fixed interval around $x_i$ or picked so that it contains the $k$ "nearest neighbors" (closest points). Consider Figure 5(b), which depicts a window of size 40 about a point $x_i = 190$. Using the $(x, y)$ data in the window (marked as squares), we could estimate $g(x_i)$ using a function that finds the center of the $y$ values at $x_i$. That function could be the mean, the median, or some fitted polynomial.

Such an approach to constructing $\hat{g}(x)$ is referred to as a *local* smoothing method. It is local in the sense that the estimate $\hat{g}(x_i)$ at $x_i$ depends primarily on observations near $x_i$. There are many such smoothers, and they are applicable in many areas of statistics; for example, they were used in Chapters 10 and 11 to smooth periodogram estimates. Particular methods can be differentiated by (1) how the neighborhood is defined (fixed interval,

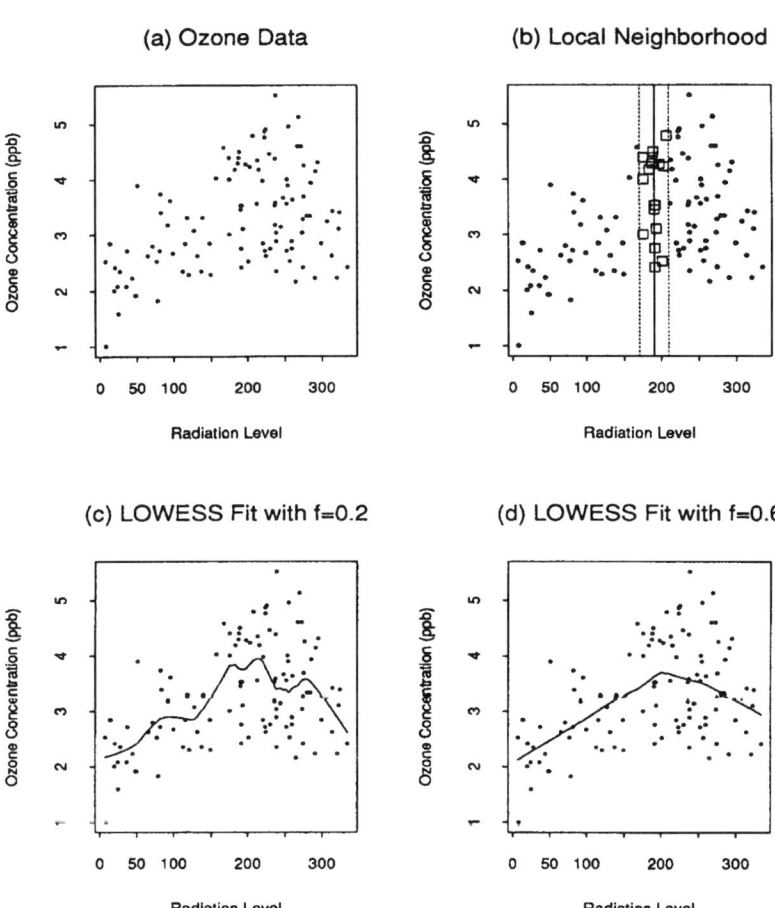

FIG. 5. Ozone data versus radiation, with an illustration of LOWESS smoothing procedure.

$k$th nearest neighbor), (2) the function of the data (e.g., mean, median, or polynomial), (3) whether observations are weighted (e.g., by distance), and (4) how they deal with outliers (robustness considerations).

For a particular smoothing method, the degree of smoothness in $\hat{g}(x)$ will depend on the value of a smoothing parameter, which controls the width of the neighborhoods. A large value of the smoothing parameter yields larger neighborhoods, which leads to a smoother fit. Smaller values of the smoothing parameter mean smaller neighborhoods and a "rougher" fit. Note that a very smooth curve, which will not fit the data well at individual points, may not capture the important features of the $y$ vs. $x$ relationship. On the other hand, a curve that fits the data well will be too "rough" to be of much use. Thus, the smoothing parameter must be selected to balance these two extremes. One way to do this is to pick the parameter so that the fitted function performs well as a predictor of future observations. This makes sense because a good predictor must not only capture the important features of a relationship, but also should not be overfit to the data. Prediction performance can be measured using the mean squared error (MSE) of prediction. One data-driven technique for selecting the smoothing parameter that attempts to minimize this MSE is *cross-validation*. The cross-validation method obtains MSE estimates through a sample reuse procedure that is similar to the bootstrap.

Good introductions to smoothing, and the selection of the smoothing parameter, may be found in the books by Härdle [22, 23] and Eubanks [24]. Other, i.e., nonlocal, smoothing methods exist; most notable are the spline methods [25]. In fitting spline models, the $\hat{g}(x_i)$ are selected by minimizing a global criterion that balances both the residual error at each point and the roughness in the curve. Again, a smoothing parameter must be chosen, and either cross-validation or generalized cross-validation may be used to select it.

**17.4.2.2 Example: LOWESS.** One widely used smoothing method, called LOWESS [26, 27], is a local ($k$th nearest neighbor) method that uses weighted, robust, polynomial fits to obtain the $\hat{g}(x_i)$ from the data in the neighborhood. Its smoothing parameter, denoted $f$, determines the fraction of the data to be included within each neighborhood.

Again, consider the data of Figure 5(a). The data are from an environmental study that looked at the relationship of air pollutant ozone and several meteorological variables in New York City [28]. Here we plot ozone vs. radiation level. While the data suggest that ozone initially increases with radiation, there appears to be a decrease in ozone at high radiation levels. However, the relationship is obscured by excessive noise in the data. Figures 5(c) and 5(d) show two possible LOWESS fits to the data, computed using the `loess` function in S. Both fits suggest that ozone increases with radiation up to a radiation level of 200; after that point, ozone is a decreasing function of radiation. However, the $f = 0.2$ fit in Figure 5(c) is decidedly rougher than the $f = 0.6$ fit in Figure 5(d), illustrating how the choice of smoothing parameter can affect the fitted model.

## 17.5 Application: Differential Equation Models

In the previous sections we have described a variety of techniques and methods that make up a toolkit to be used in structuring computations for statistical analyses. In this section we look in detail at a nonstandard statistical problem—fitting differential equation models—and show how we must draw on many of these techniques to solve that problem. In particular, the solution will require the use of numerical methods for integration and optimization, as well as computer intensive methods for smoothing and the bootstrap. In addition to discussing the general issues associated with the problem and its solution, we will illustrate the concepts by developing a new S function to handle the fitting of some simple differential equation models.

### 17.5.1 Problem Description

Differential equation models are used in many fields of applied physical science to describe the dynamic aspects of systems. The typical dynamic variable is time, and if it is the only dynamic variable, the analysis will be based on an ordinary differential equation (ODE) model. When, in addition to time, geometrical considerations are also important, partial differential equation (PDE) models are used. PDE models, together with their boundary and initial conditions, arguably constitute the most sophisticated and challenging models in today's science. Very often, however, the spatial aspects of the system under consideration can be safely ignored, and an ODE model used to describe the phenomena. In this section we will only be concerned with ODE models.

Consider three situations where the ODE models arise naturally. The first one is taken from the field of chemical engineering and deals with single-phase reaction kinetics in a well-stirred reaction vessel. In this case it can be often assumed that the essential properties of the reaction mixture such as temperature, concentrations of the reacting species, density, and other physical properties of the mixture are constant across the reactor and depend only on time. The model in this case constitutes a set of ODEs that describe dependence of concentrations on time. Usually, the temperature dependence of kinetic constants is expressed explicitly in these models. This dependence can take various forms. The Arrhenius equation, where the logarithm of a kinetic constant is a linear function of the inverse of absolute temperature, is used quite often in chemical kinetics.

The second situation, found in physical metallurgy, is very similar to the chemical kinetics problem. Here, the strength of alloys is modeled as a first-order ODE in time, although the temperature dependence of the kinetic constant is much more complex than the simple Arrhenius equation. This dependence is called the *C-curve* and, unlike the Arrhenius model, it predicts slow rates for both high and low temperatures. In chemical kinetics experiments the temperature is often held constant; in metallurgical kinetics experiments the temperature changes continuously during the heat treatment of alloys.

The third, and most complex problem, comes from the use of constitutive equations in physical metallurgy [29]. Constitutive equations describe the mechanical response of a material to applied forces, usually relating stress $\sigma$ (force per unit of cross-section area) to strain $\epsilon$ (relative elongation of the material) and to temperature. The simplest constitutive equation is Hooke's law, which states that, for a constant temperature, $\sigma = E\epsilon$ where the proportionality constant $E$ is usually called the *Young's modulus*. Hooke's law is valid over the *elastic* region of the material. However, in most applications we are interested in the region of *plastic* deformations, where material shape is permanently changed. The constitutive laws in the plastic region are sets of nonlinear differential and algebraic equations relating material properties to all possible thermomechanical histories (trajectories of temperature and strain). Given observations of such trajectories and associated material properties, we want to estimate the constitutive equation parameters.

For these models, the notion of a least squares fit to obtain estimates of the unknown model parameters is straightforward. However, the methodology is decidedly nonstandard and cannot be handled with the built-in methodology of any statistical package. In the following section we will show how a solution can be implemented by extending the S system.

### 17.5.2 Kinetics Example

**17.5.2.1 Model, Data, and Method.** The simplest nontrivial kinetic model can be described by the following first-order differential equation:

$$\dot{y} \equiv \frac{dy}{d\tau} = -k(T)y \qquad (17.4)$$

where $y$ denotes concentration, $\tau$ denotes time, and $k(T)$ is the so-called kinetic constant, which is a parametric function of absolute temperature $T$. We will assume the Arrhenius form of this parametric dependence, *i.e.*,

$$\ln k(T) = A - \frac{B}{T}$$

Note that concentration appears to be a function of two independent variables, time and temperature. However, because temperature is a function of time, time is the only real independent variable in this model.

Given observations of $y$ and $T$ over time, we want to estimate the parameters $A$ and $B$ of $k(T)$. For a constant value of $T$ the equation can be integrated analytically, producing the standard "negative exponential" decay equation, and the model fitting is straightforward. However, if the temperature follows some general time trajectory $T(\tau)$, then the model fitting is nonstandard.

As an example, assume that we have the data presented in Tables II and III. Table II and Figure 6(a) show the concentrations $y$ that were measured at times $\tau$ for three different experiments. The concentrations at time 0 are treated as known initial conditions and not as data subject to the usual sampling variability. Temperature trajectories for the experiments are shown in Table III and Figure 6(b).

An outline of the methodology follows. To fit the ODE model by least squares, we use an optimization routine to find the parameters $A$ and $B$ that minimize the sum of squared residuals. The residuals are defined as observed concentration minus the predicted concentration, and in the example we would have 11 such residuals. Computing the residuals means first computing the predicted concentrations, and that requires a numerical integration of the model equation (ODE solver). For clarity, we can use a simple ODE solver that uses the straightforward discretization

TABLE II. Raw Data for Kinetics Example

| Experiment | Time (min) | Concentration (%) |
|---|---|---|
| 1 | 0.0 | 50.00 |
| 1 | 22.0 | 49.82 |
| 1 | 73.5 | 47.56 |
| 1 | 120.0 | 44.59 |
| 2 | 0.0 | 90.00 |
| 2 | 17.5 | 90.47 |
| 2 | 38.0 | 88.90 |
| 2 | 82.0 | 72.34 |
| 2 | 120.0 | 63.34 |
| 3 | 0.0 | 80.00 |
| 3 | 15.5 | 80.13 |
| 3 | 33.0 | 79.07 |
| 3 | 74.0 | 78.14 |
| 3 | 120.0 | 54.92 |

TABLE III. Temperature Data (°C) for the Kinetics Example

| Time (min) | Experiment | | | Time (min) | Experiment | | |
|---|---|---|---|---|---|---|---|
| | 1 | 2 | 3 | | 1 | 2 | 3 |
| 0 | 25.4 | 26.1 | 24.3 | 65 | 100.2 | 128.1 | 96.8 |
| 5 | 40.2 | 35.4 | 32.6 | 70 | 102.4 | 127.3 | 105.4 |
| 10 | 53.8 | 44.8 | 40.1 | 75 | 103.0 | 125.1 | 111.7 |
| 15 | 70.5 | 55.2 | 46.2 | 80 | 101.8 | 121.6 | 117.0 |
| 20 | 84.4 | 63.9 | 53.8 | 85 | 100.6 | 118.3 | 121.1 |
| 25 | 102.3 | 74.4 | 61.2 | 90 | 99.5 | 116.9 | 125.7 |
| 30 | 104.7 | 85.0 | 62.8 | 95 | 99.1 | 117.2 | 130.8 |
| 35 | 105.2 | 96.1 | 61.4 | 100 | 99.3 | 118.7 | 133.2 |
| 40 | 103.6 | 105.2 | 59.6 | 105 | 100.8 | 120.1 | 135.4 |
| 45 | 100.9 | 113.7 | 60.3 | 110 | 100.1 | 121.5 | 136.9 |
| 50 | 98.3 | 120.8 | 68.4 | 115 | 100.2 | 122.6 | 138.5 |
| 55 | 98.5 | 123.4 | 77.5 | 120 | 105.0 | 122.7 | 138.7 |
| 60 | 100.1 | 126.6 | 85.3 | | | | |

of the differential equation. By replacing the derivative with the finite difference quotient we get

$$\frac{dy}{d\tau} = \frac{\Delta y_i}{\Delta \tau_i} = \frac{y_i - y_{i-1}}{\tau_i - \tau_{i-1}}$$

Assuming constant $\Delta \tau$ we get by substitution

$$y_i = y_{i-1} - \exp\left(A - \frac{B}{T_i}\right) \Delta \tau \, y_{i-1} \qquad (17.5)$$

Given starting values for $y$ and $T$, we can use this procedure to iteratively solve the differential equation. The time step $\Delta \tau$ needs to be reasonably small, so we do not accumulate to much discretization error. Note that the ODE solver requires temperature predictions over a time grid with intervals $\Delta \tau$. To make those predictions we must first fit smooth functions to the observed temperature trajectories.

Once we compute the estimates of $A$ and $B$, we would like to obtain standard errors for those estimates. For simpler problems we can rely on asymptotic methods; however, for more general ODE models, the bootstrap offers the only real alternative.

### 17.5.2.2 Building the S Function.
We now describe an S function to compute estimates of the parameters $A$ and $B$ for Model (17.4). The first step of our implementation is the prediction of temperatures over a fairly dense grid of equidistant time values. We can do this using the smoothing

### (a) Comparison of measured and predicted concentrations

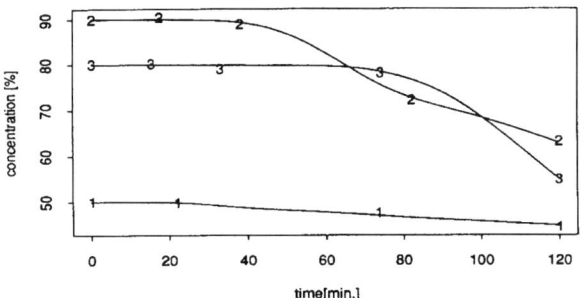

### (b) Comparison of measured and smoothed temperatures

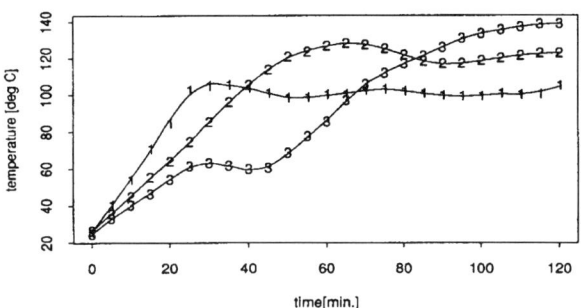

FIG. 6. Data displays for the chemical kinetics example.

function loess (described in Section 17.4.2.2). Store the temperature and time data of Table III in an S matrix called temp, where column 1 holds the times, and columns 2, 3, and 4 hold the temperatures for experiments 1, 2, and 3, respectively. Then, the following calls will smooth the data and predict the temperature values at the grid points 0.0, 0.1, . . . , 120.0 ($\Delta \tau = 0.1$) for the first experiment:

```
temp.fit <- loess(temp[,2] ~ temp[,1], data.frame(temp), span = 0.3)
temp.pred <- predict(temp.fit, seq(0, 120, by = 0.1))
```

where span holds the LOWESS smoothing parameter $f$, and the predict function is that associated with the object produced by loess. Here, temp[,1] and temp[,2] refer to columns 1 and 2 of matrix temp; the result temp.pred is a 1201 × 1 vector. Repeating this with columns 3 and 4 in place of column 2 and combining the resulting predictions produces a 1201 × 3 matrix of predictions called temp.pred.all. Figure 6(b) shows a comparison between the measured and predicted temperatures; these predictions will be used as input to the ODE solver.

Differential equation solvers generally involve iterative procedures. Because S is not very efficient for algorithms that involve large loops, we chose to write our procedure in FORTRAN, compile it, and dynamically load the resulting object into S. Then we can call the external procedure using the built-in interface to FORTRAN, an S function called .Fortran. Our simple ODE solver, based on Eq. (7.5), is coded as subroutine odegauss in Figure 7. After writing and testing the odegauss procedure outside of S, we load the object code into S using the statement dyn.load2(''odegauss.o''), making odegauss available to the .Fortran function.

We now need an S function that calls odegauss to predict concentrations $y$ at points on the time grid. The S function f.ode in Figure 8 accomplishes this. This function takes the matrix of smoothed temperature values, the indices of time points at which predictions are desired, the vector of parame-

```
      subroutine odegauss(npoint, npred, ind, initial, a, b, deltau,
     +                    temp, pred)
      integer npoint, npred, ind(npred)
      double precision initial, a, b, deltau, temp(npoint),
        pred(npred)
      integer i,j
      double precision y, t0
      data t0/273.15d0/
      j=1
      y=initial
      do 20 i=1,npred
10      y=y - exp(a - b*1.d3/(temp(j)+t0))*y*deltau
        j = j+1
        if (j.le.ind(i)) then
          goto 10
        else
          pred(i) = y
        endif
20    continue
      return
      end
```

FIG. 7. Fortran code for ODE solver.

APPLICATION: DIFFERENTIAL EQUATION MODELS 507

```
> f.odemodel <- function(meas.conc, temp, time.ind, start.val)
  {
    n. expt <- ncol(temp) - 1      # number of experiments
    temp.pred.all <- NULL          # initialize before loop
    for (i in 1:n.expt)            # get temps over time grid
    {
      temp.fit <- loess(temp[,i+1] ~ temp[,1], data.frame(temp),
        span = 0.3)
      temp.pred <- predict(temp.fit, seq(0, 120, by = 0.1))
      temp.pred.all <- cbind(temp.pred.all, temp.pred)
    }
    # fit the model
    nlsfit <- nls(~res(temp.pred.all, meas.comp, time.ind, a, b),
       start.val)
    return(nlsfit)
  }

> f.ode.resid <- function(pred.temp, meas.conc, time.ind, a, b)
  {
    ntemp <- length(meas.conc)
    pred <- ode(pred.temp, time.ind, c(a, b))
    resid <- NULL
    for (i in 1:ntemp) { resid <- c(resid, pred[[i]]
       - meas.conc[[i]]) }
    return(resid)
  }

> f.ode <- function(pred.temp, time.ind, theta)
  {
    init <- rep(100, dim(prd.temp)[2])
    delta.tau <- 0.10              # time grid interval
    ngrid <- dim(pred.temp)[1]     # number of grid points
    nexpt <- dim(pred.temp)[2]     # number of experiments
    pred <- vector("list", length = nexpt)
    for (i in 1:nexpt)
    {
      npred <- length(ind[[i]])
      pp <- vector(length = npred)
      odesolve <- .Fortran("odegauss, npoint=as.integer(ngrid),
```

FIG. 8. S functions to implement the least squares fitting of differential equation models. *Figure continues.*

```
npred=as.integer(npred), ind=as.integer(ind[[i]]),
initial=as.double(init[i]), a=as.double(theta[1]),
b=as.double(theta[2]), deltau=as.double(delta.tau),
temp=as.double(pred.temp[,i]), pp=as.double(pp))
  pred[[i]] <- odesolve$pp
 }
 return(pred)
}
```

FIG. 8. Continued

ters $(A, B)$, the values of $\Delta\tau$, and the vector of initial values as its arguments. The function loops through the number of experiments (columns of the `temp.pred.all` matrix) calling `odegauss` through the `.Fortran` function. The results are returned as a list of vectors of predictions for all the experiments. By using the list structure, we allow for the flexibility of having different number of measured concentrations for each experiment. Thus, for a given set of parameters, the `ode` function produces a set of predicted values.

These predictions are then used to calculate the residuals that will be used as input to the optimization portion of our computations. For simpler problems such as the kinetics example, we can make use of the familiar capabilities of the `nls` function. However, for complicated ODE models, such as those occurring in constitutive equation analysis, a general optimization procedure would be required.

Recall that the `nls` function will fit a nonlinear model using least squares. However, as used in Section 17.2, `nls` required a model equation as the first argument, and we cannot write the ODE model in the usual explicit notation. However, `nls` will allow us to replace the model equation with an S expression that returns the vector of residuals for given values of the parameters. Thus, we need to write a function to compute those residuals. The S function `f.ode.resid` calls `f.ode`, extracts the measured and predicted values, and forms the residual vector. Again, because our model cannot be written explicitly, we use the implicit form of the call to `nls` with `f.ode.resid`, as follows:

```
nls (~f.ode.resid(temp.pred.all, meas.conc, time.ind, a, b,
start.val))
```

where `meas.conc` and `time.ind` are lists of vectors of measured concentrations and their time indices, `a` and `b` are the Arrhenius parameters to be estimated, and `start.val` are the starting values for `a` and `b`.

All of these S statements have been collected in the S function `odemodel` (Figure 8). The call

```
> ode.fit <- f.odemodel(meas.conc, temp, time.ind, start,
start.val= list(a=19, b=9))
> summary(ode.fit)
```

produces the following output:

```
Formula: ~f.ode.resid(temp.pred.all, meas.conc, time.ind, a, b,
start.val)

Parameters:
    Value    Std. Error  t value
a 19.86160   0.902805    21.9999
b  9.98769   0.358631    27.8495

Residual standard error: 0.476353 on 9 degrees of freedom
```

The data used in this example were generated within the computer using $a = 19.4$, $b = 9.8$, and normal errors with $\sigma = 0.5$. The parameter estimates match these "true" values quite well.

By using `nls` for this simple problem, we are able to use asymptotic methods to obtain standard errors for the parameters. However, if we replace `nls` with a more general optimization routine for complex ODE models, the standard error estimates may not be easily available. In that case, we might use the bootstrap to estimate standard errors. The implementation will not be described here, but will be very similar to that described in Section 17.4.1.4.

## 17.6 Graphical Methods

### 17.6.1 Overview

In recent years the work in the field of statistical graphics has had strong links to the field of statistical computing. This is primarily because the computer has driven much of the new work in statistical graphics, with many new graphical methods being designed to exploit the capabilities of the latest hardware and software. This marriage of computing and graphics has made graphical methods a very important class of data analytic tools.

Standard graphical methods have been used throughout this volume, and examples in this Chapter include the histograms of Figure 3, and the scatterplots of Figure 5. Figure 9 and 10 show examples of less familiar graphical methods, including comparative *box* plots, the *quantile–quantile* (Q–Q) plot, and the *conditioning* plot (coplot). These particular methods perform comparative, or conditional analyses of the data. In Figure 9(a), two box plots are used to compare the distribution of tensile strengths for aluminum extrusions fabricated under four different conditions. A comparison of the distributions for two of these conditions is made in Figure 9(b) using a Q–Q plot (which is described in Chapter 7). The box plot enables a comparison of certain features of the distribution, primarily the median and the upper and lower quartiles. The Q–Q plot provides a very detailed comparison of the entire distribution. In Figure 10 we are using the coplot to investigate the spray congealing data from Chapter 9. In particular, the coplot shows how the relationship of $Y$ to $\log(X_2)$ changes with the value of $\log(X_1)$. Here, the $(Y, \log(X_2))$ pairs are separated into four groups, based on four ranges of the values of $\log(X_1)$. The pairs are plotted, with each group appearing on a separate graph. The top panel shows the four ranges, each corresponding to one of the four lower panels. Each of the lower panels uses the same horizontal and vertical scales, and a reference line has been added to clarify the changing pattern of the data points. It is clear from these plots that the relationship of $Y$ to $\log(X_2)$ remains linear, with approximately the same slope, but with an intercept that increases with the value of $\log(X_1)$.

All of these commonly used graphical methods use static displays and are referred to as *static graphical methods*. As implied by the name, static methods do not require motion to convey their information. While static methods have existed for centuries, interest in developing improved methods grew rapidly after John Tukey focused his attention on graphics in the 1960s and 1970s and wrote his book *Exploratory Data Analysis* [30]. These methods became more popular as increasing computing capabilities enabled and encouraged their widespread use.

While the computer is now the most popular medium for implementing static methods, the methods themselves transcend that medium. For that reason, we will not discuss these methods any further, and the reader is referred to the collection of excellent graphics books for descriptions and examples of the many important static graphical methods (see [26, 31–33]). Instead, we will focus the remainder of the chapter on the newer graphical methods that have been based on the real-time interaction between the analyst and a computer graphics device. Here the impact of the modern computer is most obvious, as the addition of motion and real-time interactivity has enabled data analysts to more effectively explore high-dimensional data. These *dynamic graphical methods* are discussed in the next section. While the emphasis of that section is on the graphical analysis of raw data,

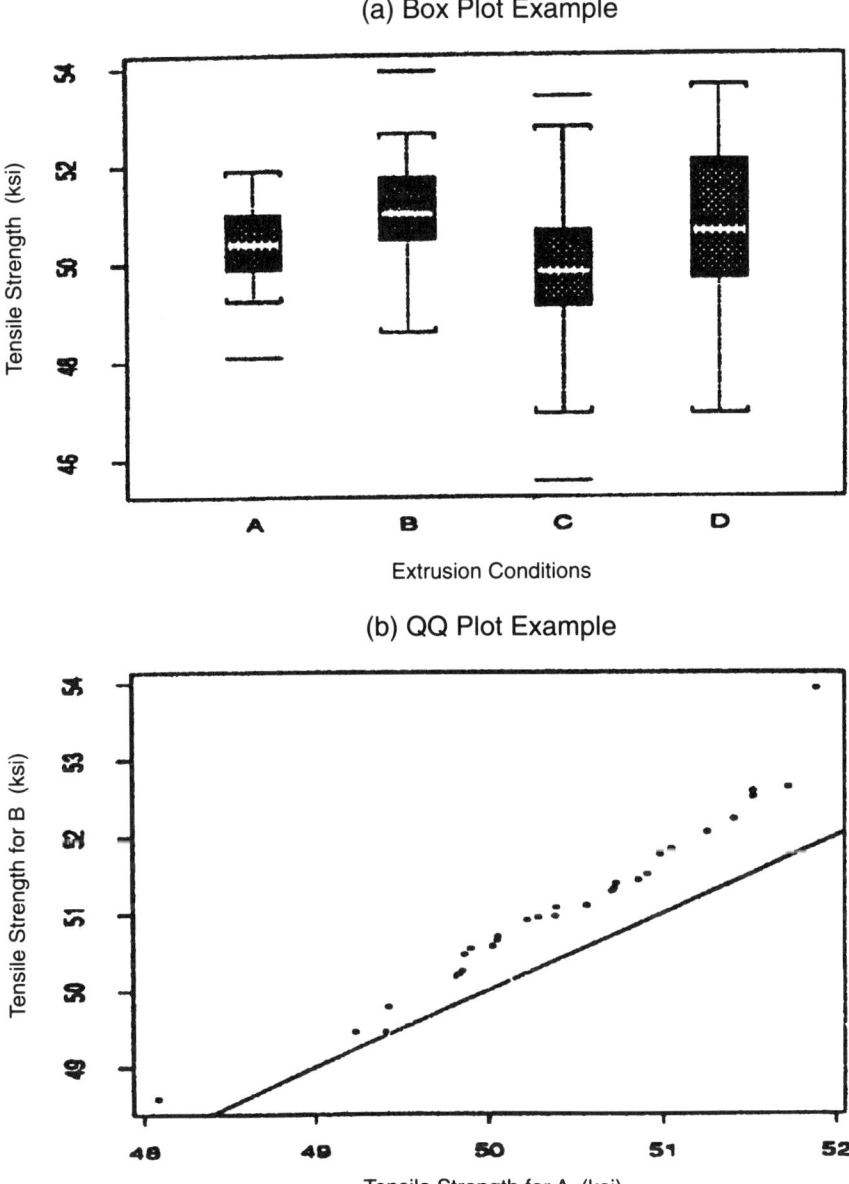

FIG. 9. Application of static graphical methods to tensile strength data.

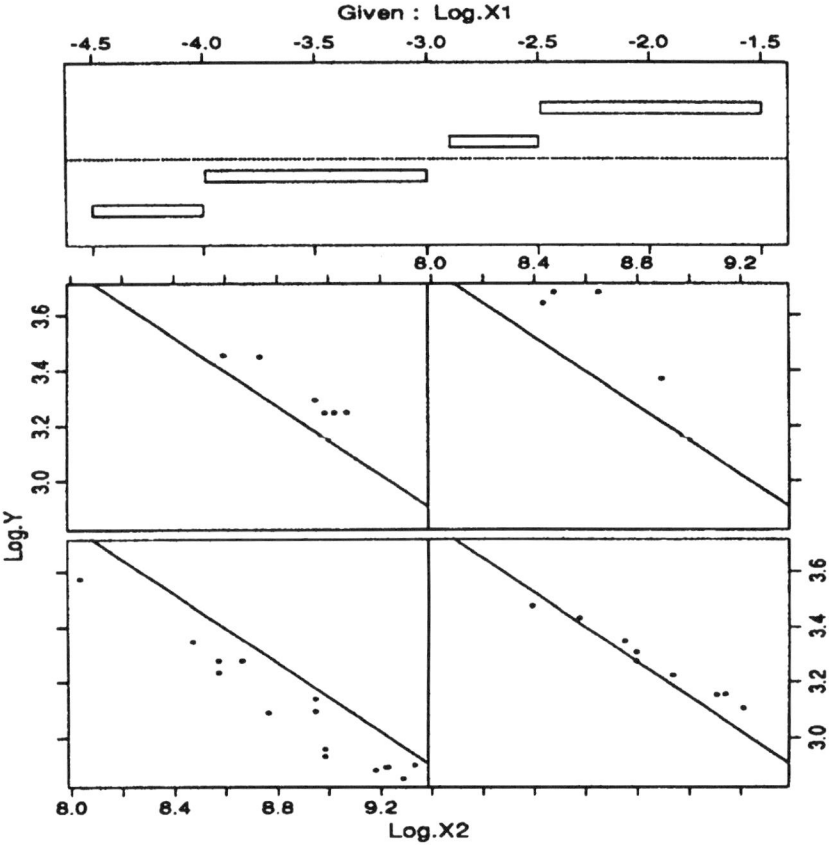

FIG. 10. Coplot for exploring the spray congealing data of Chapter 9.

static and dynamic graphical methods may also be used to analyze models that are developed and fit from that raw data. The use of graphical methods in model assessment is often referred to as *statistical visualization*. The reader is referred to [34] and [35] for more information on this topic.

### 17.6.2 Dynamic Methods

A dynamic graphical method is characterized by "direct manipulation of graphical elements on a computer graphics screen, and virtually instantaneous change of the elements" [36]. Thus, the analyst is not passively observing the graph, but rather actively interacts with the graph, instantly manipulating it using a mouse or other pointing device. This capability puts extraordinary power into the hands of the data analyst. While we cannot adequately convey that power within the pages of a book, we can at least describe some of the existing dynamic methods and their uses. In particular, we will discuss the methods of identification, linking, brushing, and rotation.

All dynamic methods act upon an initially static display, such as the one shown in Figure 11. This display of the environmental data discussed in Section 17.4.2 is generated by the S system by passing a matrix of data to the `brush` function, and consists of five separate elements: (1) histograms of each variable (bottom), (2) a scatterplot matrix showing all pairwise relationships of the four variables (upper left), (3) a three-dimensional scatterplot with rotation control panel (upper right), (4) a scrollable list of observation identifiers or case labels, here the dates of each observation (lower right), and (5) a brushing control panel (far right). Note that the three-dimensional scatter plot simply shows one two-dimensional projection of the "point cloud" of data. (The `brush()` function is part of the S-PLUS version of S [37]; SPLUS is a registered trademark of Statsci, Inc.).

This initial display provides one view of this four variable data set. Of course, many different views are required to thoroughly explore and describe the structure of the data. By interacting in real-time with this display, we can quickly produce a variety of views. For example, Figure 12 shows the same display after some manipulation, and several things have changed. We now see a number of highlighted data points in the scatterplots, the three-dimensional plot shows a different projection, some of the observation identifiers (month/day labels) are marked, and the histograms are now split. What does all this mean?

The first method at work here is *identification*. When we see a point, or group of points, on a scatter plot that are of interest, we want to associate the points with some identifying characteristic—in this case that would be the date the observation was collected. We can do that here by highlighting a particular point on a plot and seeing which label is marked on the scrolable list, or by adding a label directly to the plot itself. In Figure 12 we see a number of points highlighted on the scatterplots, and the corresponding month/day labels are marked in the scrolable list.

The next, and perhaps most important method at work here is *linking*. Note that the highlighted points in each panel are all from the same observations. That is, the highlighted points in the ozone versus wind speed scatterplot are from the same days as the highlighted points in the temperature versus radiation scatterplot. Thus, we can link the points on different scatterplots. This enables the analyst to perform a variety of conditional analyses (in the same manner as the coplot). For example, all of the highlighted

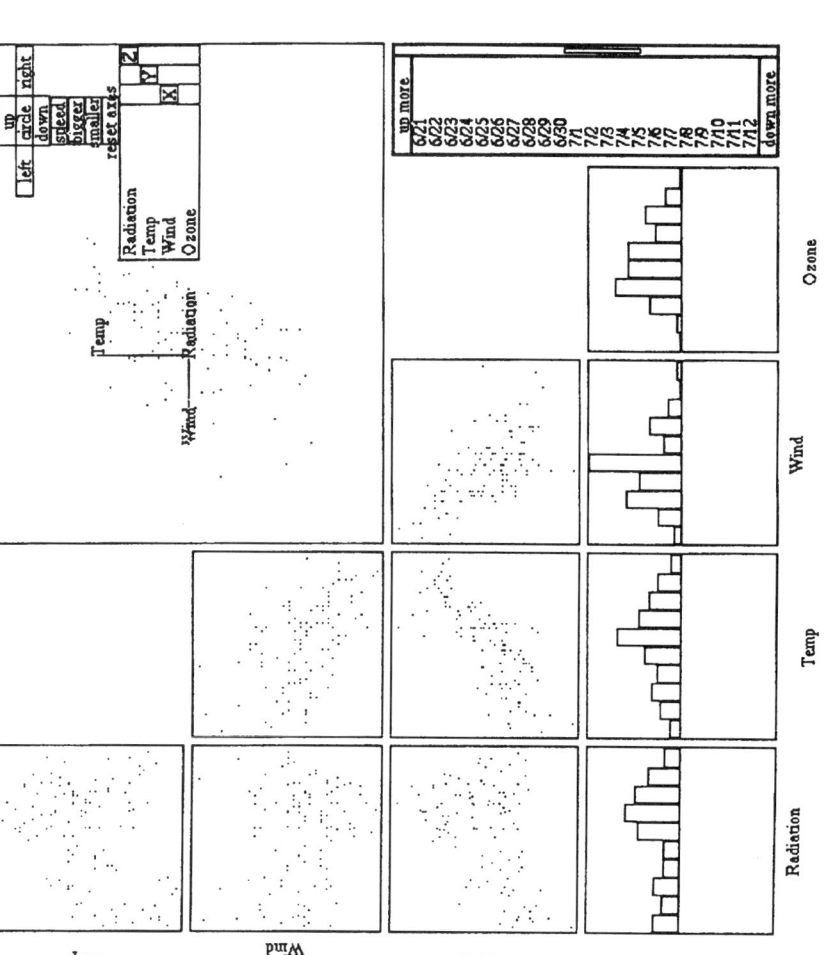

FIG. 11. Static display prior to application of dynamic methods.

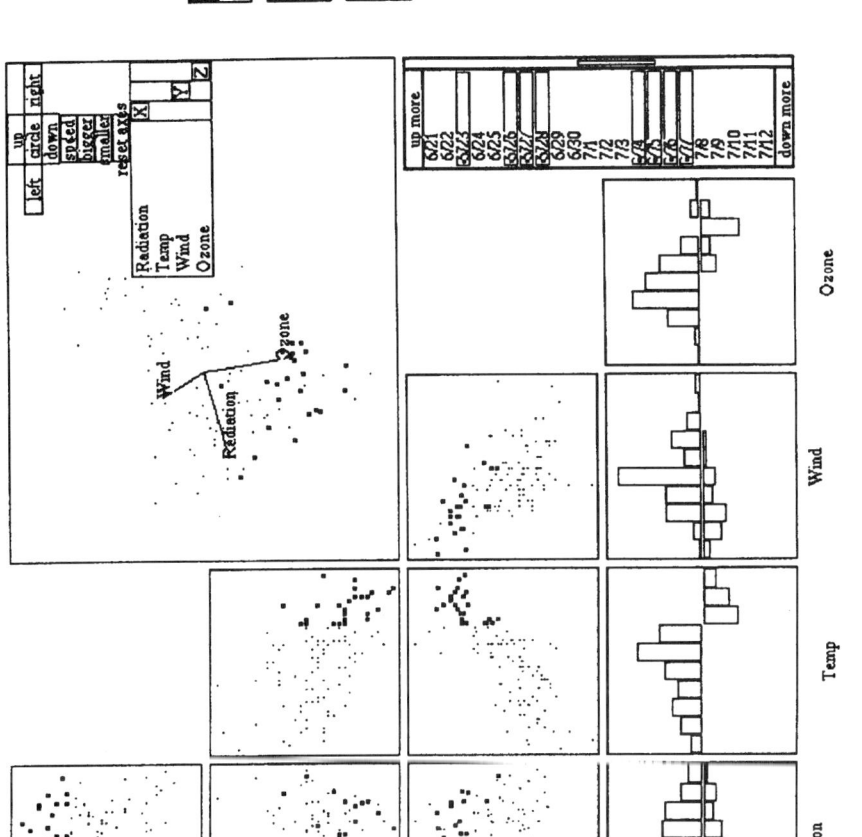

FIG. 12. Interactive display incorporating brushing and rotation.

points come from the days with the highest temperatures. By looking at the highlighted points on the ozone vs. wind speed scatterplot, we can learn how ozone and wind speed are related on hot days.

Note how the histograms are also linked to the scatterplots. Think of the highlighting operation as dividing the data into two sets of observations: highlighted, and unhighlighted. The histogram bars that extend upwards describe the distribution of the unhighlighted points. The histogram bars that extend downwards describe the distribution of the highlighted points. By placing the histograms back-to-back for each variable, we can compare the distributions of the two groups quite easily.

How do we interact with the display to identify and link points? The tool we use is a small rectangle called a *brush*. The brush is moved around the screen so as to "paint" or highlight portions of the display. The operation of the brush is defined by the control panel on the righthand side—we can set the paint mode (persistent, transient), the brush size, and the type of symbol for highlighting. By altering these parameters, we have great flexibility in implementing the identification and link methods. There is a rich collection of other brushing techniques; see [36] and [38] for more details.

Selecting the particular projection on the three-dimensional scatter plot is accomplished by *rotating* the point cloud. By rotation, we mean rapidly changing the two-dimensional views so as to achieve the perception of a rotating point cloud. The addition of motion adds a perception of depth that enables the analyst to study the trivariate structure of the data, identifying patterns and clusters that would otherwise go undetected. In this display, we combine identification with rotation to study the relationship of wind speed, radiation, and ozone concentration for the days with highest temperatures.

Of course, while Figure 12 presents a useful display, these dynamic techniques enable us to see many different versions of that display in a short time period. However, the value of these techniques lies not in the fact that many different static displays can be quickly created. Rather, it is the *animation* of the displays that enhances the user's ability to uncover interesting features of the data. For that reason the reader is strongly encouraged to try out these techniques, which are available in many popular statistical packages.

## 17.7 Conclusion

One of the clear trends in statistical computing is the way in which computers are used in the development of new statistical methods. Many classical methods were developed when computer facilities, in today's meaning of the word, did not exist. Their development relied heavily on theory, which in turn relied on assumptions assuring mathematical tractability. This process produced an impressive body of statistical knowledge that, of course, was put into computers as they become available. Modern statistical and graphical methods, on the other hand, use computers as an indispensable development tool. The bootstrap and dynamic graphical methods described in this chapter are examples of this. Another example is the whole family of Monte Carlo methods described in Chapter 5.

Two other examples of statistical methods based on computing were omitted from this review due to lack of space. One of them is known under a collective name of *Markov Chain Monte Carlo (MCMC)*. It is a general tool for simulation of complex stochastic processes useful in many types of statistical inference. The Metropolis algorithm (described briefly in Chapter 5) is a well-known example of a MCMC method. Others are Metropolis–Hastings algorithm and the Gibbs sampler algorithm. Geyer [39] presents an excellent summary of the MCMC methods. Another field of considerable importance in statistical computing is the design and analysis of computer experiments. With computational models becoming more and more complex and expensive to run, one often has to resort to experimentation to tune the parameters of the computer code. This creates a host of unique statistical issues because computer experiments are quite different from physical experiments. See Sacks, Schiller, and Welch [40] for a review of these issues.

The other clear trend in statistical computing is change. Because computer hardware and software change so quickly, we refrained from offering a comprehensive list of statistical software in Section 17.2. Such a list would become obsolete rather quickly. While one cannot predict where this change will lead, it is clear that in the future we will graphically explore larger data sets, fit more complex models, and run larger simulations with greater ease. Beyond that, we can expect that the integration of graphical user interfaces, high-level programming languages, and artificial intelligence concepts with powerful, multiprocessor hardware and fast, seamless networking will fundamentally change the way people work with the computer. These new computing environments will certainly affect the practice of statistics (see [1]), but all we can be sure of is that the future will be exciting!

## References

1. Becker, R. A., and Chambers, J. M. 1989. "Statistical Computing Environments: Past, Present, and Future." *Proceedings of the American Statistical Association, Sequicentennial Invited Paper Sessions,* pp. 245–258.

2. Becker, R. A., Chambers, J. M., and Wilks, A. R. (1988). *The New S Language: A Programming Environment for Data Analysis and Graphics.* Wadsworth and Brooks/Cole, Pacific Grove, CA.
3. Chambers, J. M., and Hastie, T. J. (1992). *Statistical Models in S.* Wadsworth and Brooks/Cole, Pacific Grove, CA.
4. SAS Institute, Inc. (1991). *SAS Language: Reference, Version 6.*
5. Tierney, L. (1990). *Lisp-Stat: An Object-Oriented Environemnt for Statistical Computing and Dynamic Graphics.* John Wiley and Sons, New York.
6. Thisted, R. A. (1988). *Elements of Statistical Computing: Numerical Computation.* Chapman and Hall, New York.
7. Stoer, J., and Bulirsch, R. (1980). *Introduction to Numerical Analysis.* Springer.
8. Golub, G. H., and Van Loan, C. F. (1983). *Matrix Computations.* Johns Hopkins University Press, Baltimore.
9. Press, W. H., Teukolsky, S. A., Vetterling, W. T., and Flannery, B. P. (1992). *Numerical Recipes in Fortran: The Art of Scientific Computing,* 2d ed. Cambridge Press, Cambridge.
10. Kincaid, D., and Cheney, W. (1991). *Numerical Analysis: Mathematics of Scientific Computing.* Brooks/Cole, Pacific Grove, CA.
11. Kennedy, W. J., and Gentle, J. E. (1980). *Statistical Computing,* Marcel Dekker, New York.
12. Luenberger, D. G. (1984). *Linear and Nonlinear Programming,* 2d ed. Addison-Wesley, Reading, MA.
13. Gill, P. E., Murray, W. M., and Wright, M. H. (1981). *Practical Optimization.* Academic Press, New York
14. Abramowitz, M., and Stegun, I. A. (eds.) (1970). *Handbook of Mathematical Functions,* Dover.
15. Efron, B. (1979). "Bootstrap Methods: Another Look at the Jacknife." *Annals of Statistics,* **7,** 1–26.
16. Efron, B. (1982). "The Jacknife, the Bootstrap, and Other Resampling Plans." SIAM, **38.**
17. Diaconis, P., and Efron, B. (1983). "Computer Intensive Methods in Statistics." *Scientific American* (May).
18. Efron, B., and Gong, G. (1983). "A Leisurely Look at the Bootstrap, the Jacknife, and Cross-Validation." *The American Statistician* **37**(1), 36–48.
19. Efron, B., and Tibshirani, R. (1986). "The Bootstrap Method for Assessing Statistical Accuracy." *Statistical Science* **1,** 54–77.
20. LePage, R., and Billard. L. (1992). *Exploring the Limits of the Bootstrap.* John Wiley and Sons, New York.
21. Mammen, E. (1992). *When Does the Bootstrap Work? Asymptotic Results and Simulations.* Springer-Verlag, New York.
22. Härdle, W. (1990). *Applied Nonparametric Regression.* Cambridge University Press, Cambridge.
23. Härdle, W. (1991). *Smoothing Techniques with Implementation in S,* Springer Verlag, New York.
24. Eubank, R. L. (1988). *Spline Smoothing and Nonparametric Regression.* Marcel Dekker, New York.
25. Wahba, G. (1990). "Spline Models for Observational Data." SIAM, **59.**
26. Cleveland, W. S. (1985). *The Elements of Graphing Data.* Wadsworth, Moneterey, CA.

27. Cleveland, W. S. (1979). "Robust Locally Weighted Regression and Smoothing Scatter Plots." *Journal of the American Statistical Association* **74,** 829–836.
28. Bruntz, S. M., Cleveland, W. S., Kleiner, B., and Warner, J. L. (1974). "The Dependence of Ambient Ozone on Solar Radiation, Wind, Temperature, and Mixing Height." *Symposium on Atmospheric Diffusion and Air Pollution,* pp. 125–128.
29. Sample, V. M., Lalli, L. A., and Richmond, O. (1992). "Experimental Aspects and Phenomenology of Internal State Variable Constitutive Model." *Third International Conference on Aluminum Alloys Physical and Mechanical Properties,* **3,** pp. 385–431. Norwegian Institute of Technology, Trondheim.
30. Tukey, J. W. (1977). *Exploratory Data Analysis.* Addison Wesley, Reading, MA.
31. Tufte, E. R. (1983). *The Visual Display of Quantitative Information.* Graphics Press, Cheshire, CN.
32. Tufte, E. R. (1990). *Envisioning Information.* Graphics Press, Cheshire, CN.
33. Chambers, J. M., Cleveland, W. S., Kleiner, B., and Tukey, P. A. (1983). *Graphical Methods for Data Analysis.* Wadsworth, Monterey, CA.
34. Newton, H. J. (ed.). (1992). *Computing Science and Statistics,* **24.** *Graphics and Visualization.* Interface Foundation of North America, Fairfax Staion, VA.
35. Cleveland, W. S. (1993). *Visualizing Data.* Hobart Press, Summit, NJ.
36. Becker, R. A., Cleveland, W. S., and Wilks, A. R. (1987). "Dynamic Graphics for Data Analysis." *Statistical Science,* **2,** 355–395.
37. Statistical Sciences, Inc. (1991). *S-PLUS User's Manual and Reference Manual.* Seattle.
38. Cleveland, W. S., and McGill, M. E. (1988). *Dynamic Graphics for Statistics.* Wadsworth and Brooks/Cole, Pacific Grove, CA.
39. Geyer, C. J. (1992). "Practical Markov Chain Monte Carlo." *Statistical Science* **4,** 473–483.
40. Sacks, J. Schiller, S. B., and Welch, W. J. (1989). "Designing for Computer Experiments." *Technometrics* **31,** 41–47.

# TABLES

## Table 1
## Pseudo-Random Digits

| | | | | | | | | | |
|---|---|---|---|---|---|---|---|---|---|
| 13018 | 75799 | 25484 | 27887 | 05657 | 17585 | 23677 | 61339 | 75801 | 90264 |
| 69405 | 38082 | 17163 | 38445 | 55760 | 04603 | 87555 | 45181 | 96531 | 43624 |
| 83879 | 12500 | 27974 | 18440 | 14473 | 13923 | 84803 | 35278 | 54146 | 27973 |
| 06245 | 35226 | 16154 | 72770 | 18215 | 29282 | 76392 | 41314 | 11633 | 01725 |
| 96511 | 51723 | 90189 | 38426 | 49708 | 93708 | 05827 | 71620 | 85457 | 20152 |
| | | | | | | | | | |
| 62365 | 52374 | 57285 | 57119 | 33860 | 08124 | 94833 | 15194 | 48205 | 94217 |
| 64674 | 56953 | 34305 | 29579 | 04012 | 03984 | 23731 | 96290 | 94477 | 62887 |
| 78387 | 84360 | 79321 | 18892 | 66826 | 56188 | 97741 | 53232 | 66983 | 46816 |
| 64786 | 82091 | 91423 | 52023 | 96178 | 78732 | 75696 | 02684 | 19958 | 82437 |
| 93059 | 58107 | 57976 | 53409 | 08548 | 70785 | 53032 | 16339 | 82171 | 31692 |
| | | | | | | | | | |
| 76281 | 51623 | 78191 | 61165 | 50730 | 02966 | 15266 | 53115 | 31430 | 22041 |
| 55607 | 26147 | 26573 | 58555 | 82331 | 01396 | 38497 | 16440 | 36152 | 74287 |
| 26391 | 88456 | 42547 | 92627 | 66327 | 76643 | 52764 | 44973 | 45715 | 58261 |
| 93740 | 96364 | 73850 | 33207 | 56661 | 97982 | 57519 | 09722 | 39756 | 09520 |
| 37863 | 87878 | 40416 | 00674 | 73815 | 79321 | 34135 | 83246 | 92762 | 30174 |
| | | | | | | | | | |
| 84148 | 59996 | 99432 | 93469 | 80734 | 68712 | 62088 | 17092 | 28257 | 39491 |
| 53789 | 65383 | 15466 | 52510 | 43784 | 72630 | 93585 | 40408 | 24389 | 03356 |
| 70332 | 01967 | 73742 | 39873 | 14690 | 91171 | 85013 | 20713 | 10084 | 39039 |
| 24048 | 59285 | 67075 | 36635 | 79929 | 54396 | 12556 | 31863 | 93225 | 14083 |
| 42571 | 10009 | 71566 | 92959 | 34078 | 47663 | 98518 | 31739 | 01542 | 70311 |
| | | | | | | | | | |
| 50803 | 29585 | 74790 | 28462 | 18281 | 17480 | 74749 | 05105 | 88624 | 23161 |
| 02580 | 22673 | 86278 | 68183 | 80298 | 21014 | 12672 | 27946 | 53877 | 86159 |
| 95167 | 06804 | 93606 | 15983 | 98253 | 35047 | 99974 | 01179 | 23388 | 42561 |
| 67515 | 75541 | 79461 | 77546 | 29489 | 83701 | 70545 | 82496 | 73528 | 53748 |
| 36274 | 80737 | 32152 | 87015 | 34085 | 33549 | 99872 | 07090 | 61385 | 12606 |
| | | | | | | | | | |
| 53181 | 23806 | 51339 | 95408 | 99140 | 99547 | 97879 | 09183 | 87482 | 98535 |
| 52011 | 91087 | 06140 | 17871 | 18213 | 76071 | 48262 | 93189 | 29450 | 11614 |
| 80052 | 25260 | 96939 | 08238 | 11590 | 80494 | 01044 | 61634 | 21388 | 50055 |
| 09803 | 67440 | 54829 | 38692 | 72249 | 35959 | 80825 | 60257 | 29356 | 36924 |
| 21327 | 84869 | 65199 | 45936 | 46229 | 37809 | 44275 | 93897 | 09957 | 26196 |
| | | | | | | | | | |
| 49274 | 83089 | 15643 | 46950 | 20005 | 98145 | 75552 | 21330 | 38037 | 09304 |
| 69474 | 94009 | 09761 | 49117 | 87545 | 23377 | 77455 | 39660 | 80301 | 14029 |
| 48419 | 18621 | 05654 | 67980 | 60638 | 36553 | 11168 | 25601 | 47395 | 59832 |
| 64825 | 58545 | 55149 | 42453 | 67663 | 66525 | 27500 | 53788 | 03420 | 57573 |
| 72094 | 59161 | 92166 | 20130 | 56189 | 48182 | 86275 | 22959 | 91920 | 67512 |
| | | | | | | | | | |
| 01984 | 64546 | 06092 | 11325 | 52336 | 14982 | 24676 | 88300 | 40774 | 30317 |
| 98018 | 02701 | 95268 | 31566 | 22730 | 87915 | 11034 | 62318 | 07530 | 31642 |
| 49699 | 01873 | 62424 | 54097 | 73628 | 43394 | 35667 | 35225 | 20975 | 18835 |
| 37366 | 96305 | 12209 | 56673 | 34152 | 03698 | 29785 | 37064 | 58687 | 62633 |
| 25066 | 11090 | 64478 | 97960 | 79166 | 17703 | 19086 | 09952 | 37811 | 26581 |

This table was created using the "RANDOM" command in the MINITAB statistical package.

## Table 2
## Standard Normal Cumulative Probabilities

$$\Phi(z) = \int_{-\infty}^{z} \frac{1}{\sqrt{2\pi}} \exp(-\frac{t^2}{2})\, dt$$

| z | .00 | .01 | .02 | .03 | .04 | .05 | .06 | .07 | .08 | .09 |
|---|---|---|---|---|---|---|---|---|---|---|
| −3.4 | .0003 | .0003 | .0003 | .0003 | .0003 | .0003 | .0003 | .0003 | .0003 | .0002 |
| −3.3 | .0005 | .0005 | .0005 | .0004 | .0004 | .0004 | .0004 | .0004 | .0004 | .0003 |
| −3.2 | .0007 | .0007 | .0006 | .0006 | .0006 | .0006 | .0006 | .0005 | .0005 | .0005 |
| −3.1 | .0010 | .0009 | .0009 | .0009 | .0008 | .0008 | .0008 | .0008 | .0007 | .0007 |
| −3.0 | .0013 | .0013 | .0013 | .0012 | .0012 | .0011 | .0011 | .0011 | .0010 | .0010 |
| −2.9 | .0019 | .0018 | .0018 | .0017 | .0016 | .0016 | .0015 | .0015 | .0014 | .0014 |
| −2.8 | .0026 | .0025 | .0024 | .0023 | .0023 | .0022 | .0021 | .0021 | .0020 | .0019 |
| −2.7 | .0035 | .0034 | .0033 | .0032 | .0031 | .0030 | .0029 | .0028 | .0027 | .0026 |
| −2.6 | .0047 | .0045 | .0044 | .0043 | .0041 | .0040 | .0039 | .0038 | .0037 | .0036 |
| −2.5 | .0062 | .0060 | .0059 | .0057 | .0055 | .0054 | .0052 | .0051 | .0049 | .0048 |
| −2.4 | .0082 | .0080 | .0078 | .0075 | .0073 | .0071 | .0069 | .0068 | .0066 | .0064 |
| −2.3 | .0107 | .0104 | .0102 | .0099 | .0096 | .0094 | .0091 | .0089 | .0087 | .0084 |
| −2.2 | .0139 | .0136 | .0132 | .0129 | .0125 | .0122 | .0119 | .0116 | .0113 | .0110 |
| −2.1 | .0179 | .0174 | .0170 | .0166 | .0162 | .0158 | .0154 | .0150 | .0146 | .0143 |
| −2.0 | .0228 | .0222 | .0217 | .0212 | .0207 | .0202 | .0197 | .0192 | .0188 | .0183 |
| −1.9 | .0287 | .0281 | .0274 | .0268 | .0262 | .0256 | .0250 | .0244 | .0239 | .0233 |
| −1.8 | .0359 | .0351 | .0344 | .0336 | .0329 | .0322 | .0314 | .0307 | .0301 | .0294 |
| −1.7 | .0446 | .0436 | .0427 | .0418 | .0409 | .0401 | .0392 | .0384 | .0375 | .0367 |
| −1.6 | .0548 | .0537 | .0526 | .0516 | .0505 | .0495 | .0485 | .0475 | .0465 | .0455 |
| −1.5 | .0668 | .0655 | .0643 | .0630 | .0618 | .0606 | .0594 | .0582 | .0571 | .0559 |
| −1.4 | .0808 | .0793 | .0778 | .0764 | .0749 | .0735 | .0721 | .0708 | .0694 | .0681 |
| −1.3 | .0968 | .0951 | .0934 | .0918 | .0901 | .0885 | .0869 | .0853 | .0838 | .0823 |
| −1.2 | .1151 | .1131 | .1112 | .1093 | .1075 | .1056 | .1038 | .1020 | .1003 | .0985 |
| −1.1 | .1357 | .1335 | .1314 | .1292 | .1271 | .1251 | .1230 | .1210 | .1190 | .1170 |
| −1.0 | .1587 | .1562 | .1539 | .1515 | .1492 | .1469 | .1446 | .1423 | .1401 | .1379 |
| −0.9 | .1841 | .1814 | .1788 | .1762 | .1736 | .1711 | .1685 | .1660 | .1635 | .1611 |
| −0.8 | .2119 | .2090 | .2061 | .2033 | .2005 | .1977 | .1949 | .1922 | .1894 | .1867 |
| −0.7 | .2420 | .2389 | .2358 | .2327 | .2297 | .2266 | .2236 | .2206 | .2177 | .2148 |
| −0.6 | .2743 | .2709 | .2676 | .2643 | .2611 | .2578 | .2546 | .2514 | .2483 | .2451 |
| −0.5 | .3085 | .3050 | .3015 | .2981 | .2946 | .2912 | .2877 | .2843 | .2810 | .2776 |
| −0.4 | .3446 | .3409 | .3372 | .3336 | .3300 | .3264 | .3228 | .3192 | .3156 | .3121 |
| −0.3 | .3821 | .3783 | .3745 | .3707 | .3669 | .3632 | .3594 | .3557 | .3520 | .3483 |
| −0.2 | .4207 | .4168 | .4129 | .4090 | .4052 | .4013 | .3974 | .3936 | .3897 | .3859 |
| −0.1 | .4602 | .4562 | .4522 | .4483 | .4443 | .4404 | .4364 | .4325 | .4286 | .4247 |
| −0.0 | .5000 | .4960 | .4920 | .4880 | .4840 | .4801 | .4761 | .4721 | .4681 | .4641 |

| z | .00 | .01 | .02 | .03 | .04 | .05 | .06 | .07 | .08 | .09 |
|---|---|---|---|---|---|---|---|---|---|---|
| 0.0 | .5000 | .5040 | .5080 | .5120 | .5160 | .5199 | .5239 | .5279 | .5319 | .5359 |
| 0.1 | .5398 | .5438 | .5478 | .5517 | .5557 | .5596 | .5636 | .5675 | .5714 | .5753 |
| 0.2 | .5793 | .5832 | .5871 | .5910 | .5948 | .5987 | .6026 | .6064 | .6103 | .6141 |
| 0.3 | .6179 | .6217 | .6255 | .6293 | .6331 | .6368 | .6406 | .6443 | .6480 | .6517 |
| 0.4 | .6554 | .6591 | .6628 | .6664 | .6700 | .6736 | .6772 | .6808 | .6844 | .6879 |
| 0.5 | .6915 | .6950 | .6985 | .7019 | .7054 | .7088 | .7123 | .7157 | .7190 | .7224 |
| 0.6 | .7257 | .7291 | .7324 | .7357 | .7389 | .7422 | .7454 | .7486 | .7517 | .7549 |
| 0.7 | .7580 | .7611 | .7642 | .7673 | .7704 | .7734 | .7764 | .7794 | .7823 | .7852 |
| 0.8 | .7881 | .7910 | .7939 | .7967 | .7995 | .8023 | .8051 | .8078 | .8106 | .8133 |
| 0.9 | .8159 | .8186 | .8212 | .8238 | .8264 | .8289 | .8315 | .8340 | .8365 | .8389 |
| 1.0 | .8413 | .8438 | .8461 | .8485 | .8508 | .8531 | .8554 | .8577 | .8599 | .8621 |
| 1.1 | .8643 | .8665 | .8686 | .8708 | .8729 | .8749 | .8770 | .8790 | .8810 | .8830 |
| 1.2 | .8849 | .8869 | .8888 | .8907 | .8925 | .8944 | .8962 | .8980 | .8997 | .9015 |
| 1.3 | .9032 | .9049 | .9066 | .9082 | .9099 | .9115 | .9131 | .9147 | .9162 | .9177 |
| 1.4 | .9192 | .9207 | .9222 | .9236 | .9251 | .9265 | .9279 | .9292 | .9306 | .9319 |
| 1.5 | .9332 | .9345 | .9357 | .9370 | .9382 | .9394 | .9406 | .9418 | .9429 | .9441 |
| 1.6 | .9452 | .9463 | .9474 | .9484 | .9495 | .9505 | .9515 | .9525 | .9535 | .9545 |
| 1.7 | .9554 | .9564 | .9573 | .9582 | .9591 | .9599 | .9608 | .9616 | .9625 | .9633 |
| 1.8 | .9641 | .9649 | .9656 | .9664 | .9671 | .9678 | .9686 | .9693 | .9699 | .9706 |
| 1.9 | .9713 | .9719 | .9726 | .9732 | .9738 | .9744 | .9750 | .9756 | .9761 | .9767 |
| 2.0 | .9773 | .9778 | .9783 | .9788 | .9793 | .9798 | .9803 | .9808 | .9812 | .9817 |
| 2.1 | .9821 | .9826 | .9830 | .9834 | .9838 | .9842 | .9846 | .9850 | .9854 | .9857 |
| 2.2 | .9861 | .9864 | .9868 | .9871 | .9875 | .9878 | .9881 | .9884 | .9887 | .9890 |
| 2.3 | .9893 | .9896 | .9898 | .9901 | .9904 | .9906 | .9909 | .9911 | .9913 | .9916 |
| 2.4 | .9918 | .9920 | .9922 | .9925 | .9927 | .9929 | .9931 | .9932 | .9934 | .9936 |
| 2.5 | .9938 | .9940 | .9941 | .9943 | .9945 | .9946 | .9948 | .9949 | .9951 | .9952 |
| 2.6 | .9953 | .9955 | .9956 | .9957 | .9959 | .9960 | .9961 | .9962 | .9963 | .9964 |
| 2.7 | .9965 | .9966 | .9967 | .9968 | .9969 | .9970 | .9971 | .9972 | .9973 | .9974 |
| 2.8 | .9974 | .9975 | .9976 | .9977 | .9977 | .9978 | .9979 | .9979 | .9980 | .9981 |
| 2.9 | .9981 | .9982 | .9983 | .9983 | .9984 | .9984 | .9985 | .9985 | .9986 | .9986 |
| 3.0 | .9987 | .9987 | .9987 | .9988 | .9988 | .9989 | .9989 | .9989 | .9990 | .9990 |
| 3.1 | .9990 | .9991 | .9991 | .9991 | .9992 | .9992 | .9992 | .9992 | .9993 | .9993 |
| 3.2 | .9993 | .9993 | .9994 | .9994 | .9994 | .9994 | .9994 | .9995 | .9995 | .9995 |
| 3.3 | .9995 | .9995 | .9996 | .9996 | .9996 | .9996 | .9996 | .9996 | .9996 | .9997 |
| 3.4 | .9997 | .9997 | .9997 | .9997 | .9997 | .9997 | .9997 | .9997 | .9997 | .9998 |

This table was created using the "CDF" command in the MINITAB statistical package.

# Table 3
## Chi-Square Distribution Quantiles

| $\nu$ | $Q(.005)$ | $Q(.01)$ | $Q(.025)$ | $Q(.05)$ | $Q(.1)$ | $Q(.9)$ | $Q(.95)$ | $Q(.975)$ | $Q(.99)$ | $Q(.995)$ |
|---|---|---|---|---|---|---|---|---|---|---|
| 1  | 0.000  | 0.000  | 0.001  | 0.004  | 0.016  | 2.706  | 3.841  | 5.024  | 6.635  | 7.879 |
| 2  | 0.010  | 0.020  | 0.051  | 0.103  | 0.211  | 4.605  | 5.991  | 7.378  | 9.210  | 10.597 |
| 3  | 0.072  | 0.115  | 0.216  | 0.352  | 0.584  | 6.251  | 7.815  | 9.348  | 11.345 | 12.838 |
| 4  | 0.207  | 0.297  | 0.484  | 0.711  | 1.064  | 7.779  | 9.488  | 11.143 | 13.277 | 14.860 |
| 5  | 0.412  | 0.554  | 0.831  | 1.145  | 1.610  | 9.236  | 11.070 | 12.833 | 15.086 | 16.750 |
| 6  | 0.676  | 0.872  | 1.237  | 1.635  | 2.204  | 10.645 | 12.592 | 14.449 | 16.812 | 18.548 |
| 7  | 0.989  | 1.239  | 1.690  | 2.167  | 2.833  | 12.017 | 14.067 | 16.013 | 18.475 | 20.278 |
| 8  | 1.344  | 1.646  | 2.180  | 2.733  | 3.490  | 13.362 | 15.507 | 17.535 | 20.090 | 21.955 |
| 9  | 1.735  | 2.088  | 2.700  | 3.325  | 4.168  | 14.684 | 16.919 | 19.023 | 21.666 | 23.589 |
| 10 | 2.156  | 2.558  | 3.247  | 3.940  | 4.865  | 15.987 | 18.307 | 20.483 | 23.209 | 25.188 |
| 11 | 2.603  | 3.053  | 3.816  | 4.575  | 5.578  | 17.275 | 19.675 | 21.920 | 24.725 | 26.757 |
| 12 | 3.074  | 3.571  | 4.404  | 5.226  | 6.304  | 18.549 | 21.026 | 23.337 | 26.217 | 28.300 |
| 13 | 3.565  | 4.107  | 5.009  | 5.892  | 7.042  | 19.812 | 22.362 | 24.736 | 27.688 | 29.819 |
| 14 | 4.075  | 4.660  | 5.629  | 6.571  | 7.790  | 21.064 | 23.685 | 26.119 | 29.141 | 31.319 |
| 15 | 4.601  | 5.229  | 6.262  | 7.261  | 8.547  | 22.307 | 24.996 | 27.488 | 30.578 | 32.801 |
| 16 | 5.142  | 5.812  | 6.908  | 7.962  | 9.312  | 23.542 | 26.296 | 28.845 | 32.000 | 34.267 |
| 17 | 5.697  | 6.408  | 7.564  | 8.672  | 10.085 | 24.769 | 27.587 | 30.191 | 33.409 | 35.718 |
| 18 | 6.265  | 7.015  | 8.231  | 9.390  | 10.865 | 25.989 | 28.869 | 31.526 | 34.805 | 37.156 |
| 19 | 6.844  | 7.633  | 8.907  | 10.117 | 11.651 | 27.204 | 30.143 | 32.852 | 36.191 | 38.582 |
| 20 | 7.434  | 8.260  | 9.591  | 10.851 | 12.443 | 28.412 | 31.410 | 34.170 | 37.566 | 39.997 |
| 21 | 8.034  | 8.897  | 10.283 | 11.591 | 13.240 | 29.615 | 32.671 | 35.479 | 38.932 | 41.401 |
| 22 | 8.643  | 9.542  | 10.982 | 12.338 | 14.041 | 30.813 | 33.924 | 36.781 | 40.290 | 42.796 |
| 23 | 9.260  | 10.196 | 11.689 | 13.091 | 14.848 | 32.007 | 35.172 | 38.076 | 41.638 | 44.181 |
| 24 | 9.886  | 10.856 | 12.401 | 13.848 | 15.659 | 33.196 | 36.415 | 39.364 | 42.980 | 45.559 |
| 25 | 10.520 | 11.524 | 13.120 | 14.611 | 16.473 | 34.382 | 37.653 | 40.647 | 44.314 | 46.928 |
| 26 | 11.160 | 12.198 | 13.844 | 15.379 | 17.292 | 35.563 | 38.885 | 41.923 | 45.642 | 48.290 |
| 27 | 11.808 | 12.879 | 14.573 | 16.151 | 18.114 | 36.741 | 40.113 | 43.195 | 46.963 | 49.645 |
| 28 | 12.461 | 13.565 | 15.308 | 16.928 | 18.939 | 37.916 | 41.337 | 44.461 | 48.278 | 50.994 |
| 29 | 13.121 | 14.256 | 16.047 | 17.708 | 19.768 | 39.087 | 42.557 | 45.722 | 49.588 | 52.336 |
| 30 | 13.787 | 14.953 | 16.791 | 18.493 | 20.599 | 40.256 | 43.773 | 46.979 | 50.892 | 53.672 |
| 31 | 14.458 | 15.655 | 17.539 | 19.281 | 21.434 | 41.422 | 44.985 | 48.232 | 52.192 | 55.003 |
| 32 | 15.134 | 16.362 | 18.291 | 20.072 | 22.271 | 42.585 | 46.194 | 49.480 | 53.486 | 56.328 |
| 33 | 15.815 | 17.074 | 19.047 | 20.867 | 23.110 | 43.745 | 47.400 | 50.725 | 54.775 | 57.648 |
| 34 | 16.501 | 17.789 | 19.806 | 21.664 | 23.952 | 44.903 | 48.602 | 51.966 | 56.061 | 58.964 |
| 35 | 17.192 | 18.509 | 20.569 | 22.465 | 24.797 | 46.059 | 49.802 | 53.204 | 57.342 | 60.275 |
| 36 | 17.887 | 19.233 | 21.336 | 23.269 | 25.643 | 47.212 | 50.998 | 54.437 | 58.619 | 61.581 |
| 37 | 18.586 | 19.960 | 22.106 | 24.075 | 26.492 | 48.364 | 52.192 | 55.668 | 59.893 | 62.885 |
| 38 | 19.289 | 20.691 | 22.878 | 24.884 | 27.343 | 49.513 | 53.384 | 56.896 | 61.163 | 64.183 |
| 39 | 19.996 | 21.426 | 23.654 | 25.695 | 28.196 | 50.660 | 54.572 | 58.120 | 62.429 | 65.477 |
| 40 | 20.707 | 22.164 | 24.433 | 26.509 | 29.051 | 51.805 | 55.759 | 59.342 | 63.691 | 66.767 |

This table was created using the "INVCDF" command in the MINITAB statistical package.

# Table 4
# t Distribution Quantiles

| $\nu$ | Q(.9) | Q(.95) | Q(.975) | Q(.99) | Q(.995) | Q(.999) | Q(.9995) |
|---|---|---|---|---|---|---|---|
| 1 | 3.078 | 6.314 | 12.706 | 31.821 | 63.657 | 318.317 | 636.607 |
| 2 | 1.886 | 2.920 | 4.303 | 6.965 | 9.925 | 22.327 | 31.598 |
| 3 | 1.638 | 2.353 | 3.182 | 4.541 | 5.841 | 10.215 | 12.924 |
| 4 | 1.533 | 2.132 | 2.776 | 3.747 | 4.604 | 7.173 | 8.610 |
| 5 | 1.476 | 2.015 | 2.571 | 3.365 | 4.032 | 5.893 | 6.869 |
| 6 | 1.440 | 1.943 | 2.447 | 3.143 | 3.707 | 5.208 | 5.959 |
| 7 | 1.415 | 1.895 | 2.365 | 2.998 | 3.499 | 4.785 | 5.408 |
| 8 | 1.397 | 1.860 | 2.306 | 2.896 | 3.355 | 4.501 | 5.041 |
| 9 | 1.383 | 1.833 | 2.262 | 2.821 | 3.250 | 4.297 | 4.781 |
| 10 | 1.372 | 1.812 | 2.228 | 2.764 | 3.169 | 4.144 | 4.587 |
| 11 | 1.363 | 1.796 | 2.201 | 2.718 | 3.106 | 4.025 | 4.437 |
| 12 | 1.356 | 1.782 | 2.179 | 2.681 | 3.055 | 3.930 | 4.318 |
| 13 | 1.350 | 1.771 | 2.160 | 2.650 | 3.012 | 3.852 | 4.221 |
| 14 | 1.345 | 1.761 | 2.145 | 2.624 | 2.977 | 3.787 | 4.140 |
| 15 | 1.341 | 1.753 | 2.131 | 2.602 | 2.947 | 3.733 | 4.073 |
| 16 | 1.337 | 1.746 | 2.120 | 2.583 | 2.921 | 3.686 | 4.015 |
| 17 | 1.333 | 1.740 | 2.110 | 2.567 | 2.898 | 3.646 | 3.965 |
| 18 | 1.330 | 1.734 | 2.101 | 2.552 | 2.878 | 3.610 | 3.922 |
| 19 | 1.328 | 1.729 | 2.093 | 2.539 | 2.861 | 3.579 | 3.883 |
| 20 | 1.325 | 1.725 | 2.086 | 2.528 | 2.845 | 3.552 | 3.849 |
| 21 | 1.323 | 1.721 | 2.080 | 2.518 | 2.831 | 3.527 | 3.819 |
| 22 | 1.321 | 1.717 | 2.074 | 2.508 | 2.819 | 3.505 | 3.792 |
| 23 | 1.319 | 1.714 | 2.069 | 2.500 | 2.807 | 3.485 | 3.768 |
| 24 | 1.318 | 1.711 | 2.064 | 2.492 | 2.797 | 3.467 | 3.745 |
| 25 | 1.316 | 1.708 | 2.060 | 2.485 | 2.787 | 3.450 | 3.725 |
| 26 | 1.315 | 1.706 | 2.056 | 2.479 | 2.779 | 3.435 | 3.707 |
| 27 | 1.314 | 1.703 | 2.052 | 2.473 | 2.771 | 3.421 | 3.690 |
| 28 | 1.313 | 1.701 | 2.048 | 2.467 | 2.763 | 3.408 | 3.674 |
| 29 | 1.311 | 1.699 | 2.045 | 2.462 | 2.756 | 3.396 | 3.659 |
| 30 | 1.310 | 1.697 | 2.042 | 2.457 | 2.750 | 3.385 | 3.646 |
| 40 | 1.303 | 1.684 | 2.021 | 2.423 | 2.704 | 3.307 | 3.551 |
| 60 | 1.296 | 1.671 | 2.000 | 2.390 | 2.660 | 3.232 | 3.460 |
| 120 | 1.289 | 1.658 | 1.980 | 2.358 | 2.617 | 3.160 | 3.373 |
| $\infty$ | 1.282 | 1.645 | 1.960 | 2.326 | 2.576 | 3.090 | 3.291 |

This table was created using "INVCDF" command in the MINITAB statistical package.

## Table 5A
## F Distribution .90 Quantiles

$\nu_1$ (Numerator Degrees of Freedom)

| $\nu_2$ | 1 | 2 | 3 | 4 | 5 | 6 | 7 | 8 | 9 | 10 |
|---|---|---|---|---|---|---|---|---|---|---|
| 1 | 39.86 | 49.50 | 53.59 | 55.84 | 57.24 | 58.20 | 58.90 | 59.44 | 59.85 | 60.20 |
| 2 | 8.53 | 9.00 | 9.16 | 9.24 | 9.29 | 9.33 | 9.35 | 9.37 | 9.38 | 9.39 |
| 3 | 5.54 | 5.46 | 5.39 | 5.34 | 5.31 | 5.28 | 5.27 | 5.25 | 5.24 | 5.23 |
| 4 | 4.54 | 4.32 | 4.19 | 4.11 | 4.05 | 4.01 | 3.98 | 3.95 | 3.94 | 3.92 |
| 5 | 4.06 | 3.78 | 3.62 | 3.52 | 3.45 | 3.40 | 3.37 | 3.34 | 3.32 | 3.30 |
| 6 | 3.78 | 3.46 | 3.29 | 3.18 | 3.11 | 3.05 | 3.01 | 2.98 | 2.96 | 2.94 |
| 7 | 3.59 | 3.26 | 3.07 | 2.96 | 2.88 | 2.83 | 2.78 | 2.75 | 2.72 | 2.70 |
| 8 | 3.46 | 3.11 | 2.92 | 2.81 | 2.73 | 2.67 | 2.62 | 2.59 | 2.56 | 2.54 |
| 9 | 3.36 | 3.01 | 2.81 | 2.69 | 2.61 | 2.55 | 2.51 | 2.47 | 2.44 | 2.42 |
| 10 | 3.28 | 2.92 | 2.73 | 2.61 | 2.52 | 2.46 | 2.41 | 2.38 | 2.35 | 2.32 |
| 11 | 3.23 | 2.86 | 2.66 | 2.54 | 2.45 | 2.39 | 2.34 | 2.30 | 2.27 | 2.25 |
| 12 | 3.18 | 2.81 | 2.61 | 2.48 | 2.39 | 2.33 | 2.28 | 2.24 | 2.21 | 2.19 |
| 13 | 3.14 | 2.76 | 2.56 | 2.43 | 2.35 | 2.28 | 2.23 | 2.20 | 2.16 | 2.14 |
| 14 | 3.10 | 2.73 | 2.52 | 2.39 | 2.31 | 2.24 | 2.19 | 2.15 | 2.12 | 2.10 |
| 15 | 3.07 | 2.70 | 2.49 | 2.36 | 2.27 | 2.21 | 2.16 | 2.12 | 2.09 | 2.06 |
| 16 | 3.05 | 2.67 | 2.46 | 2.33 | 2.24 | 2.18 | 2.13 | 2.09 | 2.06 | 2.03 |
| 17 | 3.03 | 2.64 | 2.44 | 2.31 | 2.22 | 2.15 | 2.10 | 2.06 | 2.03 | 2.00 |
| 18 | 3.01 | 2.62 | 2.42 | 2.29 | 2.20 | 2.13 | 2.08 | 2.04 | 2.00 | 1.98 |
| 19 | 2.99 | 2.61 | 2.40 | 2.27 | 2.18 | 2.11 | 2.06 | 2.02 | 1.98 | 1.96 |
| 20 | 2.97 | 2.59 | 2.38 | 2.25 | 2.16 | 2.09 | 2.04 | 2.00 | 1.96 | 1.94 |
| 21 | 2.96 | 2.57 | 2.36 | 2.23 | 2.14 | 2.08 | 2.02 | 1.98 | 1.95 | 1.92 |
| 22 | 2.95 | 2.56 | 2.35 | 2.22 | 2.13 | 2.06 | 2.01 | 1.97 | 1.93 | 1.90 |
| 23 | 2.94 | 2.55 | 2.34 | 2.21 | 2.11 | 2.05 | 1.99 | 1.95 | 1.92 | 1.89 |
| 24 | 2.93 | 2.54 | 2.33 | 2.19 | 2.10 | 2.04 | 1.98 | 1.94 | 1.91 | 1.88 |
| 25 | 2.92 | 2.53 | 2.32 | 2.18 | 2.09 | 2.02 | 1.97 | 1.93 | 1.89 | 1.87 |
| 26 | 2.91 | 2.52 | 2.31 | 2.17 | 2.08 | 2.01 | 1.96 | 1.92 | 1.88 | 1.86 |
| 27 | 2.90 | 2.51 | 2.30 | 2.17 | 2.07 | 2.00 | 1.95 | 1.91 | 1.87 | 1.85 |
| 28 | 2.89 | 2.50 | 2.29 | 2.16 | 2.06 | 2.00 | 1.94 | 1.90 | 1.87 | 1.84 |
| 29 | 2.89 | 2.50 | 2.28 | 2.15 | 2.06 | 1.99 | 1.93 | 1.89 | 1.86 | 1.83 |
| 30 | 2.88 | 2.49 | 2.28 | 2.14 | 2.05 | 1.98 | 1.93 | 1.88 | 1.85 | 1.82 |
| 40 | 2.84 | 2.44 | 2.23 | 2.09 | 2.00 | 1.93 | 1.87 | 1.83 | 1.79 | 1.76 |
| 60 | 2.79 | 2.39 | 2.18 | 2.04 | 1.95 | 1.87 | 1.82 | 1.77 | 1.74 | 1.71 |
| 120 | 2.75 | 2.35 | 2.13 | 1.99 | 1.90 | 1.82 | 1.77 | 1.72 | 1.68 | 1.65 |
| $\infty$ | 2.71 | 2.30 | 2.08 | 1.94 | 1.85 | 1.77 | 1.72 | 1.67 | 1.63 | 1.60 |

$\nu_1$ (Numerator Degrees of Freedom)

| $\nu_2$ | 12 | 15 | 20 | 24 | 30 | 40 | 60 | 120 | $\infty$ |
|---|---|---|---|---|---|---|---|---|---|
| 1 | 60.70 | 61.22 | 61.74 | 62.00 | 62.27 | 62.53 | 62.79 | 63.05 | 63.33 |
| 2 | 9.41 | 9.42 | 9.44 | 9.45 | 9.46 | 9.47 | 9.47 | 9.48 | 9.49 |
| 3 | 5.22 | 5.20 | 5.18 | 5.18 | 5.17 | 5.16 | 5.15 | 5.14 | 5.13 |
| 4 | 3.90 | 3.87 | 3.84 | 3.83 | 3.82 | 3.80 | 3.79 | 3.78 | 3.76 |
| 5 | 3.27 | 3.24 | 3.21 | 3.19 | 3.17 | 3.16 | 3.14 | 3.12 | 3.10 |
| 6 | 2.90 | 2.87 | 2.84 | 2.82 | 2.80 | 2.78 | 2.76 | 2.74 | 2.72 |
| 7 | 2.67 | 2.63 | 2.59 | 2.58 | 2.56 | 2.54 | 2.51 | 2.49 | 2.47 |
| 8 | 2.50 | 2.46 | 2.42 | 2.40 | 2.38 | 2.36 | 2.34 | 2.32 | 2.29 |
| 9 | 2.38 | 2.34 | 2.30 | 2.28 | 2.25 | 2.23 | 2.21 | 2.18 | 2.16 |
| 10 | 2.28 | 2.24 | 2.20 | 2.18 | 2.16 | 2.13 | 2.11 | 2.08 | 2.06 |
| 11 | 2.21 | 2.17 | 2.12 | 2.10 | 2.08 | 2.05 | 2.03 | 2.00 | 1.97 |
| 12 | 2.15 | 2.10 | 2.06 | 2.04 | 2.01 | 1.99 | 1.96 | 1.93 | 1.90 |
| 13 | 2.10 | 2.05 | 2.01 | 1.98 | 1.96 | 1.93 | 1.90 | 1.88 | 1.85 |
| 14 | 2.05 | 2.01 | 1.96 | 1.94 | 1.91 | 1.89 | 1.86 | 1.83 | 1.80 |
| 15 | 2.02 | 1.97 | 1.92 | 1.90 | 1.87 | 1.85 | 1.82 | 1.79 | 1.76 |
| 16 | 1.99 | 1.94 | 1.89 | 1.87 | 1.84 | 1.81 | 1.78 | 1.75 | 1.72 |
| 17 | 1.96 | 1.91 | 1.86 | 1.84 | 1.81 | 1.78 | 1.75 | 1.72 | 1.69 |
| 18 | 1.93 | 1.89 | 1.84 | 1.81 | 1.78 | 1.75 | 1.72 | 1.69 | 1.66 |
| 19 | 1.91 | 1.86 | 1.81 | 1.79 | 1.76 | 1.73 | 1.70 | 1.67 | 1.63 |
| 20 | 1.89 | 1.84 | 1.79 | 1.77 | 1.74 | 1.71 | 1.68 | 1.64 | 1.61 |
| 21 | 1.87 | 1.83 | 1.78 | 1.75 | 1.72 | 1.69 | 1.66 | 1.62 | 1.59 |
| 22 | 1.86 | 1.81 | 1.76 | 1.73 | 1.70 | 1.67 | 1.64 | 1.60 | 1.57 |
| 23 | 1.84 | 1.80 | 1.74 | 1.72 | 1.69 | 1.66 | 1.62 | 1.59 | 1.55 |
| 24 | 1.83 | 1.78 | 1.73 | 1.70 | 1.67 | 1.64 | 1.61 | 1.57 | 1.53 |
| 25 | 1.82 | 1.77 | 1.72 | 1.69 | 1.66 | 1.63 | 1.59 | 1.56 | 1.52 |
| 26 | 1.81 | 1.76 | 1.71 | 1.68 | 1.65 | 1.61 | 1.58 | 1.54 | 1.50 |
| 27 | 1.80 | 1.75 | 1.70 | 1.67 | 1.64 | 1.60 | 1.57 | 1.53 | 1.49 |
| 28 | 1.79 | 1.74 | 1.69 | 1.66 | 1.63 | 1.59 | 1.56 | 1.52 | 1.48 |
| 29 | 1.78 | 1.73 | 1.68 | 1.65 | 1.62 | 1.58 | 1.55 | 1.51 | 1.47 |
| 30 | 1.77 | 1.72 | 1.67 | 1.64 | 1.61 | 1.57 | 1.54 | 1.50 | 1.46 |
| 40 | 1.71 | 1.66 | 1.61 | 1.57 | 1.54 | 1.51 | 1.47 | 1.42 | 1.38 |
| 60 | 1.66 | 1.60 | 1.54 | 1.51 | 1.48 | 1.44 | 1.40 | 1.35 | 1.29 |
| 120 | 1.60 | 1.55 | 1.48 | 1.45 | 1.41 | 1.37 | 1.32 | 1.26 | 1.19 |
| $\infty$ | 1.55 | 1.49 | 1.42 | 1.38 | 1.34 | 1.30 | 1.24 | 1.17 | 1.00 |

This table was created using the "INVCDF" command in the MINITAB statistical package.

# Table 5B
# F Distribution .99 Quantiles

$\nu_1$ (Numerator Degrees of Freedom)

| $\nu_2$ | 1 | 2 | 3 | 4 | 5 | 6 | 7 | 8 | 9 | 10 |
|---|---|---|---|---|---|---|---|---|---|---|
| 1 | 4052 | 4999 | 5403 | 5625 | 5764 | 5859 | 5929 | 5981 | 6023 | 6055 |
| 2 | 98.51 | 99.00 | 99.17 | 99.25 | 99.30 | 99.33 | 99.35 | 99.38 | 99.39 | 99.40 |
| 3 | 34.12 | 30.82 | 29.46 | 28.71 | 28.24 | 27.91 | 27.67 | 27.49 | 27.35 | 27.23 |
| 4 | 21.20 | 18.00 | 16.69 | 15.98 | 15.52 | 15.21 | 14.98 | 14.80 | 14.66 | 14.55 |
| 5 | 16.26 | 13.27 | 12.06 | 11.39 | 10.97 | 10.67 | 10.46 | 10.29 | 10.16 | 10.05 |
| 6 | 13.75 | 10.92 | 9.78 | 9.15 | 8.75 | 8.47 | 8.26 | 8.10 | 7.98 | 7.87 |
| 7 | 12.25 | 9.55 | 8.45 | 7.85 | 7.46 | 7.19 | 6.99 | 6.84 | 6.72 | 6.62 |
| 8 | 11.26 | 8.65 | 7.59 | 7.01 | 6.63 | 6.37 | 6.18 | 6.03 | 5.91 | 5.81 |
| 9 | 10.56 | 8.02 | 6.99 | 6.42 | 6.06 | 5.80 | 5.61 | 5.47 | 5.35 | 5.26 |
| 10 | 10.04 | 7.56 | 6.55 | 5.99 | 5.64 | 5.39 | 5.20 | 5.06 | 4.94 | 4.85 |
| 11 | 9.65 | 7.21 | 6.22 | 5.67 | 5.32 | 5.07 | 4.89 | 4.74 | 4.63 | 4.54 |
| 12 | 9.33 | 6.93 | 5.95 | 5.41 | 5.06 | 4.82 | 4.64 | 4.50 | 4.39 | 4.30 |
| 13 | 9.07 | 6.70 | 5.74 | 5.21 | 4.86 | 4.62 | 4.44 | 4.30 | 4.19 | 4.10 |
| 14 | 8.86 | 6.51 | 5.56 | 5.04 | 4.69 | 4.46 | 4.28 | 4.14 | 4.03 | 3.94 |
| 15 | 8.68 | 6.36 | 5.42 | 4.89 | 4.56 | 4.32 | 4.14 | 4.00 | 3.89 | 3.80 |
| 16 | 8.53 | 6.23 | 5.29 | 4.77 | 4.44 | 4.20 | 4.03 | 3.89 | 3.78 | 3.69 |
| 17 | 8.40 | 6.11 | 5.18 | 4.67 | 4.34 | 4.10 | 3.93 | 3.79 | 3.68 | 3.59 |
| 18 | 8.29 | 6.01 | 5.09 | 4.58 | 4.25 | 4.01 | 3.84 | 3.71 | 3.60 | 3.51 |
| 19 | 8.19 | 5.93 | 5.01 | 4.50 | 4.17 | 3.94 | 3.77 | 3.63 | 3.52 | 3.43 |
| 20 | 8.10 | 5.85 | 4.94 | 4.43 | 4.10 | 3.87 | 3.70 | 3.56 | 3.46 | 3.37 |
| 21 | 8.02 | 5.78 | 4.87 | 4.37 | 4.04 | 3.81 | 3.64 | 3.51 | 3.40 | 3.31 |
| 22 | 7.95 | 5.72 | 4.82 | 4.31 | 3.99 | 3.76 | 3.59 | 3.45 | 3.35 | 3.26 |
| 23 | 7.88 | 5.66 | 4.76 | 4.26 | 3.94 | 3.71 | 3.54 | 3.41 | 3.30 | 3.21 |
| 24 | 7.82 | 5.61 | 4.72 | 4.22 | 3.90 | 3.67 | 3.50 | 3.36 | 3.26 | 3.17 |
| 25 | 7.77 | 5.57 | 4.68 | 4.18 | 3.85 | 3.63 | 3.46 | 3.32 | 3.22 | 3.13 |
| 26 | 7.72 | 5.53 | 4.64 | 4.14 | 3.82 | 3.59 | 3.42 | 3.29 | 3.18 | 3.09 |
| 27 | 7.68 | 5.49 | 4.60 | 4.11 | 3.78 | 3.56 | 3.39 | 3.26 | 3.15 | 3.06 |
| 28 | 7.64 | 5.45 | 4.57 | 4.07 | 3.75 | 3.53 | 3.36 | 3.23 | 3.12 | 3.03 |
| 29 | 7.60 | 5.42 | 4.54 | 4.04 | 3.73 | 3.50 | 3.33 | 3.20 | 3.09 | 3.00 |
| 30 | 7.56 | 5.39 | 4.51 | 4.02 | 3.70 | 3.47 | 3.30 | 3.17 | 3.07 | 2.98 |
| 40 | 7.31 | 5.18 | 4.31 | 3.83 | 3.51 | 3.29 | 3.12 | 2.99 | 2.89 | 2.80 |
| 60 | 7.08 | 4.98 | 4.13 | 3.65 | 3.34 | 3.12 | 2.95 | 2.82 | 2.72 | 2.63 |
| 120 | 6.85 | 4.79 | 3.95 | 3.48 | 3.17 | 2.96 | 2.79 | 2.66 | 2.56 | 2.47 |
| $\infty$ | 6.63 | 4.61 | 3.78 | 3.32 | 3.02 | 2.80 | 2.64 | 2.51 | 2.41 | 2.32 |

$\nu_1$ (Numerator Degrees of Freedom)

| $\nu_2$ | 12 | 15 | 20 | 24 | 30 | 40 | 60 | 120 | $\infty$ |
|---|---|---|---|---|---|---|---|---|---|
| 1 | 6107 | 6157 | 6209 | 6235 | 6260 | 6287 | 6312 | 6339 | 6366 |
| 2 | 99.41 | 99.43 | 99.44 | 99.45 | 99.47 | 99.47 | 99.48 | 99.49 | 99.50 |
| 3 | 27.05 | 26.87 | 26.69 | 26.60 | 26.51 | 26.41 | 26.32 | 26.22 | 26.13 |
| 4 | 14.37 | 14.20 | 14.02 | 13.93 | 13.84 | 13.75 | 13.65 | 13.56 | 13.46 |
| 5 | 9.89 | 9.72 | 9.55 | 9.47 | 9.38 | 9.29 | 9.20 | 9.11 | 9.02 |
| 6 | 7.72 | 7.56 | 7.40 | 7.31 | 7.23 | 7.14 | 7.06 | 6.97 | 6.88 |
| 7 | 6.47 | 6.31 | 6.16 | 6.07 | 5.99 | 5.91 | 5.82 | 5.74 | 5.65 |
| 8 | 5.67 | 5.52 | 5.36 | 5.28 | 5.20 | 5.12 | 5.03 | 4.95 | 4.86 |
| 9 | 5.11 | 4.96 | 4.81 | 4.73 | 4.65 | 4.57 | 4.48 | 4.40 | 4.31 |
| 10 | 4.71 | 4.56 | 4.41 | 4.33 | 4.25 | 4.17 | 4.08 | 4.00 | 3.91 |
| 11 | 4.40 | 4.25 | 4.10 | 4.02 | 3.94 | 3.86 | 3.78 | 3.69 | 3.60 |
| 12 | 4.16 | 4.01 | 3.86 | 3.78 | 3.70 | 3.62 | 3.54 | 3.45 | 3.36 |
| 13 | 3.96 | 3.82 | 3.66 | 3.59 | 3.51 | 3.43 | 3.34 | 3.25 | 3.17 |
| 14 | 3.80 | 3.66 | 3.51 | 3.43 | 3.35 | 3.27 | 3.18 | 3.09 | 3.00 |
| 15 | 3.67 | 3.52 | 3.37 | 3.29 | 3.21 | 3.13 | 3.05 | 2.96 | 2.87 |
| 16 | 3.55 | 3.41 | 3.26 | 3.18 | 3.10 | 3.02 | 2.93 | 2.84 | 2.75 |
| 17 | 3.46 | 3.31 | 3.16 | 3.08 | 3.00 | 2.92 | 2.83 | 2.75 | 2.65 |
| 18 | 3.37 | 3.23 | 3.08 | 3.00 | 2.92 | 2.84 | 2.75 | 2.66 | 2.57 |
| 19 | 3.30 | 3.15 | 3.00 | 2.92 | 2.84 | 2.76 | 2.67 | 2.58 | 2.49 |
| 20 | 3.23 | 3.09 | 2.94 | 2.86 | 2.78 | 2.69 | 2.61 | 2.52 | 2.42 |
| 21 | 3.17 | 3.03 | 2.88 | 2.80 | 2.72 | 2.64 | 2.55 | 2.46 | 2.36 |
| 22 | 3.12 | 2.98 | 2.83 | 2.75 | 2.67 | 2.58 | 2.50 | 2.40 | 2.31 |
| 23 | 3.07 | 2.93 | 2.78 | 2.70 | 2.62 | 2.54 | 2.45 | 2.35 | 2.26 |
| 24 | 3.03 | 2.89 | 2.74 | 2.66 | 2.58 | 2.49 | 2.40 | 2.31 | 2.21 |
| 25 | 2.99 | 2.85 | 2.70 | 2.62 | 2.54 | 2.45 | 2.36 | 2.27 | 2.17 |
| 26 | 2.96 | 2.81 | 2.66 | 2.58 | 2.50 | 2.42 | 2.33 | 2.23 | 2.13 |
| 27 | 2.93 | 2.78 | 2.63 | 2.55 | 2.47 | 2.38 | 2.29 | 2.20 | 2.10 |
| 28 | 2.90 | 2.75 | 2.60 | 2.52 | 2.44 | 2.35 | 2.26 | 2.17 | 2.06 |
| 29 | 2.87 | 2.73 | 2.57 | 2.49 | 2.41 | 2.33 | 2.23 | 2.14 | 2.03 |
| 30 | 2.84 | 2.70 | 2.55 | 2.47 | 2.39 | 2.30 | 2.21 | 2.11 | 2.01 |
| 40 | 2.66 | 2.52 | 2.37 | 2.29 | 2.20 | 2.11 | 2.02 | 1.92 | 1.80 |
| 60 | 2.50 | 2.35 | 2.20 | 2.12 | 2.03 | 1.94 | 1.84 | 1.73 | 1.60 |
| 120 | 2.34 | 2.19 | 2.03 | 1.95 | 1.86 | 1.76 | 1.66 | 1.53 | 1.38 |
| $\infty$ | 2.18 | 2.04 | 1.88 | 1.79 | 1.70 | 1.59 | 1.47 | 1.32 | 1.00 |

This table was created using the "INVCDF" command in the MINITAB statistical package.

# Index

## A

Acceptance, simulation, 451–452
ACE transformation method, 1, 273–274
Acid deposition, 376–378
Added variable plot, 270
Additive-error model, 246
$\alpha$-particles, emission, 190–191
  americium-241, 214–215
Alternating conditional expectation algorithm, 273–274
Americium-241, $\alpha$-particle emissions, 214–215
Anderson–Darling test, 202
Annealing, simulated, 440–441
Antoine's equation, 484–487
Approximation, 491
Atmospheric temperature fields, field statistics, 466–471
  data, 466
  field significance testing, 470–471
  map statistical significance estimation, 467–471
  preprocessing of data, 467
  random time series generation, 468–470
Autocovariance function, lag, 306
Autocovariance sequence, 314
Autoregressive models, 326
Autoregressive process, first-order, 463
Autoregressive spectral estimate, 327

## B

Background rejection, 441
Background signals, removal, in spectral analysis, 302–305
Bandpass filter, 295–296
Bandwidth, resolution, 323
Bayesian credibility interval, 413
Bayesian estimation, 413
Bayesian maximum a posteriori estimator, 408

Bayesian methods, 403–430
  climate change, 418, 423–425
  Gibbs sampler, 427, 429–430
  hyperparameters, 426
  hyperprior distribution, 429
  maximum entropy method, 415–416
  neutron scattering, 414–417
  nuclear magnetic resonance, 417–422
  prior distribution, 404, 407–413
    conjugate priors, 410–411
    improper priors, 409–410
    maximum entropy priors, 411–413
    noninformative priors, 408–409
    prior choice consequences, 407–408
  rocket photometry, 426–428
Bayesian spectrum analysis, 417–422
  high signal-to-noise ratio, 418–420
  low signal-to-noise ratio, 418, 421–422
Bayesian statistical inference, 403–407
  posterior distribution, 405
  prior distribution, 404
  subjective probability, 405
Bayes's theorem, 406
  continuous version, 407
  gamma posterior distribution, 410
Bayes, Thomas, 403
Benzene, vapor pressure, 485
Bernoulli trials, 16–17, 32
Bessel function, modified, 369
Binomial distribution, 17
Binomial trials, 166
Bivariate normal model method, 464
Bivariate time series, 340–346
  spectral estimation, 343–346
Block kriging, 397
Bonferroni intervals, 261
Bonferroni method, 175–176, 178
Bonferroni tests, 261
Bootstrap methods, 491–498
  applications, 495–498
  implementation, 494–495
  methodology, 492–494
  motivation, 491–492

## 532 INDEX

Bootstrap methods *(continued)*
    nonparametric bootstrap, 493–494
    parametric bootstrap, 493–494
Bootstrap sample, 493
Box plots, 510–511
Buffon's needle, 437
Burg's algorithm, 326–328

### C

Calorimeter detectors, simulation, 447
C-curve, 502
Central limit theorem, 26–27, 252
Chebyshev's inequality, 25
Chi-squared distribution, 167
Chi-squared tests, 188–192
    composite hypothesis, 189–190
    simple hypothesis, 189
    two-sampled tests, 191
Classical probability, 6
Coefficient of determination, 262
Coherence, weak periodic signals, 367–368
Coherence time, 368
Cokriging, 399
Complex coherency, 341
Composite hypothesis, 187–188
    chi-squared tests, 189–190
Compound events, probability, 12–13
Computer software, *see* Software
Computing, *see* Statistical computing
Conditional distribution, 15
Conditional probability, 11–12
Conditioning plot, 510, 512
Confidence bands, simultaneous, quantile–quantile plots with, 197–200
Confidence interval, 178, 213
    application, 169–170
    approximate, based on normal-theory method, 229
    interpreting, 167–168
    likelihood-based, 225, 226
        limited failure population model, 236–238
    relation to statistical tests, 181
    single parameter, 165–167
    simulation-based, 242
    simultaneous, 261
    Weibull, 233–235

Confidence limits for mean, 264
Confidence rectangles, 174–178
    Bonferroni method, 175–176
    grain boundary triple point, frequencies, 174–175
Confidence regions, 173–174, 213
    approximate likelihood-based, 225–226
    elliptical, 178
    normal-theory, 228–229
    Weibull, 233–235
Cook's distance, 272–273
Correlation coefficient, 23
Cospectrum, 341
Covariance, 22–25
    matrix, 382–383
Covariance function, 379–380
Cramer–von Mises test, 201–202
Critical correlation estimation, local, 463–464
Critical value, 180
Cross amplitude spectrum, 341
Cross covariance sequence, 340–341
Cross entropy, 412
Cross periodogram, 343
Cross-semivariogram, 399
Cross spectral density function, 340
Cross spectrum, 340
    multitaper estimator, 344–346
Cross-validation, 500
Cumulative distribution, definition, 14
Cumulative distribution function, 212, 217–218
    completely specified, assessing fit, 201–202
    statistical computing, 491
    Weibull, 231
Curvature array
    intrinsic, 260
    small, 263

### D

Data
    censored
        assessing distributional assumptions, 204–206
        interval-censored, 214–216
        preprocessing, 467

INDEX 533

prewhitening, 305, 320, 325–328
tapering, 300, 320–325
time series with missing, 305–309
windowing, 297–302
windows, frequency transforms, 299
dc excess, versus periodic strength, 361–364
Decision procedure, 179
Delta method, 228
Density approximation, 216
  potential problems, 216–218
Descriptive statistics, 156–158
Diagnostics, 266–277
  ACE transformation method, 273–274
  ill-conditioning, 274–276
  Larsen and McCleary plots, 269–270
  model inadequacy, 266
  multicollinearity, 274–277
  nonconstant variance, 267–268
  normal probability plot, 268–269
  outliers, 267
  plot residual versus
    each regressor, 269
    fitted value, 266–268
  plots to detect influential observations, 270–273
  plotting variables in pairs, 269
Differential equation models, 501–509
  kinetics example, 502–509
    model, data, and method, 502–504
    S function, 504–509
  problem description, 501–502
Digital filter, 283–284
  design, 292–296
Direct spectral estimate, data taper, 324
Direct spectral estimator, 321
Discrete prolate spheroidal sequence tapers, 323
  zeroth-order, 336–337
Distributional assumptions, assessment, 187–209
  chi-squared tests, 188–192
  extensions to censored data, 204–206
  formal test procedures, 201–204
    completely specified cdf, assessing fit, 201–202
    normality tests, 203–204
    parametric families, testing for, 202–203
  goodness-of-fit problem, 187–188
  quantile–quantile plots, 192–200

two-sample comparisons, 207–209
Distribution function, confidence bands, 176–177
Dominant likelihood situation, 407
Duration–bandwidth product, 323
Durbin–Watson test, 268, 276
Dynamic graphics, 512–516

## E

Efficiency, simulation, 451–452
Electromigration
  definition, 156
  failure time
    confidence band, 176–177
    linear regression, 163
    natural logs, 155–156
  statistical model, 180
Entropy, Shannon–Jaynes, 411–412, 415
Envelope delay, 341
Epoch folding, 354–355
  coherent signal, 368
Equivalent degrees of freedom, 331
Error, 245–246
  definition, 1
  estimation, 452
  instrumental, 30–31
  matrix, 23
  maximum possible, 3
  standard, 257
  statistical view, 252–253
  systematic, 28–29
Estimator
  biased, 316
  robust, 390
Euler, Leonhard, 3
Expected values, 18–19
Experimental design, 184
Experimental science, probability, 1–4
Explanatory variable, 220–221, 245

## F

Failure, definition, 156
Fast Fourier transform filtering method, 296–297

# 534 INDEX

Fejér's kernel, 317–319
Field significance, 458
 testing for
  field statistics, 465–466
  temperature correlation maps, 470–471
Field statistics, 457–478
 assessment by Monte Carlo simulation, 462–466
  critical area estimation, 465
  field significance testing, 465–466
  local critical correlation estimate, 463–464
  random time series, 462–463
  simulation of field statistics, 463
 atmosphere temperature fields, 466–471
 correlation statistic map calculation, 460–461
 cross correlation between ozone and solar flux time series, 475–477
 field significance, 458
 finite data set lengths, temporal interdependence, 459
 global ozone data fields, 471–475
 higher dimensions, 478
 known spatial degrees of freedom, 459–460
 spatial interdependence, 459
 time-lag correlation for TOMS, 457–458
Filter
 frequency response, 284–291
 prewhitening, 327–328
Filtering, Fourier, 417
Filtering time series, 283–297
 bandpass filter, 295–296
 digital filter, 283–284
  design, 292–296
 direct Fourier analysis–resynthesis, 296–297
 fast Fourier transforms, 296–297
 frequency coefficients, 285
 frequency response, 284–291
  filter coefficients, 289
  finite difference operators, 290–291
  low-pass filter, 287
  1-1-1 moving average filter, 286–289
  periodogram, 285
  spectral power densities, 285, 287–289
 Gibbs phenomenon, 292
 Lanczos window weights, 293–295
 ringing, 296

 total variance, 284–285
Finite difference operators, 290–291
Fisher information, 227
 matrix, 241
Fisher's $g$ statistic, 339
Fortran, code for ODE solver, 506
Fourier filtering, 417
Fourier frequency, independent, 366–368
Fourier representation, 313
$F$-ratio, 262
Frequency probability, 7
Frequency response, filter, 284–291
Frequency spacing, 365
$F$-statistic, 262
Functional model, 278
Fundamental frequency, 339

## G

Games, nonalgebraic solutions, 443–444
Gamma distribution, generalized, 239
Gaussian distribution, 18
 central limit theorem, 27
Gaussian noise time series, generation, 468
Gaussian random number generator, 448
Gaussian white noise process, 339
Gauss–Newton method, 255–256
Geostatistical model, 378–381
Gibbs phenomenon, 292
Gibbs sampler, 427, 429–430
Global ozone data fields, field statistics, 471–475
Global significance, 458
Grain boundary triple points, 162, 166–167
 frequencies, confidence rectangle, 174–175
Graphical methods, 509–510
Graphics
 box plots, 510–511
 brush, 515–516
 conditioning plot, 510, 512
 dynamic, 512–516
 identification, 513
 linking, 513, 516
 quantile–quantile plot, 510–511
 rotation, 515–516
 static, 510, 514
Group delay, 341

# INDEX

## H

Hanning data taper, 321–322
Hanning window, 300–301
Hardware platforms, 482
Harmonic frequency, 339
Hat matrix, 255
Hierarchical Bayes, 429
High-energy particles, interactions, 444–445
Highest posterior density, 413
High leverage point, 271
Histogram, 157–158
Hooke's law, 502
$H$-test, 360
$H$-value, maximum, 371
Hyperparameters, 426
Hyperprior distribution, 429
Hypothesis
  composite, 187–188
  simple, 187
    chi-squared tests, 189

## I

Ill-conditioning, 274–276
Independence, concept of, 12
Instrumental errors, 30–31
Integration, 490–491
Interval-censored data, 214–216
Interval estimation, 165–178
  Bonferroni method, 178
  comparing independent samples, 170–171
  confidence interval
    application, 169–170
    interpretation, 167–168
  confidence rectangles, 174–178
  confidence regions, 173–174
  prediction intervals, 171–173
  single parameter confidence intervals, 165–167
Intrinsic curvature array, 260
Inverse problem, 414
Isocorrelation contours, 380
Isotropy, 380

## J

Joint distribution, 14–15, 21

## K

Kaplan–Meier estimates, 205–206
$K$-band, 197–198, 201
Kolmogorov band, 197–198, 201
Kolmogorov–Smirnov test, two-sample, 207–209
Kolmogorov test statistic, 201
  two sample, 208–209
Kriging
  application to sulfate deposition, 397–398
  median-polish, 397
  universal, 396–397
  variance, 396–397, 400
Kullback–Leibler distance, 412

## L

Ladder of powers, 268
Lag window, 330–331
  spectral estimators, 329–333
Lanczos window weights, 293–295
Landau tail, 445
Larsen and McCleary plots, 269–270
Law of large numbers, 7, 25–26
Leakage
  periodogram, 319–320, 324–325
  smoothing window, 330
  spectral analysis, 300–301
Least squares, 245–280
  compartmental analysis of humans, 248
  diagnostics, 266–277
  error process statistical view, 252–253
  errors in regressors, 277–280
  extreme points effect, 271–272
  fitting, 254–256
  growth model, 246–247
  nonlinear, 259–261
    weighted, 391
  Ohm's law, 246

Least squares *(continued)*
  ordinary, properties, 256–257
  with and without outlier, 270
  $n$-pentane, catalytic isomerization, 247–248
  robust methods, 259
  S functions, 506–508
  solution freezing point, 248–250
  spray paint particle congealing, 250–252
  statistical inference, 261–265
  statistical modeling, 245–252
  weighted, 257–259
Left-censored observations, 218–219
  fitting Weibull, 230–234
Left truncation, 219–220
Libration, definition, 2
Light curve, 351–353
Likelihood, log, 222
Likelihood displacement, 273
Likelihood function, 164, 213–214, 406–407
  limited failure population model, 236
  maximum, 222–224, 363
  relative, 223–224
  Weibull, 232–233
Likelihood ratio, 224–225
  principle, 183
  tests, 178
    based on $F$-statistic, 262
Limited failure population model, 235–238
  likelihood-based confidence, 236–238
  likelihood function, 236
  profile likelihood functions, 236–238
Linear algebra, 490
Linear growth model, 74
Linear models, 184
Line splitting, spontaneous, 328
LOWESS, 500
  smoothing parameter, 505
Low-pass filter, 287

## M

Magnitude squared coherence, 342–346
Map statistics, *see* Field statistics
Marginal distribution, 15
Markov model, first-order, 463
Maximum entropy method, 415–416
Maximum likelihood
  estimators, 164–165
  fitting semivariogram models, 392
Mayer, Johann Tobias, 2–3
Mean, 19
Mean response, 264
Mean squared error, 160–161
Mean squared prediction error, 396
Measured amplitude with random phase method, 469–470
Median-polish algorithm, 384–385
Median-polish kriging, 397
Method of least squares, point estimation, 162–164
Method of maximum likelihood, 211–242
  applications, 211
    to exponential distribution, 221–230
      approximate confidence regions, 228–230
      approximate likelihood-based confidence regions and intervals, 225–226
      asymptotic distribution of estimators, 226–228
      asymptotic normality of estimators, 228–230
      data and model, 221–222
      Fisher information, 227
      likelihood-based confidence intervals, 225
      likelihood function, 222–224
      normal-theory method, 228–230
      profile likelihood, 224–225
  density approximation, 216
    potential problems, 216–218
  distributions with threshold parameters, 239
  estimators
    asymptotic behavior, 241
    asymptotic distribution, 226–228
    asymptotic normality, 228–230
  explanatory variables, 220–221
  fitting limited failure population model, 235–238
  generalized gamma distribution, 239
  interval-censored observations, 214–216
  left-censored observations, 218–219
    fitting Weibull, 230–234

INDEX 537

left truncation, 219–220
modeling variability, with parametric distribution, 212
nonregular models, 241–242
numerical methods, 240
point estimation, 164–165
point processes, 240
random truncation, 239
regression, 220–221
regularity conditions, 241–242
right-censored observations, 219
right truncation, 220
simulation-based confidence intervals, 242
software packages, 240
Method of moments, point estimation, 161–162
Minimum variance, 161
  unbiased estimator, 161
Moments, 19
Monte Carlo experiment, 163–164
Monte Carlo integration technique, 429
Monte Carlo simulation
  field statistics assessment, 462–466
    critical area estimation, 465
    field significance testing, 465–466
    local critical correlation estimate, 463–464
    random time series, 462–463
    simulation of field statistics, 463
  global ozone data fields, 473–475
Monte Carlo technique, estimating map's statistical significance, 467–470
Multicollinearity, 274–277
Multiplicative law, 13
Multitaper estimator, cross spectrum, 344–346
Multitaper spectral estimators, 335–338
Multivariate distribution, covariance, 22
Multivariate nonlinear model, 248
Multivariate normal distribution, 256–257

**N**

Neutron scattering, Bayesian methods, 414–417
Neymann–Pearson hypothesis testing, 182–183

Nonconstant variance, 267–268
Normal error model, 164
Normality, tests for, 203
Normal-theory method, 228
Nuclear magnetic resonance, Bayesian methods, 417–422
Nuisance parameter, 166
  planning value, 169–170
Null hypothesis, 179
Numerical mathematics, 489
  areas important in statistical computing, 490–491
Numerical methods, 488–489
  method of maximum likelihood, 240
Nyquist frequency, 309–310, 313, 316, 319

**O**

Operating systems, 482–483
Optimization, 490
Ordinary differential equation model, 501, 503, 506, 508–509
Outliers, 184, 259, 267
  least squares fit with and without, 270
Oversampling, weak periodic signals, 366–367
Ozone, correlation with solar flux, 475–477

**P**

Parameter, 245
  values, statistical inference, 264
Parameter-effects array, 260
Parameter-effects curvature, 264
Parametric families, tests for, 202–203
Partial differential equation models, 501
Partial residual plot, 270
Particle, see also α-particles
  detector, simulation, 446–447
  energy loss, simulation, 445
  multiple scattering, simulation, 445
Parzen lag window, 332–333
$n$-Pentane, catalytic isomerization, 247–248
Periodic signals, weak, 349–371
  dc excess vs. periodic strength, 361–364

Periodic signals, weak *(continued)*
  $H$-test, 360
    power of statistical tests, 360–361
    Protheroe test, 358–359
    Rayleigh test, 355–358
    $Z^2m$-test, 359–360
  independent Fourier frequencies, 366–368
  light curve, 351–353
  multiple data sets, 367–371
    coherence, 367–368
    use of all the data, 369–371
  oversampling, 366–367
  tests for uniformity of phase, 352, 354–361
    epoch folding, 354–355
  white noise, 351–352
Periodic strength, versus dc excess, 361–364
Periodogram, 285, 316–320
  bias sources, 318–319
  cross, 343
  data taper, 324
  leakage, 319–320, 324–325
  spectral window, 317
  statistical properties, 317
Phase spectrum, 341
Physics parameter, estimation, 452
π, value, estimation, 436–437
Pivots, 166, 178
Point estimation, 160–165, 213
  desirable properties, 160–161
  method of least squares, 162–164
  method of maximum likelihood, 164–165
  method of moments, 161–162
  statistic, concept of, 160
Point process, *see also* Periodic signals, weak
Point processes, 240
Poisson distribution, 17, 167, 351
Population parameters, 158
Power spectral density, calculation, time series with missing data, 305–309
Power spectral density function, *see* Spectral density function
Power transformations, 268
Practical significance, versus statistical significance, 181–182
Prediction intervals, 171–173
Predictive distribution, 413
Prewhitening, data, 305, 320, 325–328

Priors
  choice, consequences of, 407–408
  conjugate, 410–411
  improper, 409–410
  maximum entropy, 411–413
  noninformative, 408–409
Probability
  additivity, mutually exclusive events, 10–11
  definition, 4–7
    classical, 6
    frequency, 7
    quantum, 5
  distribution, 492
    random variables, 13–18
  experimental science, 1–4
  subjective, 405
Probability distribution functions, weighting schemes, 453
Probability function, 9
Probability modeling, 1–33
  modeling measurement, 27–33
    random errors, 29–33
    systematic errors, 28–29
Probability of observed data, 213–214
Probability plot, 157–158
  normal, 268–269
Probability theory, 4
  elements, 8–27
    central limit theorem, 26–27
    compound events, 12–13
    conditional probability and independence, 11–12
    covariance, 22–25
    expected values, 18–19
    law of large numbers, 25–26
    moment and mean and variance, 19–22
    random variables and probability distributions, 13–18
    sample space, 8–11
Profile likelihood, 224–225
  Weibull distribution, 234
Profile likelihood functions, 178
  limited failure population model, 236–238
Profile $t$ function, 263
Projection matrix, 255
Propagation of error, 228
Protheroe test, 358–359
$p$-value, 181–182

## Q

Quadratic approximation, 228
Quadrature spectrum, 341
Quantile–quantile plots, 192–200, 510–511
  completely specified distribution, assessing fit, 192–195
  mechanical device failure times, 192–195
  parametric families, assessing fit, 196–197
  for randomly censored data, 205–206
  with simultaneous confidence bands, 197–200
  two-sample, 207–208
Quantum probability, 5
Quasi-periodic oscillators, 368

## R

Random errors, 29–33
  instrumental, 30–31
  statistical, 31–33
Random number generator, 126
Random time series, in Monte Carlo simulation, 462–463
Random truncation, 239
Random variables, probability distributions, 13–18
Range, 157
Rayleigh powers, distribution, 369–370
Rayleigh test, 355–358
  power of, 360
Red noise spectrum, generation, 468–469
Regression, method of maximum likelihood, 220–221
Regressor
  errors, 277–280
  random, 278
  versus residual, 269
  variances, 279
Rejection region, 180, 181–182
Relative entropy, 412
Residual, 247
  versus regressor, 269
Residual plus component plot, 270
Residual sum of squares, 257
Residual sum of squares function, weighted, 391–392
Resolution bandwidth, 323
Resonance experiments, 349–350
Response variable, 245
Restricted maximum likelihood, fitting semivariogram models, 392
Right-censored observations, 219
Right truncation, 220
Ringing, 296
Rocket photometry, bayesian methods, 426–428
Roulette, nonalgebraic solutions, 441–443

## S

Sample mean, 156
Sample median, 156
Sample space, 8–11
Sample variance, 161
Sampling, 20
  correlated, 454
  distribution, 492
SAS, 484
Scientific inference, definition, 156
Second-order structure, estimation, 388–395
Semivariogram, 380–381
  estimating, 388–389
  model, fitting, 390–392
  nonparametric estimates, 389–390, 393–395
Semivariogram model, spherical, 394–395
Serial autocorrelation, 253, 259
S function, 504–509
  least squares, 506–508
Shannon–Jaynes entropy, 411–412, 415
Shannon number, 336–337
Shapiro–Wilk test, 203–204
Signal-to-noise ratio, 74
Significance test, 179–181
Simple linear regression, 162–163
Simulation, see also Monte Carlo simulation
  $K^0_1 \to \pi^+\pi^-$ decay, 447–450
  motivation for, 435–436
  particle detector, 446–447
  roulette, 441–443
  sampling any distribution, 437–440

$K^0_1 \to \pi^+\pi^-$ decay *(continued)*
   within simulations, 454–455
   traveling salesman problem, 441
   validity testing, 450–451
   variance reduction, 439
   weighting schemes, 453
  of physical systems, 433–455
   analysis of experiments, 451–453
   annealing, 440–441
   available codes, 450
   background rejection, 441
   correlated sampling, 454
   examples, simple simulations, 436–437
   games, 443–444
   high-energy particle interactions, 444–445
   history, 434–435
Smoothing
  statistical computing, 498–500
  window, leakage, 330
Software
  method of maximum likelihood, 240
  statistical, 483–484
Solar flux, correlation with ozone, 475–477
Spatial autocorrelation, 253, 259
Spatial data, statistical analysis, 375–400
  application to sulfate decomposition, 393–395
  extensions, 398–400
  geostatistical model, 378–381
  median polish, 383–385, 387–389
  second-order structure estimation, 388–395
  semivariogram, nonparametric estimation, 389–390
  spatial labels, 375
  spatial prediction, 396–398
  sulfate deposition data, 376–378
   application, 386–389
  trend surface analysis, 381–383, 386
Spatial interdependence, field statistics, 459
Spatial prediction, 396–398
Special functions, 491
Spectral analysis, 297–305, 313–346
  aliasing, 309–310
  bandwidth, 299–300
  bivariate time series, 340346
  data windowing, 297–302
  Hanning window, 300–301
  leakage, 300–301
  removal of background trends, 302–305
  smoothing, 301–302
  spectral power density, 303–304
  spectral window, 321
  univariate time series, 316–340
Spectral density function, 313–315
Spectral estimator
  lag window, 329–333
  multitaper, 335–338
  WOSA, 333–335
Spectral power density
  filter, 285
  spectral analysis, 303–304
Spectral window, 321
  periodogram, 317
Spectrum analysis, Bayesian, 417–422
S system, 484
  bootstrapping code for median, 494
  code for bootstrapping nonlinear regression coefficients, 497
  extending, 487–488
  on Unix workstation, 484–488
Standard deviation, 20, 157
Standard error, 166, 257
Static graphical methods, 510, 514
Stationarity, intrinsic, 379
Statistic, concept, 160
Statistical computing, 481–517, *see also* Graphics
  bootstrap methods, 491–498
   applications, 495–498
   implementation, 494–495
   methodology, 492–494
   motivation, 491–492
   nonparametric bootstrap, 493–494
   parametric bootstrap, 493–494
  computer-intensive methods, 491–500
  cumulative distribution functions, 491
  difference between numerical methods and numerical mathematics, 489
  differential equation models, 501–509
Statistical computing *(continued)*
   kinetics example, 502–509
   problem description, 501–502
  environments, 482–488
   hardware platforms and operating systems, 482–483
   statistical software, 483–484
   using S on Unix workstations, 484–488
  important numerical mathematics areas, 490–491

literature, 489–490
methods, 488–491
smoothing, 498–500
types, 488
Statistical decision, 179
Statistical errors, 31–33
Statistical hypothesis, 179
Statistical inference, 155–185
 basic ideas, 212–213
 Bayesian, 403–407
 descriptive statistics, 156–158
 interval estimation, *see* Interval estimation
 least squares, 261–265
 likelihood function, 213–214
 linear models, 184
 parameter values, 264
 point estimation, 160–165
 population parameters, 158
 robust estimation, 184
 sequential methods, 184–185
 statistical model, 158–159, 185
 statistical tests, 179–183
Statistical model, 157–159, *see also* Method of maximum likelihood
 diagnostics, 266–277
 electromigration, 180
 fitting, 240–241
 functional, 278
 multivariate nonlinear, 248
 nonparametric, 185
 nonregular, 241–242
 parametric, 185
 statistical inference, 158–159, 185
 structural, 278–279
 theoretical and empirical, 278
 variability, parametric distribution, 212
Statistical modeling
 basic ideas, 213–214
 least squares, 245–252
 terminology, 245–246
Statistical significance, versus practical significance, 181–182
Statistical tests, 179–183
 choosing test statistic and rejection region, 180–181
 Neymann–Pearson hypothesis testing, 182–183
 $p$-value, 181–182
 relation to confidence interval, 181
 significance test, 179–181

statistical versus practical significance, 181–182
Statistical uncertainties, estimation, 452
Statistical visualization, 512
Steady model, 74
$S$-test, 201
Structural model, 278
Sulfate
 decomposition, statistical analysis application, 393–395
 deposition
  data, 376–378
  kriging application, 397–398
  statistical analysis application, 386–389
Sum of squares for error, minimization, 163
Systematic errors, 28–29
Systematic uncertainties, estimation, 452–453

**T**

Test statistic, 180
Threshold parameters, 239
Time series
 cross correlation between ozone and solar flux, 475–477
 data preprocessing, 467
 random generation methods, 468–470
 randomly generated reference point, 465
 univariate, *see* Univariate time series
Time series analysis, 283–310
 aliasing, 309–310
 bivariate, 340–346
 data windowing, 297–302
 filtering, 283–297
 with missing data, power spectra calculation, 305–309
 spectral analysis, 297–305
Tolerance intervals, 184
Tracking detectors, simulation, 446–447
Traveling salesman problem, 441
Trend surface analysis, 381–383, 386

**U**

Unbiasedness, 161
Univariate time series, 316–340

Univariate time series *(continued)*
    data tapering, 320–325
    direct spectral estimator, 321
    discrete prolate spheroidal sequence tapers, 323
    duration–bandwidth product, 323
    periodic nonsinusoidal signal, 339
    periodogram, 316–320
    spectral peaks, evaluating significance, 338–340
    variance reduction, 329–338
        lag window spectral estimators, 329–333
        multitaper spectral estimators, 335–338
        WOSA spectral estimators, 333–335
Unix, workstation, using S, 484–488

**V**

Validity testing, simulation, 450–451
Variance, 19
Variance–covariance matrix, 256–257
Variance function, asymptotic, 228

**W**

Wald's method, 228
Weak law of large numbers, 26

Weibull, 230–234
    confidence intervals, 233–235
    confidence regions, 233–235
    cumulative distribution function, 231
    galaxy variability, 230
    likelihood function, 232–233
    limited failure population model, 235–238
Weibull distribution, three-parameter, 239
Weibull probability, 230–231
Weibull probability plot, right-censored failure data, 235
Weighting schemes, simulation, 453
Welch's overlapped segment averaging spectral estimators, 333–335
White noise, weak periodic signals, 351–352
Working hypothesis, 179
Workstation, Unix, using S, 484–488
WOSA spectral estimators, 333–335

**Y**

Young's modulus, 502

**Z**

$Z^2_m$-test, 359–360

ISBN 0-12-475973-4